Process Systems Engineering-Incubating Sustainability for Industrial Revolution 4.0

Process Systems Engineering-Incubating Sustainability for Industrial Revolution 4.0

Guest Editors

Tin Sin Lee

Thomas S.Y. Choong

MDPI

Basel • Beijing • Wuhan • Barcelona • Belgrade • Novi Sad • Cluj • Manchester

Guest Editors
Tin Sin Lee
Department of Chemical Engineering
Universiti Tunku Abdul Rahman
Kajang
Malaysia

Thomas S.Y. Choong
Department of Chemical and Environmental Engineering
Universiti Putra Malaysia
Serdang
Malaysia

Editorial Office
MDPI AG
Grosspeteranlage 5
4052 Basel, Switzerland

This is a reprint of the Special Issue, published open access by the journal *Processes* (ISSN 2227-9717), freely accessible at: https://www.mdpi.com/journal/processes/special_issues/29SJS28Y95.

For citation purposes, cite each article independently as indicated on the article page online and as indicated below:

Lastname, A.A.; Lastname, B.B. Article Title. *Journal Name* **Year**, *Volume Number*, Page Range.

ISBN 978-3-7258-6586-4 (Hbk)
ISBN 978-3-7258-6587-1 (PDF)
https://doi.org/10.3390/books978-3-7258-6587-1

Contents

About the Editors

Lee Tin Sin

Lee Tin Sin is an academician from Universiti Tunku Abdul Rahman. He has been involved in rubber processing, biopolymer, nanocomposite, and polymer synthesis with more than 100 publications of journal papers, book chapters, and conferences. He was the recipient of the Society of Chemical Engineers Japan Award for Outstanding Asian Researcher and Engineer 2018 for his contribution to polymer research.

Thomas S.Y. Choong

Thomas S.Y. Choong is an academician from Universiti Putra Malaysia. He is actively involved in chemical and environmental engineering research. He is currently a fellow of the Institution of Engineers, Malaysia. He has been involved in the research area of chemical engineering simulation, palm oil production, synthesis, optimization, and safety. He was the organizing committee of renowned such as Asian Pacific Confederation of Chemical Engineering and Asian Process System Engineering Conferences.

processes

MDPI

Editorial

Special Issue on "Process Systems Engineering—Incubating Sustainability for Industrial Revolution 4.0"

Tin Sin Lee [1,*] and Thomas Shean Yaw Choong [2,*]

1 Department of Chemical Engineering, Lee Kong Chian Faculty of Engineering and Science, Universiti Tunku Abdul Rahman, Bandar Sungai Long, Kajang 43000, Selangor, Malaysia
2 Department of Chemical and Environmental Engineering, Faculty of Engineering, Universiti Putra Malaysia, Serdang 43400, Selangor, Malaysia
* Correspondence: leets@utar.edu.my (T.S.L.); csthomas@upm.edu.my (T.S.Y.C.)

Process systems engineering (PSE) plays a crucial role in enhancing the efficiency and sustainability of industrial systems by employing advanced methodologies to optimize processes. PSE allows for the meticulous design, modeling, and control of complex industrial systems, ensuring that every component operates at peak efficiency. With the advent of artificial intelligence (AI), these capabilities have been significantly augmented. AI-driven modeling and analysis tools enable more accurate predictions, better decision-making, and seamless integration of various subsystems, leading to improved overall performance [1]. By leveraging AI, PSE can now tackle increasingly complex challenges, making it possible to manage and optimize large-scale industrial operations with unprecedented precision and efficiency.

In an era defined by rapid technological advancement and global sustainability imperatives, the role of engineering and scientific innovation has become increasingly central to shaping a resilient, intelligent, and environmentally responsible future. The industrial landscape of the 21st century demands not only efficiency and productivity but also sustainability, adaptability, and ethical responsibility. Hence, the research contributions presented in this Special Issue embody the evolving intersection of advanced technologies, sustainable practices, and safety-driven design philosophies. Together, these works reflect the essence of transformative research—bridging theory and application, human and machine, science and society [2]. Simultaneously, the global push towards carbon reduction has intensified the focus on environmental considerations within industrial operations. As industries are being held to stricter environmental standards, there is a growing need to evaluate and optimize processes not just for economic gain but also for their environmental impact. This holistic approach requires the analysis of industrial systems through multiple approaches, including energy consumption, waste generation, and emissions. PSE provides the tools to perform such multi-faceted analyses, enabling industries to balance economic performance with environmental responsibility. By integrating these methodologies, PSE not only helps industries to meet their carbon reduction targets but also ensures that these efforts are economically viable and technologically feasible. This systemic approach is essential for driving the transition towards more sustainable industrial practices while maintaining competitiveness in a rapidly evolving market [3].

This Special Issue brings together a carefully selection from the collection of studies presented at the PSE ASIA 2024—11th Asian Symposium on Process Systems Engineering at Penang, Malaysia, representing the diverse and interconnected fields that define modern engineering and industrial evolution. The featured articles traverse three overarching groups: (i) sustainability and environmental engineering, (ii) digitalization and Industry 4.0 transformation, and (iii) artificial intelligence, safety, and risk management. Each theme contributes to a shared goal—to redefine the future of industrial systems through innovation that is both intelligent and sustainable.

The first set of papers highlights the importance of environmental sustainability and resource optimization as foundational pillars of future-ready industries. The article "*Analyzing the Impact of Orifice Size and Retention Time in Private Tanks on Water Quality Indicators*" provides critical insights into chlorine decay in water distribution networks which is significantly affected by the presence of storage tanks, particularly due to the orifice size and retention time, which influence both hydraulic flow behavior and water residence time. This study introduces a novel simulation framework that integrates pressure-driven analysis with a first-order kinetic model for chlorine decay, implemented using the WQnetXL tool and validated through simulations in EPANET. Complementing this, "*Esterification of Kenaf Core Fiber as a Potential Adsorbent for Oil Removal from Palm Oil Mill Effluent (POME)*" explores the potential of esterified kenaf core (EKC) fiber as an oil adsorbent for oil removal from POME, optimized using a full central composite design (CCD) within the response surface methodology (RSM) framework. The optimum conditions achieved 76% oil removal efficiency, with a 1:0.5 ratio of mercerized kenaf core to stearic acid (MKC:SA), 15 wt% of catalyst, and 1 h reflux time during the esterification process. On the other hand, the paper "*Scaling Up a Heater System for Devulcanization of Off-Spec 2 Latex Waste: A Two-Phase Feasibility Study*" explores the feasibility of devulcanizing off-spec latex waste using a two-phase approach comprising laboratory and pilot-scale trials. Findings from the laboratory phase guided the design and operation of a pilot-scale process using a retrofitted waste rubber machine. Results showed comparable devulcanization efficiency at both scales, with gel contents of 52.5% (lab) and 55.2% (pilot). GPC analysis indicated increased molecular weight and reduced polydispersity index (2.266 to 1.7601), signifying more uniform molecular weight distribution and effective crosslink scission. The successful scale-up demonstrates the potential for industrial application of latex waste devulcanization. Meanwhile, "*Emerging Advances in Sustainable Manufacturing*" is a systematic review of the latest trends in sustainable manufacturing over the past five years, exploring future developments in technologies, methods, and strategies. Utilizing the PRISMA (Preferred Reporting Items for Systematic Reviews and Meta-Analyses) methodology, adapted to engineering, this review ensures transparency and reproducibility throughout the identification, screening, eligibility, and inclusion stages of the review process. Collectively, these papers reinforce the importance of sustainability—not as an auxiliary consideration, but as the vital framework through which future industrial and environmental systems to evolve.

The second cluster of papers focuses on digital transformation and intelligent automation, capturing the essence of Industry 4.0. The paper "*Implementation of a Sustainable Framework for Process Optimization Through the Integration of Robotic Process Automation and Big Data in the Evolution of Industry 4.0*" demonstrates the integration of Robotic Process Automation (RPA) and Big Data within a sustainable framework for process optimization in the context of Industry 4.0. As industries strive to enhance operational efficiency while maintaining sustainability, the need for innovative solutions has become crucial. The research applies the PICO methodology (Population, Intervention, Comparison, Outcome) to assess the impact of combining these technologies on process optimization and sustainability. "*Trends in Sustainable Inventory Management Practices in Industry 4.0*" further extends this narrative to examine 52 recently published papers on sustainable inventory management in Industry 4.0, intending to bridge theory and practice through a compre-

 https://doi.org/10.3390/pr14010036

hensive literature review. By analyzing the latest advancements discussed over the past two years, covering 2024 and 2025, key trends shaping the field are identified, highlighting existing gaps that may require further exploration. Focusing on this time frame is particularly relevant because it reflects how companies have recently started using artificial intelligence more practically to support sustainability goals. Adding a human-centered perspective, "*Prototype of a Multimodal Platform Including EEG and HRV Measurements Intended for Neuroergonomics Applications*" offers an innovative approach to understanding human–machine interaction through physiological measurements. This work introduces prototype testing to demonstrate the system's ability to detect stress and drowsiness. Along with other indicators such as body temperature, heart rate (HR), and SpO_2 levels, the system incorporates electroencephalography (EEG) and heart rate variability (HRV). Together, these contributions present a compelling vision of how digitalization and human intelligence can be harmonized—forming the foundation of smart, adaptive, and sustainable industrial ecosystems.

The third research group brings into focus artificial intelligence, cybersecurity, and safety management—domains that are increasingly critical as industries become more interconnected and data-driven. The study "*A Type-2 Fuzzy Logic Expert System for AI Selection in Solar Photovoltaic Applications*" integrates fuzzy logic and AI to support energy decision-making, illustrating how intelligent systems can guide sustainable energy adoption. The review provides details on the advantages, limitations, and optimal use cases of various review techniques, such as Artificial Neural Networks, Fuzzy Logic, Convolutional Neural Networks, Long-Short Term Memory, Support Vector Machines, Decision Trees, Random Forest, k-Nearest Neighbors, and Particle Swarm Optimization. "*MAL-XSEL: Enhancing Industrial Web Malware Detection with an Explainable Stacking Ensemble Model*" addresses the growing challenge of cybersecurity in industrial networks, emphasizing the importance of transparency and explainability in AI-driven detection systems. MAL-XSEL explicates the model predictions through Shapley additive explanations (SHAP) and local interpretable model-agnostic explanations (LIME), which enable security analysts to validate how the detection logic works and prioritize the features contributing to the most critical threats. Evaluated on two benchmark datasets, MAL-XSEL outperformed conventional machine learning models, achieving top accuracies of 99.62% (ClaMP dataset) and 99.16% (Malware-DataSet). The paper "*Artificial Intelligence in Manufacturing Industry Worker Safety: A New Paradigm for Hazard Prevention and Mitigation*" pioneers the application of AI for proactive safety management, highlighting the potential of machine learning in protecting human lives in industrial environments. Meanwhile, "*A Comparative Review of IG-541 System Use in Total Flooding Application for Energized Electrical Fire*" and "*A Two-Level Facility Layout Design Method with the Consideration of High-Risk Facilities in Chemical Industries*" underscore the critical role of design optimization and fire safety strategies in managing industrial risks. Another notable paper focuses on "*Multi-objective Large-scale ALB Considering Position and 2 Equipment Conflicts Using an Improved NSGA-II*". An improved NSGA-II algorithm is developed by incorporating artificial immune algorithm concepts and neighborhood search. A dominance rate and concentration-based selection strategy, along with refined crossover and mutation operators, enhances search efficiency. Case studies show that the improved algorithm achieves the same optimal number of workstations as the traditional NSGA-II but improves workload balance by 5%, reduces costs by 2%, and yields 76% higher hyper-volume and 133% more Pareto front solutions. Collectively, these studies represent the emerging paradigm of intelligent safety systems—where automation not only enhances productivity but also safeguards people, assets, and the environment.

 https://doi.org/10.3390/pr14010036

Taken together, these papers illustrate a shared vision: the integration of sustainability, intelligence, and safety as the new triad of engineering excellence. They collectively reflect how research communities are responding to the challenges of climate change, industrial safety, and digital transformation—not in isolation, but through interconnected innovation. The convergence of environmental science, artificial intelligence, and human-centered design marks a profound shift toward the holistic evolution of industrial systems.

In conclusion, this Special Issue stands as the spirit of innovation and collaboration that defines today's engineering research. The diversity of topics—spanning water quality, renewable energy, process automation, human–machine interfaces, and safety systems—reflects the multidisciplinary fabric of sustainable industrial development. On behalf of the editorial team, we extend sincere appreciation to all authors, reviewers, and conference organizers for their invaluable contributions. It is our hope that this collection of works will inspire further interdisciplinary research, foster meaningful collaborations, and ignite new ideas toward building an intelligent, safe, and sustainable future for all.

Author Contributions: Investigation, T.S.L.; Writing—Original draft preparation, T.S.L.; Writing—Review and editing, T.S.L. and T.S.Y.C. All authors have read and agreed to the published version of the manuscript.

Conflicts of Interest: The authors declare no conflicts of interest.

List of Contributions

1. Xu, G.; Xu, S.; Wang, Y. A Two-Level Facility Layout Design Method with the Consideration of High-Risk Facilities in Chemical Industries. *Processes* **2025**, *13*, 161. https://doi.org/10.3390/pr13010161.
2. Alias, N.; Abdullah, L.; Yaw, T.; Jamil, S.; Ting, T.; Asis, A.; Lee, C.; Adeyi, A. Esterification of Kenaf Core Fiber as a Potential Adsorbent for Oil Removal from Palm Oil Mill Effluent (POME). *Processes* **2025**, *13*, 463. https://doi.org/10.3390/pr13020463.
3. Loo, K.; Lee, T.; Bee, S. A Comparative Review of IG-541 System Use in Total Flooding Application for Energized Electrical Fire. *Processes* **2025**, *13*, 485. https://doi.org/10.3390/pr13020485.
4. Patrício, L.; Varela, L.; Silveira, Z. Implementation of a Sustainable Framework for Process Optimization Through the Integration of Robotic Process Automation and Big Data in the Evolution of Industry 4.0. *Processes* **2025**, *13*, 536. https://doi.org/10.3390/pr13020536.
5. Aljuaid, A. Prototype of a Multimodal Platform Including EEG and HRV Measurements Intended for Neuroergonomics Applications. *Processes* **2025**, *13*, 1074. https://doi.org/10.3390/pr13041074.
6. Carpitella, S.; Izquierdo, J. Trends in Sustainable Inventory Management Practices in Industry 4.0. *Processes* **2025**, *13*, 1131. https://doi.org/10.3390/pr13041131.
7. Khurram, M.; Zhang, C.; Muhammad, S.; Kishnani, H.; An, K.; Abeywardena, K.; Chadha, U.; Behdinan, K. Artificial Intelligence in Manufacturing Industry Worker Safety: A New Paradigm for Hazard Prevention and Mitigation. *Processes* **2025**, *13*, 1312. https://doi.org/10.3390/pr13051312.
8. Hemdan, E.; Alshathri, S.; Elwahsh, H.; Ghoneim, O.; Sayed, A. MAL-XSEL: Enhancing Industrial Web Malware Detection with an Explainable Stacking Ensemble Model. *Processes* **2025**, *13*, 1329. https://doi.org/10.3390/pr13051329.
9. Pérez-Briceño, C.; Ponce, P.; Mei, Q.; Fayek, A. A Type-2 Fuzzy Logic Expert System for AI Selection in Solar Photovoltaic Applications Based on Data and Literature-Driven Decision Framework. *Processes* **2025**, *13*, 1524. https://doi.org/10.3390/pr13051524.
10. Polo, S.; Rubio, E.; Ayllón, J.; de Agustina, B. Emerging Advances in Sustainable Manufacturing. *Processes* **2025**, *13*, 1549. https://doi.org/10.3390/pr13051549.
11. Rizvi, S.; Rustum, R. Analyzing the Impact of Orifice Size and Retention Time in Private Tanks on Water Quality Indicators in Distribution Networks. *Processes* **2025**, *13*, 1674. https://doi.org/10.3390/pr13061674.

12. Alias, S.; Ramarad, S.; Ng, L.Y.; Murugappan, V.A.; Low, J.B.C.; Leng, F.P.; Leam, J.J.; Ng, D.K.S. Scaling up a Heater System for Devulcanization of Off-Spec Latex Waste: A Two-Phase Feasibility Study. *Processes* **2025**, *13*, 4062.
13. Li, H.; Gao, Y.; Kong, F.l.; Zhang, Z.; Song, G. Multiobjective Largescale Assembly Line Balancing with Parallel Stations and Equipment Constraints using an Improved NSGA-II. *Processes* **2025**, *13*, 3574.

References

1. Khoo, T.L.; Lee, T.S.; Bee, S.-T.; Ma, C.; Zhang, Y.-Y. A Comparative Review of Large Language Models in Engineering with Emphasis on Chemical Engineering Applications. *Processes* **2025**, *13*, 2680. [CrossRef]
2. Zhao, K.; Che, X.; Wei, C.; Tang, Z.; Yu, H.; Wang, D.; Wang, J.; Zhang, L. The Molecular Modeling, Simulation, and Design of Base Oils and Additives in Lubricating Oils: A Review. *Processes* **2024**, *12*, 2407. [CrossRef]
3. Briceño, C.P.; Ponce, P.; Fayek, A.R.; Anthony, B.; Bradley, R.; Peffer, A.; Meier, A.; Mei, Q. Optimizing Solar PV Deployment in Manufacturing: A Morphological Matrix and Fuzzy TOPSIS Approach. *Processes* **2025**, *13*, 1120. [CrossRef]

 https://doi.org/10.3390/pr14010036

processes

MDPI

Review

A Comparative Review of IG-541 System Use in Total Flooding Application for Energized Electrical Fire

Kheng Hooi Loo [1,*], Tin Sin Lee [1,*] and Soo Tueen Bee [2]

1 Department of Chemical Engineering, Lee Kong Chian Faculty of Engineering and Science, Universiti Tunku Abdul Rahman, Bandar Sungai Long, Kajang 43000, Selangor, Malaysia
2 Department of Mechanical and Material Engineering, Lee Kong Chian Faculty of Engineering and Science, Universiti Tunku Abdul Rahman, Bandar Sungai Long, Kajang 43000, Selangor, Malaysia; beest@utar.edu.my
* Correspondence: khenghooi.loo@ten.com (K.H.L.); leets@utar.edu.my (T.S.L.)

Abstract

Clean agent fire suppression systems are commonly used to protect areas containing valuable or critical equipment, especially in data centers and electrical substations, where traditional fire suppression methods are less effective or pose additional risks. This review evaluates the IG-541 fire suppression system as an alternative to halocarbon-based agents like HFC-227ea and FK-5-1-12, which are being phased out under environmental regulations, focusing on their application in energized electrical fires. IG-541 offers environmental advantages, including zero ozone depletion potential, no global warming potential, and negligible atmospheric lifetime, making it compliant with stringent environmental regulations. This review compares IG-541 with halocarbon agents across parameters such as extinguishing efficacy, safety considerations, environmental impacts, cost impacts, and system design considerations. Key findings underscore IG-541's effectiveness in reducing fire damage without producing harmful by-products or exacerbating climate change. Furthermore, the study highlights the regulatory frameworks influencing the phase-out of halocarbon agents and the transition toward environmentally sustainable alternatives. While IG-541 emerges as a promising replacement for halocarbon agents, further exploration into its application in varied fire scenarios and energy-intensive environments is recommended to optimize its deployment.

Keywords: clean agent; IG-541; HFC-227ea; FK-5-1-12; halocarbon agents; energized electrical fires; total flooding application

1. Introduction

Clean agent fire suppression systems are commonly used in the building industry to safeguard facilities, including computer rooms, data centers, switch rooms, motor control centers, process equipment, and other critical assets. These systems are typically implemented to secure high-value assets, ensure company continuity, and safeguard people and processes. These systems are ideal for protecting delicate equipment, documents, or artefacts since they leave no residue after discharge. Clean agent fire suppression systems consist of extinguishing agent storage cylinders connected to a network of discharge pipework and nozzles that deliver the extinguishing agent to the room being protected. They are often equipped with dedicated fire detection and control systems that provide an automatic discharge if a fire incident happens [1]. The tremendous development in computing demand for artificial intelligence has accelerated the global expansion of data centers. According to predictions by the International Energy Agency (IEA), energy consumption in data centers could double by 2026 due to the rise in power-intensive workloads, such as AI and cryptocurrency mining [2]. Therefore, selecting clean agent fire extinguishing systems to safeguard the essential and costly equipment in data centers is of utmost importance to the owners. This is being done to prevent any potential harm to the assets and to avoid any interruptions in operations. In addition to data centers, most of the industrial and manufacturing facilities, as well as the oil and gas industry, in use today run on electricity through an electrical substation. A high-voltage electric power system facility called a substation is an essential element of any industry. These are the places in the power network where busbars and transformers are linked for the purpose of power distribution and transmission via switchgears and circuit breakers. Arcs from motors, switches, and electric igniters are just a few of the sources of ignition that can occur in an electrical substation. The associated industry's frequent electrical fire events in substations, which cause injuries, fatalities, and property damage, must be prevented to protect people and property [3]. The advantages of clean agent fire suppression systems, particularly their effectiveness in suppressing energized electrical fires, have made them a preferred choice for installation in substations, ensuring the safety of both people and property.

Automatic sprinkler systems are the most common type of fixed fire protection system and have a long history of proven reliability. However, they are not suitable for certain hazards, such as live electrical equipment, flammable liquid fires, and any risks that would react violently with water. Dry chemical systems are effective against flammable liquid fires, including spray fires, but they provide minimal cooling and become ineffective once the powder settles. Additionally, powders are not recommended for use in occupied spaces due to post-fire cleanup issues. Low and medium-expansion foam systems are particularly effective for liquid pool fires. The foam forms a barrier between the fire and its oxygen supply while also cooling the fuel. However, since foams are aqueous solutions, they should not be used for risks that would react violently with water. While these traditional fire suppression systems perform well in specific applications, they are not sufficiently effective in some scenarios to replace the use of halon [4]. The purpose of this review is to assess the impact and effectiveness of the IG-541 agent fire suppression system as an alternative to the halocarbon agent-based fire suppression system, given the possibility that end users will eventually phase out the halocarbon agent entirely as part of the ESG (Environmental and Social Governance) approach to comply with the local standards and regulations. This review was conducted over various aspects, including performance in an electrical fire, safety considerations, environment impacts, economic considerations, and system design comparison.

2. Review Methodologies

This review systematically investigates IG-541 system use in total flooding application focusing on suppressing the energized electrical fire, providing a comprehensive comparison and analysis of the literature published between 1994 and 2024. The methodology employed was designed to ensure scientific rigor, comprehensive coverage, and relevance of the reviewed material.

Database selection: The review utilizes well-established academic databases to collect relevant literature. Major reference databases selected for the research include the National Institute of Standards and Technology (NIST), the National Fire Protection Association

(NFPA), the International Organization for Standardization (ISO), the United Nations Environment Programme (UNEP), and Scopus.

Search strategy: Specific and comprehensive search terms were developed, including keywords such as "IG-541", "FM200", "Novec 1230", "FK-5-1-12", "Inergen", "hfc-227ea", "total flooding application", "energized electrical fire", "halocarbon agent", "inert gas". The iterative search process ensured comprehensive coverage of the relevant literature.

Inclusion criteria: Studies were included based on their publication in peer-reviewed journals, conference proceedings, research reports, and book chapters. The selection criteria also required that these studies be published within the last 30 years, written in English, and directly related to the research topic.

Article selection: The initial search resulted in 418 articles, which were then evaluated for methodological quality and content relevance. During this process, duplicates and studies not related to the topic were excluded.

The selected articles were thoroughly analyzed and synthesized to offer a comprehensive and up-to-date overview of the state of the art on the topic of clean agent fire suppression systems in total flooding applications for energized electrical fires. This review paper makes significant contributions to the field of sustainable fire safety by providing an unparalleled breadth and depth of analysis.

3. Energized Electrical Fires Overview

Understanding the chemistry between fire and electricity is essential, especially when examining how electrical fires form. Electrical fire is a fire that is directly caused by the flow of electric current or a discharge of static electricity if it is not the result of normal and purposeful heat produced by an electrical device. Electric current is a form of energy resulting from the movement of electrons, while static electricity involves electric charges that accumulate on the surface and remain stationary rather than flowing as a continuous electric current [5]. Heat energy may be produced from electrical energy in several ways, including heat generated by lightning, heat from arcing, resistance heating, dielectric heating, induction heating, static electricity heating, and leakage current heating. Arc/spark and hot components are two broad categories of potential sources of ignition in electrical systems. An arc or spark is an electric discharge that typically produces a popping sound. This phenomenon occurs when the air or gas between two charged conductors becomes highly ionized, leading to a breakdown that allows current to flow through a visible, luminous channel. On the other hand, a hot component refers to an element that releases enough energy to ignite a surrounding combustible mixture when exposed to any ignition sources. The discharge of capacitive circuits, the interrupting (opening) of inductive circuits, the opening or closing of resistive circuits with slow, intermittent interruption increasing the igniting capacity (hazard), and hot temperature sources can all act as ignition sources due to the electrical system. The mechanisms for ignition can be found in fuses, relay contacts, switch contacts, short circuits caused by damage or component failure, and arcing over between conductors or components [6]. According to NFPA 921, for ignition to occur from an electrical source, the electrical wiring, equipment, or component must be energized by the building's wiring, an emergency system, a battery, or some other source. Additionally, sufficient heat and temperature to ignite a close combustible material must have been produced by electrical energy at the point of origin by the electrical source. The ignition by electrical energy involves transferring sufficient heat to a fuel (i.e., competent ignition source) by passage of electrical current to ignite material that is close [7]. Electrical fires are a significant type of structure fire, but the mechanisms that lead to their ignition are often uniquely specialized and require specific research. Key areas needing investigation include the physical mechanisms of ignition from electrical current, the minimum voltage, current, or power levels necessary for an electrical fire to occur, the time frame for ignition, electrical fires related to medium- and high-voltage circuits, and the fire hazard or risk associated with code-compliant electrical installations [8].

Some electrical faults associated with wiring or wiring devices lead to fire situations that can involve a two-step sequence: first, overheating, then arcing and igniting. For example, a wire may get heated due to excessive current or a poor connection. This degradation can result in a short circuit, especially when the wire bends or passes over a metal edge [9]. The thermal conductivity of the wire core significantly influences the spread rate, potentially having an impact comparable to that of the wire's dimensions. Generally, a wire with a smaller diameter and higher conductance (due to a larger wire core and greater thermal conductivity) will experience faster flame spread, indicating a higher fire hazard, although it may be more challenging to ignite. During flame spread, molten polymer dripping is observed in thicker-coated wires. This dripping occurs more frequently in wires with higher conductivity and at lower pressures [10]. Solid polymers do not burn directly; their volatile decomposition products ignite in the gas phase when mixed with oxygen. Ignition occurs when these gases reach a lower flammability limit for the particular fuel–air mixture. When the surface temperature of plastics at ignition is close to the thermal degradation temperature, the thermal degradation reactions on the plastic surface occur more quickly than the rate at which heat is absorbed. As a result, the heat transfer process primarily controls the burning rate, heat release rate, and smoke production during flaming combustion. The three coupled processes needed for flaming combustion are shown in Figure 1: (1) heating the polymer; (2) thermal decomposition of the solid polymer to gaseous fuel; and (3) ignition and combustion of the fuel gases in the air. Heat is applied to the polymer surface via an ignition source (such as arc/spark and hot component) or thermal feedback of radiant energy from the flame, which results in the thermolysis of the polymer molecules' primary chemical bonds. Low-molar-mass degradation products evaporate, and when they react with air (oxygen) in the combustion zone above the surface, heat is released along with carbon dioxide, water, and incomplete combustion products such as soot, mineral acids, carbon monoxide, and unburned hydrocarbons [11]. The clean agent fire suppression system is introduced to break the fire triangle at one or more places to prevent burning. Halocarbon agents disrupt the uninhibited chain reaction of combustion by absorbing the thermal energy of a fire, preventing the combustion reaction from continuing. Inert agents reduce the oxygen level to a point where the fire can no longer sustain itself.

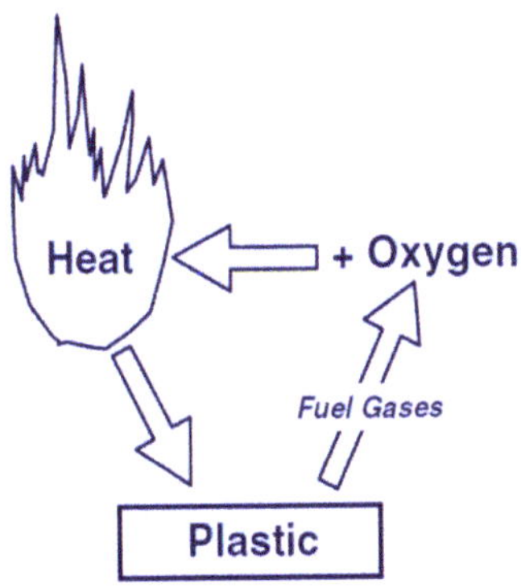

Figure 1. The fire triangle. Heating of the plastic generates volatile thermal degradation products (fuel gases) that mix with air, forming a combustible mixture. Ignition of the combustible mixture releases heat that continues the burning process [11].

Compartment fire development can be divided into four stages: incipient, growth, fully developed, and decay. Flashover is not a stage itself but a rapid transition between the growth and fully developed stages. The progression of an incipient fire is largely influenced by the fuel's characteristics and configuration, making it a fuel-controlled fire, as shown in Figure 2a. The air in the compartment provides sufficient oxygen for the fire to continue developing. If there is enough oxygen, additional fuel will ignite, increasing the fire's heat release rate. As the gases in the compartment heat up, they expand and increase pressure within the confined space. The fire can grow further through flame spread or by igniting other fuels in the compartment. Flashover is the sudden transition from the growth stage to a fully developed fire, resulting in total surface involvement of all combustible materials in the compartment. Flashover conditions are defined in various ways, and it is important to note that flashover does not always occur; sufficient fuel and oxygen are required for it to happen. In the post-flashover stage, energy release is at its peak but is generally limited by ventilation. A compartment fire may enter the decay stage as the available fuel is consumed or due to limited oxygen [12]. Although some oxygen remains, visible flames diminish, and the fire continues to smolder. High heat and smoke conditions persist, creating the potential for a backdraft. A backdraft is a smoke explosion that can occur when additional oxygen is introduced into a smoldering fire, causing the heated gases to ignite explosively. Warning signs of a possible backdraft include heavy, dense smoke with no visible flames in a tightly closed space, black smoke pushing out around closed doors or window frames, smoke-stained glass, and pulsating pressure from the fire, as well as the reversal of air movement pulling smoke back into the building through a doorway [13]. An idealized curve of a ventilation-controlled fire is depicted in Figure 2b. Similar to a fuel-controlled fire, it begins with ignition and progresses through the growth stage. However, as high heat-of-combustion fuels burn within the nearly airtight compartment, the fire enters a decay stage due to the limited availability of oxygen necessary for combustion. As the oxygen supply diminishes, the fire's heat-release rate and the gas temperatures within the compartment decrease. If a door or window is opened while the fire is still burning, albeit at a reduced intensity, and additional fuel is present, the influx of outside air can cause a rapid increase in the fire's heat-release rate. This can generate enough energy to cause a flashover in the room, a phenomenon known as a ventilation-induced flashover. Once sufficient oxygen is available, the fire can reach the fully developed stage and may become fuel-controlled again until it eventually decays or until suppression by the fire department [14].

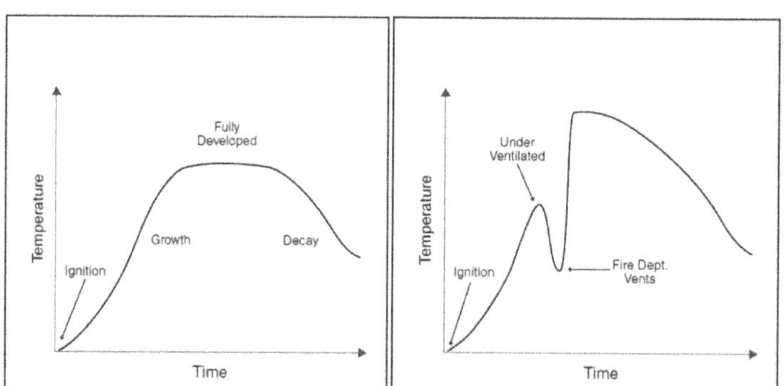

(a) (b)

Figure 2. Idealized fire behavior graph: (**a**) typical fuel-controlled fire; (**b**) typical ventilation-controlled fire [14].

The typical four stages of fire growth for an electrical fire in a data center, along with the possible responses, are listed below [15]:

(1) Incipient stage—Overheating of equipment/circuits and trace levels of combustion gasses equivalent to the lowest amount detectable by an aspirating system without other detection. The possible response will involve alertness and action by the occupant, accompanied by the activation of a pre-alarm signal.
(2) Smoldering (visible smoke) stage—Increased burning detectable by human smell triggers spot-type smoke detection and activates the highest alert level for aspirating systems. The possible response will involve action by the occupant and pre-alarm signal activation.
(3) Flaming stage—Pyrolysis and flaming combustion activate multiple spot-type detectors, leading to increased room temperature and the development of an upper gas layer. The possible responses include activating the fire alarm, shutting down the HVAC system, and initiating either the clean agent system countdown or releasing the solenoid valve in the pre-action system.
(4) Fire growth/spread stage—The rapid production of large quantities of smoke quickly activates multiple spot detectors, leading to a rapid acceleration in heat release and the activation of the nearest sprinkler. The possible responses include activating the fire alarm, triggering the sprinkler or gaseous system discharge, and closing the fire/smoke dampers.

The four stages of fire growth for an electrical fire in a data center can also be applied to other electrical-related facilities, such as substations and telecommunications, as these environments primarily contain cables and electrical equipment. Clean agent fire suppression systems can be designed to activate early, during the flaming stage of a fire, ensuring minimal fire damage since the clean agent does not harm assets or occupants. In contrast, pre-action sprinkler systems only activate when the sprinkler heads reach a certain temperature, resulting in more downtime compared to clean agent systems and causing greater fire damage to the building.

While it is generally recommended to disconnect electrical equipment from its power source before attempting to extinguish a fire, equipment owners may opt to keep the equipment running due to considerations such as the risk of casualties from power shutdown, hazards to people or property posed by the fire, economic losses from interrupted functionality, or financial losses from facility damage. Therefore, there are situations in which the design of fire suppression systems is predicated on the presence of electrical equipment that has been activated. Unfortunately, there is currently no accepted test procedure to determine the exact quantity of agent required to put out flames in situations where the input of energy from an electrical source might increase combustion. Various test methods have been employed to simulate the effects on energized electrical equipment. These methods include tests designed to suppress flames during realistic electrical failure events involving representative polymeric materials, as well as tests that aim to control key parameters such as external heat flux (EHF) [16]. Several studies have been conducted to simulate the impact of different clean agent gas concentrations on energized polymeric materials and metal surfaces. The test findings typically demonstrate that higher agent concentrations are necessary in the presence of energy input from various sources [17–21]. Unlike NFPA 2001 [22], which clearly defines energized electrical fire as a Class C fire and specifies the minimum design concentration for each clean agent gas, ISO 14520-1 [23] provides a guideline for applications using energizing electrical equipment because it recognizes the more challenging nature of energized electrical fires. Under NFPA 2001's Section 5.4.2.5, the Class A minimum flame extinguishing concentration gives a safety factor of 1.35, which yields the minimum design concentration for a Class C hazard. In contrast, energized electrical fires have been designated as higher-hazard class A hazards because they exhibit the characteristics mentioned in ISO 14520-1's caution statement in 7.5.1.3. As a result, the minimum design concentration for an energized electrical fire shall be higher than that of surface class A or 95% of the class B minimum design concentrations. As shown in Table 1, ISO 14520 specifies a more cautious and higher minimum design concentration for each clean agent gas than NFPA 2001. It was explicitly stated in ISO-14520 that the minimum design concentration of higher hazard class A should be applied to the following conditions: cable bundles greater than 100 mm in diameter; cable trays with a fill density greater than 20% of the tray cross-section; horizontal or vertical stacks of cable trays (closer than 250 mm) apart; and equipment energized during extinguishment period where the collective power consumption exceeds 5 kW. In contrast, NFPA 2001 merely indicates that a higher extinguishing concentration may be necessary for these conditions.

Table 1. Comparison of NFPA 2001 and ISO 14520 minimum design concentrations for each clean agent gas on class C fire.

Agent	Minimum Design Concentration (vol%)	
	NFPA 2001 [22]	ISO 14520
FK-5-1-12	4.5	5.6 [24]
HFC-125	9.0	11.5 [25]
HFC-227ea	7.0	8.5 [26]
HFC-23	20.3	16.3 [27]
IG-541	38.5	41.7 [28]
IG-55	42.7	45.1 [29]
IG-100	41.9	41.5 [30]

4. Clean Agent Gaseous Fire Suppression System Overview

Fire protection applications can generally be classified into two main categories: (1) those that permit the use of water-based sprinklers and (2) special hazards that necessitate the use of alternative fire extinguishing agents, such as carbon dioxide, halon, halon replacements, dry chemicals, wet chemicals, or foams. From the 1920s to the 1960s, carbon dioxide was the predominant gaseous fire suppression agent. However, beginning in the 1960s, halon-based systems became extensively utilized. The application of carbon dioxide is now largely restricted due to factors that limit its effectiveness and the health risks it poses [31]. The extinguishing mechanism of carbon dioxide involves displacing oxygen to a level below that required for combustion. Consequently, carbon dioxide is not recommended for use as a fire suppressant in occupied spaces, as the concentration needed for fire extinguishing is lethal to humans. Halons 1301, 1211, and 2402 are highly effective and important fire-extinguishing chemicals, but they are also potent ozone-depleting substances (ODSs). Currently, the availability of halons depends entirely on recovered and recycled materials. Defined in Group II of Annex A of the Montreal Protocol, halons are a class of bromine-containing halogenated chemicals that have been, and in some cases continue to be, used as gaseous extinguishing agents in various fire and explosion protection applications [32]. In 1987, the Montreal Protocol began restricting the production and use of halons due to their significant environmental impact, particularly their role in ozone depletion and global warming. Consequently, halon production was phased out in Montreal Protocol countries by 1994, though it was allowed in Article 5(1) countries until 2009 [33]. The discontinuation of halon production has significantly affected the fire and explosion protection industry. Halons were known for being clean, non-conductive, safe for humans, and highly effective. However, identifying appropriate substitutes for their various applications continues to be a challenge for fire protection experts. To reduce reliance on halons, the use of conventional non-halon fire protection materials has been encouraged. Additionally, new clean agent replacement chemicals and alternative technologies have been rapidly introduced [34]. According to the Halons Technical Options Committee, several gaseous fire extinguishing agent technologies have been commercialized as alternatives to Halon 1301 for use in total flooding applications that are suitable for occupied spaces, as shown in Table 2. Many studies have been conducted to compare the performance of alternative agents to Halon 1301 in various contexts [35–38]. As shown in Table 2, the clean agents can be categorized into two main types: halocarbons and inert gases. Both types are used in total flooding extinguishing systems, where the agent is discharged throughout an entire protected area to quickly reach the minimum design concentration needed for fire extinguishment. The minimum design concentrations vary depending on the international standards selected, the type of agent, and the specific hazard to be protected. HFC-227ea and FK-5-1-12 are the most commonly used halocarbon agents in the market for combating energized electrical fires in total flooding applications. Therefore, this review will compare these agents with IG-541, a widely used inert gas, from various aspects.

Table 2. Alternative gaseous fire extinguishing agent to halons for suitable use in total flooding applications [39].

Agent	Composition
	Inert Gases
IG-01	Argon, Ar
IG-100	Nitrogen, N_2
IG-541	Nitrogen, 52 vol%; Argon, 40 vol%; Carbon dioxide, 8 vol%
IG-55	Nitrogen, 50 vol%; Argon, 50 vol%
	Hydrofluorocarbons
HFC-125	C_2HF_5—Pentafluoroethane
HFC-23	CHF_3—Trifluoromethane
HFC-227ea	CF_3CHFCF_3—1,1,1,2,3,3,3-heptafluoropropane
HFC-236fa	CF_3CHFCF_3—1,1,1,2,3,3,3-heptafluoropropane
HFC Blend B	HFC-143a, CH_2FCF_3, 1,1,1,2-tetrafluoroethane, 86 wt%, HFC-125, C_2HF_5, pentafluoroethane, 9 wt%, Carbon dioxide, CO_2, 5 wt%
	Perfluoroketone
FK-5-1-12	$CF_3CF_2C(O)CF(CF_3)_2$—Dodecafluoro-2-methylpentan-3-one
	Hydrochlorofluorocarbons
HCFC Blend A	HCFC-22, $CHC1F_2$—Chlorodifluoromethane, 82 wt%, HCFC-124, $CHC1F$-CF_3, 1-Chloro-tetrafluoroethane, 9.5 wt%, HCFC-123, $CHC1_2$-CF_3, 1,1-dichloro-trifluoroethane, 4.75 wt%, Isopropenyl-1-methylcyclohexane, 3.75 wt%

A clean agent gaseous fire suppression system is an engineered set of components designed to detect and extinguish fire quickly while alerting occupants. It works by filling the protected area with a gas or chemical extinguishing agent to prevent extensive damage [40]. As illustrated in Figure 3, the typical elements and components are as follows:

(1) Discharge nozzle—It is used to distribute the extinguishing agent into the protected area.

(2) Piping system—It is used to transport the extinguishing agent (inert gas or halocarbon agent) from its storage container to the discharge nozzle.

(3) Control Panel—It integrates all devices and displays their operational status and condition.
(4) Discharge or warning alarms—Electronic devices that provide an audible or visual alarm when detected.
(5) Hazard warning or caution signs—It must be posted at the entrance to, and inside, areas protected by fixed extinguishing systems.
(6) Automatic fire detection devices—A device that detects fire and generates an alarm signal.
(7) Manual discharge station—A device that provides a way to manually discharge the fire extinguishing system.
(8) Storage containers and extinguishing agent—The storage system discharges agent into the piping and through the discharge nozzles when activated by a manual or automatic device.

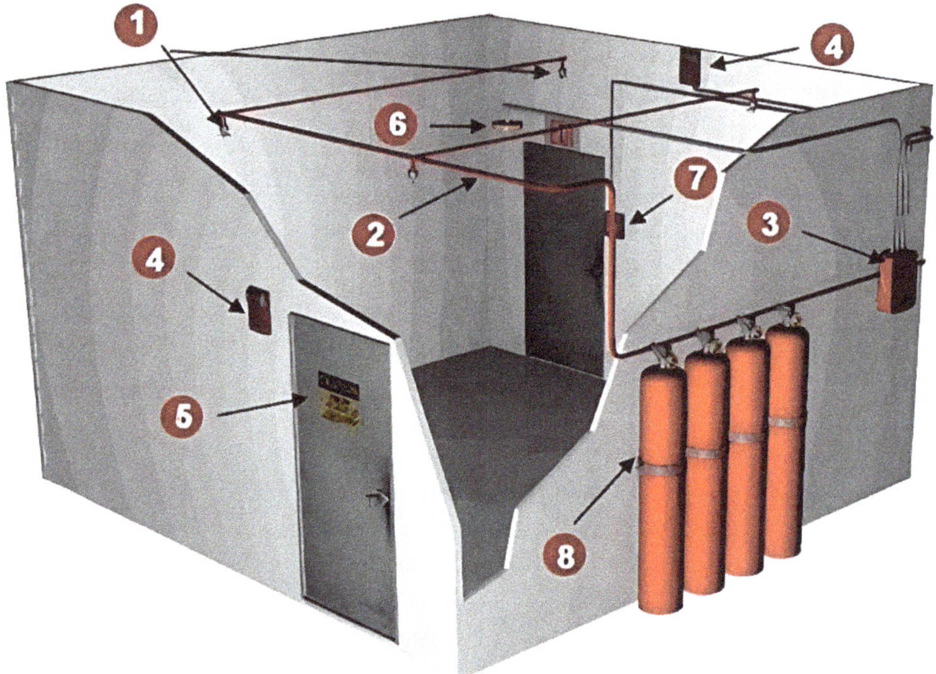

Figure 3. Typical arrangement of the clean agent gaseous fire suppression system and the components are as follows: (1) discharge nozzles; (2) piping system; (3) control panel; (4) discharge or warning alarms; (5) hazard warning or caution signs; (6) automatic fire detection devices; (7) manual discharge station; (8) storage containers and extinguishing agent [40].

In the event of a fire, clean agent systems are activated by a suppression releasing panel that detects the fire using an automatic detection system. Once a fire has been detected, a releasing sequence starts, often with a delay to allow occupants to evacuate. Notification appliances in the protected area sound an alarm for a predetermined duration before the system is activated. The gas is released from the cylinders by the releasing panel via an electronic signal to a solenoid valve on the storage cylinders. The gas flows through the piping system and is discharged through open nozzles to flood the protected enclosure [41]. The clean agent gaseous fire extinguishing system can also be configured with a selector valve system to protect multiple hazard areas using a shared cylinder bank. In this setup, the manifold is equipped with normally closed selector valves, each featuring its own actuator and solenoid valve assembly. When activated, these valves release the clean agent into the designated hazard area. A portion of the total clean agent cylinders can be allocated to a specific hazard, facilitated by the use of non-return (check) valves in the cylinder pilot line. To manage pressure, a restrictor is installed at the outlet of each selector valve, ensuring controlled flow into the distribution piping. Optional isolation (lockout) valves may also be installed upstream of the selector valves. These valves must remain supervised in an open position and should only be closed during system servicing or routine maintenance [42].

4.1. Halocarbon Agent

These chemical agents contain chlorine, fluorine, or iodine, either individually or in combination. The categories of these agents include hydrochlorofluorocarbons (HCFCs), hydrofluorocarbons (HFCs), perfluorocarbons (PFCs), and fluoroidocarbons (FICs). They share several common characteristics with halons: they are all electrically nonconductive, clean agents and exist as liquefied gases or compressed liquids. However, these agents vary significantly in terms of ozone depletion potential, toxicity, volume requirements, cost, environmental impact, and availability. For instance, PFCs are recommended only for use in situations where no other agent meets the technical performance or safety requirements. All of these systems require precise design to avoid decomposition and prevent the production of HF. It is important to note that these agents can be stored and discharged using systems similar to those used for Halon 1301 [43]. Halocarbon systems are effective in protecting against Class A, Class B, and Class C fires. Generally, all halocarbons extinguish fires through a similar mechanism. As a halocarbon gas approaches a fire, it absorbs significant amounts of heat energy from the combustion process, thereby lowering the surface temperatures of combustibles below their ignition point. Additionally, halocarbons release highly reactive halogen ions that combine with free hydrogen, which is a secondary fuel source released during a hydrocarbon fire. This elimination of free hydrogen helps to prevent the fire from spreading while also cooling the temperatures at the heat source. The suitability of a halocarbon fire extinguishing agent for use in occupied areas (i.e., areas with people) or solely in unoccupied areas (i.e., areas without people) is determined by its toxicity [44].

Hydrofluorocarbons (HFCs) are second-generation substitutes for compounds that contribute to human-caused stratospheric ozone depletion. Since they do not contain chlorine, bromine, or iodine, their ozone depletion potentials are nearly zero. With the phase-out of chlorofluorocarbons (CFCs) and their first-generation replacements of hydrochlorofluorocarbons (HCFCs), HFCs have gained increasing significance [45]. In the early 1990s, Great Lakes Chemical Corporation filed a patent for HFC-227ea to serve as a replacement for Halon 1301 in special hazards fire suppression systems [46]. Then, DuPont Fluoroproducts reached a patent cross-licensing agreement with Great Lakes Chemical Corporation (GLCC) regarding the use of HFC-227ea in fire protection products, marketed under the trademarks FE-227 and FM200, respectively. HFC-227ea, also known as 1,1,1,2,3,3,3-heptafluoropropane with the structural formula CF3CHFCF3, has been listed as an acceptable substitute for Halon 1301 and for new applications in the US EPA's SNAP program for both total flooding and streaming systems [47]. While HFC-227ea contains fluoride, there are currently no outright bans on its use. However, some countries have implemented restrictions that limit the sale of HFC-227ea fire protection systems. Due to its global warming potential being greater than 1, HFC-227ea is subject to F-Gas regulations.

The primary extinguishing mechanism of HFC-227ea involves heat removal as the agent decomposes in the presence of flames, with some additional chemical activity potentially occurring. Since HFC-227ea is heavier than air, uncontrolled leakage at low levels can negatively impact the maintenance of agent concentration after discharge [48]. Table 3 presents the chemical and physical properties of HFC-227ea, as documented by PubChem.

Table 3. Chemical and physical properties of HFC-227ea [49].

Properties	
Chemical name	1,1,1,2,3,3,3-Heptafluoropropane
Chemical formula	C_3HF_7
Molecular structure	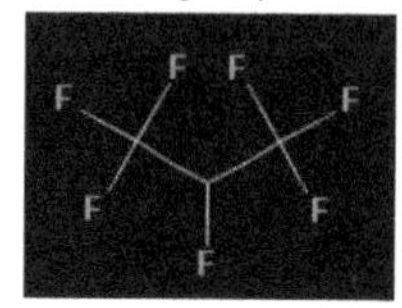
PubChem CID	67940
CAS number	431-89-0
Molecular weight	170.03 g/mol
Color/form	Colorless gas
Odor	Light, ethereal odor
Boiling point	−16.34 °C
Melting point	−126.8 °C
Density	1.394 g/cu cm at 25 °C
Critical density	0.582 g/cu cm
Vapor density	4.2 at 20 °C (Air = 1)
Vapor pressure	390 kPa at 20 °C
Viscosity	0.0132 mPa.s at 25 °C (gas); 0.184 mPa.s (liquid)
Latent heat of vaporization	132.7 kJ/kg
Refractive index	1.407 at 25 °C/D

In 2004, the 3M Company introduced Novec 1230 fluid as a next-generation halon replacement with a low environmental impact and a high margin of safety [50]. The complete chemical name for Novec 1230 fluid is dodecafluoro-2-methylpentan-3-one with the structural formula $CF_3CF_2C(O)CF(CF_3)_2$. In the ASHRAE nomenclature, which is used in the NFPA and ISO 14520 clean agent standards, it is designated as FK-5-1-12. This substance is also referred to as "waterless water" or "dry water". FK-5-1-12 was introduced as an environmentally friendly alternative to HFC-based gaseous agents like FM200 (HFC-227ea), FE13 (HFC-23), and ECARO (HFC-125), which are being phased out under the EU's F-gas Regulation. FK-5-1-12 extinguishes fires primarily through cooling. Due to its high density, it is important to consider the design of the compartment where it will be used, particularly with regard to low-level uncontrolled ventilation paths [51]. It has proven to be an excellent alternative to ozone-depleting compounds, offering an ideal combination of performance, safety, environmental benefits, and commercial acceptance [52]. As a fluoroketone agent, it represents a valuable class of compounds capable of producing nonflammable molecules with short atmospheric lifetimes, making it a sustainable alternative to halons in fire protection. The brief atmospheric lifetimes of these compounds lead to low global warming potentials (GWPs). These environmental characteristics are significantly different from those of halons and their first-generation replacements, such as HFCs and PFCs [53]. It combines with air to form a gaseous mixture upon discharge. The heat capacity of this agent/air combination is significantly higher than that of air by itself. This gas combination will absorb more energy or heat for every degree that its temperature changes because of its increased heat capacity. The agent/air combination absorbs enough heat at the system design concentration to disturb the circumstances necessary for combustion to take place. The presence of the agent increases the heat loss from the fire to its surroundings. This causes the combustion zone to cool to the point where the fire is extinguished [54]. FK-5-1-12 has sufficient vapor pressure and a low heat of vaporization, allowing it to achieve and maintain gas concentrations well above the levels needed for effective fire extinguishing. Adiabatic calculations indicate that the room air and incoming liquid agent provide enough sensible heat to vaporize the agent sufficiently for extinguishment. This is similar to HFC-227ea, with comparable energy requirements for evaporation [55]. Table 4 presents the chemical and physical properties of FK-5-1-12 as documented by PubChem and the 3M company. Among halocarbon agents, FK-5-1-12 is undoubtedly the best choice for use in total flooding applications due to its low global warming potential (GWP) and zero ozone depletion potential (ODP), as indicated in Table 5. Additionally, there is no international mandate for its phase-out. However, 3M has proactively decided to cease production because the PFAS substances present in the agents could potentially affect human health and the environment, as proposed by the European Chemicals Agency (ECHA).

Table 4. Chemical and physical properties of FK-5-1-12 [54,56].

Properties	
Chemical name	1,1,1,2,2,4,5,5,5-nonafluoro-4-(trifluoromethyl)pentan-3-one
Chemical formula	$C_6F_{12}O$, $CF_3CF_2C(O)CF(CF_3)_2$
Molecular structure	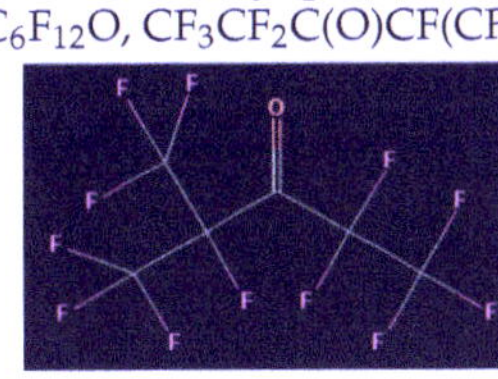
PubChem CID	2782408
CAS number	756-13-8
Molecular weight	316.04 g/mol
Physical state	Liquid
Color	Colorless
Boiling point	49.2 °C
Melting point	−108.0 °C
Density, saturated liquid	1.60 g/mol
Density, gas @ 1 atm	0.0136 g/mol
Heat of vaporization @ boiling point	88.0 kJ/kg
Critical density	639.1 kg.m^3
Liquid viscosity @ 0 °C/25 °C	0.56/0.39 centistokes
Vapor pressure	0.404 bar at 25 °C
Vapor density	11.6 (Air = 1)

Table 5. Potential environmental impacts for the halocarbon agents used in total flooding applications [22].

Agent	Global Warming Potential (GWP) [1]	Ozone Depletion Potential (ODP)
HFC-125	3170	0
HFC-23	12,400	0
HFC-227ea	3350	0
HFC-236fa	8060	0
HFC Blend B	1400	0
FK-5-1-12	<1	0
HCFC Blend A	1500	0.048

[1] GWP is reported over a 100-year integrated time horizon.

4.2. Inert Gas Agent

Most halon replacements, such as halocarbon agents, have a global warming potential and may be subject to future regulations under the Kyoto Protocol. Additionally, the use of these replacements for fire suppression often results in the production of toxic or corrosive by-products, like hydrogen fluoride and carbonyl fluoride, raising concerns about the safety of humans and property [57]. Therefore, it is highly desirable to find alternatives with performance characteristics similar to those of halocarbons. The latest commercially available inert gas systems use nitrogen gas, which serves both as the fire suppression agent and the pressurization agent. Nitrogen gas has been a staple in the fire suppression industry, primarily used as a pressurization gas in total flood–fire suppression systems and in portable fire extinguishers [58]. Inert gases, often known as noble gases or rare gases, include helium, neon, argon, krypton, xenon, and radon. These gases are chemically inert, which means they do not react with other elements or compounds. The term "inert gaseous fire extinguishing agents" refers to pure naturally occurring gases present in the earth's atmosphere (e.g., argon, neon, or helium) or a blend of inert gases with others, such as an argon–nitrogen combination. These substances are kept as non-liquefied gases. The term "passive agent" is sometimes used to describe inert gas extinguishing agents due to their method of fire suppression. These agents reduce the amount of oxygen available, making the fire insufficient to sustain combustion. Although CO_2 is not technically an inert gas, it is used in fire safety and has comparable qualities [59]. Unlike halocarbon agents, inert gas agents do not disrupt the chemical chain reaction or the cooling of a fire. According to the study by Rattananon and Patvicaichod, increasing the concentration of inert agents does not contribute to lowering the temperature inside the room [60]. There are four types of inert gas agents approved for use in total flooding applications in accordance with NFPA 2001. IG-541 is a mixture of three inert gases: nitrogen (52%), argon (40%), and carbon dioxide (8%). IG-01 consists solely of argon, an unblended inert gas. IG-100 represents nitrogen, another unblended inert gas. IG-55 is a 50/50 blend of two inert gases, nitrogen and argon.

Table 6 summarizes the chemical and physical properties of IG-541. IG-541 is an abundant, non-corrosive gas that neither supports combustion nor reacts with most substances. It consists solely of naturally occurring gases that have no adverse effects on the ozone layer or the environment as a whole. Unlike halons and halocarbon agents, IG-541 does not contain halogen components such as chlorine, fluorine, bromine, or iodine, resulting in zero ozone depletion potential. Furthermore, IG-541 is not classified as a chemical toxin and poses no mutagenic, teratogenic, or carcinogenic risks. It also does not affect the central nervous system or sensitize the heart, unlike the existing halons and halocarbon agents [61]. IG-541 offers a significant advantage over other types of inert gases due to its inclusion of carbon dioxide, which enhances respiratory characteristics and allows humans to survive in oxygen-deficient environments. The carbon dioxide in IG-541 increases blood flow and oxygen delivery to the brain, counteracting the effects of carbon monoxide, which reduces the amount of oxygen available to the brain by binding with hemoglobin. Typical individuals will not experience adverse effects from exposure to IG-541, even over extended periods. Additionally, individuals with pre-existing cardiac or pulmonary conditions are not negatively affected by IG-541 during evacuation and can safely exit the environment without impairment [62]. IG-541 extinguishes fires by reducing the oxygen content to a level that no longer supports combustion. When discharged into a room, IG-541 introduces a specific mixture of gases that enable a person to breathe in a low-oxygen environment. This mixture actually enhances the body's ability to absorb oxygen. Normally, a room's atmosphere contains about 21% oxygen and less than 1% carbon dioxide. If the oxygen content drops below 15%, most common materials will not burn. IG-541 reduces the oxygen content to around 12.5% while increasing the carbon dioxide content to approximately 3%. This rise in carbon dioxide boosts a person's respiration rate and the body's capacity to absorb oxygen [63].

Table 6. Chemical and physical properties of IG-541 [64].

Properties	
Trade name	Inergen
Chemical formula	52% N_2, 40% Ar, 8% CO_2
CAS number	7727-37-9, 7440-37-1, 124-38-9
Molecular weight	34.08 g/mol
Physical state	Gas
Color	Colorless
Odor	Odorless
Vapor density at 1 atm	1.416 kg/m^3 at 20 °C
Relative density at 1 atm	1.18 at 20 °C

4.3. Regulations and Standards

Halocarbon emissions contribute to ozone layer depletion and climate change. Therefore, preventing these emissions through the halocarbon agent regulatory framework shown in Figure 4 offers the dual benefits of protecting the ozone layer and mitigating climate change. International regulations and national government actions may affect the support for existing fire protection systems that use halocarbons and their future availability. Most countries have incorporated the requirements of ISO 14520 and NFPA 2001 standards into their local regulations to govern the type and usage of halocarbon agents. Both ISO 14520 and NFPA 2001 outline the minimum requirements for the design, installation, approval, and maintenance of total flooding and local application fire extinguishing systems that use halocarbon and inert gas agents. Any halocarbon agent intended for recognition or inclusion in this standard must first undergo an evaluation process equivalent to that used by the U.S. Environmental Protection Agency (EPA) Significant New Alternatives Policy (SNAP) Program for total flooding agents [22,23]. EPA's Significant New Alternatives Policy (SNAP) identifies and evaluates substitutes for end-uses that have traditionally relied on ozone-depleting substances (ODSs). These substitutes are assessed based on environmental and health risks, considering factors such as ozone depletion potential, global warming potential, toxicity, flammability, and exposure potential. The lists of acceptable and unacceptable substitutes for total flooding applications are updated multiple times each year [65]. The Montreal Protocol mandates a gradual reduction in the consumption and production of various ozone-depleting substances (ODSs), with distinct timelines for developed and developing countries. Under this protocol, hydrochlorofluorocarbons (HCFCs) used in fire protection are set to be phased out by 2020 in developed countries and by 2030 in developing countries. Hydrofluorocarbons (HFCs), introduced as non-ozone-depleting alternatives to facilitate the phase-out of HCFCs, are also subject to phase-down. Developed countries are expected to begin reductions in 2019, while developing countries will start by freezing HFC consumption levels in 2024, with some nations following in 2028 [33]. The Kyoto Protocol identifies carbon dioxide, methane, nitrous oxide, and the fluorochemicals HFCs, PFCs, and SF6 as the main gases contributing to the greenhouse effect and potentially subject to emission controls. It mandates that developed countries first halt the growth of their greenhouse gas emissions and then reduce them to 5.2% below 1990 levels. The Doha Amendment to the Kyoto Protocol established a second commitment period from 2013 to 2020. Initially entering into force in 2005 and ratified by 192 Parties, the Kyoto Protocol has since been superseded by the Paris Agreement. Unlike the Kyoto Protocol, which required only developed countries to reduce emissions, the Paris Agreement acknowledges that climate change is a global issue and calls on all countries to set their own emissions targets [66]. In the European Union, Regulation (EU) No. 517/2014 (known as the F-Gas Regulation) establishes rules to protect the environment by reducing emissions of fluorinated greenhouse gases. To achieve this, it establishes rules for the containment, use, recovery, and destruction of these gases, along with related ancillary measures. It also sets conditions for the market placement of specific products and equipment that contain or rely on fluorinated greenhouse gases, imposes conditions on certain uses of these gases, and establishes quantitative limits for the market placement of hydrofluorocarbons [67]. On 7 February 2023, the European Chemicals Agency (ECHA) proposed a ban on an entire class of per- and polyfluoroalkyl compounds. This restriction aims to reduce emissions of PFASs and their degradation products, which can persist in the environment longer than any other human-made chemicals. Additional concerns include their bioaccumulation, mobility, long-range transport potential (LRTP), accumulation in plants, global warming potential, and (eco)toxicological effects [68]. As part of this shift, 3M announced on 20 December 2022 that it would cease all PFAS manufacturing by the end of 2025, including the production of Novec 1230 (FK-5-1-12). While NOVEC 1230 does not deplete the ozone layer and has a low global warming potential (GWP), it contains PFAS substances that may pose risks to human health and the environment [69]. Despite 3M's exit, other manufacturers may continue to produce the FK-5-1-12 agent, which is identical to NOVEC 1230, as the 3M patent expired in July 2020.

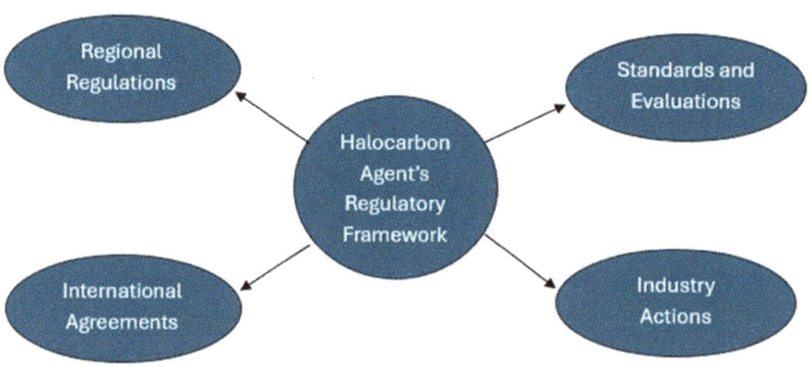

Figure 4. Diagram of the regulatory framework for halocarbon agents.

Figure 5 illustrates the regulatory framework of clean agent suppression systems since the regulation of these systems is required to ensure their effectiveness, safety, and environmental protection. Third-party certification is a conformity assessment process conducted by an independent body, separate from both the supplier and customer organizations. This process ensures that products and services comply with specified standards and other normative documents. The advantages of third-party certification include risk reduction, prevention of costly errors, and time savings for specifiers and regulators. For manufacturers and installers, the benefits include increased global sales, enhanced product or service value, and reduced liability [70]. Under Underwriters Laboratories, UL certification includes UL 2166, the Standard for Halocarbon Clean Agent Extinguishing System Units [71], which details the requirements for the construction and operation of halocarbon clean agent fire extinguishing system units. Additionally, UL 2127, the Standard for Inert Gas Clean Agent Extinguishing System Units [72], specifies the requirements for the construction and operation of inert gas clean agent fire extinguishing system units. Both standards of systems must be installed, inspected, tested, and maintained in accordance with NFPA 2001 and the National Fire Code of Canada. Under Factory Mutual Approval, approval is conducted according to FM 5600, the Examination Standard for Clean Agent Extinguishing Systems [73]. This standard outlines the testing and certification requirements for fixed fire extinguishing systems that use vaporizing liquid or inert gas clean agents for total flooding protection. During testing, the full systems, including their design, installation, operation, and maintenance instructions, are evaluated to ensure they perform effectively under worst-case conditions. Under the Loss Prevention Certification Board, the requirements for fire testing of fixed gaseous fire extinguishing systems are outlined in the LPS 1230 standard [74]. This standard evaluates the performance of halon alternative fixed extinguishing systems, including non-liquefied inert and halocarbon gas types, in accordance with the ISO 14520 standard.

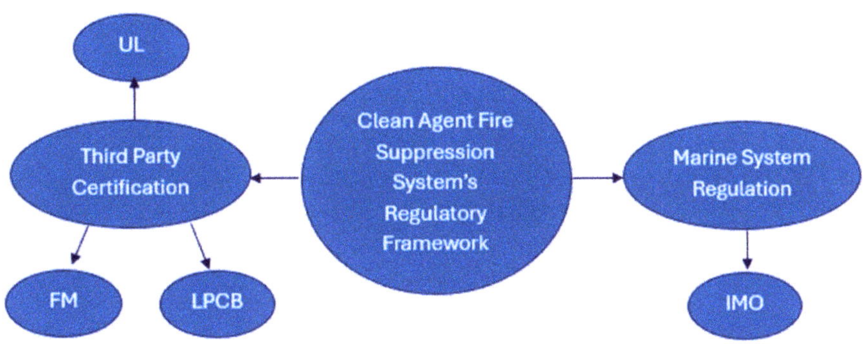

Figure 5. Diagram of the regulatory framework for clean agent fire suppression system.

In marine systems, various regulatory bodies and standards govern the requirements for clean agent fire extinguishing systems to ensure vessel health and safety. These regulations provide guidelines for the design, installation, and maintenance of these systems. Most marine regulatory bodies follow the International Maritime Organization's standards, specifically the International Convention for the Safety of Life at Sea (SOLAS) [75], which outlines minimum safety standards to be adhered to for constructing and operating merchant ships. Since its inception in 1974, the SOLAS Convention has undergone several revisions and amendments. In 1998, the Maritime Safety Committee revised the guidelines for fixed gas fire-extinguishing systems used in category A machinery spaces and cargo pump rooms, requiring them to demonstrate the same reliability as specified in SOLAS regulations II-2/7 and II-2/63. These systems must also be tested to ensure they can extinguish various types of fires that may occur in a ship's engine room [76]. Further amendments were made to MSC/Circ.848 with the issuance of MSC.1/Circ.1267, which outlines the potential risks and mitigation measures associated with fluorocarbon fire extinguishing agents. During fires, hydrogen fluoride (HF) vapor can be produced from these agents, leading to health issues such as respiratory and eye irritation. Control stations and manned locations during a fire should ensure that HF and HCl levels remain below 5 ppm, and other hazardous products should be kept below harmful concentrations for the duration of exposure. For all ships, the fire-extinguishing system design manual should include procedures to control decomposition products like HF vapor generated from fluorocarbon agents, as longer exposure of the agent to high temperatures can produce higher concentrations of these gases, which can impair escape [77]. In 2000, the Maritime Safety Committee amended Regulation 10 4.1.3 of SOLAS Chapter II-2 under Resolution MSC.99(73) to prohibit the use of fire extinguishing systems employing Halon 1211, Halon 1301, Halon 2402, and perfluorocarbons on all new constructions and new installations on existing vessels [78].

5. Performance in Energized Electrical Fire

As mentioned earlier, there is no established test method to accurately determine the exact amount of agent required to extinguish an energized electrical fire. This section will focus on the extinguishing performance of IG-541, FK-5-1-12, and HFC227-ea in energized electrical fires, using various types of testing. Generally, for protecting electrical equipment, such as servers and cables, the design concentration for an ordinary combustible (Class A) fire should be used according to the relevant standard when the equipment and cables are de-energized. However, if the electrical equipment and cables are not immediately de-energized and there is a delayed power disconnect, the design concentration for an energized electrical fire (Class C) should be applied as specified by the relevant standard [79]. FM Global conducted research to determine the minimum extinguishing concentrations required for various clean agents when dealing with energized electrical equipment. This research involved a series of tests using FM Global's fire propagation apparatus (FPA), as shown in Figure 6. In these tests, a sample of polymethyl methacrylate (PMMA) was placed in the FPA and exposed to a specific concentration of a clean agent. An 1100 W electrical arc was then introduced to see if the PMMA would ignite. If ignition occurred, the concentration of the clean agent increased in subsequent tests until the PMMA no longer ignited. The concentration at which ignition could no longer be achieved was identified as the minimum design concentration necessary for extinguishing fires in high-energy electrical environments. Table 7 presents the minimum design concentration of the IG-541 and halocarbon agents needed to suppress the energized electrical fire. If the electrical equipment is not de-energized before the clean agent concentration drops below these minimum design concentrations, reignition is anticipated [80]. FM Global's testing indicates that higher agent concentrations are necessary for high-energy arcing faults exceeding 480 V. However, it is crucial to assess and limit the use of these higher concentrations in typically occupied areas to minimize exposure duration, especially when these concentrations exceed the agent's NOAEL (no observed adverse effect level). When evaluating the effectiveness of suppressing energized electrical fires exceeding 480 V, G-541 demonstrates greater efficiency compared to halocarbon agents. IG-541 requires approximately a 50% increase in concentration, whereas FK-5-1-12 requires about a 110% increase, and HFC-227ea requires roughly a 70% increase to achieve the same suppression results.

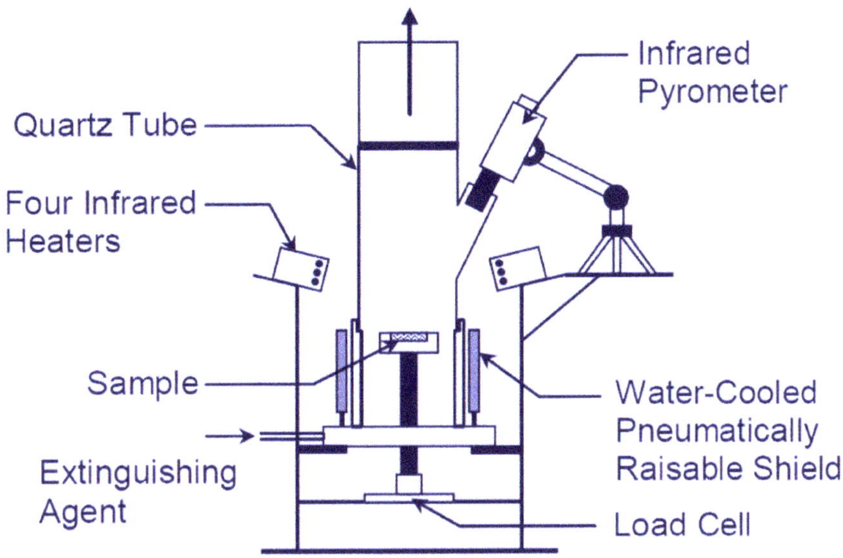

Figure 6. Diagram of the fire propagation apparatus [81].

Table 7. Minimum design concentration for IG-541 and halocarbon agents in Class C fires across different test methods.

Agent	Minimum Design Concentration on Class C Fire (vol%)	Reference
IG-541	38.5 (≤480 V), 57.0 (>480 V)	FM Global [79]
HFC-227ea	7.0 (≤480 V), 12.0 (>480 V)	
FK-5-1-12	4.7 (≤480 V), 10.0 (>480 V)	
IG-541	49.0 (48 watts), 56.1 (192 watts)	G. Bengston and R. Niemann [82]
HFC-227ea	8.0 (48 watts), 9.3 (192 watts)	
FK-5-1-12	4.5 (48 watts), 5.1 (192 watts)	

Nieman et al. [82] conducted experiments using nickel–chromium resistance wire powered by 48 watts and 192 watts of DC electricity to ignite PMMA samples, as shown in Figure 7, testing the extinguishing effectiveness of agents such as FC-3-1-10, HFC-227ea, and HFC-23, with Halon 1301 serving as a baseline. They concluded that clean agent extinguishing and inerting concentration values for Class "C" energized fires should be based on the energy levels to be protected. Higher energy circuits necessitate higher agent concentrations to ensure adequate fire suppression. To prevent reflash or reignition of Class "C" energized fires, clean agent concentration values should be designed to include the test inerting value plus a minimum 10% safety factor. The increased levels of HF generated when using clean agent inerting concentrations to prevent reignition or reflash may necessitate limiting the time that electronic circuits can remain energized to prevent corrosion damage to sensitive electronic equipment. Subsequently, Bengston and Niemann [83] extended these experiments to include other agents such as IG-541, FK-5-1-12, HFC-125, and HFC-236ea. Their findings were consistent with previous experiments, and they additionally found that the design concentration standards of ISO-14520 for Class

"A" fires are insufficient to ensure suppression, as they do not adequately prevent reflash and reignition. Bengston and colleagues [84] discovered that FK-5-1-12 shows significant potential as an alternative to the replacement of Halon 1301, such as HFC agents for use in equivalent concentrations. Its performance profile suggests that FK-5-1-12 could be a major advancement in efforts to eliminate or replace global warming substances (PFCs and HFCs) in total flooding applications.

Smith et al. [85] evaluated the effectiveness of HFC-227ea and L-15566 (the experimental product number for FK-5-1-12) in extinguishing continuously energized fires using both the conductive heating test protocol and a modified version of this protocol. The conductive heating test protocol involved testing with two different cable types, each heated with 1000 watts of energy while operating on 240 VAC. Upon sustained ignition, the extinguishing agent was discharged at the desired design concentration and monitored for 10 min post-discharge. In contrast, the modified conductive heating test protocol introduced a continuous ignition source throughout the test. This was achieved by placing a 6000 V, 20 mA electrical arc above the cable specimens, simulating conditions where conductive heating of cable connections occurs near an electrical arc. This setup allowed for the evaluation of scenarios where an arc develops between electrical equipment in close proximity to heated cables. The results from the conductive heating test indicated that agent concentrations lower than standard design values could extinguish a specific fire scenario. However, under the modified protocol with an electric arc, these concentrations were less effective in extinguishing the fire and preventing reignition. This modified protocol highlights the need for higher design concentrations in situations where internal heating may occur, and there is a possibility of an external ignition source being nearby. Previous testing of L-15566 (FK-5-1-12) at a 3.5 vol% concentration resulted in reignition during the modified test. Bengston et al. [86] updated the examination by increasing the concentrations to 3.9%, 4.5%, and 5.1%, which successfully prevented reignition. They concluded that higher agent concentrations were necessary to prevent reignition/reflash with both cables. Nonetheless, FK-5-1-12 was able to extinguish and prevent reignition/reflash at much lower concentrations compared to HFC-227ea. McKenna et al. [87] conducted a study on the effectiveness of HFC-227ea in achieving fire control using the ohmic heating test, conductive heating test, and printed wiring board (PWB) failure test. The test samples, including cables and PWBs, were selected based on their common presence in the telecommunications industry. The study found that fires initiated by and involving energized electrical circuits could be controlled by HFC-227ea at concentrations below 7%. Most of the test fires were completely extinguished within 15 s of agent discharge. In all tests, the circuits remained energized for 5 min after agent discharge, and no re-ignition was observed. In the PWB tests, the initiating arc continued to propagate after agent discharge, with intermittent flickering flames observed attached to the arc. These intermittent flames were immediately extinguished.

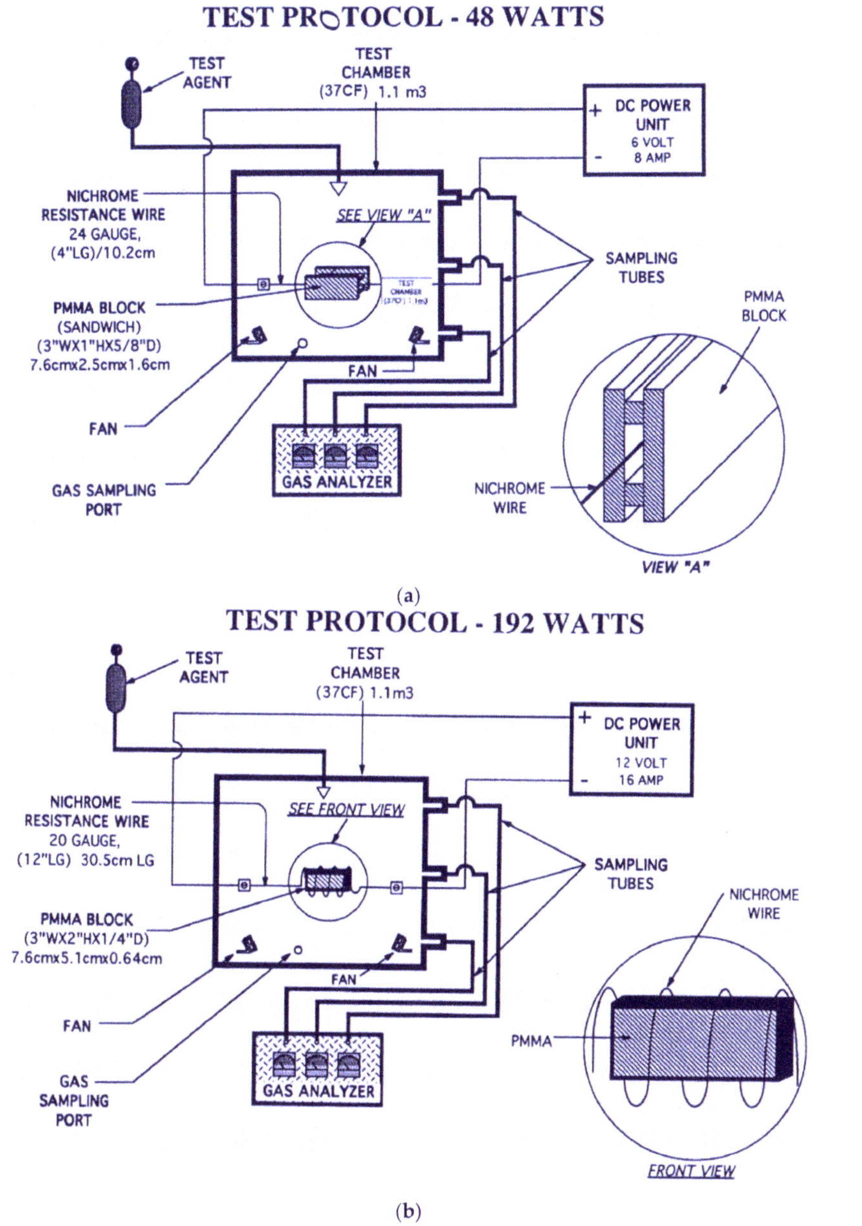

Figure 7. Testing protocol involving nickel–chromium resistance wire wrapped around PMMA samples ignited using different DC power levels: (**a**) 48 watts; (**b**) 192 watts [82–84].

Braun and colleagues [88] conducted tests using a short-duration auto-ignition temperature method with six clean agents: N2, IG-541, HFC-23, HFC-227ea, FC-2-1-8, and FC-3-1-10, as shown in Figure 8. The nickel metal foil used in the experiments was cut, notched, and folded, with the shiny side exposed to the fuel–air–agent mixture and the reverse side in contact with two thermocouples. The foil was heated using two 20-volt power supplies connected in parallel, providing a maximum current of 40 amps, capable of reaching temperatures up to 1000 °C. Initial experiments were conducted with each new foil using only an ethene–air mixture, without any agent. Subsequent experiments introduced increasing concentrations of fire suppressant agents. The study found that the introduction of clean agents generally required higher foil surface temperatures to prevent ignition than when no agents were present. The experimental data indicate a correlation between the number of fluorine atoms in the agent and its effectiveness in inhibiting hot surface ignition. Notably, in the presence of an electrically heated metal surface, a total flooding suppressant design concentration of 120% of the cup burner value was often insufficient to prevent the reignition of flammable gases from nearby pyrolyzing polymers. The volume fraction of agent needed to prevent ignition of a stoichiometric ethene–air mixture was found to be 1.1 to 5.9 times greater than the amount required in the cup burner test apparatus [89]. Smith and Rivers [90] used the same six clean agents to test the radiant-enhanced extinguishment device (REED), as shown in Figure 9 to determine the minimum extinguishing concentration of agents at specified radiation levels. The main components of the REED include a circular brass base, two concentric Pyrex tubes, a brass shield, a glass bead mixing chamber, a fuel platform, a conical heater, and three calibrated rotameters for monitoring gas flow. The tests were conducted at energy levels of 0, 20, 40, and 60 kW/m^2. The fuels chosen for the study were PMMA, grey PVC, natural ABS, and polypropylene. The specimens, which were in rod form, had aluminum foil wrapped around their bottom and curved surfaces to minimize edge effects and contain dripping. The extinguishing volume fractions were determined by gradually increasing the amount of extinguishing media until the flames from the samples were extinguished. The requirement for higher concentrations for fuels exposed to continuously energized electrical sources was reaffirmed. For energy levels above 20 kW/m^2, the design concentrations for N2, IG-541, and HFC-227ea would exceed their respective NOAEL thresholds.

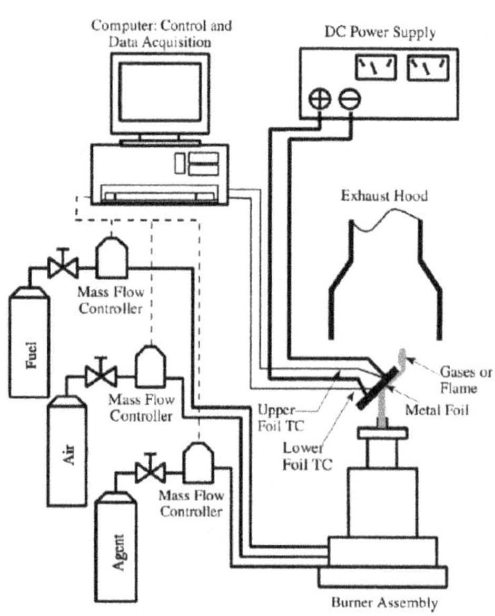

Figure 8. Schematic drawing of the setup for the short-duration auto-ignition temperature method [88,89].

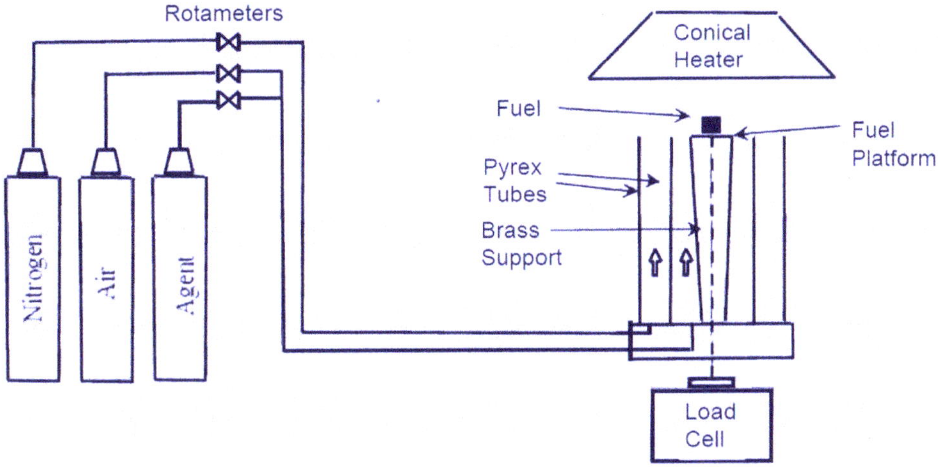

Figure 9. Schematic drawing of the setup for the radiant enhanced extinguishment device (REED) [90].

In conclusion, IG-541 and halocarbon agents demonstrated comparable effectiveness in extinguishing fires, although higher agent concentrations were required to extinguish fires caused by higher energy levels in energized electrical equipment. Therefore, IG-541 is a viable replacement for halocarbon agents if they are completely phased out from the market. Further research is needed to examine the impact of higher energy levels and various types of materials on the extinguishing concentration value of IG-541. This information is essential for determining the maximum allowable exposure time in occupied spaces.

6. Environment Factor

Three key environmental characteristics are of particular interest in evaluating IG-541 and halocarbon agents as potential halon replacements are shown below:

(1) Ozone Depletion Potential (ODP): ODP quantifies a chemical's capacity to destroy stratospheric ozone. It is expressed as ozone depletion per unit mass of the substance emitted relative to a standard of CFC-11.

(2) Global Warming Potential (GWP): GWP measures the impact of a chemical's emission on global warming relative to a reference gas, typically carbon dioxide (CO_2).

(3) Atmospheric Lifetime: This metric reflects the persistence of a chemical in the atmosphere. Chemicals with longer atmospheric lifetimes tend to contribute more significantly to global warming and may pose concerns regarding unexpected long-term environmental effects.

ODPs, GWPs, and atmospheric lifetimes are calculated but cannot be measured. HFCs are gaining increased attention as substitutes for ozone-depleting substances because, unlike HCFCs, they do not deplete the ozone layer and have shorter atmospheric lifetimes compared to PFCs. However, there remains significant concern about the role of HFCs in contributing to global warming [91]. As demonstrated in Table 8, IG-541 and halocarbon agents do not have the potential to deplete the ozone layer because they do not contain bromine or chlorine, which are responsible for the destruction of stratospheric ozone by other ozone-depleting substances. However, it is important to understand how ozone-depleting agents that contain bromine or chlorine, such as halons and HCFCs, can damage the stratospheric ozone.

Table 8. Environment characteristics of IG-541 and halocarbon agents: ODP, radiative efficiency, GWPs, and atmospheric lifetime.

Agent	ODP [22]	Radiative Efficiency ($W\ m^{-2}\ ppb^{-1}$) [92]	GWP-20 [1] [92]	GWP-100 [1] [92]	GWP-500 [1] [92]	Atmospheric Lifetime (Years) [92]
HFC-227ea	0	0.273	5850	3600	1100	36
FK-5-1-12	0	0.028	0.411	0.114	0.033	0.019
IG-541	0	0	0	0	0	0

[1] GWP is reported over a 20-year, 100-year, and 500-year time horizon.

Ozone, a naturally occurring gas in our atmosphere with the chemical formula O_3, exists predominantly in two regions of the atmosphere. Approximately 10% of Earth's ozone is found in the troposphere, which extends from the surface up to about 10–15 km in altitude. The remaining 90% is situated in the stratosphere, the atmospheric layer between the top of the troposphere and an altitude of about 50 km. Stratospheric ozone is naturally produced through chemical reactions involving solar ultraviolet (UV) radiation and oxygen molecules, which constitute about 21% of the atmosphere. Initially, solar UV radiation breaks apart an oxygen molecule (O_2) into two separate oxygen atoms (2O). Subsequently, each of these highly reactive oxygen atoms combines with an oxygen molecule to form an ozone molecule (O_3) in a binding reaction. These reactions continuously take place whenever solar UV radiation is present in the stratosphere. Consequently, the highest ozone production occurs in the tropical stratosphere. In the overall process, three oxygen molecules plus sunlight react to form two ozone molecules [93]

$$O_2 + UV \rightarrow O + O \quad (1)$$

$$O + O_2 \rightarrow O_3 \quad (2)$$

$$\text{Overall reaction: } 3\ O_2 + UV \rightarrow 2O_3 \quad (3)$$

Although the UV radiation dissociates the ozone molecule, ozone can reform through the following reactions, resulting in no net loss of ozone:

$$O_3 + UV \rightarrow O_2 + O \quad (4)$$

$$O + O_2 \rightarrow O_3 \quad (5)$$

Ozone is also destroyed by the following reaction:

$$O + O_3 \rightarrow O_2 + O_2 \quad (6)$$

These reactions are referred to as the Chapman reactions. As altitude increases, the reaction between oxygen molecules and oxygen atoms slows down, whereas the reaction between ozone molecules and UV radiation accelerates [94]. Halons, HCFCs, and other halocarbons that contain chlorine or bromine have been demonstrated to contribute to the depletion of stratospheric ozone. While natural phenomena can lead to temporary ozone loss, chlorine and bromine released from human-made compounds like CFCs are now recognized as the primary cause of this depletion. These compounds are highly stable and do not readily break down in the lower atmosphere, allowing them to persist for extended periods. Due to their stability, these halogenated compounds eventually reach the stratosphere, where ultraviolet (UV) radiation breaks them down, releasing free chlorine atoms [95]. The chlorine atoms released from these compounds by photolysis and oxidation react rapidly with ozone, initiating a catalytic cycle through the following reaction:

$$Cl + O_3 \rightarrow ClO + O_2 \quad (7)$$

$$ClO + O \rightarrow Cl + O_2 \quad (8)$$

$$O_3 + O \rightarrow 2\,O_2 \quad (9)$$

Stratospheric bromine is generated through the photolysis and oxidation by OH of brominated compounds released at the surface. Brominated hydrocarbon compounds, such as halons, are produced solely by industrial activities. The most efficient catalytic cycle involving bromine atoms is one that couples bromine and chlorine radicals.

$$BrO + ClO \rightarrow Br + ClO_2 \quad (10)$$

$$ClO_2 + M \rightarrow Cl + O_2 \quad (11)$$

$$Br + O_3 \rightarrow BrO + O_2 \quad (12)$$

$$Cl + O_3 \rightarrow ClO + O_2 \quad (13)$$

$$2\,O_3 \rightarrow 3\,O_2 \quad (14)$$

In conclusion, the relative significance of chlorine and bromine in the stratosphere is determined by their concentrations and the reaction rates of their specific catalytic cycles. Consequently, this process is effective only at altitudes above 30 km, where the concentration of atomic oxygen (O) is sufficient. At lower altitudes, the destruction of odd oxygen occurs through different catalytic cycles that do not require atomic oxygen [96].

The quantity present in the atmosphere must be monitored annually over a specific time period, as the global warming potential (GWP) is a cumulative effect calculated over that duration. These calculations account for the atmospheric lifetime of the chemical, as well as its infrared-absorbing properties, which also influence its GWP. GWPs are used to estimate the future global warming impact of a substance over a given time frame by multiplying the weight of the greenhouse gas by its GWP for that specific time horizon. The resulting value can be compared with others to determine which substance will have the least or greatest impact over that period. Emitting a large amount of low-GWP greenhouse gas may result in less global warming than a small release of high-GWP greenhouse gas over a certain time horizon. Different time horizons can yield different outcomes [97]. The global warming potential (GWP) estimates the cumulative warming effect of a unit mass of a specific greenhouse gas in the current atmosphere compared to that of carbon dioxide (CO_2). Carbon dioxide, known as reference gas, is given a GWP value of 1, regardless of the time horizon or other factors that may affect the GWPs of other greenhouse gases. GWPs are formulated based on the concept of radiative forcing, which links radiative forcing to climate response. This metric allows for the comparison and aggregation of the climate effects of different greenhouse gases (GHGs). A higher GWP indicates a greater potential for a GHG to cause global warming. The GWP value is influenced by two main factors: the ability of the GHG to trap heat in the atmosphere, described by its radiative forcing or radiative efficiency, and the atmospheric lifetime of the GHG, described by its decay rate. The radiative efficiency of a gas is not constant, so changes in the concentration of the reference gas CO_2 affect the GWPs of all other GHGs. GWPs are calculated for specific time horizons, with the Intergovernmental Panel on Climate Change (IPCC) providing values for 20, 100, and 500 years. Therefore, the choice of time horizon affects the relative importance of different gases when using the GWP metric [98]. The most commonly used time periods are either 20 years (GWP-20) or 100 years (GWP-100), being the norm used to report aggregated GHG emissions, including some IPCC communications [99]. Hydrofluorocarbons (HFCs) are being used as replacements for many chlorofluorocarbons (CFCs) and hydrochlorofluorocarbons (HCFCs) that are being phased out under the Montreal Protocol on Substances that Deplete the Ozone Layer. Unlike these ozone-depleting substances (ODS), HFCs do not harm the ozone layer. However, they are extremely potent greenhouse gases (GHGs), with a global warming potential up to thousands of times greater than that of carbon dioxide. The use of HFCs is currently increasing more rapidly than any other category of GHGs. Projections indicate that HFC use could increase by as much as 30-fold by 2050, potentially contributing up to 0.1 °C to the global average temperature rise by mid-century and increasing up to fivefold, potentially contributing up to 0.5 °C by 2100 [100]. Although HFCs are harmless to the stratospheric ozone layer and generally have lower radiative efficiencies compared to the most common ozone-depleting substances (ODSs), they are potent greenhouse gases. As a result, HFCs were included in the list of substances regulated by the 1997 Kyoto Protocol under the United Nations Framework Convention on Climate Change (UNFCCC). Later, specific HFCs were incorporated into the Montreal Protocol framework through the Kigali Amendment in 2016. HFC-227ea is present in the atmosphere at lower abundances and contributes relatively little to the overall radiative forcing of HFCs. It is utilized as a fire suppressant, mainly serving as a replacement for halons in both streaming and total flooding applications [101]. It is crucial to understand the naturally occurring greenhouse effect and identify the greenhouse gases, such as HFCs, that contribute to it. Solar radiation enters the Earth's atmosphere from space, with a portion being absorbed by the Earth's surface. The Earth then re-emits this radiation back toward space, but its properties change from high-frequency solar radiation to lower-frequency infrared radiation. Greenhouse gases, which are transparent to solar radiation, effectively absorb infrared radiation. Consequently, this radiation, which would have otherwise escaped into space, is retained, leading to a warming of the atmosphere [102]. Therefore, the high radiative efficiency of HFC-227ea leads to a higher global warming potential compared to FK-5-1-12 and IG-541, as shown in Table 8.

With the phase-out of halons, HCFCs, and HFCs, FK-5-1-12 is becoming a popular alternative as an environmentally friendly fire suppression agent due to its negligible global warming potential (GWP). A study by Taniguchi et al. found that FK-5-1-12 is removed from the environment through photolysis, which occurs over approximately 1–2 weeks. Its global warming potential is negligible due to its short atmospheric lifetime. When FK-5-1-12 undergoes photolysis in air, it produces $CF_3C(O)F$ and COF_2. $CF_3C(O)F$ is absorbed into rain, clouds, or seawater, where it hydrolyzes to form trifluoroacetic acid, while COF2 hydrolyzes to produce CO_2 and HF [103]. A study conducted by MIT explored the atmospheric loss mechanisms of FK-5-1-12, demonstrating that while the chemical does not react with hydroxyl radicals (OH), it undergoes significant degradation when exposed to ultraviolet (UV) light. The authors evaluated the UV cross-section of FK-5-1-12 and found a maximum absorption wavelength of 306 nm. Photolysis in the lower atmosphere will act as a substantial sink for this chemical due to its high absorbance at wavelengths over 300 nm [104]. The scientists came to the following conclusion: "In fact, the absorption spectrum is comparable to that of acetaldehyde" [105], a species with a roughly 5-day lifespan against solar photolysis [106]. Hence, they anticipate that FK-5-1-12's atmospheric lifetime against solar radiation will be between three and five days because of its bigger absorption cross-sections. According to recent laboratory observations, FK-5-1-12's photodissociation rate is comparable to acetaldehyde's, within an experimental error [107]. Jackson and colleagues [108] examined the potential roles of photolysis, hydrolysis, and hydration in determining the environmental fate of FK-5-1-12. They found that hydration is not a significant pathway for FK-5-1-12. Although the hydrolysis rate constant is much higher than that for photolysis, the typically low levels of liquid water in the atmosphere make hydrolysis less impactful. Even during nighttime, the production of PFPrA and HFC-227ea would be minimal, indicating that photolysis is the predominant atmospheric fate of FK-5-1-12. Therefore, photolysis will always be more dominant than hydrolysis due to the low amount of liquid water in the atmosphere, which outweighs the higher hydrolysis rate constant. Even during heavy rain at night, significant hydrolysis is unlikely to occur. Nevertheless, the FK-5-1-12 underwent a re-assessment using the IPCC 2021 evaluation method and was subsequently reported in the IPCC WGI Sixth Assessment Report. By applying the IPCC 2021 evaluation method with a 100-year integration time horizon and a radiative forcing of 0.028 Wm-2ppb-1, a global warming potential (GWP) value of 0.114 was determined [92]. As a result, its relatively low GWP value makes it significantly more environmentally friendly than HFC-227ea, although it still cannot compare to IG-541, which has a GWP value of zero.

The atmospheric lifetime of HFC-227ea, a hydrogen-containing fluorocarbon, is predominantly determined by its reaction with hydroxyl (OH) radicals, while other degradation pathways, such as photolysis and reactions with O_3, HO_2, or NO_3, are negligible. The rate coefficient used to calculate the tropospheric lifetime of HFC-227 suggests it is approximately 40 years. The main products of HFC-227ea's tropospheric oxidation are identified as trifluoroacetylfluoride (CF_3COF) and, by implication, carbonylfluoride (COF_2). Both products are expected to undergo rapid hydrolysis in the tropospheric cloud environment, resulting in trifluoroacetic acid (CF_3COOH), CO_2, and HF [109]. CF_3COOH is not toxic to animals but does exhibit a mild herbicidal effect. The concentration of CF_3COOH anticipated in rainfall due to the atmospheric degradation of HFCs is significantly lower than levels known to affect plant systems. The hydrolysis of COF_2 and CF_3-OH produces HF and CO_2, which do not pose environmental concerns [110]. Prior to the research conducted by Zellner et al., Nelson et al. [111] determined that the atmospheric lifetime of HFC-227ea is 42 years, using a similar approach that evaluated its reaction with OH radicals as the dominant tropospheric sink. In contrast, oxidation by hydroxyl radicals and wet deposition/rain out are ineffective at removing FK-5-1-12 from the atmosphere. The primary atmospheric sink for FK-5-1-12 is photolysis. FK-5-1-12 has been shown to photolyze rapidly in the atmosphere, resulting in a short atmospheric lifetime. The degradation products from the atmospheric decomposition of FK-5-1-12 already exist in the environment and are not expected to be harmful. Despite its rapid degradation in the environment, FK-5-1-12 remains sufficiently stable for use as a clean extinguishing agent [112]. D'Anna et al. [113] conducted a photolysis study under natural sunlight conditions in a simulation chamber and estimated that the atmospheric lifetime of FK-5-1-12 is approximately one week. Despite FK-5-1-12 having a large integrated absorption cross-section, its short atmospheric lifetime results in a negligible global warming potential. In conclusion, according to the IPCC documented values, the atmospheric lifetimes for HFC-227ea and FK-5-1-12 are 36 years and 0.019 years, respectively, which align with the researchers' evaluations. HFC-227ea has a significantly longer atmospheric lifetime compared to FK-5-1-12, indicating that HFC-227ea is more persistent in the atmosphere. This persistence contributes to the growing concern regarding its impact on global warming. In contrast, because of its mixture of naturally occurring gases, IG-541 has excellent environmental credentials and contains no harmful hydrogen fluoride-producing chlorofluorocarbons (CFCs). It is also considered an inert gas with no adverse effect on the ozone layer, zero global warming potential, and no atmospheric lifetime. The suppression agent is not regarded to be harmful to the environment and does not come under the 'F-gas Regulations' because it contains no fluorinated greenhouse gases [114]. Although IG-541 contains 8% CO_2, which is known to contribute to global warming, this CO_2 is not artificially produced like in halocarbon agents; instead, it is directly sourced from the environment. As such, IG-541 is a highly environmentally friendly option, especially for total flooding applications and suppressing energized electrical fires, compared to halocarbon agents.

It is worth noting that 3M discontinued manufacturing of FK-5-1-12 owing to the presence of PFAS substance, even though the PFAS contained in FK-5-1-12 is not on the EU regulation's banned list. Per- and polyfluoroalkyl substances (PFAS) are a class of synthetic organic compounds characterized by a wide range of structures, properties, uses, bioaccumulation potentials, and toxicities. Despite their diversity, all PFAS share the common feature of containing perfluoroalkyl groups that are highly resistant to environmental and metabolic degradation. Consequently, most PFAS are either non-degradable or eventually transform into stable terminal products, which are still classified as PFAS [115]. The European Chemicals Agency (ECHA) [68] has proposed an unprecedented ban on the entire class of per- and polyfluoroalkyl compounds. If approved, this prohibition will impact over 7 million such substances, as reported by PubChem [116]. This will undoubtedly affect the FK-5-1-12 agent if regulatory action is taken because it is a fully fluorinated ketone and fits under the previously stated broad definition of "PFAS." Fire Suppression Systems Association (FSSA) expressed a divergent perspective and strongly urged ECHA to exclude FK-5-1-12 agents from the ban list. FK-5-1-12, in contrast to aqueous film-forming foam (AFFF), has the potential to partition into the groundwater during end use. Upon discharge from the fire suppression system, it rapidly transforms into gas and completely breaks down in the atmosphere within approximately one week. Thus, FK-5-1-12 is not considered persistent in the environment [117]. Furthermore, the Environment Protection Agency (EPA) has classified FK-5-1-12 as an acceptable use under the Significant New Alternatives Policy (SNAP) program for total flooding applications. This is because it does not present a higher overall environmental and human health risk compared to other approved alternatives for this specific purpose [118].

7. Safety Consideration

The toxicity of a clean agent fire suppression system in a total flooding application is a major safety factor, particularly in the normally occupied space. Table 6 presents data regarding the toxicological and physiological impacts of IG-541 and halocarbon agents, as defined by NFPA 2001. It also includes the safety margin of each agent in comparison to the minimum design concentration specified by NFPA 2001 and ISO 14520 standards. The no observable adverse effect level (NOAEL) is the highest concentration at which no adverse physiological or toxicological effect has been observed. The lowest observable adverse effect level (LOAEL) is the lowest concentration at which an adverse physiological or toxicological effect has been observed. Cardiac sensitization and hydrogen fluoride (HF) formation from thermal decomposition products are the primary acute toxicity effects of the halocarbon agents. LC50 is the concentration lethal to 50 percent of the rat population during a 4 h exposure, specifically for halocarbon agents. It is a statistically derived concentration that, over a specified exposure period, is expected to result in the death of 50% of the test animals within a given timeframe. This term is relevant to experiments involving inhalation exposure and aquatic toxicity testing [119]. Test Guideline No. 403 (TG 403) is designed to gather sufficient information on the acute toxicity of a test substance to enable its classification and provide lethality data for one or both sexes, as needed for quantitative risk assessments. This guideline presents two testing methods. The first method is a traditional protocol where groups of animals are exposed to either a limit concentration (limit test) or a series of concentrations in a stepwise manner for a predetermined duration, typically 4 h, although other exposure durations may be used to meet specific regulatory requirements. The second method is a (C × t) protocol, where groups of animals are exposed to either a single limit concentration or a series of multiple concentrations over various durations [120]. The study, conducted in accordance with TG 403 guidelines, involved exposing rats to a concentration of 788,696 ppm of the test material. The substance was continuously administered in oxygen to the test chamber for a duration of 4 h. Following exposure, the rats were monitored for a period of 14 days and then subjected to terminal necropsy. The results indicate that HFC-227ea has an acute inhalation LC50 greater than 79 vol% in rats [121]. This finding is consistent with the NFPA documented value of a concentration greater than 80 vol%. As shown in Table 9, the acute inhalation LC50 of FK-5-1-12 is greater than 10 vol%, whereas the acute inhalation LC50 of HFC-227ea is greater than 80 vol%. This indicates that HFC-227ea has lower acute inhalational toxicity compared to FK-5-1-12.

Table 9. Toxicity information and safety margin for IG-541 and halocarbon agents use in occupied space for total flooding application.

Agent	LC50 (vol%) [22]	NOAEL (vol%) [22]	LOAEL (vol%) [22]	Class C Fire Minimum Design Concentration (vol%) [1]	Safety Margin (%) Compared to NOAEL [2]	Safety Margin (%) Compared to LOAEL [3]
HFC-227ea	>80.0	9.0	10.5	7.0–8.5	6–29	24–50
FK-5-1-12	>10.0	10.0	>10.0	4.5–5.6	79–122	79–122
IG-541	-	43.0	52.0	38.5–41.7	3–12	25–35

[1] The range of minimum design concentrations for Class C fires is determined by using the NFPA standard as the lower limit and the ISO 14520 standard as the upper limit. [2] The range of the safety margin is determined by comparing the Class C fire minimum design concentration with the NOAEL, ranging from the lower percentage of the ISO 14520 standard to the higher percentage of the NFPA standard. [3] The range of the safety margin is determined by comparing the Class C fire minimum design concentration with the LOAEL, ranging from the lower percentage of the ISO 14520 standard to the higher percentage of the NFPA standard.

Cardiac sensitization is the condition in which the heart becomes temporarily more susceptible than usual to the arrhythmogenic effects of endogenous catecholamines when exposed to high concentrations of numerous halogenated and unsubstituted hydrocarbons. It is measured in dogs after they have been exposed to halocarbon agents for 5 min. At a 5 min interval, an external dose of adrenaline is given, and any resulting cardiac sensitization in the dog is documented. Five minutes of exposure is typically selected in advance of the adrenaline challenge dose to simulate the exposure duration involved in "sniffing" situations. The maximum blood level is typically attained after about five minutes of exposure to a given chemical concentration [122]. The lowest dose at which a cardiac response is observed is referred to as the LOAEL (lowest observed adverse effect level). The next lowest dose tested, at which no positive response is observed, is referred to as the NOAEL (no observed adverse effect level). These endpoints are determined in animals that have already been primed for a cardiac response due to the high level of epinephrine challenge. Without an external epinephrine challenge, the concentration required to evoke a response would need to be much higher. The EPA applies the LOAEL and NOAEL endpoints from these dog tests directly to humans, recognizing the high sensitivity of the test procedure. Therefore, the LOAEL and NOAEL levels established by this test procedure are very conservative estimates of the levels that would trigger a cardiac response without an external source of epinephrine [123]. In summary, the NOAEL levels for FK-5-1-12 and HFC-227ea, as shown in Table 9, indicate that cardiac sensitization occurs at 10 vol% and 9 vol%, respectively. This results in a safety margin of 6 to 29 percent for HFC-227ea and 79 to 122 percent for FK-5-1-12 when compared to the minimum design concentration for Class C fires as specified in NFPA 2001 and ISO 14520. The LOAEL for FK-5-1-12 mirrors the safety margin of the NOAEL, as its level is above 10 vol%. For HFC-227ea, the LOAEL occurs at 10.5 vol%, yielding a higher safety margin of 24 to 50 percent compared to the NOAEL. In conclusion, FK-5-1-12 is safer for humans, providing a higher safety margin at both LOAEL and NOAEL levels compared to HFC-227ea. It also ensures minimal hazardous effects on human health in normally occupied spaces with total flooding applications to protect energized electrical equipment.

One scientific approach used to better understand the health effects of chemicals is known as physiologically based pharmacokinetic (PBPK) modeling. A PBPK model relates the amount of chemical exposure to the concentration of the chemical in the blood and organs over time. For instance, PBPK models can help determine whether a toxic level of a chemical would be present in the blood or an organ of a person after exposure to a certain amount of the chemical through drinking water. These models are developed using mathematical values (referred to as "parameters") and equations that describe various characteristics and processes of the body, such as body weight, blood flow rate, and metabolism rate. The purpose of PBPK models is to interpret animal toxicity data in the context of human health, a process known as extrapolation. By adjusting the physiological parameters in a PBPK model that describes a chemical in a laboratory animal species, the model can be adapted for humans. This allows for more accurate and human-relevant predictions [124]. The time for safe human exposure to HFC-227ea at specified concentrations is detailed in NFPA 2001's Table 4.3.2.3 (c). According to this table, exposure to HFC-227ea is considered safe for at least five minutes at a concentration of 10.5%, which corresponds to the LOAEL level. It is important to evaluate the duration of safe exposure in relation to the concentration of the agent required to extinguish a fire. By comparing the concentration for safe five-minute exposure to the recommended design concentration of the agent, we can gain insights into safety margins. For HFC-227ea, the safe exposure concentration is 10.5%, while the recommended design concentration for Class C fires is 7.0%. This indicates that even though the safe exposure concentration is at the LOAEL, there is a 3.5% difference between the safe exposure concentration and the recommended design concentration for HFC-227ea [125]. In contrast, since no PBPK model has been conducted for FK-5-1-12, it is not possible to make a comparison between the concentration for safe five-minute exposure and the recommended design concentration of the agent based on the PBPK model.

The extent of thermal decomposition products generated when a halocarbon agent interacts with a flame front depends on three critical factors that affect the time required to extinguish the fire. These factors are the ratio of the fire size to the enclosure volume, the volumetric concentration of the agent, and the discharge time. Testing of thermal decomposition products has demonstrated that significant concentrations of HF can be produced when fluorinated halon replacement agents are used for fire suppression. Typically, HF concentrations for these agents are between 2 and 10 times higher than those for Halon 1301. It is important to note that these peak HF concentrations generally decrease quickly. Tests have indicated that HF concentrations can drop to half of their peak levels within 5 min after the agent is discharged [126]. Hydrogen fluoride (HF) is a colorless, fuming liquid or gas with a strong, irritating odor. The following acute health effects may occur immediately or shortly after exposure to HF. It can severely irritate and burn the eyes, potentially causing permanent damage. Contact with HF can cause irritation and severe burns to the skin and deep tissues. Burns may develop hours after contact, even if no pain is initially felt. HF can irritate the nose and throat. Inhaling HF can irritate the lungs, leading to coughing and/or shortness of breath. Higher exposures may result in a build-up of fluids in the lungs (pulmonary edema), a medical emergency characterized by severe shortness of breath. Exposure to HF can cause headaches, dizziness, nausea, and vomiting. Very high exposure can lead to fluoride poisoning, which includes symptoms such as stomach pain, weakness, convulsions, collapse, and potentially death. The following chronic health effects can occur sometime after exposure to HF and may last for months or years. Repeated inhalation of HF may cause bronchitis, characterized by coughing, phlegm, and/or shortness of breath. HF may damage the liver and kidneys. Long-term exposure to very high concentrations can cause fluoride deposits in the bones and teeth, a condition known as "Fluorosis" (changes in bone structure). This can result in bone pain, fractures, disability, and mottling of the teeth [127]. The Occupational Safety and Health Administration (OSHA) defines the immediate hazardous to life or health (IDLH) for HF concentration as 30 ppm, while the legal airborne permissible exposure limit (PEL) is 3 ppm averaged over an 8 h work shift [128].

It is important to note that HF gases will react with moisture to form hydrofluoric acid on the surfaces of electronic equipment. The corrosion damage caused by HF gas and acid, along with smoke and fire effluents, can lead to metal loss, electrical current leakage, and increased contact resistance. These damages may result in serious electrical short circuits or cause the electronic equipment to receive false data, disrupting electronic systems and potentially leading to system failure. The damage caused by HF, smoke, and fire effluents affects not only the equipment located near the fire but also remote equipment. Even if there is no immediate reaction on the electronic equipment from HF exposure, long-term damage due to HF corrosion is possible. The corrosion process on electronic equipment does not cease once the fire is extinguished. The extent of corrosion damage depends on factors such as HF concentration, temperature, and relative humidity in the enclosure, the duration of exposure, the nature of the combustible material involved, and the properties of the exposed equipment [129]. Kim et al. [130] conducted a study on the use of HFC-227ea as a total flooding–fire suppression agent for in-cabinet cable fires and a small liquid fuel pool fire in a simulated electronic equipment room, focusing on the measurement of acidic gas products. The study found that the extinguishment time of a cable bundle fire is affected by the cabinet's ventilation conditions, which in turn influence the levels of acidic gases like HF. In a closed fire cabinet, HF concentrations peaked at 800 ppm, whereas in an open or ventilated cabinet, they stayed below 100 ppm due to quicker extinguishment. There was minimal migration of HF to adjacent non-fire cabinets, and HF levels in the test room generally remained below 100 ppm, which is considered safe for short-term exposure. However, in the case of large fires, HF concentrations in the test room reached hazardous levels, averaging 2000 ppm over 10 min, with substantial HF transport to non-fire cabinets. This situation poses a risk of corrosion to electronic equipment and health hazards to individuals during large fires when using HFC-227ea. Therefore, early fire detection is essential to control small fires before they grow, ensuring the safety of both people and property.

Enlow [131] illustrated the complete combustion of the FK-5-1-12 agent, which results in the production of hydrogen fluoride and carbon dioxide:

$$C_6F_{12}O + 2.5O_2 + 6\ H_2O \rightarrow 6\ CO_2 + 12\ HF \tag{15}$$

The combustion of the FK-5-1-12 agent will, therefore, generate 12 volumes of hydrogen fluoride per molecule basis. Under dry conditions, the combustion of the FK-5-1-12 agent results in incomplete combustion and the production of carbonyl fluoride. Carbonyl fluoride is a highly reactive gas that reacts with available water to produce carbon dioxide and hydrogen fluoride via the following reaction:

$$CF_2O + H_2O \rightarrow CO_2 + 2\ HF \tag{16}$$

The magnitude of the HF concentration for a "normal" sized room is illustrated in Figure 10. It is a useful tool for estimating the thermal decomposition products for a specific room size. For example, to maintain HF concentrations below an acceptable range in a 1000 ft^3 (28 m^3) room, fire detection should be activated before the room reaches 3.7 kW in size. The severity of a fire is relative to the size of the enclosure. Increasing both the fire size and the enclosure size may not result in higher concentrations of HF during extinguishing. However, increasing the fire size while maintaining the enclosure size has been shown to increase HF production. Consequently, it is expected that larger fire sizes (one order of magnitude larger within the same enclosure) than those used in this study would have resulted in higher HF production [132]. Ditch et al. [133] demonstrated that FK-5-1-12 exhibits comparable performance to currently available halon alternatives in thermal decomposition product (TDP) testing. The experiments confirmed its effectiveness in extinguishing fires fueled by various fuels under a wide range of conditions. The results revealed a correlation between HF production and factors such as fire size, discharge time, and agent concentration. Notably, shorter exposure times to fire consistently resulted in lower TDP generation. Figure 11 depicts the dangerous toxic loads (DTLs) for humans based on Meldrum's analyses and compares the level of HF measured during extinguishment of fire less than 25 kW [134]. The DTL corresponds to exposure levels that would result in significant distress for all exposed personnel. The testing results reveal that levels of HF produced in the referenced fires, where the FK-5-1-12 agent is used as an extinguishant, fall significantly below the DTL curve. As a result, the FK-5-1-12 agent can be designed to operate optimally for personnel protection and sensitive assets, reducing the generation of HF that would exacerbate the potential hazard posed by a fire [135]. Thus, the design of FK-5-1-12 systems should focus on quick fire detection and rapid discharge to minimize the risk of excessive HF exposure and maximize the safety of both personnel and sensitive assets.

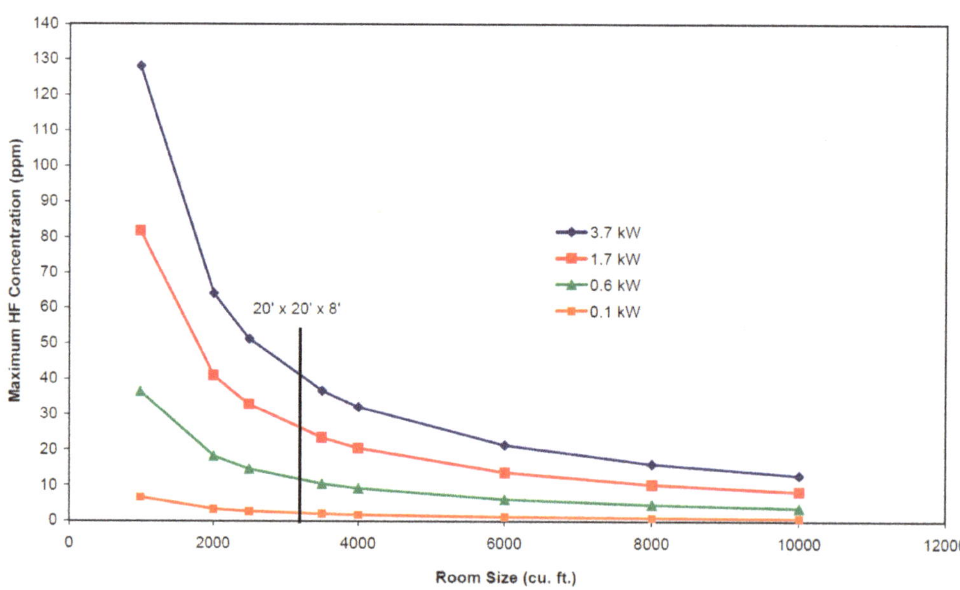

Figure 10. Effect of room size on HF production for FK-5-1-12 at 4.9 vol% concentration [132].

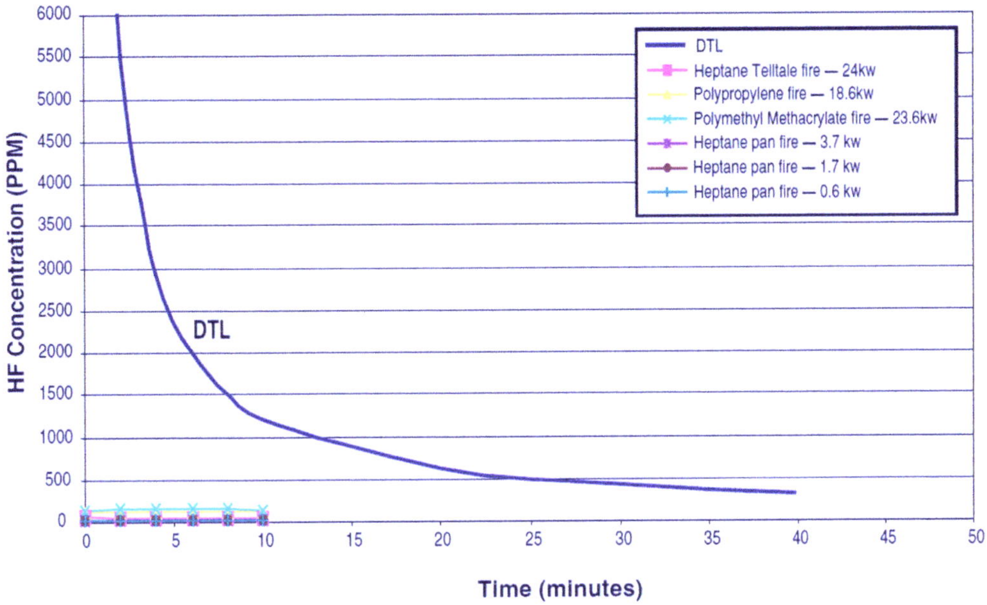

Figure 11. Time-dependent HF concentrations related to toxicological endpoints [135].

The thermal decomposition of HFCs in fire suppression systems can produce high concentrations of HF gas. Depending on the specific HFC chemistry and the size of the fire, the IDLH levels of HF gas can be reached in under 45 s. This hazard is particularly dangerous when an HFC suppression system is used indoors. The creation of toxic byproducts during fire scenarios presents additional risks to first responders and any nearby personnel beyond the immediate danger of the fire itself, and they may be unaware of these risks [136]. Hynes et al. [137] conducted a shock tube study on the pyrolysis of CF_3CHFCF_3 within a temperature range of 1200–1500 K and at pressures of 16–18 atm. The study utilized gas chromatography (GC-MS) and FTIR spectroscopy for product analysis, aiming to develop a comprehensive chemical mechanism to describe the decomposition of CF_3CHFCF_3. Experimental results and modeling indicated that the primary initiation reaction involved bond fission ($CF_3CHFCF_3 \rightarrow CF_3 + CF_3CHF$). Additionally, the abstraction of the secondary hydrogen atom by fluorine atoms was identified as a significant reaction pathway. In contrast, the 1, 2 HF elimination process ($CF_3CHFCF_3 \rightarrow CF_3CF{=}CF_2 + HF$) was found to occur at a slower rate. Copeland et al. [138] investigated the thermal decomposition of HFC-227ea at low pressure, highly diluted in argon, across a temperature range of 600–2000 °C using photoelectron spectroscopy. The primary decomposition reaction at lower temperatures (600–900 °C) was identified as the thermodynamically favored reaction producing $CF_3CF{=}CF_2$ and HF. At elevated temperatures (900–1200 °C), the decomposition pathway yielding C_2F_4 and CF_3H became more significant. No evidence was found for the reaction $CF_3CHFCF_3 \rightarrow CF_3CHF + CF_3$, which was anticipated based on a shock tube study conducted at much higher pressures, nor for $CF_3CHFCF_3 \rightarrow CF_3CF + CF_3H$. For the latter reaction, it is likely that CF_3CF is converted into C_2F_4 under the experimental conditions before photoionization occurs in the ionization region of the photoelectron spectrometer. At even higher temperatures, C_3F_6 was observed to decompose into C_2F_4 and CF_2, while C_2F_4 further decomposed into CF_2. The report by Andersson et al. examines the production of HF and COF_2 during experiments with halocarbon agents HFC-227ea and FK-5-1-12. It was found that at lower application rates, most fluorine from HFC-227ea ends up as HF. As the concentration approaches extinguishing levels, the proportion of COF_2 increases, but total fluorine recovery, including HF, remains below 40% at a 90% application rate. FK-5-1-12 shows lower fluorine recovery as HF and COF_2 compared to HFC-227ea. The yield of HF decreases for both agents as the application rate increases, with HFC-227ea producing a high yield of HF at low rates, while FK-5-1-12 yields significantly less HF at similar rates. No data are available for high application rates for both agents. The yield of COF_2 increases with higher application rates for all agents. At low rates, the extinguishing agent combusts more, transferring fluorine to HF. At higher rates, less combustion occurs, favoring COF_2 formation over HF. Effective fire extinguishing requires sufficient agent quantity and distribution to minimize harmful combustion products [139].

Since inert gases are not suitable for assessing cardiac sensitization like halocarbon gases, the values for the IG-541 agent in Table 6 are derived from the physiological effects in humans in a hypoxic environment and equivalent to the NOAEL and LOAEL values provided for the FK-5-1-12 agent. The IG-541 agent's 43% and 52% design concentrations correspond to NOAEL and LOAEL values, which equate to 12% minimum oxygen concentration for the no-effect level and 10% minimum oxygen concentration for the low-effect level, respectively. The primary health risk associated with the use of inert agents is hypoxia, which occurs when there is an insufficient level of oxygen reaching the body's tissues and organs. Hypoxia can result from inhaling air with a low oxygen concentration. IG-541 extinguishes fires by reducing oxygen levels to a point where combustion cannot continue. However, this creates a hypoxic environment that can be harmful to humans. The extent of potential harm depends on oxygen concentration and the duration of exposure to the hypoxic atmosphere. At very low oxygen levels (below 8 percent), exposure can be potentially fatal [140]. Acknowledging the presence of CO_2 in IG-541 and its beneficial effects on blood oxygenation and cerebral blood flow, the New Extinguishants Advisory Group (NEAG) has established the following guidelines for the use of IG-541:

(1) The oxygen concentration must not drop below 10% at any time while people are present.
(2) For oxygen concentrations between 10% and 12%, exposure should be limited to no more than 2 min.
(3) For oxygen concentrations between 12% and 15%, exposure should be limited to no more than 20 min.
(4) The final CO_2 levels should be maintained between 2.5% and 5%.

It is important to note that in a fire situation, elevated CO_2 levels can help maintain normal cognitive function due to their positive effects on blood oxygen saturation and cerebral blood flow, which can aid in evacuating the room. However, the increased ventilation and potential reduction in breath-hold time induced by higher CO_2 levels may also enhance the body's uptake of toxic materials from the fire [141]. In comparison to the NFPA 2001 guidelines for human exposure limitations, as shown in Table 10, there are notable differences in the exposure limits for oxygen concentrations between 10% and 12%. While NFPA 2001 allows an exposure limit of up to 3 min, the NEAG guideline restricts this to no more than 2 min. Additionally, NFPA 2001 permits up to 5 min of human exposure for oxygen concentrations above 12%, whereas the NEAG guideline extends the exposure limit to 20 min for oxygen concentrations between 12% and 15%. When suppressing Class C fires in accordance with NFPA 2001 and ISO 14520 standards, the safety margin relative to the NOAEL value of the IG-541 agent is reduced to only 3% to 12%, despite the improved breathing effects due to the addition of CO_2 in the IG-541 agent. Although the safety margin can be increased from 25% to 35% when compared to the LOAEL value of the

IG-541 system, the personnel exposure time limit must be reduced to 3 min in normally occupied spaces, as indicated in Table 10.

Table 10. Human exposure limitations for IG-541 and halocarbon agents used in total flooding application as specified in NFPA 2001 [22].

Agent Design Concentration (vol%)	Occupancy	Exposure Time Limit	Remarks
		HFC-227ea	
≤9.0	Normally occupied	Up to 5 min	No PBPK data needed
>9.0 and ≤10.5	Normally occupied	Up to 5 min	PBPK data
>10.5 and ≤11.0	Not occupied	Up to 1.13 min	PBPK data
>11.0 and ≤11.5	Not occupied	Up to 0.60 min	PBPK data
>11.5 and ≤12.0	Not occupied	Up to 0.49 min	PBPK data
		FK-5-1-12	
≤10.0	Normally occupied	Up to 5 min	No PBPK data needed
>10.0	Not occupied	Up to 30 s	No PBPK data available
		IG-541	
≤43.0	Normally occupied	Up to 5 min	≥12% oxygen concentrations
>43.0 and ≤52.0	Normally occupied	Up to 3 min	<12.0% and ≥10.0% oxygen concentrations
>52.0 and ≤62.0	Not occupied	Up to 30 s	<10.0% and ≥8.0% oxygen concentrations
>62.0	Not occupied	No exposure allowed	<8.0 oxygen% concentrations

Table 10 presents the human exposure limits for IG-541 and halocarbon agents, as specified in NFPA 2001. These limits specify the agent design concentration in relation to the type of occupancy and the allowable exposure time, specifically for use in total flooding applications. To prevent unnecessary exposure to the agents, both inert and halocarbon agents have a maximum exposure time of 5 min. Additionally, unprotected personnel are prohibited from entering a protected space during or after agent discharge. The FK-1-12 agent, with a design concentration above 10%, is permitted on the condition that personnel may evacuate the area within 30 s and that it is utilized exclusively in unoccupied spaces. Table 11 summarizes the overall effects of decreased oxygen concentration. Inert gases such as nitrogen, argon, and helium are odorless, colorless, and tasteless, making them particularly dangerous because they provide no warning of their presence or the resulting decrease in oxygen levels. For individuals unaware of their surroundings, the onset of asphyxiation can occur without warning symptoms and may happen very quickly, sometimes in just seconds, if oxygen levels are extremely low. As such, you may not realize you are in danger until it is too late [142]. According to the NFPA standard, as shown in Table 10, human exposure to IG-541 agents is limited to concentrations below 43%, which corresponds to an oxygen concentration of 12%, for no more than 5 min in normally occupied spaces. When gases other than oxygen are introduced or mixed with the air we breathe, the oxygen concentration decreases, leading to oxygen deficiency. If the IG-541 agent causes this deficiency, a person's physical and mental performance can decline without realizing it. Fainting can occur unexpectedly when oxygen levels drop to around 10%. If oxygen concentration falls below 10%, death from asphyxia can occur within minutes unless resuscitation is administered promptly [143]. Therefore, Table 10 shows the installation of IG-541 in a normally occupied space is prohibited if the agent design concentration exceeds 52% to ensure that it is safe for human exposure. At a design concentration exceeding 62%, the installation of the IG-541 agent is restricted to an unmanned room and human exposure is prohibited due to its equivalence to an oxygen concentration of less than 8%. Profound fainting upon exposure to oxygen concentrations below 8% can lead to severe consequences, such as brain damage in individuals, even if they are promptly rescued. In conclusion, the HF gas generated in fires as a breakdown product of the halocarbon agents can cause toxic effects on humans and corrosion damage to electronic equipment. However, the IG-541 does not decompose measurably during the extinguishment of a fire. When comparing the NOAEL and LOAEL values, the design concentration of the FK-5-1-12 agent has a much larger safety margin than the IG-541 and HFC-227ea.

Table 11. The overall effect of reduced O_2 concentrations [142].

O_2 (vol%)	Effects and Symptoms
18–21	No discernible symptoms can be detected by the individual. A risk assessment must be undertaken to understand the causes and determine whether it is safe to continue working.
11–18	Reduction of physical and intellectual performance without the sufferer being aware.
8–11	Possibility of fainting within a few minutes without prior warning. Risk of death below 11%.
6–8	Fainting occurs after a short time. Resuscitation possible if carried out immediately.
0–6	Fainting almost immediate. Brain damage, even if rescued.

8. Economic Factor

It is imperative for business owners who seek to prioritize safety within budget constraints to understand the several factors that affect the cost of installing a clean agent fire suppression system. The main considerations that impact the economic aspect of implementing the system are building size and layout, system type and complexity, compliance requirement, installation labor, and material cost, as well as maintenance and testing. Gordon and Preston [144] evaluated the comparative costs of the IG-541 system against alternative gaseous fire suppression systems for hydroelectric generator applications. The cost of IG-541 as a gaseous agent is significantly lower than that of HFC-227ea and FK-5-1-12, as it uses naturally occurring atmospheric gases. However, IG-541 requires a larger quantity of gas compared to halocarbon agents. When comparing FK-5-1-12 to HFC-227ea, FK-5-1-12 is typically about 10% more expensive. The equipment cost of the IG-541 system is higher due to the substantial number of bottles and extensive manifolding required, whereas halocarbon systems involve far fewer valves. Annual maintenance costs for IG-541 and halocarbon systems are similar, driven by the testing requirements outlined in relevant standards and codes. However, over a ten-year period, maintenance costs can vary significantly due to gas bottle retesting expenses, which are considerably lower—by a factor of 5 to 10 times—for systems with fewer bottles. Capital budget differences between systems are narrowing due to increased competition and the availability of similar generic products. Additional factors such as manifold configurations and redundancy requirements also impact system costs. Robin [145] evaluated the quantities of inert gas and halocarbon agents required to extinguish Class A and Class B fires in a 100 m^3 enclosure. The findings showed that HFC-227ea requires the least mass for extinguishment compared to IG-541 and

FK-5-1-12. While FK-5-1-12 requires a larger mass than HFCs, it also has a higher cost per unit mass compared to both HFCs and inert gases, making FK-5-1-12 the least cost-effective option among clean agent systems. Inert gas agents, due to their physical properties, cannot be compressed into a liquid state and must be stored as high-pressure gases. This necessitates high-pressure storage cylinders and piping, significantly increasing the cost of inert gas systems. Additionally, their low volumetric efficiency and inability to store liquids require a larger number of cylinders, leading to increased storage space, a larger system footprint, and higher overall costs. In contrast, halocarbon agents, except HFC-23, can be stored as liquids in standard low-pressure cylinders, allowing a greater mass of agents to be stored in a given cylinder volume. This reduces the number of cylinders and the associated storage requirements, making these systems more space-efficient and cost-effective. As a result, system costs for inert gas agents escalate more rapidly with increasing system size compared to halocarbon systems.

The IPCC [146] documented Wickham's study, which compared the average values of space, weight, and cost for IG-541 and halocarbon agent systems against a Halon 1301 system. The comparison focused on protecting room volumes ranging from 500 m^3 to 5000 m^3 in compliance with IMO regulations for shipboard machinery spaces, as detailed in Table 12. The weight includes the storage containers and their contents, excluding piping and hangers. Footprint refers to the area occupied by the agent containers, defined by the dimensions of the cylinder bank. Cube represents the volume occupied by the containers, calculated as the footprint area multiplied by the cylinder height, measured to the top of the valves. System cost includes the agent, storage containers, actuators, brackets, hoses, valves, controls, alarms, and other components sold by manufacturers to distributors or installers. It excludes distribution piping, supports, fittings, electrical components, installation labor, packing, and shipping costs. The analysis revealed that HFC-227ea systems had the lowest average values for weight, footprint, cube, and overall system cost. However, IG-541 systems had the lowest agent cost compared to FK-5-1-12 and HFC-227ea. While FK-5-1-12 systems required less weight, footprint, and cube than IG-541 systems, IG-541 maintained a cost advantage in terms of agent pricing over FK-5-1-12. Further details on the comparison of IG-541 and halocarbon agent systems for 500 m^3, 1000 m^3, 3000 m^3, and 5000 m^3 volumes, focusing on weight, footprint, cube, and cost, are available in Wickham's report [147]. In terms of system weight, the HFC-227ea system is the most weight-efficient, with the IG-541 system being three times heavier. FK-5-1-12 and HFC-227ea systems are nearly identical in weight. Regarding system footprint, the IG-541 system requires significantly more floor space compared to halocarbon agent systems, with cubic space requirements closely aligning with the footprint. When comparing costs, HFC-227ea emerges as the most cost-effective option. However, cost evaluations can vary significantly when considering factors such as selector valve systems, bulk storage configurations, and combined-use systems. The IG-541 system is frequently configured with selector valve arrangements, allowing a single agent supply to protect multiple hazards. This configuration is highly cost-effective, as the expense of the agent supply can be distributed across several applications. In both land-based and marine systems, there is a point at which transitioning from high-pressure cylinder storage to low-pressure bulk storage becomes more economical due to the increasing agent weight. Additionally, some regulatory bodies and local authorities mandate combined system use to ensure comprehensive protection for equipment and room spaces. For instance, SOLAS regulations mandate a local application water system for hazardous machinery spaces alongside a total flooding system for machinery spaces in marine environments.

Table 12. Comparison of average values in the 500 to 5000 m^3 range, using the Halon 1301 system as the basis [138].

Agent	Concentration (vol%)	Weight (kg/m^3)	Footprint (10^4 m^2/m^3)	Cube (10^4 m^3/m^3)	System Cost (USD/m^3)	Agent Cost (USD/m^3)
HFC-227ea	8.7	1.1	6.8	13.1	28.05	19.08
FK-5-1-12	5.5	1.2	7.3	13.8	35.98	26.59
IG-541	40.0	4.3	28.2	56.6	34.07	9.62

The IPCC [148] also documented a comparison of primary clean agent fire extinguishing systems currently in use, under development, or demonstrated for fixed systems suitable for occupied spaces. The analysis highlights relative cost differences among the options. FK-5-1-12 has the highest initial investment cost, as well as higher annual service and end-of-life recovery costs, compared to IG-541 and HFC-227ea. Annual service costs account for replacing 2% of the agent lost per year. For halocarbon agents, the end-of-life value is positive, as recovered agents can be recycled and resold. This cost recovery is equivalent to 50% of the agent's initial cost, enabling reuse in new systems or replenishment of existing ones. Wolf's [149] cost study provides a more comprehensive analysis than Wickham's, as it includes installation labor and recharge costs. The cost comparison for IG-541 and halocarbon agents, considering new system installation and one or two recharge events, is shown in Table 13 for 4000 and 6000-cubic-foot systems. The analysis indicates that for smaller spaces, the cost of a new system installation is similar for IG-541 and halocarbon agents, assuming no discharges over the system's lifespan. For larger spaces, the initial cost of installing an IG-541 system is significantly higher if no discharges occur. If there is one recharge during the system's life, the total cost for small spaces remains comparable across all agents. For large spaces with one recharge, the costs for IG-541 and FK-5-1-12 are similar, while HFC-227ea is less expensive. For smaller systems, if there are two discharges, FK-5-1-12 becomes more expensive than both IG-541 and HFC-227ea. In larger systems with two recharges, IG-541 and FK-5-1-12 have the highest costs. Additionally, Bank et al. [150] conducted a life-cycle assessment comparing the environmental impacts of HFC-227ea and IG-541 in a typical UK-based fire suppression system. The study revealed that approximately six times the amount of IG-541 is required compared to HFC-227ea to achieve equivalent performance. Consequently, six cylinders of IG-541 would be needed in place of a single HFC-227ea cylinder.

Table 13. System cost comparisons of IG-541 and halocarbon agents for the 4000 and 60,000 cubic-foot spaces considering new installations with and without recharges [149].

Agent	New System Installation	New System Installation and One Recharge	New System Installation and Two Recharges
Cost Over Life of Small System—4000 Cubic Foot (USD)			
IG-541	42,000	45,088	48,176
FK-5-1-12	36,000	51,696	57,840
HFC-227ea	35,000	40,276	45,552
Cost Over Life of Large System—60,000 Cubic Foot (USD)			
IG-541	220,000	246,700	273,400
FK-5-1-12	130,500	198,640	266,796
HFC-227ea	109,500	162,580	215,660

In conclusion, each fire suppression system offers distinct advantages, making the installation costs of IG-541 and halocarbon systems generally comparable. Among the options, the HFC-227ea system is the least expensive for initial installation compared to IG-541 and FK-5-1-12. However, the high agent cost of FK-5-1-12 leads to higher overall costs than IG-541 and HFC-227ea when future recharge expenses are factored in. The IG-541 system requires a significantly larger footprint and greater weight than halocarbon systems due to its high volumetric requirements, necessitating a larger number of cylinders. Inert gas systems are commonly designed to protect multiple areas, using a cylinder bank with enough capacity to safeguard the largest enclosure. Directional valves are employed to direct the gas to the appropriate hazard area, with gas supplied either through individual piping from the cylinder bank or via a high-pressure manifold extending throughout the protected building [151]. While the FK-5-1-12 system also supports selector valve configurations, its 10 s maximum discharge period limits its application to nearby hazards unless the protected rooms are closely situated. In data centers and substations, raised floors are often built to provide a flexible and efficient environment for managing cables and cooling systems. According to NFPA 75, A.9.1.1.3 notes that halocarbon agents should not be used to protect the space below a raised floor unless the space above the floor is also protected. If a fire exists in the equipment above the raised floor, agents below the extinguishing concentration may be exposed, and halocarbon agents may undergo significant decomposition. An inert gas system must be used to protect the space under a raised floor if only the space is to be protected by a total flooding system [152].

9. System Design Consideration

Table 14 provides a comparison of IG-541 and halocarbon agents based on system characteristics and design requirements. IG-541 is stored as a compressed gas in steel cylinders at storage pressures of either 200 bar or 300 bar. In contrast, halocarbon agents are stored as compressed liquids in steel containers, super-pressurized with nitrogen to 25 bar, 42 bar, or 50 bar to improve discharge flow characteristics. Consequently, the minimum design working pressure for system piping upstream of the pressure reducer must align with the storage pressure. This necessitates the use of higher-grade piping materials for IG-541 systems compared to halocarbon agent systems. For halocarbon agent systems, the maximum permitted discharge time is 10 s, which is defined as the period during which all liquid agents pass through the nozzle. The total discharge time is extended due to the expulsion of nitrogen and agent vapor from the system. The 10 s discharge time limit for halocarbon agents aims to achieve four key objectives [153]:

(1) Ensure high flow rates through nozzles for adequate mixing of the agent with air inside the enclosure.
(2) Provide sufficient velocity through pipes to maintain a homogeneous flow of liquid and vapor.
(3) Limit the formation of agent thermal decomposition products.
(4) Minimize direct and indirect fire damage, especially in rapidly developing fire scenarios.

Among these objectives, minimizing the formation of agent thermal decomposition products is the most critical concerning discharge. For Class C fires, the IG-541 system has a discharge time requirement of 120 s. The primary reasons for this limitation, despite IG-541 producing no decomposition products, are the following:

(1) To limit direct and indirect fire damage.
(2) To minimize the duration that the fire burns in a depleted oxygen atmosphere.

Table 14. System characteristic and design requirements comparison between IG-541 and halocarbon agents.

Agent	Stored Pressure (bar)	Discharge Time (s) for Class C Fire	Pipe Run Distance	Visibility After Discharge	Pressure Effect	Combustion Product Ventilation	Concentration Hold Time	Agent Recovery, Recycling, and Reclamation
HFC-227ea	25/42/50	10	Short	Low	Positive + Negative	Yes	Weak	Yes
FK-5-1-12	25/42/50	10	Short	Low	Positive + Negative	Yes	Weak	Yes
IG-541	200/300	120	Long	High	Positive	No	Good	No

The IG-541 agent is stored and discharged as a gas, which minimizes energy loss due to phase change and results in lower friction losses. This enables the IG-541 agent to travel longer distances through smaller-diameter pipes. In contrast, the halocarbon agent stores the agent as compressed liquid within the cylinder, pressurized with dry nitrogen to propel the agent through the piping system. The liquid is transformed into a gas as it travels under pressure through the discharge piping due to generated friction. Compressed liquids do not flow as well as inert gases since their pressure in the pipe is lower. Consequently, the halocarbon system's discharge pipe length is more constrained than the IG-541 system. Discharging halocarbon agents creates the appearance of fog, which may reduce visibility. However, this typically clears quickly and should not obstruct personnel's ability to safely escape the protected area. There is no significant condensation of air moisture during the IG-541 agent discharge; therefore, the gases do not reduce escape route visibility [154].

The effectiveness of a gaseous total flooding fire suppression system relies on retaining the air–extinguishant mixture within the protected area for a certain period. This requires minimizing gas exchange (leakage) between the enclosure and the outside environment, necessitating a high degree of enclosure integrity. If the enclosure is too tightly sealed during extinguishant discharge, the resulting pressure change could exceed the structural limits of its surfaces, such as windows and doors. Conversely, if there is too much vent area, rapid gas exchange will shorten the retention time of the extinguishant. Therefore, gaseous fire suppression systems must manage pressure within the protected area during discharge and ensure the extinguishant–air mixture is retained for a specified period afterward. Pressure venting of enclosures is crucial for designing a clean agent fire prevention system because it reduces the forces caused by changes in enclosure pressure when gaseous agents are released into an enclosure, hence preventing the likelihood of structural integrity failure. Furthermore, the operation of pressure vent or extract systems requires the removal of mixtures of air/gaseous media, combustion products, and extinguishant breakdown products from a protected enclosure into the environment [155]. Figure 12 depicts typical pressure excursions on the structure of the protected enclosure during inert gas and halocarbon agent discharge. The peak positive and negative pressures generated during the discharge of halocarbon fire suppression systems can potentially damage the structural components of the protected area, such as walls, studs, and windows, if not properly installed. If these structural components are damaged during system discharge, it can no longer be assured that the enclosure will maintain the necessary concentration of the agent long enough to effectively suppress the fire. While the choice of agent is a significant factor, it is only one of many that influence the magnitude of pressure peaks within the protected space. Other critical factors include agent concentration, wall construction, the size and location of enclosure leakage areas, fire size, humidity, and retention time, all of which play a role in the overall pressure dynamics. The discharge characteristics of the halocarbon agent system typically include an increase in negative pressure at the beginning of the discharge, which is partly caused by the rapid temperature decrease, followed by a rise in positive pressure. Therefore, a dual-flow damper with both a positive (outward) and a negative (inward) opening damper is needed [156]. The initial negative pressure spike results from the cooling effect due to the agent's high heat capacity, which absorbs sensible heat during its rapid vaporization upon discharge. The subsequent positive pressure spike occurs because of the relatively rapid introduction of the gaseous agent and propellant mass into the fixed volume of the enclosure [157]. The discharge of inert gas agents is characterized by a rapid increase in pressure to a peak level, followed by a relatively slow decrease over time. For this type of discharge, an outward-opening damper is needed, as positive pressure is consistently generated during an IG-541 system discharge. The pressure dynamics of inert gas agent discharges are notably different from those of halocarbon-based agent discharges. The thermodynamic properties of a halocarbon-based agent determine the extent of heat transfer during discharge, resulting in varying pressure dynamics for different halocarbon-based agents [158]. The IG-541 system's constant flow design prevents peak pressure increases during discharge and introduces lower pressure into the piping system. The valve is specially designed to evenly distribute the gas flow into the protected area, reducing the need for venting and minimizing the risk of over-pressurization. This design offers an excellent alternative to standard high-pressure inert gas systems, which typically have a distinctive peak flow and pressure spike, necessitating larger, higher-specification pipework and a greater venting area [159].

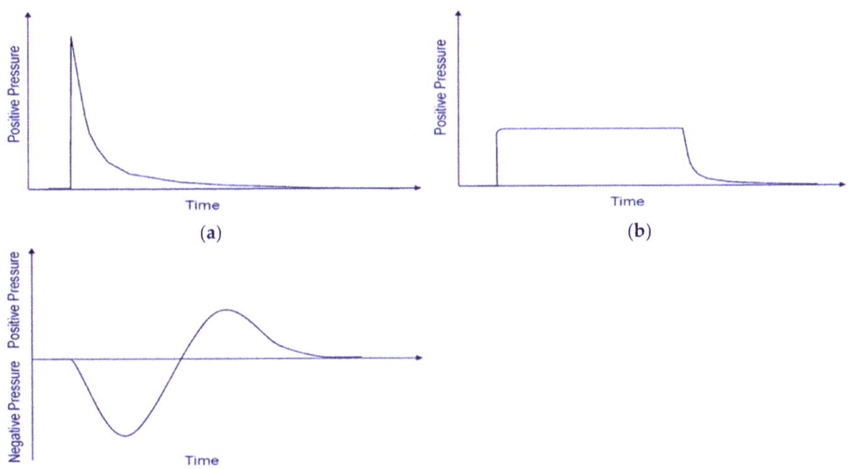

Figure 12. Typical pressure excursions that would occur during discharge. (**a**) Inert gas (**b**) Inert gas (constant flow). (**c**) Halocarbon agent [155].

Mechanical extraction systems may be necessary for both halocarbon and inert gas agents, albeit for different reasons. Halocarbon agents can decompose when exposed to flames and hot surfaces, producing acidic byproducts that may cause corrosion if not promptly removed. Therefore, mechanical extraction is essential when using halocarbon agents. In contrast, inert gases do not generate acidic decomposition products, but mechanical extraction is still preferred to eliminate smoke and soot deposits dispersed throughout the room by the force of the inert gas discharge. To prevent the fire from re-igniting, the extinguishing gas should be retained for a period of time after discharge, which typically requires a well-sealed enclosure. Upon fire detection, the compressed agent, whether a halocarbon or an inert gas, is released into the enclosure, causing a peak pressure of approximately 250 to 1250 Pascals for a brief moment. The exact pressure depends on the total leakage area of the enclosure. Once the enclosure is fully flooded with the clean agents, they will begin to leak out at a rate primarily determined by the leakage area in the lower part of the enclosure. The distribution of the remaining agent will either stay consistent throughout the enclosure due to continuous mixing or form an interface with the air above and the agent below, which will gradually descend over time [160]. The IG-541 has a density similar to that of air [63], resulting in slower leakage and allowing the gaseous agent to be stored for longer periods. In contrast, the FK-5-1-12 has a higher density than air [161], leading to a weak concentration hold time and faster leakage. Although the density of HFC-227ea [162] is lower than FK-5-1-12, it is still higher than IG-541, which also results in a weak concentration hold time and faster leakage. Consequently, a higher volume of halocarbon agent is required to compensate for the faster leakage while maintaining the minimum hold time.

To minimize the environmental impact of halocarbon agents, recovery and recycling should be performed safely and responsibly, adhering to established quality standards. The FIA [163] provides guidance outlining the criteria for recycled halocarbon agents and their use in discharged or service-exchange containers. When containers are returned from service, the agent should be reclaimed during servicing or decommissioning using a process that reverses the original filling procedure. Closed-loop recovery systems, which prevent contact with ambient air, are preferred to minimize losses and avoid contaminant introduction. If the halocarbon agent requires reconditioning to meet the required standards, it should undergo filtering, drying, distillation, or other appropriate processes before reuse. After processing, the agent must be tested to ensure compliance with relevant standards. Agents that fail to meet the required purity levels should not be reused and must be destroyed in an environmentally friendly manner or returned to the manufacturer in accordance with applicable regulations. Halocarbon fire extinguishing agents should only be reused if they have been properly recovered, tested, and processed to meet the necessary quality requirements. The FSTOC [164] has established standard methods for the treatment of HFC-227ea. Reuse involves transferring an agent cylinder from one application to another without additional processing. Recovery entails removing the agent, in any condition, from an extinguishing system cylinder and storing it in an external container without testing or further processing. Recycling focuses on cleaning recovered agents for reuse without meeting all the stringent requirements for reclamation. This process typically includes removing nitrogen, pressurizing the agent, as well as reducing moisture, particulate matter, and non-volatile residue content to meet relevant standards. Reclamation, the preferred method, involves reprocessing the agent to achieve the highest level of purity specified by applicable standards. This process requires specialized separation or distillation equipment to eliminate impurities, including other halocarbon contaminants. The HTOC [165] recommends strategies to reduce halocarbon agent releases that exceed those required for fire protection and explosion prevention. Key recommendations include adopting recycling equipment to recover surplus or reusable fire extinguishing agents, discontinuing the use of fire extinguishants as test gases in discharge testing for fire protection systems, and amending regulations that mandate such practices. Additional measures involve utilizing well-managed central storage facilities for halocarbon extinguishant reserves and avoiding the release of system cylinder contents into the atmosphere during equipment servicing. Since the composition of IG-541 consists of naturally occurring atmospheric gases, it can be safely released into the atmosphere without environmental concerns. Furthermore, atmospheric discharge is a cost-effective option for disposing of IG-541 during the decommissioning of fire extinguishing systems.

10. Practical Implementation

10.1. Past Electrical Fire Incidents

The research report conducted from 2017 to 2021 by U.S. Fire Departments identified equipment or heat source failure as a leading cause of structure fires in industrial and manufacturing properties. Electrical distribution, lighting, and power transfer equipment were frequently involved in ignitions, with wiring and related equipment accounting for 12% of these fires. A significant 24% of structure fires in industrial properties were attributed to equipment or heat source failures, including electrical and mechanical malfunctions or heat sources too close to combustibles. Notably, arcing, unclassified heat from powered equipment, radiated or conducted heat from operating equipment, and sparks or embers were common heat sources [166]. Additionally, a separate report on non-home electrical fires from 2012 to 2016 revealed that electrical distribution and lighting equipment were often implicated in these fires. About 46% of these fires resulted from electrical failures or malfunctions, with arcing being a significant contributor, highlighting the dangers posed by unintended electrical discharges between conductors [167]. Uptime Institute's outages database shows that fires in data centers are relatively rare and have a minimal impact on operations, accounting for only 1% of significant outages [168]. While the frequency of fires is not increasing relative to the number of data centers or IT loads, they still pose a serious risk. In the case of the OVHcloud data storage center fire on 9–10 March 2021, the fire started in the UPS2 area and spread rapidly across the SBG2 building. The lack of a VESDA (very early smoke detection apparatus) system and an automatic fire suppression system, such as water or gas, was a significant factor that allowed the fire to escalate. In the aftermath, the Bureau of Investigation and Analysis of Industrial Risks recommended the implementation of inert gas systems as a potential fire suppression solution for such facilities [169]. The incident underlines the importance of advanced fire suppression systems, particularly in high-risk areas like data centers and substations, where electrical equipment failure can have catastrophic consequences. Inert gas systems, such as IG-541, could have potentially mitigated the fire's impact by providing rapid, efficient fire suppression without causing harm to sensitive electronic equipment.

10.2. Case Studies of IG-541 System

Several prominent organizations in different industries have turned to the IG-541 fire suppression system to protect their critical infrastructure and sensitive equipment, demonstrating its effectiveness and environmental advantages. The world's largest media conglomerates in Southern California, USA, selected the IG-541 system to safeguard their data processing network from both fire and collateral damage. This solution was crucial in protecting multi-million-dollar network processors within their data processing hub. The system met the client's stringent requirements by safeguarding sensitive electronic equipment and valuable data in an environmentally friendly manner. The client specifically sought an alternative to the ozone-depleting chemical suppression agents previously in use, and IG-541, made up of atmospheric gases, provided the ideal solution without environmental harm [170]. Similarly, Blix Solutions AS chose IG-541 to align with their commitment to sustainability and environmental responsibility. The IG-541 system, which is composed of naturally occurring gases like nitrogen, argon, and carbon dioxide, offers a non-toxic and environmentally friendly fire suppression solution. This makes it an ideal choice for data centers where protecting sensitive electronic equipment is crucial. This decision was especially beneficial for protecting sensitive electronic equipment in data centers, where water-based systems might cause severe damage. The rapid detection and suppression capabilities of the IG-541 system also minimize the spread of fire, safeguard operations, and ensure business continuity while maintaining the company's commitment to reducing its carbon footprint [171]. Sindal Biogas, focused on green transition goals, also selected IG-541 to protect their technical rooms, which house critical components like switchboards and servers. These rooms are prone to fire risks that could lead to significant production downtime. As a biogas plant plays a crucial role in sustainability, Sindal Biogas prioritized an environmentally neutral solution that avoids harmful chemicals like PFAS. IG-541's environmentally friendly nature reassured the company and improved operational stability and risk management, also positively impacting their relationship with their insurance provider [172]. Copenhagen Airport has relied on IG-541 for over 30 years to protect critical IT and technology rooms. The system has been deployed successfully in two minor incidents, including a short circuit and a fan heater fire. In both cases, IG-541 effectively extinguished the fire without causing damage to buildings or IT equipment, ensuring no cleanup was required and maintaining operational continuity [173]. E.ON's Kemsley Combined Heat and Power (CHP) plant uses IG-541 to safeguard control and electrical switch rooms. This system protects both personnel and equipment, automatically deploying when fire is detected in the protected zones. In a recent incident, the system successfully extinguished a fire in a switchgear room, preventing potential damage and allowing production to continue without disruption. In a notable incident, the IG-541 system automatically triggered when a fire broke out in a switchgear room. The system efficiently extinguished the fire, averting a potentially dangerous situation. Importantly, the IG-541 agent did not interfere with any electrical equipment, and no cleanup was necessary, allowing the plant to resume production immediately. This case exemplifies the system's ability to maintain safety and ensure operational continuity without causing any disruptions [174]. These case studies highlight the versatility and reliability of IG-541 in protecting critical infrastructure, offering both safety and operational continuity while maintaining environmental responsibility.

10.3. Effectiveness and Limitations of IG-541 System

The IG-541 system is recognized for its effectiveness as a pre-engineered fire suppression solution and offers the flexibility to suppress various classes of fires. As outlined in NFPA 2001's Section 4.2.1.1 [22], all pre-engineered systems shall be installed to protect hazards within the limitations that have been established by the listing. Those consisting of system components designed to be installed according to pre-tested limitations by a testing laboratory. Automatic extinguishing units incorporate special nozzles, flow rates, methods of application, nozzle placement, actuation techniques, piping materials, discharge times, mounting techniques, and pressurization levels that could differ from those detailed elsewhere in this standard. A key advantage of these systems is their ability to provide early fire detection and rapid response. Depending on the installed detection package, the system can activate immediately or with a programmed delay, ensuring prompt containment of fires at their earliest stage. Additionally, pre-engineered systems are safe and user-friendly, as they operate automatically without requiring human intervention, though a manual override option is available for emergencies. Their pre-tested design ensures reliable performance, meeting NFPA standards and effectively extinguishing specific fire types in specialized environments. Another significant benefit is their ability to eliminate the fuel source of a fire automatically, particularly in cases involving electrical equipment, where the fuel or electrical source might otherwise continue to feed the flames [175]. The

IG-541 system, in particular, has demonstrated its effectiveness in various fire scenarios. It can suppress Class C fires effectively while also showing strong performance against Class A and Class B fires. Laboratory evaluations confirm its efficacy through standardized testing protocols, such as the wood crib test for Class A fires and the cup burner method for Class B fires [176]. In real-world applications, the IG-541 system has gained popularity for its reliability in protecting against these types of fires, as demonstrated in various projects, including machinery spaces, museums, libraries, and heritage sites [177–179]. This adaptability highlights the IG-541 system's ability to adjust design concentrations to meet the requirements for suppressing different fire classes, making it a versatile and efficient fire suppression solution.

The IG-541 fire suppression system has notable limitations, particularly in its inability to suppress fires involving specific materials and the potential effects of acoustical noise. Clean agent fire suppression systems shall not be used on fires involving specific materials unless the agents have been tested and approved by the authority having jurisdiction. These materials include chemicals or mixtures capable of rapid oxidation in the absence of air, such as cellulose nitrate and gunpowder. Reactive metals, including lithium, sodium, potassium, magnesium, titanium, zirconium, uranium, and plutonium, are also unsuitable for clean agent suppression due to their unique chemical properties. Additionally, metal hydrides and chemicals capable of autothermal decomposition, such as certain organic peroxides, pyrophoric materials, and hydrazine, present significant risks when interacting with clean agents [22]. Li-ion batteries, widely used in data centers and substations as backup power for uninterruptible power supply (UPS) systems, pose specific challenges despite their advantages, including high energy density, longer lifespan, and lower maintenance. One key risk is thermal runaway, where a lithium-ion cell enters an uncontrollable, self-heating state due to damage, overcharging, or exposure to high temperatures. This process involves exothermic reactions releasing heat that accelerates the reaction if not dissipated effectively. The IG-541 system, while capable of reducing oxygen concentration to suppress fires, cannot halt thermal runaway as the reaction is primarily driven by internal chemical processes rather than oxygen availability. Consequently, the fire may re-ignite after suppression. Addressing this limitation requires fire suppression methods that incorporate active cooling to mitigate the risks of Li-ion battery fires effectively [180]. The potential effects of acoustical noise produced by activation of the IG-541 system shall be considered where such systems are used in an occupancy containing noise-sensitive equipment. Several reports have been recorded on a yearly basis about IG-541 gas discharges conducting to faults in hard disk drives (HDDs) operating in data centers protected by such a solution from fire due to the level of noise produced. The sound power level emitted by nozzles can vary depending on their height and the humidity profile at the measurement site. To minimize noise impact, it is recommended to position nozzles on the facets of the concrete supporting walls, ensuring they do not face HDD racks directly or other nozzles [181]. The sound output of fire suppression systems is dependent on many factors. The factors that influence this include discharge duration, peak agent flow rate, valve technology, and many others. Traditional inert gas suppression systems often use metering orifices, with nozzle sound power levels sometimes exceeding 145 dB. The acoustic nozzle is a substantial advancement in nozzle technology for data center markets, translating into a solution that can reduce sound exposure to sensitive HDDs and requires significantly fewer nozzles and piping. For optimal protection, Johnson Controls recommends maintaining sound pressure levels at 110 dBZ or below across the 500 Hz–10k Hz frequency range at HDD locations [182].

10.4. Future of IG-541 System

The future trend of IG-541 systems is being shaped by technological advancements, stricter fire safety standards, environmental priorities, and the specialized needs of critical infrastructure that involve energized electrical fires. IG-541 offers key benefits such as being electrically non-conductive, leaving no residue, and having zero ozone depletion potential (ODP) and global warming potential (GWP). These characteristics make it particularly valuable for high-risk and specialized applications. The rising global demand for fire protection systems like IG-541 is driven by the growth of data storage facilities and electrical power supply facilities. IG-541's non-conductive and environmentally friendly properties make it an ideal choice for these applications. High-tech industries, such as automation and electronics manufacturing, are also expected to adopt IG-541 due to its ability to protect sensitive equipment without leaving residue or conducting electricity. IG-541 systems offer flexibility through integration with other suppression agents and smart technologies. Combining IG-541 with water sprinklers, water mist systems, foam, or chemical agents enhances fire suppression effectiveness, especially for complex hazards like Li-ion batteries or high energy electrical equipment. Integrating Internet of Things (IoT) technology enables real-time monitoring and smart activation, improving system responsiveness and minimizing false alarms. Artificial intelligence (AI) further boosts system readiness by optimizing design and providing predictive insights through AI-based fire behavior modeling. Environmentally, IG-541 stands out for its sustainability. Composed of naturally occurring gases, it reduces carbon footprints while meeting environmental regulations and green building certifications. Its eco-friendly profile supports companies in improving their environmental, social, and governance (ESG) performance, demonstrating commitment to sustainability and responsible operations. By blending environmental responsibility, technological adaptability, and robust fire suppression capabilities, IG-541 systems are positioned as a future-proof solution for modern fire safety challenges.

11. Conclusions and Future Works

The evolution of fire suppression technologies is critical in addressing the growing complexities of protecting high-value assets, such as data centers and electrical substations, while meeting stringent environmental and safety standards. This study comprehensively evaluates IG-541, an inert gas mixture, as a cleaner and safer alternative to halocarbon-based agents like HFC-227ea and FK-5-1-12 in energized electrical fire suppression. IG-541's composition offers unique advantages, including efficient fire suppression through oxygen reduction, absence of harmful by-products, and a zero impact on ozone depletion and global warming. These attributes make it a strong candidate for adoption in industries prioritizing environmental sustainability and regulatory compliance. In comparison, halocarbon agents, while effective, face increasing scrutiny due to their significant global warming potential, extended atmospheric lifetimes, and toxic decomposition by-products, which present risks to both human health and sensitive equipment. The study also emphasizes the importance of safety in fire suppression, particularly in occupied spaces. IG-541 offers a significant safety advantage due to its inert nature and lack of hazardous decomposition products. Halocarbon agents, on the other hand, can produce hydrogen fluoride (HF) during decomposition, posing risks to human health and electronic equipment. IG-541 also maintains respiratory-friendly conditions, allowing safe evacuation during its deployment. The regulatory landscape is increasingly favoring sustainable fire suppression solutions. International agreements like the Montreal Protocol and local environmental standards are driving the phase-out of halocarbon agents. IG-541 meets stringent safety and environmental requirements outlined by standards such as NFPA 2001 and ISO 14520. This positions it as an ideal solution for future applications, especially in critical industries like data centers, substations, and telecommunications facilities. In summary, IG-541 presents a transformative opportunity for the fire suppression field by balancing high performance, environmental responsibility, and safety. Its adoption can lead the industry toward a future where fire protection supports both operational resilience and global sustainability goals.

Future Works

While IG-541 shows great promise as a sustainable and efficient alternative to halocarbon-based fire suppression agents, further research and development are essential to fully optimize its effectiveness across diverse fire scenarios. A critical area of study involves evaluating IG-541's extinguishing performance in high-energy electrical fires. As discussed earlier, higher agent concentrations are required to suppress fires involving energized electrical equipment at elevated energy levels. If IG-541 fails to completely remove heat from the equipment, re-ignition may occur due to the continuous presence of intense heat and potential arcing. Additional research should focus on understanding the persistent fire hazards in such scenarios, including ignition and combustion behaviors, as well as the overall fire suppression process. Key factors to consider include compartment fire conditions (e.g., temperature and pressure), flame propagation to adjacent electrical components (such as cables, switchgear, and transient combustibles), and the production of ionized soot, which could significantly contribute to secondary arcing. Further studies should also explore the potential ignition of secondary combustibles, fire growth dynamics, and the intensity of the blaze following an enduring fire event [183]. Assessing the effectiveness and timing of fire suppression interventions will also be crucial in enhancing the reliability of IG-541 in these challenging conditions. Battery systems, critical for backup power in data centers and substations, present additional challenges, as IG-541 is less effective against fires involving reactive metals like Li-ion batteries. Research should focus on improving suppression performance by integrating pressure relief mechanisms and combining IG-541 with cooling-based methods, such as water [184]. Additionally, further safety evaluations are necessary to assess the impact of IG-541 in occupied spaces, particularly as higher agent concentrations are required for high-energy fire scenarios and the protection of complex hazards such as battery systems. Ensuring occupant safety includes understanding the physiological effects of prolonged exposure to reduced oxygen levels. Additionally, safety protocols should be developed to enhance evacuation procedures and implement system fail-safes in the event of an IG-541 discharge within hazardous areas. An optimal IG-541 concentration, delivered through a muffled nozzle system, has been selected for a dual cab fire inerting system based on its effectiveness in ensuring safe evacuation times [185]. To maximize safety and efficiency during fire emergencies, operator training and task synchronization should accompany the system's implementation. Furthermore, an early fire detection system is essential in IG-541 systems to prevent oxygen levels from dropping dangerously low in confined spaces. If early suppression is not feasible, sufficient evacuation time must be ensured [186].

A comprehensive cost–benefit analysis should be conducted to evaluate the long-term economic viability of IG-541 systems compared to halocarbon and traditional fire suppression agents. This analysis should incorporate lifecycle studies of materials and system components to solidify IG-541's environmental and operational benefits. These recommendations aim to refine IG-541's application, reinforcing its position as a sustainable and reliable solution for fire protection in the face of future challenges. In fire safety engineering, a cost–benefit analysis provides a structured approach to determining whether the benefits of a fire safety measure outweigh its costs. Various reference methodologies can be applied to conduct this assessment effectively [187,188]. Further research is needed to reduce the acoustic noise generated during the activation of the IG-541 system to minimize the risk of damage to sensitive equipment and implement effective noise control measures. This includes investigating the combined effect of multiple nozzles—both standard and silent—on noise levels, particularly their impact on HDD performance. Additionally, expanding the range of tested silent nozzles, including designs with sound-absorbing layers, could further mitigate noise-related issues [189]. Additionally, research should focus on optimizing nozzle quantity and spacing, avoiding very short discharge times, and eliminating the use of pneumatic sirens to reduce the impact of sound levels on HDD functionality [190]. Moreover, further research should focus on minimizing or eliminating IG-541 system leakage through ventilation or structural improvements to mitigate over-pressure risks that could damage structures and equipment. Ventilation systems play a crucial role not only in reducing the impact of IG-541 over-pressure but also in optimizing the quantity of required agent needed to achieve the desired oxygen concentration. Studies should investigate IG-541's extinguishing efficiency across different fire sizes and the thermal pressure effects caused by fires in enclosed spaces. In addition to mechanical ventilation, natural ventilation such as open doors, windows, and leaks also influences airflow and system performance [191,192]. Additionally, research should assess the required IG-541 quantity in scenarios with larger building openings to determine system effectiveness [193]. Furthermore, leakage performance between raised floors and equipment should be analyzed based on opening ratios to verify the reliability of IG-541 when used exclusively for raised floor protection [194].

Author Contributions: Conceptualization, K.H.L.; writing—original draft preparation, K.H.L.; writing—review and editing, K.H.L. and T.S.L.; supervision, K.H.L.; project administration, K.H.L. and S.T.B. All authors have read and agreed to the published version of the manuscript.

Funding: This research received no external funding.

Data Availability Statement: No new data were created or analyzed in this study.

Conflicts of Interest: The authors declare no conflicts of interest.

References

1. Staines, B. Fire Protection in the Building Industry Identifying Risks and Understanding Safety Measures. Available online: https://www.fpib.com.au/wp-content/uploads/building_industry_article.pdf (accessed on 25 December 2024).
2. International Energy Agency. *Electricity 2024—Analysis and Forecast to 2026*; IEA: Paris, France, 2024; Volume 170.
3. Amer, S.; Rashed, J.; Elayan, W.; Almazrouei, K.; AlShabibi, A. Managing Electrical Fire Risks in Substations: A Framework to Protect Lives and Assets. In Proceedings of the 5th European International Conference on Industrial Engineering and Operations Management, Rome, Italy, 26–28 July 2022; pp. 2160–2181.
4. Kim, A. Recent Development in Fire Suppression Systems. In Proceedings of the 5th AOSFST, Newcastle, Australia, 3–6 December 2001; pp. 12–27.
5. Babrauskas, V. Electrical Fires. In *SFPE Handbook of Fire Protection Engineering*, 5th ed.; Hurley, M., Ed.; SFPE: Gaithersburg, MD, USA, 2016; pp. 662–704, ISBN 978-1-4939-2564-3.
6. Geoffrey, B.; Derek, C.; Vijayaraghavan, M.; Steve, M. Electrical Energy, Ignition and Flammability. In *Practical Electrical Equipment and Installations in Hazardous Areas*, 1st ed.; Mackay, S., Ed.; Newnes: Oxford, UK, 2005; pp. 10–26, ISBN 0-7506-6398-7.
7. National Fire Protection Association. *NFPA 921: Guide for Fire and Explosion Investigations*; NFPA: Quincy, MA, USA, 2024.
8. Babraukas, V. Research on Electrical Fires: The State of the Art. In Proceedings of the Fire Safety Science—Proceedings of the Ninth International Symposium, Karlsruhe, Germany, 21–26 September 2008; pp. 3–18.
9. Babrauskas, V. How Do Electrical Wiring Faults Lead To Structure Ignitions? In Proceedings of the Fire and Materials 2001 Conference, San Francisco, CA, USA, 22–24 January 2001; pp. 39–51.
10. Huang, X. Ignition and Spread of Electrical Wire Fires. Master's Thesis, University of California, San Diego, CA, USA, 2012.

11. Lyon, R.E. Plastics and Rubber. In *Handbook of Materials for Fire Protection*; Harper, C.A., Ed.; McGraw-Hill: New York, NY, USA, 2004; pp. 121–172, ISBN 0-07-143330-9.
12. Hartin, E. Fire Development and Fire Behavior Indicators. Available online: https://cfbt-us.com/pdfs/FBIandFireDevelopment.pdf (accessed on 25 December 2024).
13. Davie County, NC, Fire Marshal's Office. *Stages of Fire Growth*; DCFM: Mocksville, NC, USA, 2018.
14. Madrzykowski, D. Fire Dynamics: The Science of Fire Fighting. *Int. Fire Serv. J. Leadersh. Manag.* **2013**, *7*, 7–15.
15. Donohue, S.S.; Suski, M.; Chen, C. Fire Protection and Life Safety Design in Data Centers. In *Data Center Handbook: Plan, Design, Build, and Operations of a Smart Data Center*, 2nd ed.; Geng, H., Ed.; John Wiley & Sons, Inc.: Hoboken, NJ, USA, 2021; pp. 533–550, ISBN 9781119597544.
16. Linteris, G.T. Clean Agent Suppression of Energized Electrical Equipment Fires. *Fire Technol.* **2011**, *47*, 1–68. [CrossRef]
17. Driscoll, M.R.; Rivers, P.E. Clean Extinguishing Agents and Continuously Energized Circuits: Recent Findings. In Proceedings of the Halon Options Technical Working Conference Proceedings, Albuquerque, NM, USA, 6–8 May 1997; pp. 129–140.
18. Hamins, A.; Borthwick, P. Suppression of Ignition over a Heated Metal Surface. *Combust. Flame* **1998**, *112*, 161–170. [CrossRef]
19. Niemann, R.; Bayless, H. Update on the Evaluation of Selected NFPA 2001 Agents for Suppressing Class "C" Energized Fires. In Proceedings of the Halon Options Technical Working Conference Proceedings, Albuquerque, NM, USA, 12–14 May 1998; pp. 293–295.
20. Donnelly, M.K.; Grosshandler, W.L. *Suppression of Fires Exposed to an External Radiant Flux*; NIST Interagency/Internal Report-6827; National Institute of Standards and Technology: Gaithersburg, MD, USA, 2001; Volume 31. [CrossRef]
21. Flamm, D.; Bengston, G.; Niemann, R. Continuing the Examination and Comparison of Existing Halon Alternatives in Preventing Re-ignition on Continuously Energized Fires. In Proceedings of the Halon Options Technical Working Conference Proceedings, Albuquerque, NM, USA, 24–26 May 2005; pp. 1–7.
22. *NFPA 2001*; Standard on Clean Agent Fire Extinguishing Systems. National Fire Protection Association (NFPA): Quincy, MA, USA, 2022.
23. *ISO 14520-1*; Gaseous Fire-Extinguishing Systems—Physical Properties and System Design—Part 1: General Requirements. International Organization for Standardization (ISO): Geneva, Switzerland, 2022.
24. *ISO 14520-5*; Gaseous Fire-Extinguishing Systems—Physical Properties and System Design—Part 5: FK-5-1-12 Extinguishant. International Organization for Standardization (ISO): Geneva, Switzerland, 2019.
25. *ISO 14520-8*; Gaseous Fire-Extinguishing Systems—Physical Properties and System Design—Part 8: HFC 125 Extinguishant. International Organization for Standardization (ISO): Geneva, Switzerland, 2019.
26. *ISO 14520-9*; Gaseous Fire-Extinguishing Systems—Physical Properties and System Design—Part 9: HFC 227ea Extinguishant. International Organization for Standardization (ISO): Geneva, Switzerland, 2019.
27. *ISO 14520-10*; Gaseous Fire-Extinguishing Systems—Physical Properties and System Design—Part 10: HFC 23 Extinguishant. International Organization for Standardization (ISO): Geneva, Switzerland, 2019.
28. *ISO 14520-15*; Gaseous Fire-Extinguishing Systems—Physical Properties and System Design—Part 15: IG-541 Extinguishant. International Organization for Standardization (ISO): Geneva, Switzerland, 2015.
29. *ISO 14520-14*; Gaseous Fire-Extinguishing Systems—Physical Properties and System Design—Part 14: IG-55 Extinguishant. International Organization for Standardization (ISO): Geneva, Switzerland, 2015.
30. *ISO 14520-13*; Gaseous Fire-Extinguishing Systems—Physical Properties and System Design—Part 13: IG-100 Extinguishant. International Organization for Standardization (ISO): Geneva, Switzerland, 2015.
31. Stratopheric Protection Division. *Carbon Dioxide as a Fire Suppressant: Examining the Risks*; EPA430-R-00-002; United States Environmental Protection Agency: Washington, DC, USA, 2000; Volume 54.
32. Halons Technical Options Committee. *Fire Protection Alternatives to Halons, HCFCs and HFCs*; United Nations Environment Programme: Nairobi, Kenya, 2018; Volume 73, ISBN 978-9966-076-52-6.
33. Ozone Secretariat. *Handbook for the Montreal Protocol on Substances that Deplete the Ozone Layer*; United Nations Environment Programme: Nairobi, Kenya, 2019; Volume 942, ISBN 978-9966-076-59-5.
34. UNEP Division of Technology, Industry and Economics. *Eliminating Dependency on Halons: Self-Help Guide for Low Consuming Countries*; United Nations Environment Programme: Paris, France, 1999; Volume 130, ISBN 92-807-1783-9.
35. Hughes, W.J. *Options to the Use of Halons for Aircraft Fire Suppression Systems—2012 Update*; DOT/FAA/AR-11/31; United States Department of Transportation: Washington, DC, USA, 2012; Volume 102.
36. Xing, H.; Lu, S.; Yang, H.; Zhang, H. Review on Research Progress of $C_6F_{12}O$ as a Fire Extinguishing Agent. *Fire* **2022**, *5*, 50. [CrossRef]

37. Hiltz, J.A. *Acid Gas (Hydrogen Fluoride) Production from Halon 1301 and Alternative Gaseous Fire Suppressants: A Review*; DRDC-RDDC-2015-R036; Defence Research and Development Canada—Atlantic Research Centre: Dartmouth, NS, Canada, 2015; Volume 36.
38. Wickham, R.T. *Status of Industry Effects to Replace Halon Fire Extinguishing Agents*; Wickham Associates: Stratham, NH, USA, 2002; Volume 54.
39. Halons Technical Options Committee. *New Technology Halon Alternatives*; United Nations Environment Programme: Nairobi, Kenya, 2007; Volume 28.
40. Occupational Safety and Health Administration. Evacuation Plans and Procedures eTools. Available online: https://www.osha.gov/etools/evacuation-plans-procedures/emergency-standards/fixed-extinguishing (accessed on 30 January 2025).
41. O'Connor, B. Clean Agent System Basics. Available online: https://www.nfpa.org/news-blogs-and-articles/blogs/2022/05/06/clean-agent-system-basics (accessed on 30 January 2025).
42. Shukla, M.; Tiwary, A.; Yadav, A.; Khan, M.; Badodia, P.K. An Innovative Designing of a Clean Agent Flooding System for Server Room of an Educational Institute. *Int. J. Adv. Res. Sci. Commun. Technol.* **2021**, *7*, 7.
43. UNEP Division of Technology, Industry and Economics. *Standards and Codes of Practice to Eliminate Dependency on Halons: Handbook of Good Practices in the Halon Sector*; United Nations Environment Programme: Paris, France, 2001; Volume 80, ISBN 92-807-1988-1.
44. Pollution Prevention Program—Federal Programs Division. *Pollution Prevention Fact Sheet #14: Alternatives to Halon & Other Halocarbon Fire Extinguishing Agents*; Environment Canada Environment Protection Branch: Nepean, ON, Canada, 2000; Volume 7.
45. Laube, J.C.; Martinerie, P.; Witrant, E.; Blunier, T.; Schwander, J.; Brenninkmeijer, C.A.M.; Schuck, T.J.; Bolder, M.; Rockmann, T.; Ven de Veen, C.; et al. Accelerating Growth of HFC-227ea (1,1,1,2,3,3,3-heptafluoropropane) in the Atmosphere. *Atmos. Chem. Phys.* **2010**, *10*, 5903–5910. [CrossRef]
46. Likubo, Y.; Robin, M.L. Fire Extinguishing Methods and Blends Utilizing Hydrofluorocarbons. U.S. Patent 5,124,053, 23 June 1992.
47. DuPont Fluoroproducts. *DuPont FM-200 Fire Extinguishing Agent: Technical Information*; DuPont: Wilmington, DE, USA, 2009.
48. RISC Authority. *Active Fire Protection Guide FM200*; Fire Protection Association: Moreton-in-Marsh, UK, 2022.
49. PubChem. 1,1,1,2,3,3,3-Heptafluoropropane. Available online: https://pubchem.ncbi.nlm.nih.gov/compound/67940 (accessed on 30 January 2025).
50. Rivers, P.E.; Minday, R.M.; Behr, F.E.; Vitcak, D.R.; Flynn, R.M.; Costello, M.G.; Parent, M.J.; Owens, J.G.; Zhang, Z. Use of Fluorinated Ketones in Fire Extinguishing Compositions. European Patent 1261398 B1, 29 September 2004.
51. RISC Authority. *Active Fire Protection Guide: Migration of NOVEC 1230 and HFC Gaseous Extinguishing Agents to Environmentally*; Fire Protection Association: Moreton-in-Marsh, UK, 2024.
52. Rivers, P.E. Advancement in Sustainable Fire Suppression Development C_6F-Ketone: A Novel New Halon Replacement Alternatives to HFCS and PFCS. In Proceedings of the Halon Options Technical Working Conference Proceedings, Albuquerque, NM, USA, 24–26 April 2001; pp. 341–348.
53. Owens, J.G. Physical and Environmental Properties of a Next Generation Extinguishing Agent. In Proceedings of the Halon Options Technical Working Conference Proceedings, Albuquerque, NM, USA, 30 April–2 May 2002.
54. 3M Company. *3M Novec 1230 Fire Protection Fluid: Technical Data*; 3M Company: Saint Paul, MN, USA, 2022.
55. Rivers, P.E.; Schmeer, J.S.; Owens, J.G. Consideration of Liquid to Gas Phenomena for a Sustainable, High-Boiling, Tropodegrable Halocarbon. Available online: https://www.nist.gov/system/files/documents/el/fire_research/R0301569.pdf (accessed on 25 December 2024).
56. PubChem. 3-Pentanone, 1,1,1,2,2,4,5,5,5-nonafluoro-4-(trifluoromethyl)-. Available online: https://pubchem.ncbi.nlm.nih.gov/compound/2782408 (accessed on 20 January 2025).
57. Su, J.Z.; Kim, A.K.; Crampton, G.P.; Liu, Z. Fire Suppression with Inert Gas Agents. *J. Fire Prot. Eng.* **2001**, *11*, 72–86. [CrossRef]
58. Moore, T.A.; Yamada, N. Nitrogen Gas as a Halon Replacement. In Proceedings of the Halon Options Technical Working Conference Proceedings, Albuquerque, NM, USA, 12–14 May 1998.
59. Fire Protection Association of Australia. *Gaseous Fire Suppression Systems in Australia*; FPA Australia: Victoria, Australia, 2018; Volume 40.
60. Rattananon, N.; Patvichaichod, S. Performance Analysis of Inert Gas Systems in Main Distribution Board Room Using Fire Dynamics Simulation. *IOP Conf. Ser. Mater. Sci. Eng.* **2020**, *715*, 012006. [CrossRef]
61. Riley, J.F. Inergen a Breathable Gaseous Extinguishing Agent. Available online: https://www.nist.gov/document/r0301006pdf#:~:text=The%20composition%20of%20INERGEN%20consists,and%20Carbon%20Dioxide%20(8%25) (accessed on 25 December 2024).
62. Tyco Fire Protection Products. *The Physiology of INERGEN Fire Extinguishing Agent*; Tyco Fire Products LP: Marinette, WI, USA, 2016.

63. Ansul. *Inergen 200-Bar Systems—Design, Installation, Recharge and Maintenance Manual*; Tyco Fire Protection Products: Marinette, WI, USA, 2012.
64. *Inergen*; CAS RN: 7727-37-9, 7440-37-1, 124-38-9; Fire Eater, Hillerød, Denmark, 16 February 2022. Available online: https://partner.fire-eater.com/wp-content/uploads/2022/03/SDS-IG541-ENG-2022-02-16.pdf (accessed on 25 December 2024).
65. U.S. Environment Protection Agency. Substitutes in Total Flooding Agents. Available online: https://www.epa.gov/snap/substitutes-total-flooding-agents (accessed on 25 December 2024).
66. United Nations. Marking the Kyoto Protocol's 25th Anniversary. Available online: https://www.un.org/en/climatechange/marking-kyoto-protocol%E2%80%99s-25th-anniversary (accessed on 25 December 2024).
67. European Union. Regulation (EU) No 517/2014 of the European Parliament and of the Council of 16 April 2014 on Fluorinated Greenhouse Gases and Repealing Regulation (EC) No 842/2006 Text with EEA Relevance. Available online: https://eur-lex.europa.eu/legal-content/EN/TXT/?uri=celex:32014R0517 (accessed on 25 December 2024).
68. European Chemicals Agency. *Annex XV Restriction Report—Per~ and Polyfluoroalkyl Substances (PFASs)*; ECHA: Helsinki, Finland, 2023; Volume 224.
69. 3M Company. 3M to Exit PFAS Manufacturing by the End of 2025. Available online: https://news.3m.com/2022-12-20-3M-to-Exit-PFAS-Manufacturing-by-the-End-of-2025 (accessed on 25 December 2024).
70. Loss Prevention Certification Board. *LPCB Red Book: Part: 4 Fixed Fire Fighting Systems*; Bre Global Ltd.: London, UK, 2024; Volume 75.
71. Underwriters Laboratories Inc. *UL 2166: Halocarbon Clean Agent Extinguishing System Units*; UL: Northbrook, IL, USA, 2017.
72. Underwriters Laboratories Inc. *UL 2127: Inert Gas Clean Agent Extinguishing System Units*; UL: Northbrook, IL, USA, 2024.
73. FM Approvals LLC. *FM 5600: Examination Standard for Clean Agent Extinguishing System*; FM Global: West Glocester, RI, USA, 2023.
74. Loss Prevention Certification Board. *LPS 1230: Requirements for Fire Testing of Fixed Gaseous Fire Extinguishing Systems*; Bre Global Ltd.: London, UK, 2014; Volume 26.
75. International Maritime Organization. *International Convention for the Safety of Life at Sea, 1974*; IMO: London, UK, 1974.
76. International Maritime Organization. *MSC/Circ.848: Revised Guidelines for the Approval of Equivalent Fixed Gas Fire-Extinguishing Systems, As Referred to in SOLAS 74, for Machinery Spaces and Cargo Pump Rooms*; IMO: London, UK, 1998.
77. International Maritime Organization. *MSC.1/Circ.1267: Amendments to Revised Guidelines for the Approval of Equivalent Fixed Gas Fire-Extinguishing Systems, As Referred to in SOLAS 74, for Machinery Spaces and Cargo Pump-Rooms (MSC/Circ.848)*; IMO: London, UK, 2008.
78. International Maritime Organization. *Resolution MSC.99(73): Adoption of Amendments to the International Convention for the Safety of Life at Sea, 1974, As Amended*; IMO: London, UK, 2000.
79. Factory Mutual Insurance Company. *Property Loss Prevention Data Sheets: Data Centers and Related Facilities*; FM Global: West Glocester, RI, USA, 2022.
80. Factory Mutual Insurance Company. *Property Loss Prevention Data Sheets: Halocarbon and Inert Gas (Clean Agent) Fire Extinguishing Systems*; FM Global: West Glocester, RI, USA, 2023.
81. Chaos, M. Spectral Aspects of Bench-scale Flammability Testing: Application to Hardwood Pyrolysis. In Proceedings of the Fire Safety Science—Proceedings of the Eleventh Internation Symposium, Christchurch, New Zealand, 10–14 February 2014; pp. 165–178.
82. Niemann, R.; Bayless, H.; Craft, C. Evaluation of Selected NFPA 2001 Agents for Suppressing Class "C" Energized Fires. In Proceedings of the Halon Options Technical Working Conference Proceedings, Albuquerque, NM, USA, 7–9 May 1996; pp. 399–412.
83. Bengston, G.; Niemann, R. Update in the Evaluation of Selected NFPA 2001 Agents for Suppressing Class "C" Energized Fires. Available online: https://www.nist.gov/system/files/documents/el/fire_research/R0501593.pdf (accessed on 25 December 2024).
84. Bengston, G.; Flamm, J.; Niemann, R. Update on the Evaluation for Selected NFPA 2001, Agents for Suppressing Class "C" Energized Fires Featuring C6F-Ketone. Available online: https://www.nist.gov/system/files/documents/el/fire_research/R0201330.pdf (accessed on 25 December 2024).
85. Smith, D.M.; Niemann, R.; Bengston, G. Examination and Comparison of Existing Halon Alternatives and New Sustainable Clean Agent Technology in Suppressing Continuously Energized Fires. In Proceedings of the Halon Options Technical Working Conference Proceedings, Albuquerque, NM, USA, 24–26 April 2001; pp. 288–298.
86. Bengston, G.; Flamm, J.; Niemann, R. Update on the Examination and Comparison of Existing Halon Alternatives and New Sustainable Clean Agent Technology in Suppressing Continuously Energized Fires. Available online: https://nist.gov/system/files/documents/el/fire_research/R0201322.pdf (accessed on 25 December 2024).

87. McKenna, L.A.; Gottuk, D.T.; DiNenno, P.J. Extinguishment Tests of Continuously Energized Class C Fires. In Proceedings of the Halon Options Technical Working Conference Proceedings, Albuquerque, NM, USA, 12–14 May 1998; pp. 80–90.
88. Braun, E.; Womeldorf, C.A.; Grosshandler, W.L. Suppression Concentration of Clean Agents Exposed to a Continuously Energized Heated Metal Surface. *Fire Saf. J.* **1999**, *33*, 141–152. [CrossRef]
89. Braun, e.; Womeldorf, C.; Grosshandler, W.L. Determination of Suppression Concentration for Clean Agents Exposed to a Continuously-Energized Heated Metal Surface. In Proceedings of the Halon Options Technical Working Conference Proceedings, Albuquerque, NM, USA, 6–8 May 1997; pp. 149–161.
90. Smith, D.M.; Rivers, P.E. Effectiveness of Clean Agents on Burning Polymeric Materials Subjected to an External Energy Source. In Proceedings of the Halon Options Technical Working Conference Proceedings, Albuquerque, NM, USA, 21–29 April 1999; pp. 522–533.
91. Tapscott, R.E.; Speitel, L.C. Halon Replacement Options for Aircraft. In Proceedings of the International Aircraft Fire and Cabin Safety Research Conference, Atlantic City, NJ, USA, 16–20 November 1998.
92. Smith, C.; Nicholls, Z.R.J.; Armour, K.; Collins, w.; Forster, P.; Meinshausen, M.; Palmer, M.D.; Watanabe, M. The Earth's Energy Budget, Climate Feedbacks, and Climate Sensitivity Supplementary Material. In *Climate Change 2021: The Physical Science Basis. Contribution of Working Group I to the Sixth Assessment Report of the Intergovernmental Panel on Climate Change*; IPCC: Geneva, Switzerland, 2021.
93. Salawitch, R.J.; McBride, L.A.; Thompson, C.R.; Fleming, E.L.; McKenzie, R.L.; Rosenlof, K.H.; Doherty, S.J.; Fahey, D.W. *20 Questions & Answers About the Ozone Layer 2022 Update, Scientific Assessment of Ozone Depletion: 2022*; World Meteorological Organization: Geneva, Switzerland, 2022; Volume 81, ISBN 978-9914-733-98-3.
94. Ibanez, J.G.; Hernandez-Esparza, M.; Doria-Serrano, C.; Fregoso-Infante, A.; Singh, M.M. Halogenated Hydrocarbons and the Ozone Layer Depletion. In *Environmental Chemistry*; Springer: New York, NY, USA, 2008; pp. 115–121, ISBN 978-0-387-49493-7.
95. Sivasakthivel, T.; Siva Kumar Reddy, K.K. Ozone Layer Depletion and Its Effects: A Review. *Int. J. Environ. Sci. Dev.* **2011**, *2*, 30–37. [CrossRef]
96. Bekki, S.; Lefevre, F. Stratospheric Ozone: History and Concepts and Interactions with Climate. *Eur. Phys. J. Conf.* **2009**, *1*, 113–136. [CrossRef]
97. DiNenno, P.J.; Forssell, E.W. Clean Agent Total Flooding Fire Extinguishing Systems. In *SFPE Handbook of Fire Protection Engineering*, 5th ed.; Hurley, M.J., Ed.; SFPE: Gaithersburg, MD, USA, 2016; pp. 1483–1530, ISBN 978-1-4939-2564-3.
98. United Nations Framework Convention on Climate Change. *Information on Global Warming Potentials*; FCCC/TP/2004/3; UNFCCC: Bonn, Germany, 2004; Volume 12.
99. Cady, R.A. *A Literature Review of GWP: A Proposed Method for Estimating Global Warming Potential (GWP) of Short-Lived Climate Pollutants Like Methane*; Global Dairy Platform: Rosemont, IL, USA, 2020; Volume 37.
100. Andersen, S.O.; Brack, D.; Depledge, J. *A Global Response to HFCs Through Fair and Effective Ozone and Climate Policies*; Chatham House: London, UK, 2014; Volume 8.
101. World Meteorological Organization. *Scientific Assessment of Ozone Depletion: 2022*; GAW Report No. 278; WMO: Geneva, Switzerland, 2022; Volume 520, ISBN 978-9914-733-97-6.
102. Rio, R.D. *Environmental Impact Report: 3.5 Climate Change and Greenhouse Gases*; Nevada County: Nevada County, CA, USA, 2012; Volume 26.
103. Taniguchi, N.; Wallington, T.J.; Hurley, M.D.; Guschin, A.G.; Molina, L.T.; Molina, M.J. Atmospheric Chemistry of $C_2F_5C(O)CF(CF_3)_2$: Photolysis and Reaction with Cl Atoms, OH Radicals, and Ozone. *J. Phys. Chem. A* **2003**, *107*, 2674–2679. [CrossRef]
104. Guschin, A.G.; Molina, L.T.; Molina, M.J. *Atmospheric Chemistry of L-15381, L-15566 and L-14703 and Integrated Band Strengths of L-14374, L-14375, L-14752, L-13453 and L-14703*; 3M Company: Saint Paul, MN, USA, 1999.
105. Finlayson-Pitts, B.J.; Pitts, J.N. *Atmospheric Chemistry: Fundamentals and Experimental Techniques*; John Wiley & Sons: New York, NY, USA, 1986; ISBN 978-0471882275.
106. Seinfeld, J.H.; Prandis, S.N. *Atmospheric Chemistry and Physics*; John Wiley & Sons: New York, NY, USA, 1998.
107. Plummer, G. *Laboratory Measurements and Calculations Related to the Photo-Disassociation of L-15566 in the Earth's Lower Atmosphere*; 3M Company: Saint Paul, MN, USA, 2001.
108. Jackson, D.A.; Young, C.J.; Hurley, M.D.; Wallington, T.J.; Mabury, S.A. Atmospheric Degradation of Perfluoro-2-methyl-3-pentanone: Photolysis, Hydrolysis and Hydration. *Environ. Sci. Technol.* **2011**, *45*, 8030–8036. [CrossRef]
109. Zellner, R.; Bednarek, G.; Hoffmann, A.; Kohlmann, J.P.; Mors, V.; Saathoff, H. Rate and Mechanism of the Atmospheric Degradation of 2H-Heptafluoropropane (HFC-227). *Ber. Bunsenges. Phys. Chem.* **1994**, *98*, 141–146. [CrossRef]

110. Mogelberg, T.E.; Sehested, J.; Bilde, M.; Wallington, T.J.; Nielsen, O.J. Atmospheric Chemistry of CF3CFHCF3 (HFC-227ea): Spectrokinetic Investigation of the $CF_3CFO_2{\bullet}CF_3$ Radical, Its Reactions with NO and NO_2, and Fate of the $CF_3CFO{\bullet}CF_3$ Radical. *J. Phys. Chem.* **1996**, *100*, 8882–8889. [CrossRef]
111. Nelson, D.D.; Zahniser, M.S.; Kolb, C.E. OH Reaction Kinetics and Atmospheric Lifetimes of CF_3CFHCF_3 and CF_3CH_2Br. *Geophys. Res. Lett.* **1993**, *20*, 197–200. [CrossRef]
112. Owens, J.G. Understanding the Stability and Environmental Characteristics of a Sustainable Halon Alternative. Available online: https://www.nist.gov/system/files/documents/el/fire_research/R0301570.pdf (accessed on 25 December 2024).
113. D'Anna, B.; Sellevag, S.T.; Wirtz, K.; Nielsen, C.J. Photolysis Study of Perfluoro-2-methyl-3-pentanone under Natural Sunlight Conditions. *Environ. Sci. Technol.* **2005**, *39*, 8708–8711. [CrossRef]
114. RISC Authority. *Active Fire Protection Guide Inergen/IG541*; Fire Protection Association: Moreton-in-Marsh, UK, 2022.
115. Cousins, I.T.; DeWit, J.C.; Gluge, J.; Goldenman, G.; Herzke, D.; Lohmann, R.; Ng, C.A.; Scheringer, M.; Wang, Z. The High Persistence of PFAS is Sufficient for Their Management as a Chemical Class. *Environ. Sci. Process. Impacts* **2020**, *22*, 2307–2312. [CrossRef] [PubMed]
116. Schymanski, E.L.; Zhang, J.; Thiessen, P.A.; Chirsir, P.; Kondic, T.; Bolton, E.E. Per- and Polyfluoroalkyl Substances (PFAS) in PubChem: 7 Million and Growing. *Environ. Sci. Technol.* **2023**, *57*, 16918–16928. [CrossRef]
117. Fire Suppression Systems Association. *FSSA ECHA Science*; FSSA: Rockville, MD, USA, 2023.
118. U.S. Environmental Protection Agency. *Protection of Stratospheric Ozone: Notice 17 for Significant New Alternatives Policy Program*; 40 CFR Part 82; Federal Register: Washington, DC, USA, 2002.
119. European Centre for Ecotoxicology and Toxicology of Chemicals. *Monograph No. 6: Acute Toxicity Tests LD50 (LC50) Determinations and Alternatives*; ECETOC: Brussels, Belgium, 1985; Volume 42.
120. Organisation for Economic Co-operation and Development. *OECD Guidelines for the Testing of Chemicals: Test Guideline No. 403 Acute Inhalation Toxicity*; OECD: Paris, France, 2024; Volume 20.
121. National Occupational Health and Safety Commission. *Full Public Report: Heptafluoropropane (HFC-227ea)*; EX/5 (NA/164); Australian Industrial Chemicals Introduction Scheme: Sydney, NSW, Australia, 1999; Volume 20.
122. European Centre for Ecotoxicology and Toxicology of Chemicals. *Technical Report No. 105: Evaluation of Cardiac Sensitisation Test Methods*; ECETOC: Brussels, Belgium, 2009; Volume 97.
123. Vinegar, A.; Jepson, G.W. Epinephrine Challenge for Cardiac Sensitization Testing Vs. Endogenous Epinephrine: A Perspective. In Proceedings of the Halon Options Technical Working Conference Proceedings, Albuquerque, NM, USA, 27–29 April 1999; pp. 567–572.
124. Office of Research and Development. *Physiologically-Based Pharmacokinetic (PBPK) Models: Scientific Models to Help Evaluate Health Effects of Chemicals*; U.S. EPA: Washington, DC, USA, 2018.
125. Vinegar, A.; Jepson, G.W. Physiologically Based Modelling of Halon Replacements for Human Safety Evaluation. Available online: https://www.nist.gov/document/vinegar-toxicological-assessment-human-health-consequences-associated-inhalation-halon (accessed on 25 December 2024).
126. Peatross, M.J. A Review of Thermal Decomposition Product Testing of Halocarbon Fire Suppression Agents. In Proceedings of the Halon Options Technical Working Conference Proceedings, Albuquerque, NM, USA, 12–14 May 1998; pp. 244–253.
127. NJ Health. *Right to Know, Hazardous Substance Fact Sheet: Hydrogen Fluoride*; New Jersey Department of Health: Trenton, NJ, USA, 2017.
128. Occupational Safety and Health Administration. OSHA Occupational Chemical Database: Hydrogen Fluoride. Available online: https://www.osha.gov/chemicaldata/622 (accessed on 25 December 2024).
129. Liu, Z.G.; Kim, A.K. *A Literature Review of the Impact of Thermal Decomposition Products Generated by Halon Replacements on Electronic Equipment*; National Research Council Canada: Ottawa, ON, Canada, 1999; Volume 32.
130. Kim, J.; Su, J.; Crampton, G.; Kanabus-Kaminska, M. Measurement of Thermal Decomposition Products of HFC-227ea in an Electronic Facility. In Proceedings of the Halon Options Technical Working Conference Proceedings, Albuquerque, NM, USA, 24–26 April 2001; pp. 314–320.
131. Enlow, M.A. *Hanger Fire Suppression Utilizing Novec 1230*; AFCEC-CX-TY-TP-2018-0007; Vulcan Research and Controls, LLC: Panama City, FL, USA, 2018; Volume 41.
132. Ditch, B.D. Thermal Decomposition Product Testing with 1,1,1,2,2,4,5,5,5 Nonafluoro-4-trifluoromethyl Pentan-3-one (C_6F-Ketone) During Fire Extinguishing. Master's Thesis, Worcester Polytechnic Institute, Worcester, MA, USA, 2002.
133. Ditch, B.D.; Rivers, P.E.; Thomas, S.D. Thermal Decomposition Product Testing with C_6F-Ketone. In Proceedings of the Halon Options Technical Working Conference Proceedings, Albuquerque, NM, USA, 24–26 April 2001; pp. 349–354.
134. Meldrum, M. *Toxicology of Substances in Relation to Major Hazards: Hydrogen Fluoride*; Health and Safety Executive Information Centre: Bootle, UK, 1993.

135. 3M Company. *Thermal Decomposition Considerations with 3M Novec 1230 Fire Protection Fluid*; 3M Company: Saint Paul, MN, USA, 2004.
136. Krebs, R.; Owens, J.; Luckarift, H. Formation and Detection of Hydrogen Fluoride Gas During Fire Fighting Scenarios. *Fire Saf. J.* **2022**, *127*, 103489. [CrossRef]
137. Hynes, R.G.; Mackie, J.C.; Masri, A.R. A Shock Tube Kinetic Study on the Reaction of C_3F_6 and H Atoms at High Temperatures. *Proc. Combust. Inst.* **2000**, *28*, 1557–1562. [CrossRef]
138. Copeland, G.; Lee, E.P.F.; Dyke, J.M.; Chow, W.K.; Mok, D.K.W.; Chau, F.T. Study of 2-H-Heptafluoropropane and Its Thermal Decomposition Using UV Photoelectron Spectroscopy and ab Initio Molecular Orbital Calculations. *J. Phys. Chem. A* **2010**, *114*, 3540–3550. [CrossRef] [PubMed]
139. Andersson, B.; Blomqvist, P.; Dederichs, A. *Thermal Breakdown of Extinguishing Agents*; Report 3137; Lund University: Lund, Sweden, 2008; Volume 84.
140. Harrington, J.L. *Clean Agent Extinguishing Systems: Revised Rules for Safe Human Exposure*; NFPA: Quincy, MA, USA, 2024.
141. New Extinguishants Advisory Group. *A Review of the Toxic and Asphyxiating Hazards of Clean Agent Replacements for Halon 1301*; United Kingdom Halon Alternatives Group: Weybridge, UK, 2000.
142. Safety Advisory Group. *Safety Newsletter: Campaign Against Asphyxiation*; SAG NL N° 77/03/E; European Industrial Gases Association: Brussels, Belgium, 2003.
143. Safety Advisory Group. *Hazards of Inert Gases*; IGC Doc 44/00/E; European Industrial Gases Association: Brussels, Belgium, 2000.
144. Gordon, G.; Preston, M. Use of IG-541 (Inergen) Gaseous Fire Extinguishing Systems in Hydro Generators in New Zealand. In Proceedings of the EEA Conference & Exhibition, Auckland, New Zealand, 23–24 June 2011; pp. 1–14.
145. Robin, M.L. *DuPont Fire Extinguishants: The Roles of Hydrofluorocarbons in Global Fire Protection*; DuPont: Wilmington, DE, USA, 2009.
146. Technology and Economic Assessment Panel. *Special Report on Safeguarding the Ozone Layer and The Global System: Issues Related to Hydrofluorocarbons and Perfluorocarbons*; IPCC: Geneva, Switzerland, 2018; Volume 485.
147. Wickham, R.T. *Review of the Use of Carbon Dioxide Total Flooding Fire Extinguishing Systems*; Wickham Associates: Stratham, NH, USA, 2003.
148. Technology and Economic Assessment Panel. *Special Report on Safeguarding the Ozone Layer and The Global System: Issues Related to Hydrofluorocarbons and Perfluorocarbons, Summary for Policymakers and Technical Summary*; IPCC: Geneva, Switzerland, 2018; Volume 96.
149. Wolf, K. *Developing a California Inventory for Industrial Applications of Perfluorocarbons, Sulfur Hexafluoride, Hydrofluorocarbons, Nitrogen Trifluoride, Hydrofluoroethers and Ozone Depleting Substances*; Agreement Number 07-313; Institute for Research and Technical Assistance: San Francisco, CA, USA, 2011; Volume 107.
150. Banks, R.E.; Clarke, E.K.; Johnson, E.P.; Sharratt, P.N. Environment Aspects of Fluorinated Materials: Part 3: Comparative Life-cycle Assessment of the Impacts Associated with Fire Extinguishants HFC-227ea and IG-541. *Process Saf. Environ. Prot.* **1998**, *76*, 229–238. [CrossRef]
151. Tyco Fire Protection Products. *Protection of Data Centers: Economic and Environmental Consideration of the Alternative Technologies Available*; Tyco Fire Products LP: Marinette, WI, USA, 2015.
152. National Fire Protection Association. *NFPA 75: Standard for the Fire Protection of Information Technology Equipment*; NFPA: Quincy, MA, USA, 2024.
153. DiNenno, P.J.; Taylor, G.M. Halon and Halon Replacement Agents and Systems. In *Fire Protection Handbook*, 20th ed.; Cote, A.E., Grant, C.C., Hall, J.R., Solomon, R.E., Eds.; NFPA: Quincy, MA, USA, 2008; Volume I, pp. 93–121, ISBN 978-0-87765-758-3.
154. Taheri, P. An Intergrated Approach from the Preliminary Design to the Final Proposal of Gaseous Fire Suppression Systems. Master's Thesis, Politecnico di Milano, Milan, Italy, 2015.
155. Fire Industry Association. *Guidance on the Pressure Relief and Post Discharge Venting of Enclosures Protected by Gaseous Fire Fighting Systems*; FIA: Hampton, UK, 2012.
156. Fire Protection Association New Zealand. *FPANZ Code of Practice for Gaseous Fire Suppression Systems*; FPANZ: North Harbour, New Zealand, 2016.
157. 3M Company. *Preserving Room Integrity During Clean Agent Discharges*; 3M Company: Saint Paul, MN, USA, 2009.
158. Robin, M.L.; Forssell, E.W.; Sharma, V. Pressure Dynamics of Clean Agent Discharges. Available online: https://www.nist.gov/system/files/documents/el/fire_research/R0501582.pdf (accessed on 25 December 2024).
159. Tyco Fire Protection Products. *INERGEN Gaseous Fire Suppression Systems with i-Flow Technology*; Tyco Fire Product LP: Marinette, WI, USA, 2012.
160. Stagg, A. *Clean Agent Enclosure Design for ISO 14520 & AS 4212*; Fire Protection Technologies: Melbourne, Australia, 2013.

161. 3M Company. *3M Novec 1230 Fire Protection Fluid*; CAS RN: 756-13-8; 16-3425-2; rev 34; 3M Company: Saint Paul, MN, USA, 21 March 2024. Available online: https://multimedia.3m.com/mws/mediawebserver?mwsId=SSSSSuUn_zu8l00x482xNx29Mv70k17zHvu9lxtD7SSSSSS-- (accessed on 25 December 2024).
162. DuPont Fluoroproducts. *DuPont FM-200 (HFC-227ea) Fire Extinguishing Agent: Properties, Uses, Storage, and Handling*; DuPont: Wilmington, DE, USA, 2009.
163. Fire Industry Association. *Guidance Note: Recycled Halocarbon Extinguishing Agents*; FIA: Hampton, UK, 2020.
164. Fire Suppression Technical Options Committee. *2022 Assessment Report*; UNEP: Nairobi, Kenya, 2022; Volume 263, ISBN 978-9914-733-96-9.
165. Halons Technical Options Committee. *Emission Reduction Strategies for Halons and Other Halogenated Gaseous Extinguishing Agents*; UNEP: Nairobi, Kenya, 2018; Volume 37, ISBN 978-9966-076-53-3.
166. McGree, T. *Fires in Industrial and Manufacturing Properties*; NFPA: Quincy, MA, USA, 2023.
167. Campbell, R. *Non-Home Electrical Fires*; NFPA: Quincy, MA, USA, 2019.
168. Donnellan, D.; Lawrence, A.; Bizo, D.; Judge, P.; O'Brien, J.; Davis, J.; Smolaks, M.; Williams-George, J.; Weinschenk, R. *Uptime Institute Global Data Center Survey 2024*; Uptime Institute: New York, NY, USA, 2024; Volume 31.
169. Bureau of Investigation and Analysis of Industrial Risk. *Investigation Report on the Fire in the OVH Data Storage Center*; BEA-RI: La Defense Cedex, France, 2022; Volume 43.
170. Johnson Controls. *Case Study: Data Processing Network for Media Conglomerate*; Johnson Controls: Milwaukee, WI, USA, 2020.
171. Fire Eater. Blix Solution AS. Available online: https://fire-eater.com/blix-solutions-as/ (accessed on 30 January 2025).
172. Fire Eater. Sindal Biogas Plant. Available online: https://fire-eater.com/sindal-biogas-plant/ (accessed on 30 January 2025).
173. Fire Eater. Copenhagen Airport. Available online: https://fire-eater.com/copenhagen-airport/ (accessed on 30 January 2025).
174. Fire Eater. E.ON Kemsley. Available online: https://fire-eater.com/portfolio-item/e-on-kemsley/ (accessed on 30 January 2025).
175. Beranek, J. *An Overview of Pre-engineered Fire Suppression Systems and Industry Changes That May Affect Your Property*; Fire Equipment Manufacturer's Association: Cleveland, OH, USA, 2007.
176. Nicolas, J.S. The Inergen Approach. In Proceedings of the Halon Options Technical Working Conference Proceedings, Albuquerque, NM, USA, 11–13 May 1993; pp. 369–405.
177. Fire Eater. Strom P. Museum. Available online: https://fire-eater.com/storm-p-museum/ (accessed on 30 January 2025).
178. Fire Eater. Fosseholm Manor. Available online: https://fire-eater.com/fosseholm-manor/ (accessed on 30 January 2025).
179. Johnson Controls. *Case Study: Abraham Lincoln Presidential Library and Museum*; Johnson Controls: Milwaukee, WI, USA, 2020.
180. Fire Protection Association. Suppressing Battery Hazards. Available online: https://www.thefpa.co.uk/news/suppressing-battery-hazards (accessed on 30 January 2025).
181. Mihalache, C.; Bigan, C.; Panduru, V.; Tsakiris, C. Modelling of Noise Reduction for Datacentre Buildings Fire Protection with Inert Gas Systems. In Proceedings of the MATEC Web of Conferences, Sibiu, Romania, 5–7 June 2019; pp. 1–10.
182. Johnson Controls. *Impact of Sound on Computer Hard Disk Drives and Risk Mitigation Measures*; Johnson Controls: Milwaukee, WI, USA, 2018.
183. Brown, J.W.; Nowlen, S.P.; Wyant, F.J. *High Energy Arcing Fault Fires in Switchgear Equipment, A Literature Review*; SAND2008-4820; Sandia National Laboratories: Livermore, CA, USA, 2009; Volume 86.
184. Fire Eater. *Inert Gas Fire Extinguishing Systems and Li-Ion Batteries*; Fire Eater: Hillerød, Denmark, 2023.
185. Rosa, M.I.D.; Litton, C.D. Effectiveness of Various Concentrations of an Inert Gas Mixture for Preventing and Suppressing Mining Equipment Cab Fires: Development of a Dual Cab Fire Inerting System. *Fire Technol.* **2007**, *43*, 29–44. [CrossRef]
186. Su, J.Z.; Kim, A.K.; Liu, Z.; Crampton, G.P. Inergen Fire Suppression Testing. In Proceedings of the Halon Options Technical Working Conference Proceedings, Albuquerque, NM, USA, 27–29 April 1999; pp. 336–342.
187. Hasofer, M.; Thomas, I. Cost Benefit Analysis of a Fire Safety System Based on the Life Quality Index. In Proceedings of the Fire Safety Science—Proceedings of the Ninth International Symposium, Karlsruhe, Germany, 21–26 September 2008; pp. 969–980.
188. Lucherini, A.; Ni, S.; Unobe, D.; Gernay, T.; Chaudhary, R.K.; Peng, M.; Coile, R.V. Application of Cost-Benefit Analysis Methodology for Safet Measures in Structural Fire Engineering. In Proceedings of the The 13th International Conference on Structures in Fire, Coimbra, Portugal, 19–21 June 2024; pp. 1–12.
189. Strianese, M.; Torricelli, N.; Tarozzi, L.; Santangelo, P.E. Experimental Assessment of the Acoustic Performance of Nozzle Designed for Clean Agent Fire Suppression. *Appl. Sci.* **2023**, *13*, 186. [CrossRef]
190. Tyco Fire Protection Products. *INERGEN Acoustics Testing Position Paper*; Tyco Fire Products LP: Marinette, WI, USA, 2012.
191. Hu, X.; Kraaijeveld, A.; Log, T. Numerical Investigation of the Required Quantity of Inert Gas Agents in Fire Suppression Systems. *Energies* **2020**, *13*, 2536. [CrossRef]
192. Hu, X.; Kraaijeveld, A. Experimental and Numerical Investigation of Extinguishing Effectiveness of Inert-Gas Agents in a Leaky Enclosure. *Energies* **2022**, *15*, 4323. [CrossRef]

193. Atukpawu, I.E. Investigation of Effects of Enclosure Opening on Inert Gas Suppression System. Master's Thesis, Western Norway University of Applied Sciences, Bergen, Norway, 2022.
194. Kim, M.; Kim, B.; Kim, M. Analysis Methodology of Clean Agent Fire Extinguishing System Discharge Test for NPP MCR Raised Floor. In Proceedings of the Transactions of the Korean Nuclear Society Autumn Meeting, Yeosu, Republic of Korea, 25–26 October 2018; pp. 1–3.

processes

MDPI

Review

Artificial Intelligence in Manufacturing Industry Worker Safety: A New Paradigm for Hazard Prevention and Mitigation

Minahil Khurram [1], Catherine Zhang [2], Shalahudin Muhammad [3], Hitesh Kishnani [3], Kimi An [2], Kalana Abeywardena [1], Utkarsh Chadha [4,*] and Kamran Behdinan [4,*]

[1] The Edward S. Rogers Sr. Department of Electrical & Computer Engineering, University of Toronto, Toronto, ON M5S 3G8, Canada
[2] Department of Mechanical and Industrial Engineering, University of Toronto, Toronto, ON M5S 3G8, Canada
[3] Department of Civil & Mineral Engineering, Faculty of Applied Sciences and Engineering, University of Toronto, Toronto, ON M5S 1A4, Canada
[4] Advanced Research Laboratory for Multifunctional Lightweight Structures (ARL-MLS), Department of Mechanical and Industrial Engineering, University of Toronto, Toronto, ON M5S 3G8, Canada
* Correspondence: utkarsh.chadha@mail.utoronto.ca (U.C.); behdinan@mie.utoronto.ca (K.B.)

Abstract

The phenomenal rise of artificial intelligence (AI) in the last decade, and its evolution as a versatile addition to various fields, necessitates its usage for novel purposes in multi-dimensional fields like the manufacturing industry. Even though AI has been rigorously studied for process optimization, wastage reduction, and other quintessential aspects of the manufacturing industry, there has been limited focus on worker safety as a theme in the current literature. Safety standards contribute to worker safety, but there is no *one-size-fits-all* approach in these standards or policies, which warrants evaluation and integration of new ideas and technologies to reach the closest to ideal standards. This includes but is not limited to health, regulation of operations, predictive maintenance, and automation and control. The rise of Industry 4.0 and the migration towards Industry 5.0 facilitate easy integration of advanced technologies like AI into the manufacturing industry with real-time predictive capabilities, and this can help reduce human errors and mitigate hazards in processes where sensitivity is crucial or hazards are frequent. Keeping the future outlook in focus, AI can contribute to training workers in risk-free environments, promote engineering education for easy adaptation to new technology, and reduce resistance to changes in the industry. Furthermore, there is an urgent need for standards and regulations to govern and integrate AI technologies judiciously into the manufacturing industry, which holds AI models and their creators accountable for their decisions. This could further extend to preventing the adversarial use of new technology. This study exhaustively discusses the potential and ongoing contributions of this technology to the safety of workers in the manufacturing industry.

Keywords: artificial intelligence; worker safety; manufacturing industry; engineering education; hazard prevention; occupational hazards

1. Introduction

The recent boom in artificial intelligence has been a boon for various sectors to improve their efficiencies and untangle the complex threads of questions that were previously difficult to answer. With the multidimensional application of artificial intelligence being implemented in new industries, its nascent stages allow for further exploration of new paradigms such as worker safety in the manufacturing industry using these advanced technologies [1]. In the last few decades, AI has evolved significantly in the field of manufacturing beyond basic automation to a more data-driven approach; that is, optimization, error reduction, etc., to achieve higher efficiency [2,3]. Incorporation of AI in manufacturing started in the 1980s to design specific programmatic logic controllers (PLCS), which were later employed in automotive manufacturing plants to improve assembly lines through repetitive and less labor-intensive tasks [3]. By the mid-2000s, computer vision gained traction with advances in machine learning algorithms for predictive maintenance and product optimization [4]. This was the phase in which the field of artificial intelligence made significant progress in the manufacturing industry and its general development, as static automation was gradually abandoned for a more dynamic predictive system capable of learning and optimizing processes. However, data scarcity and computational power limited the advancements of these AI models until the rapid employment of the Internet of Things (IoT) in the mid to late 2000s, which led to the integration of cyber-physical systems in industrial settings, thus giving rise to the term "Industry 4.0", or the fourth industrial revolution [5]. This is where modern AI models emerge that can handle huge amounts of data with reduced skepticism regarding their computational power, efficiency, or access to advanced hardware requirements.

According to Deloitte, due to predictive maintenance based on ML, machine downtime has decreased by 20–30%, and equipment life has increased by 20–40% [6]. As shown in Figure 1, predictive maintenance has evolved significantly. Reactive maintenance, which allowed machines or equipment to fail, was followed by preventive maintenance to prevent or delay unanticipated breakdowns. Now, moving towards more predictive maintenance with the use of modern technology, such as the integration of sensors, allows real-time monitoring of machines and equipment in the manufacturing industry, significantly reducing risks but still requires human supervision [6]. This is further advanced using predictive models and algorithms that use sensor data to automatically take actions that models are supposed to take, slightly reducing human supervision and improving manufacturing efficiency [7]. Moreover, current AI models generate simulations, that is, digital twins, which virtually replicate current systems that allow for simulation modeling and better decision making [7,8].

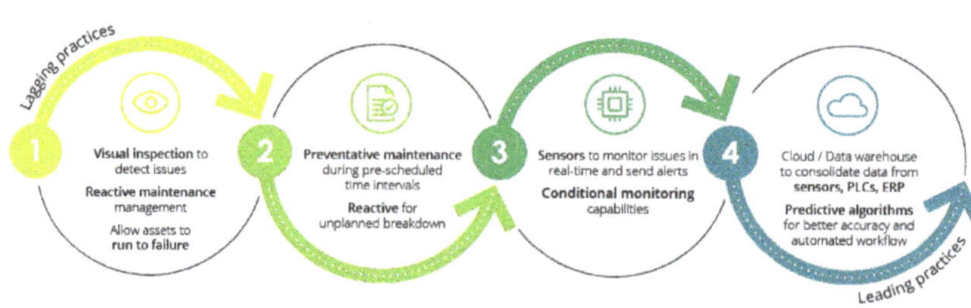

Figure 1. Evolution of predictive maintenance in the manufacturing industry. Credit to Deloitte [6].

Modern AI vision systems in manufacturing have combined deep learning and predictive modeling to the degree of accuracy where flaw detection exceeds/meets a human level [9]. Thus, manufacturing industries have seen a massive increase in workforce dynamics, i.e., human–AI collaboration, as opposed to the predominant manufacturing in the 1980s [10]. Looking ahead, AI is increasingly positioned as a cornerstone for manufacturing due to its rapid advancements, reinforcement learning, and industry acceptance we see today. While all the other uses of AI in manufacturing are crucial, worker safety has not been directly addressed in such research work.

Safety has been subconsciously discussed as an indirect benefit of using AI, which overshadows a very important aspect, i.e., human capital and its safe utilization. The manufacturing industry is essential, but it also comes with risks and hazards [11–13]. Some standards have been in place to ensure safety in such industries, but how AI could transform or significantly impact this is discussed in this study exhaustively. AI as a technology is in its nascent stages, which means that there are still some gray areas for making laws and policies and deciding the remit of utilization as well as development [14].

1.1. Research Objectives and Methodology

In this section, we outline the key objectives of this study and the methodology used to conduct the review. The aim is to provide a comprehensive and structured analysis of the role AI plays in enhancing worker safety in the manufacturing industry. Given the limited existing literature and the focus of most studies on product outputs or processes, this research aims to fill the gap by focusing on human safety.

1.1.1. Objectives

The primary objective of this review article is to explore and analyze the emerging role of AI in improving worker safety within the manufacturing sector. While the current literature is sparse and often focused on product or process outcomes, this study emphasizes the importance of worker safety in the context of AI integration. The goal is to provide a detailed source of information on the use of AI models for enhancing safety and to critically assess the current standards and policies. Although some countries and institutions have established laws or frameworks to regulate AI, there remains a need for a universal framework that applies not just to manufacturing or worker safety but to AI in broader contexts [15,16].

To gather relevant information, this study used reliable sources such as industry blogs, government websites, and scholarly research articles. These sources were accessed primarily through platforms like Scopus and Google Scholar, ensuring the inclusion of authoritative content on AI policies, models, manufacturing cases, industry hazards, and related topics.

1.1.2. Research Questions

To guide this review and address key gaps in the existing literature, the following research questions were posed:

- How is AI currently applied to worker safety in manufacturing environments?
- What are the strengths and limitations of these AI-based safety systems?
- What regulatory, ethical, and technical challenges must be addressed for the large-scale adoption of AI in manufacturing?
- What role can policy frameworks play in shaping responsible AI integration in the manufacturing sector?

1.1.3. Methodology

This study adopts a qualitative, narrative review method as limited empirical research is available on the specific topic of AI and worker safety in manufacturing. The methodology focuses on synthesizing information from peer-reviewed journal articles, industry white papers, technical reports, policy documents, and standards databases. A systematic search was conducted across a range of academic and industry databases, including Google Scholar, Scopus, IEEE Xplore, ScienceDirect, and Springer Nature Link.

The inclusion criteria for the sources were as follows:

1. Published between 2018 and 2025.
2. Focused on applications in the manufacturing sector.
3. Address issues related to safety, ethics, or governance in AI deployment.
4. Written in English.

The exclusion criteria were as follows:

1. Sources focused on non-manufacturing sector use cases.
2. Duplicates or secondary sources lacking credibility.
3. Non-peer-reviewed sources (unless government- or standards-based, and select preprints on reliable servers).

The selected sources were thoroughly analyzed to identify recurring themes, emerging technologies, and governance gaps, and to compare traditional safety practices with AI-based solutions. The insights gained from this analysis were used to inform the overall findings of the review.

2. Understanding Manufacturing Hazards

Hazards in the manufacturing industry can be characterized by an unexpected event posing a safety threat from routine processes. A system with many moving parts, workers, and often heavy machinery has many possible sources of occupational user or equipment incidents [17–19]. This range of occurrences is surprisingly vast, spanning acute injuries to slowly manifested chronic conditions or infectious agents. For example, we can point out common incidents of bodily damage such as contusions, cuts, or abrasions in subtractive manufacturing while working with automated or operated machinery [20,21]. Meanwhile, in additive manufacturing, the proliferation of atomized chemicals and fine particles poses possible chronic health problems for workers who are exposed regularly [22].

Hazard identification and mitigation is not only important to worker health and safety but also a critical component in a successful work culture. Diagnosing and preventing such incidents saves money through both reduced defects or repairs while maintaining a trusting relationship with workers and stakeholders. This contributes to a work culture that is built upon the general sentiment of workers. A positive safety culture will encourage continued behavioral safety, while a weak safety culture may result in less belief in following occupational safety. By cultivating a strong workplace culture valued around hazard identification and safety, the workers involved are more likely to have better overall health and morale [23].

2.1. Manufacturing Safety Standards

While workplace safety principles in manufacturing typically follow the same philosophy internationally, the precise jurisdiction differs from country to country. The basic anatomy of these principles are based on the International Labour Organization and ISO 45001, as shown in Figure 2 [24–27].

Workplaces tend to utilize any version of this holistic approach to safety management, as seen from the ILO, to have a cyclical method to manage workplace conduct. By thoroughly educating workers about their workplace hazards, many proceed with confidence that their employer is practicing proper hazard mitigation and are encouraged to do the same [28]. When properly reported and noted, observations and analysis of the provided incident's reaction will be scrutinized and improved for the next occurrences.

These axioms of occupational safety act as an instruction manual for workplace conduct while demonstrating further effectiveness when supported using a positive safety culture that encourages worker compliance [29]. The development of these factors in conjunction has shown high efficacy in boosting productivity and worker morale.

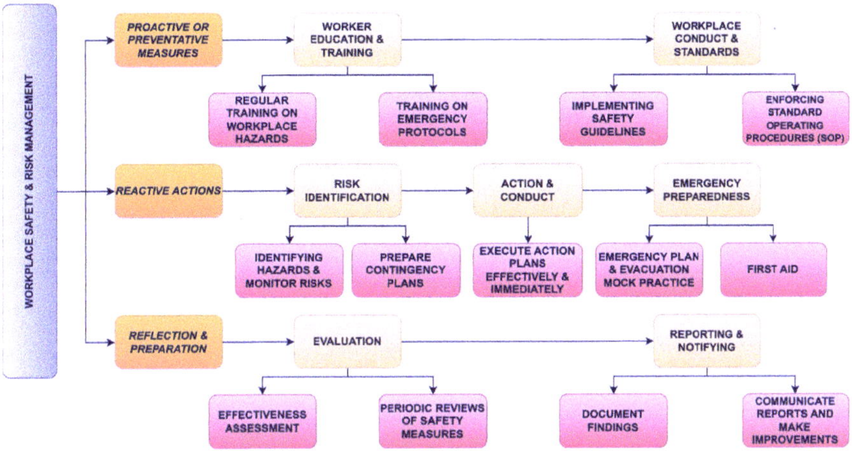

Figure 2. Steps for workplace safety and risk management, based on [24–26].

2.2. Productivity Methodologies in Manufacturing

Hand-in-hand with occupational health and safety, workplaces have developed productivity principles to optimize and improve their output. Over the last half-century, methods using statistical and process analysis to reduce waste have been widely used, leading to the prevalence of techniques such as Six Sigma and Lean in the manufacturing industry. While both promote output efficiency, they are differentiated by the sources of process waste they address. A successful combination of such methodologies proves to provide quantitative and qualitative improvements to the output or workplace culture [30].

Lean manufacturing is implemented as a process that achieves waste reduction through reconsidering the perspective on inventory. Products are manufactured directly due to consumer demand, greatly reducing inventory waste [31]. This is built upon Kaizen, a philosophy that works by mapping the value stream. It encourages rapid, cyclical analysis of a manufacturing process, where workers continuously improve upon areas of waste [32,33]. Lean methodology is closely related to just-in-time manufacturing, originating from Toyota's aim to cut down inventory levels. By allocating resources to the exact place and time, productivity is greatly encouraged and an economic advantage is created. Six Sigma aims to address waste reduction by minimizing the areas of error and discrepancies throughout the manufacturing process [34]. Through the use of data collection and statistical analysis, root causes of source variations are identified. The goal is to achieve a 'defect-free workplace', defined as less than 3.4 defects per million opportunities. It is a highly quantifiable method in comparison to Lean manufacturing (Figure 3) [35].

In the manufacturing industry, a trend of combining Lean and Six Sigma methodologies has been noticed in the form of Lean Six Sigma (LSS). With the same goal, both methods have eventually merged to encompass a further holistic perception of the manufacturing value stream. The implementation of Lean methods will reduce process waste and improve efficiency, while Six Sigma aids in perfecting it by smoothing out discrepancies [32]. LSS offers a problem-solving framework to businesses, making it a popular choice to provide a comprehensive approach to systemic improvement.

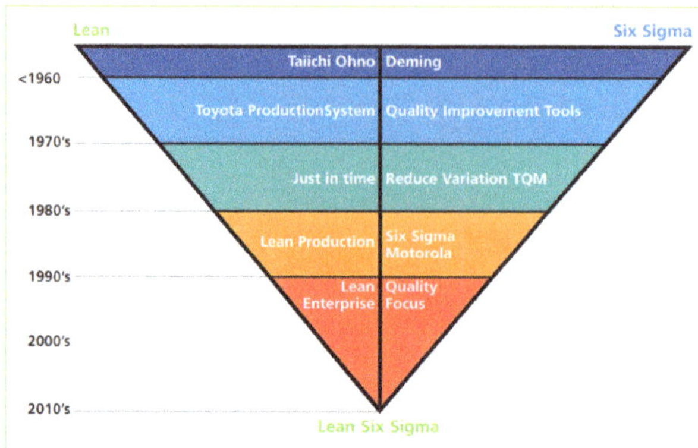

Figure 3. The development and compatibility of Lean manufacturing and Six Sigma over time. Credit to the English National Health Service (NHS) [32].

2.3. *Traditional Approaches to Hazard Prevention in Manufacturing and Their Limitations*

Traditional hazard prevention approaches in manufacturing have laid the foundation for workplace safety practices and worker protection [36–40]. Some traditional methods of preventing such dangers are safety training and education, personal protective equipment (PPE), routine safety inspections, incident investigations and reports, and safety regulations. However, despite their effectiveness in limiting some hazards, their varied limitations are evident in today's continuously changing industrial world, as represented in Figure 4 [41–45]. Table 1 further expands on the types of hazards that are experienced in a general manufacturing industry and the type of intervention that is conventionally used followed by its current limitations with evolving industry needs and advancements. These traditional methods have limitations, as the manufacturing industry has been evolving rapidly within the last few decades.

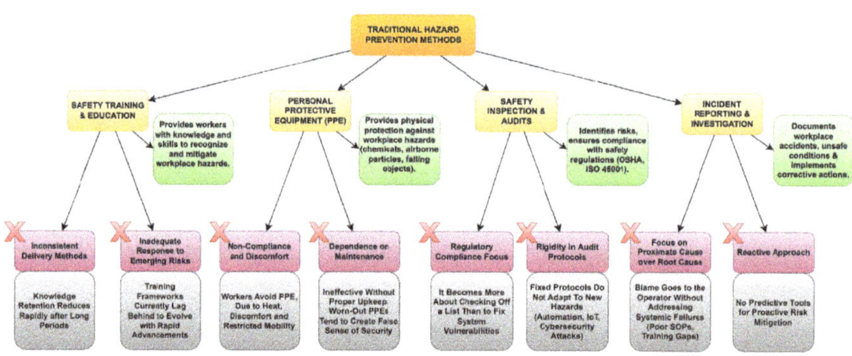

Figure 4. Traditional approaches in manufacturing for hazard prevention and their disadvantages in the current pace of developments [36–45].

An important study conducted by Dyreborg et al. [45] emphasizes that safety interventions implemented at a group or organizational level tend to be more effective than targeting individual behavior alone. The study findings suggest that engineering-based controls that do not depend on the active choice of the worker or the 'decision to use' have been shown to limit workplace injuries [45]. To overcome these limitations of 'behavior-only' strategies, the investigation suggested the integration of multifaceted safety strategies that span all levels of the organization, especially when engineering controls are included [45].

Table 1. Exhaustive overview of the types of hazards faced in the manufacturing industry.

Hazard Type	Specific Details of the Hazards Reported	Description	Traditional Method of Prevention	Limitation	Ref.
Chemical hazards	Toxic substances	These substances lead to acute or chronic poisoning, if exposed	- Use of personal protective equipment (PPE) - Implementation of proper ventilation systems - Regular safety training for personnel	Lack of hazard awareness stemming from poor training regarding health and safety in a particular setting leads to increased risk	[46,47]
	Flammable and explosive chemicals	Fire and explosion hazards	- Proper storage and handling procedures - Use of fire-resistant containers and cabinets - Installation of fire suppression systems	Current prevention strategies may not fully address the risk of fire or explosion in all scenarios	[46,47]
	Carcinogenic substance	Some chemicals used in manufacturing processes may have carcinogenic properties	- Substitution with less hazardous alternatives when possible - Strict exposure controls and monitoring	Long-term effects of exposure to carcinogens may not be immediately apparent, making prevention challenging	[47]
	Corrosive substances	Can cause severe burns and tissue damage upon contact	- Use of appropriate PPE - Installation of emergency eyewash stations and showers	Accidental spills or splashes can still occur, potentially causing immediate harm before preventive measures can be activated	[46,47]

Table 1. *Cont.*

Hazard Type	Specific Details of the Hazards Reported	Description	Traditional Method of Prevention	Limitation	Ref.
Physical hazards	Noise	Prolonged exposure to noise at or above 85dBA can lead to permanent hearing loss, tinnitus, and difficulty understanding speech in noise	- Implementing noise control measures - Using hearing protection devices - Establishing noise exposure limits	Defining "hazardous noise" based on sound level alone is insufficient	[46,48,49]
	Vibration	Hand–arm vibration (HAV) can lead to various occupational health hazards for workers	- Using anti-vibration gloves and necessary protective equipment - Regular tool maintenance - Incorporating short breaks during work and avoiding prolonged exposure - Prompt reporting of Hand–Arm Vibration (HAV) symptoms	Symptoms may not appear until after significant exposure (typically 2000 hours), which can delay early intervention and prevention efforts	[48,50,51]
	Ionizing radiation	Ionizing radiation exposure is a frequent occupational hazard	- Implementing distance, time, and shielding measures - Adhering to exposure limits	The current system of dose limitation may not fully address the optimization of protection in all scenarios	[46,52]
	Heat	Heat stress is associated with a spectrum of heat-related illnesses, including heat stroke, which can lead to death	- Implementing acclimatization programs - Training workers and supervisors about symptom recognition and first aid - Engineering and administrative controls to reduce heat stress - Medical surveillance	- Current occupational heat index guidance might not be sufficiently protective - Current guidelines may not fully address all risk factors	[46,48,53]

Table 1. *Cont.*

Hazard Type	Specific Details of the Hazards Reported	Description	Traditional Method of Prevention	Limitation	Ref.
Physical hazards	Mechanical maintenance hazards	Operation and maintenance of machinery can lead to injuries such as entanglement, crushing, or impact	- Establishing and following sufficient safe work procedures at an organizational level - Installing safety devices such as guards, barriers, and emergency stop mechanisms - Scheduling regular preventive maintenance for machines	- Inadequate training or oversight increases the risk of accidents - Regular maintenance and safety upgrades can be financially burdensome for companies, leading to neglect in some cases	[48,54,55]
	Operational hazards	Industrial operations expose workers to hazardous materials, ergonomic risks, and unsafe work practices that can lead to injuries or health issues	- Developing clear standard operating procedures for safe operation practices - Providing PPE - Conducting regular safety audits	- Workers may not always adhere to SOPs or use PPE correctly - Rapid changes in operational conditions can make it difficult to maintain effective safety measures - Limited resources may hinder the implementation of safety programs	[48,54,55]
	Electrical hazards	Electrical hazards involve risks associated with electrical systems and equipment	- Educating workers about electrical safety - Conducting routine checks on electrical systems and equipment - Installing GFCIs	- Older electrical systems may not meet current safety standards - Variability in adherence to safety protocols can lead to dangerous situations - The intricate nature of modern electrical systems makes thorough inspections challenging	[45,48,56]
	Fire and explosion risks	Flammable materials, improper storage practices, and inadequate fire prevention measures in industrial settings	- Providing education on fire prevention strategies and emergency response procedures - Ensuring that flammable materials are stored safely according to regulations - Installing smoke detectors and sprinkler systems to detect and respond to fires quickly	- Negligence or lack of awareness among workers - Improperly maintained and inadequate-sized fire suppression systems - Keeping up with evolving fire safety regulations can be resource-intensive for companies	[45,57]

Table 1. *Cont.*

Hazard Type	Specific Details of the Hazards Reported	Description	Traditional Method of Prevention	Limitation	Ref.
Ergonomic hazards	Repetitive motions	Repetitive motion causes muscle fatigue and can eventually result in long-term damage to workers	- Implementing mechanical lifting aids - Changing task design (from carrying to pushing/pulling) - Balancing work demands and pace - Varying tasks	Effectiveness of current prevention methods may not be accurate	[58–60]
	Awkward postures	Awkward postures are risk factors for neck/shoulder pain (NSP) and low back pain (LBP)	- Redesigning workspaces to avoid stooped or twisted positions - Limiting shelf height and raising worker positions - Providing unrestricted workspace	- Any exposure to awkward postures can increase pain, making complete prevention challenging - Current assessment methods often rely on self-reporting, which can be subjective and prone to recall bias	[61]
	Heavy lifting	Occupational lifting, especially for extended durations and with higher loads, increases the risk of long-term sickness absence (LTSA)	- Using mechanical lifting aids - Teaching proper lifting techniques - Limiting maximum lifting weight	- There is no universally accepted method of determining safe lifting capacity for workers - The effectiveness of teaching "proper" lifting techniques has been questioned in some studies	[62,63]
Psychosocial hazards	Work-related stress	Stress is an individual's response to high-intensity work, affecting cognitive, physical, mental, and emotional status. It can lead to mental health problems such as anxiety, depression, and burnout.	- Clearly defining roles and responsibilities - Improving communication between workers and managers - Employee Assistance Programs (EAPs)	- Underutilization of EAP services and data by companies - Difficulty in reaching all workforce segments with stress reduction programs	[64–66]

As shown in Table 1, these hazards are unique to every manufacturing process, each with its respective reactive and proactive preventative measures. There are many approaches to these risks, with many workflows, maintenance, regulatory precautions, and methodologies having been developed over the years to reduce the possibility of incidents. One of the quintessential duties of employers is to provide a safe and productive working environment, which is achieved through the research of an appropriate combination of these measures. As a result of implementing a customized incident mitigation method, hazard-induced employee injury, chronic health issues, psychological effects, and deaths can be reduced in the manufacturing workspace. By reactively tracking incidents, proactive anticipation can be formed from the occurrence patterns that slowly manifest. In the current industrial revolution, we are noticing the migration of this general process from pen and paper to digital data collection and analysis using programmable models. By exploring the status quo in hazard mitigation methodology, we can analyze the future of integrating artificial intelligence and machine learning into manufacturing safety [23].

2.4. Comparison of Traditional and AI-Based Safety Methods

Traditional workplace safety in the manufacturing industry relies on well-established protocols such as routine safety audits and checks, use of personal protective equipment (PPE), manual reporting of incidents, and hazard sign placements [67,68]. As comprehensively explained in Figure 4, the traditional methods have their own set of flaws despite setting the foundations for occupational safety [69]. These are often reactive, static, and are limited by human factors such as oversights, fatigue, or inconsistency in implementation [70].

With the rise of Industry 4.0 and AI integration, human involvement is being reduced in areas where constant attention is needed to prevent serious hazards, as we discuss in Section 3. Real-time data analysis allows for proactive safety monitoring, while predictive algorithms help ensure timely maintenance and readiness to manage any changes in the system [71]. Moreover, AI-driven decision support systems can automatically adjust and control system settings, maintaining safe operational parameters and following predefined safety guidelines, ensuring steady and reliable control in industrial environments.

These systems can process massive volumes of data from devices like sensors, cameras, logs to detect any anomalies, predict equipment failure, and also monitor worker behavior in real-time. These benefits of AI, alongside the traditionally established safety methods, can facilitate a safer work environment.

3. AI Techniques Transforming Safety Standards

The manufacturing industry presents numerous safety challenges, from equipment malfunctions to hazardous working conditions. As AI continues to evolve, its ability to enhance workplace safety becomes increasingly significant. By analyzing vast amounts of data, automating safety monitoring, and predicting potential risks, AI-driven solutions are transforming traditional safety practices [72,73]. This section explores key AI techniques that improve worker safety. Section 3.1 examines how machine learning-based predictive analysis helps identify risks before they lead to accidents. Section 3.2 delves into the role of computer vision in real-time monitoring, enabling the detection of unsafe behaviors and hazardous conditions. Section 3.3 discusses how natural language processing enhances safety communication and decision-making in manufacturing environments. Finally, Section 3.4 illustrates a few real-world case studies of AI being used in the manufacturing industry.

3.1. Predictive Analytics

Predictive analytics uses statistical techniques and machine learning (ML) to analyze historical data and forecast future outcomes [74]. In the manufacturing industry, predictive analysis is instrumental in optimizing production workflows, reducing downtime, and enhancing overall efficiency. A particularly vital application is improving worker safety by identifying potential hazards before they pose risks to worker well-being. By leveraging real-time sensor data, equipment logs, and historical incident records, predictive models can detect patterns indicating potential failures or unsafe conditions, enabling proactive interventions [71]. As illustrated in Figure 5, different ML approaches are utilized in predictive analysis depending on the availability of labeled data and the specific application.

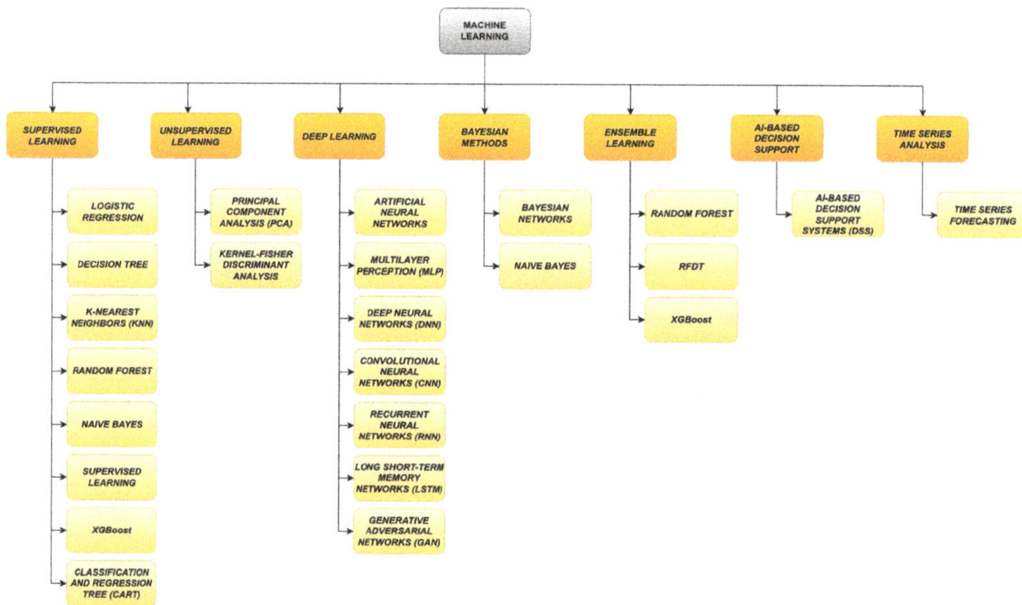

Figure 5. Taxonomy of AI models employing machine learning techniques.

3.1.1. Machine Learning Techniques for Predictive Analytics

Supervised learning is widely utilized when input–output relationships are available in existing data. For example, models trained on labeled data—such as sensor readings linked to past failures—can identify causal associations that lead to hazardous conditions. Classification methods detect defective products or hazardous situations in real-time, while regression models analyze continuous variables such as temperature, pressure, and machine vibration to determine the remaining useful life of industrial equipment, which allows for timely maintenance and failure prevention [71,75,76].

Unsupervised learning plays a crucial role in anomaly detection [71], detecting irregularities in machine behavior or environmental conditions without requiring predefined labels. Clustering techniques group similar operational patterns, aiding in detecting deviations that may signal impending failures or unsafe conditions [74]. Dimensionality reduction methods streamline complex sensor data, preserving essential information while enhancing computational efficiency [76–78]. Semi-supervised learning, which combines aspects of supervised and unsupervised learning, is particularly beneficial when labeled data are scarce, as it allows models to extract useful patterns from large pools of unlabeled data [71,74].

Reinforcement learning enhances adaptability by dynamically optimizing decision-making strategies based on real-time feedback [74]. Unlike traditional predictive models that rely solely on historical data, reinforcement learning systems adjust their strategies continuously. For example, industrial robots trained with reinforcement learning navigate factory environments safely, while adaptive control systems fine-tune machinery operations to minimize risks based on evolving conditions [71]. Commonly used models in predictive analysis include Artificial Neural Networks (ANNs), Support Vector Machines (SVMs), and Random Forests (RFs) [74,79,80]. RFs, in particular, are highly effective due to their capability in handling high-dimensional manufacturing data and capturing complex relationships between variables.

3.1.2. Enhancing Worker Safety Through Predictive Analytics

Worker safety in manufacturing is a critical concern, with risks arising from material handling, machine operation, and repetitive tasks. Predictive analytics mitigates these risks by analyzing workplace data, classifying accidents, and predicting injury severity, allowing for early interventions and informed decision-making [73].

A fundamental application of predictive analysis in worker safety is the classification of workplace accidents using historical incident reports and real-time sensor data. K-Nearest Neighbors (KNN) identifies patterns in reported incidents by clustering similar accidents based on shared characteristics such as task type, environmental conditions, and injury severity [73]. When a new incident is reported, KNN compares it to existing clusters using distance metrics like Euclidean or Manhattan distance, signaling a high probability of serious consequences if it aligns with past severe accidents. Logistic regression estimates injury severity by analyzing relationships between factors such as task complexity, worker fatigue, and environmental conditions, classifying outcomes such as minor, moderate, or severe injury [81]. The model highlights influential risk factors, enabling manufacturers to prioritize safety interventions. Furthermore, Random Forest (RF) improves prediction accuracy through ensemble learning, combining multiple decision trees trained on random subsets of data. This approach enhances robustness, reduces variance, and mitigates overfitting. Feature importance analysis within RF further identifies key factors influencing accident risks, guiding targeted safety measures [81,82].

Another crucial step in improving worker safety is identifying the groups at greatest risk and understanding common injury causes. In the study conducted by Kahkhi, F.D. et al. [82], data on injury type, claim status, body part affected, and injury cause are analyzed using a naive Bayes model to classify incidents and estimate injury likelihood. Random Forest Decision Trees (RFDT) are then applied to refine predictions and provide insights into the importance of different injury risk factors. To address ergonomic risks, deep learning-based 3D pose estimators are developed to monitor worker movements in real-time, detecting unsafe postures and alerting workers to prevent musculoskeletal disorders [75]. Similarly, predictive models evaluate the impact of workplace conditions on specific organs, such as liver function. In the study conducted by Ni, L. et al. [83], logistic regression and a restricted cubic spline model were used to assess the non-linear relationship between years of service and liver abnormalities, while a neural network refined the predictions, enhancing health risk assessments.

3.1.3. Other Applications of Predictive Analytics in Worker Safety

Predictive Analysis is further utilized in specific applications that relate to worker safety: the operation of specific tools used in industry, working environment conditions, and production.

Predictive Maintenance and Fault Diagnosis: Predictive maintenance minimizes equipment failures that could lead to worker injuries. In the metal-cutting industry, neural networks predict power consumption based on cutting parameters, optimizing tool performance and preventing malfunctions [84]. Predictive analytics is also applied in manufacturing system control, quality assurance, and defect mitigation, using Bayesian networks and ANNs to diagnose faults in industrial equipment such as power transformers and transceiver stations [85]. In smelting operations, a multivariate time series deep learning model predicts furnace temperatures in electric arc furnaces, improving process stability and preventing hazardous temperature fluctuations [86].
Workplace Environmental Monitoring: Predictive analytics enhances workplace conditions by ensuring optimal temperature, humidity, and air quality. HVAC systems use predictive models to detect faults and optimize energy consumption [87,88]. Deep neural networks are used to predict HVAC failures, maintaining safe temperature ranges in heat-intensive work environments [81,87]. Additionally, ANN-MLP algorithms monitor and predict smoke emissions from malfunctioning machines, triggering a tiered alert system to prevent worker exposure to hazardous fumes [89].
Sustainable Manufacturing Practices: Sustainability in manufacturing is another area where predictive analysis is valuable. Multi-criteria decision models use regression analysis and ANNs to optimize material consumption, energy efficiency, recyclability, and production costs, promoting sustainable and safer manufacturing practices [90].

3.1.4. Limitations and Biases in Predictive Analytics in Manufacturing

While predictive analysis offers significant advantages in manufacturing, particularly for safety, several limitations can impact its effectiveness. A key challenge is *bias–variance tradeoff* [91]. Models with high bias tend to oversimplify the data, which results in underfitting [92], where the model misses critical patterns, such as interactions between machinery, environmental factors, and human behavior. In contrast, models with high variance may overfit data [93], detecting irrelevant patterns and generating false alarms, like incorrect posture detections, that cause worker distrust. This phenomenon in machine learning models is illustrated in Figure 6.

When models generalize too much, they can miscalculate hazards, overlook irregular patterns, and delay responses. This can extend to inaccurate health monitoring and missing irregularities in worker vitals such as blood pressure or temperature, potentially leading to severe outcomes like strokes or heart complications [94]. In addition, such models may not detect new, unseen faults, such as gas leaks or electrical problems, causing delays in emergency responses [92].

Another limitation is the *quality of data* [95]. Predictive models in manufacturing rely on accurate and comprehensive data. However, real-world data are often incomplete, noisy, or inconsistent due to sensor errors or insufficient historical records. Without proper data preprocessing, which involves cleaning and formatting raw data, predictive models can be inaccurate and unreliable [96]. Ensuring data quality is foundational to the model's success, especially when the safety of workers is at stake.

Manufacturing processes are dynamic, with frequent changes in equipment, materials, and production methods. This introduces the challenge of *concept drift* [97,98]. As these changes occur, predictive models can quickly become outdated if they are not continuously updated through life-long learning [99]. Without regular retraining and monitoring, models may lose accuracy and fail to reflect the current state of operations, potentially creating unsafe conditions for workers.

In conclusion, while predictive analysis holds the promise of enhancing safety and efficiency in manufacturing, addressing these limitations, such as ensuring data quality, managing the bias–variance tradeoff, and updating models regularly, is crucial to maintaining their effectiveness and safeguarding workers' well-being.

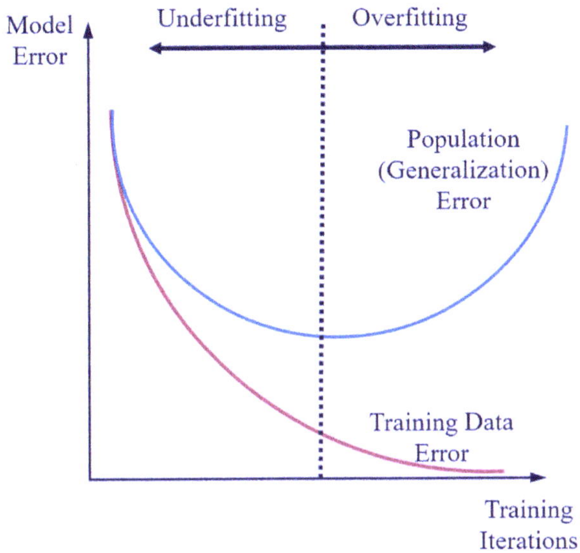

Figure 6. Illustration of overfitting and underfitting [92].

3.2. Real-Time Intelligent Surveillance Using Computer Vision

Predictive safety analytics is crucial in preventing workplace accidents, as discussed in Section 3.1. Traditional sensor-based safety solutions, such as pressure sensors, environmental monitors, and wearable devices, are commonly used in industrial settings, but they come with significant limitations [100,101]. While these systems can detect specific hazards, they often miss critical visual cues and cannot fully capture the complexity of dynamic industrial environments. Moreover, wearable safety solutions like smart personal protective equipment (PPE) and electromyograms (EMGs) can be financially burdensome for large workforces [102,103]. They require constant monitoring, regular upkeep, and strict worker compliance, making them cumbersome to implement and maintain at scale [103].

To overcome these challenges, computer vision has emerged as a scalable, non-intrusive solution. It continuously analyzes workplace conditions and enhances safety protocols without the need for constant physical maintenance. Using technologies like CCTV footage, thermal imaging, infrared sensors, depth cameras (LiDAR and time-of-flight), X-ray scans, and drone-captured aerial views, computer vision provides a comprehensive way to monitor entire workspaces [103]. This approach can detect potential hazards that might go unnoticed by traditional sensors, significantly reducing human error and enabling proactive safety interventions.

Central to modern computer vision advancements are Convolutional Neural Networks (CNNs), which extract hierarchical features from images for automated analysis [104]. These models have been widely used in safety applications, but their effectiveness depends on the complexity of the task at hand. For simple tasks like fire detection [105,106], CNN-based image classification models work well. These models classify an image as "fire" or "no fire", enabling rapid hazard response. However, in industrial settings, where workers, machinery, and dynamic interactions coexist, more complex analyses are required. A simple classifier might identify a hazard but fail to recognize specific risks, such as distinguishing between workers and machinery detecting unsafe postures, or localizing where on the work floor it materializes [107].

To address this, deep learning-based object detection models have been developed. Models like Faster R-CNN [108,109], Mask R-CNN [110], YOLO (You Only Look Once) [111], and SSD (Single-Shot Multibox Detector) [112] can detect and localize multiple objects within a scene, generating bounding boxes around them for precise tracking and hazard identification. In manufacturing environments, object and activity detection enhances safety by monitoring worker movements and identifying potential dangers [13]. For example, worker–machine proximity monitoring uses detection models to differentiate workers from heavy machinery like forklifts or robotic arms [103]. If a worker gets too close to operating equipment, the system can issue real-time alerts, preventing collisions [113]. Similarly, unsafe posture recognition systems can identify hazardous activities, such as improper lifting techniques or unsafe bending, and prompt corrective actions to reduce the risk of injury.

Computer vision also supports PPE compliance tracking by continuously verifying whether workers are wearing the required safety gear, such as helmets or gloves. If a worker is non-compliant, the system can trigger an immediate alert to ensure the safety of the worker and those around them. Beyond detection, tracking algorithms monitor worker movements over time, predicting potential hazards before accidents happen. For example, if a worker repeatedly enters restricted zones (i.e., perimeter intrusion systems [114]) or neglects safety procedures, the system can escalate warnings or alert supervisors for intervention. A practical implementation of computer vision for worker safety follows a structured workflow, as indicated in Figure 7.

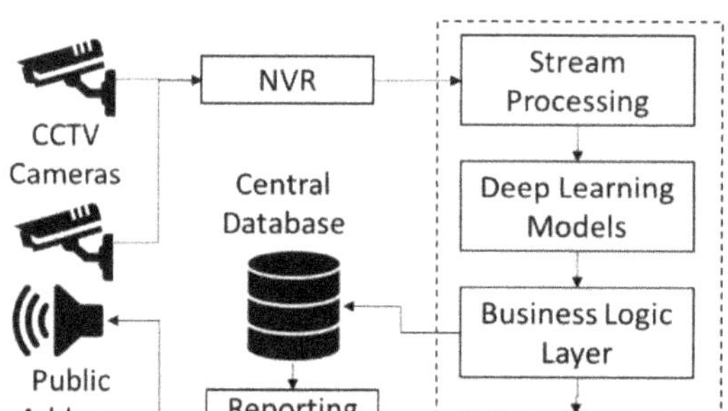

Figure 7. Solution architecture diagram to monitor and alert worker activity [13].

First, CCTV footage is captured and processed through a Network Video Recorder (NVR). The video stream is then analyzed using deep learning models to identify potential hazards. In the risk detection and business logic layer, AI models assess risks based on predefined criteria, categorizing them from "Very Unlikely" to "Very Likely" and from "Negligible" to "Severe" following a risk assessment matrix shown in Figure 8. If a hazard is detected, the system triggers alerts to notify workers and supervisors where the responses are tailored based on the severity of the situation [103]. Therefore, by eliminating the need for costly and maintenance-heavy wearable devices, computer vision offers a more efficient, scalable solution that continuously monitors workers, machinery, and the environment.

Real-time hazard detection, such as unsafe proximity to machines, improper posture, or PPE compliance, enables immediate alerts and corrective actions. This proactive approach not only reduces the risk of accidents but also ensures a smoother, more cost-effective implementation of safety measures across large and dynamic industrial settings.

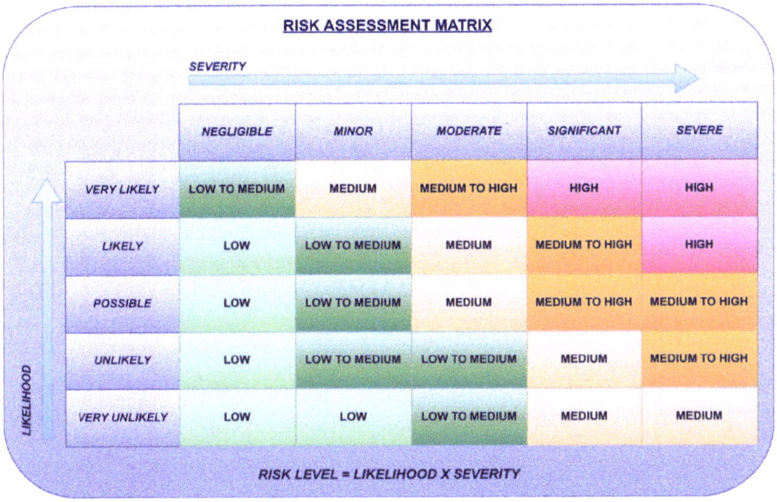

Figure 8. Risk assessment matrix used for decision-making in computer vision. Adapted and recreated with permission from [103].

While CV brings substantial benefits to workplace safety by enabling real-time monitoring and proactive hazard detection, its application in worker surveillance introduces critical privacy, ethical, and technical challenges [115]. Many CV systems rely on extensive visual datasets that may include images of workers captured without explicit consent, raising serious privacy concerns [116]. These datasets often lack adequate representation across ethnic and religious groups, leading to biased models that can skew safety assessments and result in unfair or discriminatory monitoring [117]. Such practices may contribute to a stressful work environment, where heightened anxiety and reduced trust impair essential safety-related functions such as situational awareness, attention to detail, and reflexes.

In addition to these privacy and ethical concerns, deploying CV in dynamic industrial environments presents several technical challenges [118]. Visual obstructions—such as machinery or other workers blocking the camera's view—can hinder detection accuracy, while variable lighting, glare, dust, and steam can degrade image quality [119–121]. Real-time applications also require low-latency processing, which can be difficult to achieve in resource-constrained settings. Furthermore, CV models often need retraining to adapt to new factory layouts, equipment, or worker behaviors, making scalability a challenge [122, 123]. Integration with legacy systems adds additional complexity. These limitations highlight the need for robust, adaptable, and ethically sound CV systems capable of addressing the complexities of modern manufacturing environments.

3.3. Enhancing Workplace Safety Through NLP-Powered Communication

The rapid advancements in Natural Language Processing (NLP) have allowed machines and computers to automatically analyze and classify text-based documentation, extracting valuable information for applications such as job codings, espouse assessment, data analysis, and risk valuation. Innovations in this area, such as the development of learning models like bidirectional transformers, have greatly reduced the reliance on manual input in text-processing tasks [124,125].

A major application of NLP in safety and manufacturing domains is in its ability to process vast amounts of unstructured data, including incident reports [126], maintenance work orders [127], and other important documents, transforming the data into structured, machine-readable formats. This ability allows researchers, supervisors, and safety professionals to effectively manage extensive datasets and recognize actionable insights, including root causes of incidents, that may otherwise remain unrecognized without NLP tools [128]. A specific example of the usage of NLP in worker safety is in classifying risk variables associated with multiple sclerosis musculoskeletal disorders (MSDs) for treatment and prevention. This is performed using advanced pre-trained sentence transformers (Euclidean, Bray–Curtis, and Mahalanobis), cosine similarity, and distance metrics to classify 25 MSD risk factors [129].

A standard process of an NLP workflow is shown in Figure 9. Typically, it begins with preprocessing, where textual documents are cleaned and prepared. This involves removing unwanted characters, eliminating stop words, applying lemmatization (for example, grouping words like 'best' and 'good'), and stemming (reducing words to their root forms). Following preprocessing, vectorization techniques such as Word2Vec or TD-IDF are used to convert text into numerical formats that machine learning algorithms can interpret and utilize [130]. Advanced models like BERT and GPT-3 then analyze this structured numerical data to identify patterns and extract meaningful insights for the reader [130]. A specialized branch of NLP for technical and engineering-related data is known as Technical Language Processing (TLP). It focuses on applying NLP methods to analyze such specialized data, where the sequence of words plays an important role in determining context and meaning, particularly in manufacturing. TLP systems are designed to emphasize word order due to its influence on analytical outcomes. These systems classify specific terms, such as 'leaks', into distinct components or problem categories, which are then classified and prioritized [127].

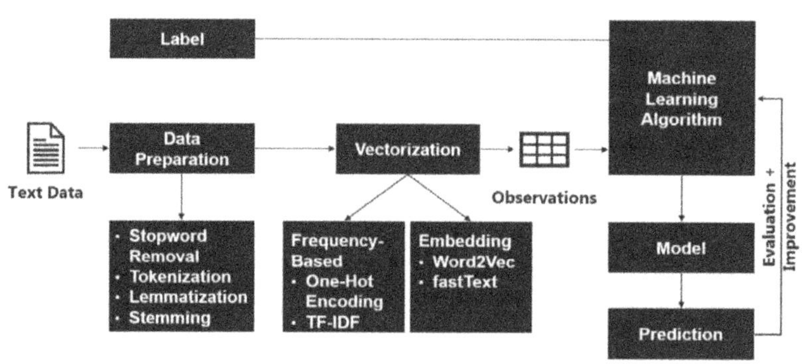

Figure 9. Standard process for applying NLP on textual data in manufacturing. Adapted with permission from [130].

Similar to the limitations discussed in Section 3.1.4, predictive analysis tools that rely on NLP for safety communications or training must be carefully managed [131]. Biases in training data—such as those related to culture, gender, or outdated safety protocols—can result in inaccurate or unsafe recommendations [132]. In manufacturing, these errors can lead to miscommunications, unsafe practices, and heightened exposure to workplace hazards. Therefore, ensuring that NLP models are free from bias, regularly updated, and aligned with the latest safety standards is crucial for maintaining the effectiveness and reliability of predictive safety systems [133].

3.4. Real Industrial Case Studies

This section highlights selected real-world industrial cases where AI technologies have been applied in manufacturing.

3.4.1. Case Study 1: Collaborative Robots *(Ford Motor Company, Tesla, General Motors Company)*

With the rise of Industry 4.0, companies such as Ford, Tesla, and General Motors have integrated AI-powered collaborative robots (CoBots) [134,135] into their production lines. Unlike traditional robots, CoBots work directly alongside human workers. Ford, for example, has deployed over 100 CoBots in 24 of its facilities [134], reflecting the growing use of these machines to improve productivity and safety. The CoBot market, valued at $649 million in 2019, is projected to grow at 49% annually through 2025 [136].

CoBots are designed to operate in hazardous conditions and for extended periods, minimizing the risk of workplace injuries [137]. By alleviating human workers from physically demanding and monotonous tasks, safety, efficiency, and productivity are accelerated [138]. Moreover, Tesla is investigating exoskeletons and wearable sensors to improve posture and reduce musculoskeletal strain [139], further promoting the safety and well-being of their workforce.

3.4.2. Case Study 2: AI-Facilitated CCTV Infrastructure *(SeeWise.AI, Intenseye, Linfox)*

Companies like SeeWise.AI and Intenseye use AI-powered CCTV and computer vision to improve safety compliance and regulatory oversight in manufacturing [140,141]. These systems enable real-time monitoring of PPE, automated safety inspections, hazardous area monitoring, and behavior analysis. Intenseye, for example, reported identifying 200 times more safety hazards using AI than manual observation [141]. Major manufacturers, including Coca-Cola, PEPSICO, Panasonic, Hyundai, Daikin, Honda, Koctas, Amazon, Coats, and Amcor, have adopted these technologies [142,143], though specific details of their use cases are not always available. Koctas, however, reported an 89% reduction in safety risks related to the absence of hard hats due to AI-enabled monitoring [143,144].

Furthermore, Linfox, an Australian logistics company, uses AI-enhanced CCTV to boost worker safety and security in its warehouses [145]. The system includes features such as material handling equipment (MHE) proximity alerts and their optimized use, footpath compliance tracking, distracted pedestrian detection, alcohol testing, and vehicle road-worthiness checks [145], improving safety and operational efficiency.

3.4.3. Case Study 3: Nvidia Virtual Factories

Virtual factories are digital replicas of real manufacturing environments, enabling the modeling, simulation, analysis, and optimization of production processes [146]. These models allow for accurate simulations of hazardous scenarios, such as machine fires or structural collapses, improving risk prediction and reducing human exposure to dangerous areas [147]. ML and 3D modeling enable the simulation of autonomous mobile robots (AMRs) and CoBots working alongside human workers. Like CoBots mentioned in Section 3.4.1, AMRs perform physically demanding tasks and address labor shortages. AMRs can operate in high-risk environments, including high-temperature, radioactive, or chemically contaminated zones, minimizing human exposure and preventing illness. With AI, ML, physics-based simulations, sensors, and mechatronics, these robots exhibit human-like perception, mobility, dexterity, cognition, and whole-body control [148–150].

AI agents are a powerful technology for automating processes and enhancing worker safety by executing complex tasks and high-level objectives, improving efficiency [151].

These agents use technologies like LLMs, RAG, vector databases, APIs, NLP, and advanced programming frameworks. LLMs process large volumes of safety reports, inspection logs, and records, offering concise summaries or alerts for safety officers. RAG enables dynamic access to up-to-date safety documents, ensuring fast interventions to prevent worker injuries from escalating into serious incidents [152]. RAG also supports adherence to safety procedures, such as forklift operation guidelines or OSHA heat exposure limits [153].

NVIDIA Omniverse is a platform that combines APIs and software development kits (SDKs) with OpenUSD rendering and generative physical AI to create digital workflows in industrial environments [154]. Companies like Continental [155], Rockwell Automation [156], and Wistron [157] have adopted this technology, leading to significant improvements. For example, Wistron reported a 51% increase in worker efficiency, a 50% reduction in process times, and a 40% decrease in defect rates through real-time monitoring and layout optimization [158].

This discussion is further expanded in Table 2, where real industrial cases are summarized with their applications, specific areas that they cover, benefits, and limitations.

Table 2. AI examples in manufacturing safety.

AI Model Used	Application	Specific Usage in the Industry	Benefits for Manufacturing Industry	Limitations	Ref.
Convolutional Neural Networks (CNNs) and Decision Trees (DTs)	Fault detection	Industrial cold forging	Superior performance in detecting faults using vibration test data	Requires high-quality vibration data for training	[159]
Long Short-Term Memory Networks (LSTM) with Improved Bat Algorithm (VSSBA)	Cloud manufacturing scheduling	Fast prediction of scheduling results	Enables efficient scheduling in cloud-based manufacturing environments	Complexity in implementation and tuning	[159]
Process Mining Techniques	Productivity improvement	Make-to-stock manufacturing	Dynamically maps and analyzes complex manufacturing processes	Requires extensive process data for accurate analysis	[159]
Classification and Regression Tree (CART) algorithm	Line feeding mode selection	Predicting optimal line feeding mode for components	Improves efficiency in component delivery to production lines	Accuracy depends on the quality and quantity of input data	[159]
Principal Component Analysis (PCA)	Machine monitoring and process optimization	Detecting incipient patterns in machinery and identifying important process factors	Enables early fault detection and process improvement	May oversimplify complex relationships in data	[160]
Autoencoders (AE)	Feature selection and dimensionality reduction	Diagnosis of product defects and event detection	Improves efficiency in data processing and pattern recognition	Requires careful tuning to avoid loss of critical information	[160]
Support Vector Machines (SVMs)	Quality prediction and wear prediction	Defect detection in products and machinery wear prediction	Effective for classification and regression tasks in manufacturing	Low efficiency with large-scale input data	[160]
Association rule-based Learning	Relationship identification in manufacturing systems	Depicting relationships between shop floor indicators and causes of action	Uncovers hidden correlations in datasets for improved management strategies	May generate a large number of rules, requiring careful interpretation	[160]
Kernel–Fisher Discriminant Analysis (KFD)	Classification and regression in manufacturing	Defect detection and quality prediction at the product level	Effective for high-dimensional data analysis	Computationally intensive for large datasets	[160]
Generative Adversarial Networks (GANs)	Predicting machine failures through data augmentation	Stimulating different scenarios to predict system failures	Improves accuracy by depicting realistic situations	Risk of generating unrealistic simulations/ unreliable data can degrade model performance	[161–163]
Bayesian networks (BNs)	Risk analysis and prediction	Predict high-risk environments through variable interactions, e.g., chemical reactants	Improved risk assessment and communication, as it provokes proactive risk management	Highly dependent on accurate data, which can be challenging to obtain from high-risk scenarios	[164–166]
Random Forest (RF)	Predictive maintenance and anomaly detection	Monitors sensor data, e.g., temperature, to predict equipment failure	Prevents unexpected breakdowns that can cause injuries and damage, as it provokes efficient maintenance	Interpretability in setting a benchmark	[167,168]

Table 2. *Cont.*

AI Model Used	Application	Specific Usage in the Industry	Benefits for Manufacturing Industry	Limitations	Ref.
Hybrid CNN and RNN	Anomaly detection and human activity recognition	Analyses of real-time footage to monitor/detect worker safety.	Provides alerts to supervisors to facilitate corrective actions while also providing early detection of potential hazards.	High quality and quantity data required to train such models + computational cost is very high for real-time use	[169–171]
Fuzzy logic Systems	Risk assessment and safety control	Predict risk level through collisions and conditions in dynamic environments.	Handle uncertainty and are flexible to employ	Complex and hard to generalize cases and data, thus making them hard to train	[172,173]

4. Predictive Maintenance and Risk Mitigation

Predictive maintenance (PdM) is a proactive strategy that leverages predictive analytics, as discussed in Section 3.1, to analyze historical performance data and real-time asset information to identify operational anomalies and potential equipment defects, enabling timely repairs before failures occur. PdM aims to optimize maintenance schedules, reducing unplanned outages, and minimizing unnecessary preventive maintenance costs [159,174]. PdM can be categorized into five types [175,176]:

1. Corrective Maintenance (CM): Restores equipment upon fault detection.
2. Preventative Maintenance (PM): Analyzes historical data to minimize breakdowns.
3. Risk-Based Maintenance (RM): Prioritizes assets posing the greatest risk in case of failure.
4. Condition-Based Maintenance (CBM): Uses sensor data to trigger maintenance upon performance decline.
5. Predetermined Maintenance (PtM): Follows manufacturer guidelines and historical data to schedule maintenance.

AI-driven PdM minimizes human error by automating the analysis of complex, high-volume data from monitored assets. By processing vast amounts of sensor and performance data, AI models are capable of identifying underlying patterns that are often obscured to humans due to limitations in concurrent data processing [177]. This capability enhances the speed and accuracy of monitoring, reduces human intervention, and minimizes operational disruptions. As a result, PdM enables early detection of anomalies, preventing costly downtime. AI also optimizes maintenance schedules, prioritizes critical issues, and autonomously recommends or executes maintenance actions based on real-time insights, improving both the reliability and cost-effectiveness of operations [178].

Root cause analysis (RCA) is a critical preliminary step in PdM, which identifies the underlying causes of failures, thereby optimizing the maintenance strategy [179]. Machines maintained by integrating PdM with RCA operate more efficiently and for extended periods, as this combined approach minimizes the need for frequent repairs [179]. Another important PdM approach, Reliability-Centered Maintenance (RCM), is widely used in industries such as nuclear and aviation. RCM determines the most effective maintenance strategy by combining PdM, PM, and CM while evaluating factors such as equipment condition, age, technical feasibility, and economic efficiency. RCA is crucial in informing these decision-making processes [180].

In addition to machinery, PdM can be applied to the well-being of workers in high-stress environments such as manual or semi-automatic assembly lines. Wearable devices that monitor physiological data—such as heart rate variability, skin respiration rate, skin conductance, and electromyography—can detect stress levels and help manage worker well-being. This application of AI contributes to both productivity and safety, reducing the impact of stress, burnout, and physical strain [160]. Moreover, computer vision-based solutions contribute to PdM by detecting potential malfunctions in machinery before they occur. Using image or video data captured through continuous monitoring, techniques like edge detection, Gray Level Co-Occurrence Matrix (GLCM), and CNN models help identify early signs of issues. This not only prevents costly breakdowns but also reduces material waste, further demonstrating the impact of AI in enhancing operational efficiency [159,179,181,182].

The terms 'risk mitigation' and 'predictive analytics' are often used interchangeably, but in this paper, we define risk mitigation as the processes employed by assets and machinery to prevent failures. Reactive risk mitigation is a widely used approach that relies on post-event analysis and training based on historical data [180]. In contrast, a proactive risk mitigation model harnesses predictive analytics and ML to detect and prevent risks before they occur. ML enhances system performance by identifying patterns and mapping correlations in data, allowing for early intervention [180]. Figure 10 illustrates an architecture that senses asset conditions and determines the most appropriate PdM technique. The scanning, sensing, and final decision-making processes are facilitated through the Internet of Things (IoT), while AI and ML algorithms are used to predict and recommend the optimal PdM strategy [181].

In conclusion, the integration of predictive maintenance (PdM) with advanced technologies such as AI, machine learning, and IoT offers significant improvements in operational efficiency and risk mitigation. By automating data analysis and identifying patterns that are often invisible to humans, AI enhances both the speed and accuracy of maintenance processes, reducing downtime and operational costs. Additionally, the application of PdM to both machinery and worker well-being exemplifies its versatility in optimizing not only asset management but also human resource efficiency. As PdM continues to evolve, its ability to predict, prevent, and respond to issues in real-time will undoubtedly lead to more resilient, cost-effective, and sustainable operations across various industries.

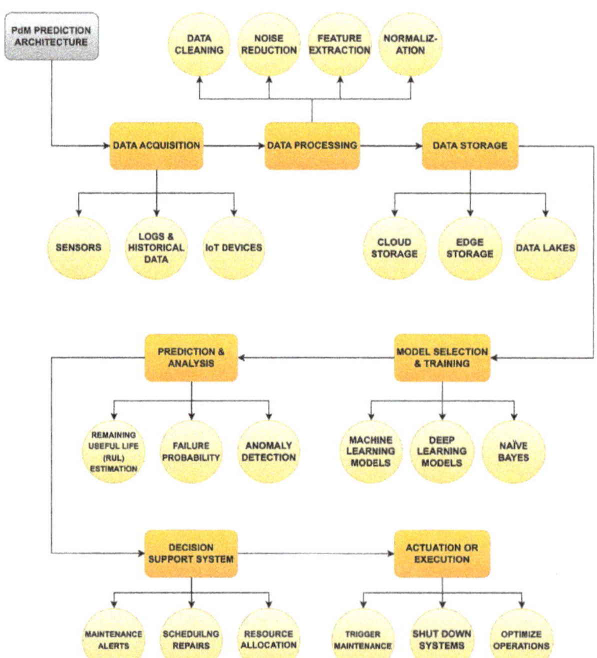

Figure 10. PdM prediction architecture [179–182].

5. Real-Time Hazard Detection and Response

Despite the quality of workplace conduct training, there will remain a factor of unavoidable risk in some activities. This is anticipated in occupational safety and health standards, where the common preliminary step of protocols is the ability to predict risks and identify anomalies. These can be categorized into short, acute incidents or long, accumulated injuries and conditions. Due to the repetitive or complex nature of some manufacturing actions, or the desire to shorten protocol activities to save time, some dangerous situations are created unintentionally by a worker [183]. In certain cases, only the symptoms of an incident can be recognized instead of the root issue. The ability for a worker to diagnose a hazard in any of these scenarios is supported through multiple methods, but unusual contingencies or unseen factors may always arise [184].

5.1. Ensuring Worker Health and Safety

A significant factor of hazard diagnostics is the high readiness of information regarding common incidents for workers in the surroundings. This can be seen in the usage of tailored Safety Data Sheets, which serve to provide rapid identification and appropriate protocols, with additional ancillary information such as emergency contacts [185]. Another example of easily comprehensible and intuitive hazard information can be seen in the form of a qualitative risk matrix, as similarly shown in Figure 8. By weighing the likelihood against the danger of an incident, workers can determine the appropriate reactive protocol. This eliminates the possibility of weighing low-danger hazards similarly to a higher-danger threats [184].

Though both of these methods come with some pitfalls of over-simplicity, the purpose they serve in a complex workplace is vital to quick emergency judgment and protocol [186]. With access to this information through rapid-identification methods such as color coding, workers can drastically reduce action time that is instead used for emergency aid. With this time, the next step of hazard control and containment is fundamental in applicable situations [187]. In a contingency, there may be a lack of time for a worker to report and await commands from a supervisor. In such cases, a Right to Refuse Dangerous Work (aka. Stop Work Authority) may be enacted by an employee to make the executive decision to terminate a process or refuse a task [188]. Though this may be exploited due to risk ambiguity, the ability of workers to act with independent judgment is crucial to saving time in dangerous scenarios [189].

The implementation of AI can supplement the decision-making and risk evaluation step for workers. Pulling from a database of past occurrences, a comprehensive understanding of workplace conditions and the presented risks can be created. AI models can be further developed to act as an emergency shutdown protocol, providing quicker response times than workers.

5.2. Maintaining Operational Efficiency

In an age where a continuous manufacturing flow is required to maintain high outputs, operational efficiency is valued on the same level as other resources. With an uninterrupted production rate, a company will grow in competitiveness and profitability [190]. This is emphasized in certain productivity methodologies. Each step of a process can be seen as a 'peak' or a 'trough' when analyzed as outputs over time, where the goal of implementing such methodology is to smooth out these efficiency bumps [191]. In an industry landscape where these minuscule effects create noticeable disruptions in outputs, the rapid elimination or suppression of hazards and incidents is crucial to keeping a company profitable and efficient.

AI may provide operational efficiency through both hazard detection and process efficiency. Similarly to incident detection, process discrepancies are easily monitored by adaptive AI using access to a database of past occurrences [183].

5.3. Preventing Costly Accidents

In manufacturing disasters, there is often a trend of deficient hazard detection. By detecting the risk of an incident before its occurrence, a company can be saved from damages or loss of resources. In some cases, this risk may not pose an immediate threat but rather pose a different danger further down the process. With timely hazard detection, risks are identified and addressed before the manifestation of danger. Catastrophic incidents in the manufacturing industry may stem from various deficiencies; in hazard detection, housekeeping, organizational, or operational common oversights, which have been comprehensively expanded in Table 3.

One such example of a catastrophic disaster is the Hayes Lemmerz Aluminum Dust explosion in 2003 [192]; an accumulation of aluminum dust within the workspace resulted in a dust deflagration. Due to poor hazard identification and housekeeping, the danger had not been appropriately handled. Another example of disaster is the Chevron Refinery Fire in 2012; a ruptured pipe released gas that formed a flammable vapor cloud, resulting in ignition and fire. Past recommendations to improve the piping had not been considered, with a history of a lack of safety inspections [193]. As a result of organizational and operational deficiencies, a proper examination of the pipe system and therefore risk identification and reaction had not occurred. To prevent occupational accidents and hazards from occurring and affecting workers, a monitoring and detection system must exist to give an early warning message before the occurrence of disaster [194].

Table 3 shows how AI introduces promising tools for risk mitigation in the manufacturing industry and improving efficiency, with possible usages in every corner of prevention. Unseen equipment deterioration may be supplemented with AI vision and monitoring to catch possible failures.

Table 3. Learnings from the common deficiencies in the manufacturing industry and how AI could generally regulate them.

Classification of Deficiencies	Example Specification	Description	Possible Complications	How AI, in General, could Regulate Such Factors	Limitations of AI in This Case	Ref.
Housekeeping deficiencies	Workplace dust buildup	Excess dust buildup accumulating around in the work environment.	• Respiratory conditions (e.g., lung cancer, asthma) • Combustible dust explosion hazard; deflagration • Clogging and interruption of mechanisms; overheating	• Monitoring equipment quality; equipment wear may result in higher dust output or act as an ignition source • Environmental surveillance	• High resource usage • Worker compliance issues	[195–198]
	Unclean Surfaces	Poor cleanup or housekeeping may lead to unclean surfaces. This may include greasy, wet, dusty, unsanitary surfaces, etc.	• Trip hazards • Exposure to excess hazardous substances (e.g., chemical, bacterial) • Puncture, scratch hazards from messy workspace	• Environmental surveillance • Worker/workplace recommendations	• High resource usage • Worker compliance issues—possible distrust in work culture	[199]
Operational	Failure in hazard identification	Lack of or inappropriate risk assessment	• Further development of risk into higher danger threat • Unpreparedness during occurrence • Legal violations	• Database-supported monitoring to provide comprehensive risk assessment	• Possible AI biases • Over-reliance on AI safeguards	[200]
	Automatnineon Errors	An automation error may arise from factors such as mechanical, programming, calibration, etc.	• Loss in resources; operational costs, materials, time. • Threat to worker safety • Machine safeguards do not fully reduce risks	• AI implemented into emergency automatic shutdown systems • Providing a further, comprehensive safeguard system than other methods	• In automatic AI shutdowns, AI is given the same executive power as a worker or higher • Possible AI biases • Overreliance on AI safeguards	[200–202]

Table 3. *Cont.*

Classification of Deficiencies	Example Specification	Description	Possible Complications	How AI, in General, could regulate such factors	Limitations of AI in this case	Ref.
Organizational	Deviation from protocol	Any action, such as skipping a step in a safety procedure, is not acceptable according to organizational standards	• Violation of safety standards • Encouraging a false norm for what is acceptable • Higher risk of incident	• Worker Surveillance • Process optimization, reducing worker desire for short-cuts	• Worker compliance issues—possible distrust in work culture • Possible AI biases	[203,204]
	Equipment failure	Sudden or gradual equipment failure	• Hidden equipment failure may go undetected, causing further complications • Resource and money loss; repairs are often cheaper than new equipment	• Equipment wear monitoring	• Overreliance on AI safeguards • Generalization of AI database; further monitoring of equipment conditions is needed	[205]
Worker health	Worker exhaustion	Factors such as overworking, monotonous work, or physical exertion may result in exhaustion	• Effects associated with fatigue; impaired judgment, poor reaction time • Higher risk of incident • Decreased productivity	• Biometric monitoring; techniques such as skin conductance or EEG activity • Worker fatigue anticipation	• Human physiology is a highly complicated subject to predict accurately • Implementation of such sensors can be considered invasive	[206–209]
	Chronic hazards	Long-term exposure to hazards; chemicals, dust, non-ergonomic worker practices	• Workers developing undetected conditions • Long-term health deterioration • Safety legislation violations	• Environmental monitoring; detecting sources of possible long-term complications • Monitoring and analysis of human movement	• High resource usage for full environmental surveillance • High complexity of human movement [3] • Need for customization of accident presentation in different workplaces	[209–212]

5.4. AI-Powered Surveillance and Anomaly Detection

Early monitoring systems play a crucial role in mitigating risks to humans, property, and the environment by detecting anomalies and dangerous events in real-time [213]. Traditional surveillance often struggles with high false positive rates due to manual monitoring and simple automation rules. However, the integration of advanced AI models like YOLOv8 and Inflated 3D (I3D) has demonstrated improvements in accuracy and speed, enabling more efficient processing of complex CCTV surveillance data. The YOLOv8 model excels in real-time object detection, processing video frames rapidly with high precision. It is particularly useful in dynamic environments, such as factories or construction sites, where quick responses are critical. YOLOv8 can detect suspicious objects, like unattended bags or misplaced equipment, which may signal a potential hazard. When paired with advanced tracking algorithms, YOLOv8 can monitor the movement of these objects or individuals across video frames, allowing the system to flag unusual patterns of activity, such as a worker entering a restricted area or engaging in unsafe behavior. I3D, on the other hand, enhances the capabilities of traditional CNNs by inflating their 2D architecture into 3D. This transformation enables I3D to capture both spatial and temporal features simultaneously, providing a deeper understanding of video content. The ability to analyze patterns over time makes I3D a powerful tool for detecting complex behaviors and suspicious activities. For instance, it can identify evolving hazards, such as a gas leak developing over several frames, or recognize unsafe worker behaviors that unfold over time, providing a more comprehensive approach to surveillance [214].

Anomaly detection in surveillance systems can be classified into two main approaches: density- and distance-based detection, and model-based detection. The first approach focuses on identifying deviations by analyzing the relationships between data points, while the second involves building a model of normal behavior and flagging deviations from this baseline. Both methods are essential in worker safety applications, where real-time monitoring can detect unsafe behaviors or environmental hazards. For instance, anomaly detection can identify a worker entering a restricted area, failure to wear safety equipment, or dangerous environmental changes like gas leaks or temperature spikes [214].

The Local Outlier Factor (LOF) algorithm is widely used for density- and distance-based anomaly detection. It identifies outliers, such as workers exhibiting unusual behavior or sensor data showing abnormal environmental conditions. LOF's ability to adapt to various data distributions makes it highly effective in safety-critical surveillance applications [215]. Other models like SVMs, RFs, and GMMs also play a key role in detecting anomalies by estimating probabilities of unusual behavior, such as a sudden fall or injury. RFs, in particular, are adept at handling high-dimensional safety data, such as video feeds or sensor data from wearables, quickly identifying patterns that signal potential risks to workers [216–218].

In modern manufacturing and supply chains, ensuring the security of sensitive data is crucial for smooth operations [219]. The increasing use of ML models that rely on personal data—like ID numbers, mobile numbers, and medical records—has heightened the importance of protecting this information [220]. To address these concerns, ML-driven anomaly detection models are being employed to enhance network security by monitoring network traffic and identifying anomalies, such as unauthorized access or fraud [221]. RFs have proven to be particularly effective in these scenarios, offering high accuracy and robustness to outliers, making them ideal for detecting anomalies in complex network environments [222].

The use of AI-powered anomaly detection not only strengthens real-time surveillance systems but also helps mitigate risks by enabling proactive measures in worker safety, environmental hazard detection, and network security. By analyzing data from various sources, including CCTV, thermal imaging, and wearable sensors, AI models can provide real-time insights, ensuring a safer and more secure working environment.

5.5. Automated Emergency Protocols and Alerts

ML can improve real-time decision-making and be used to develop robust and adaptive emergency protocols and response systems. Creating an ML emergency response system requires several steps, each with its respective algorithms and learning style, including the following, which are further expanded in Figure 11:

1. Classification of risk assessment [223]
2. Clustering to find patterns [224]
3. Crisis escalation prediction model [225]
4. Reinforcement learning for dynamic decision making [226]

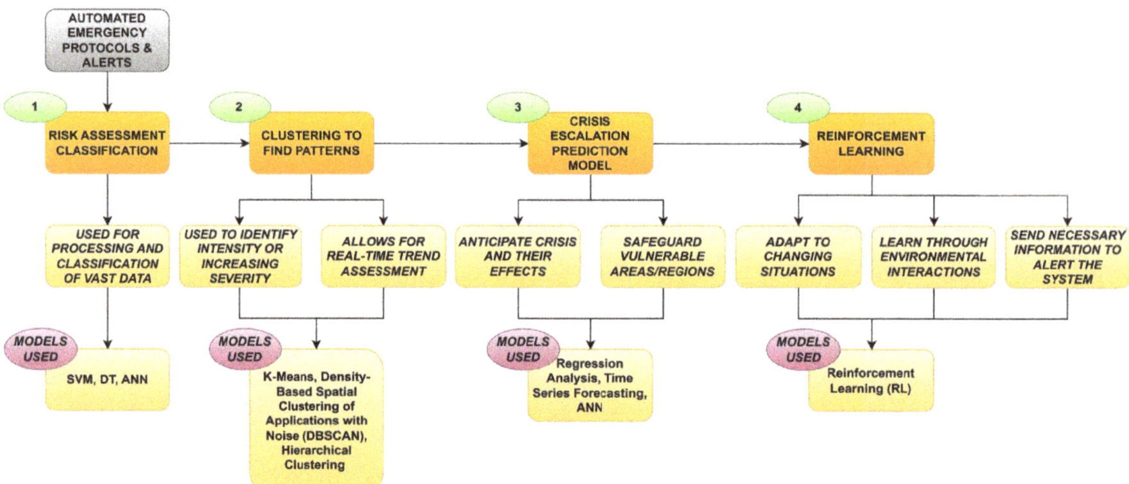

Figure 11. Automated emergency protocol and alerts system steps [223–226].

A computer vision-based architecture for alerts is proposed in Section 3.2. Alternatively, a proactive monitoring and alerting system through AI-powered cloud monitoring can be used for the automation of alerts. AI uses ML, NLP, and predictive analytics for real-time monitoring and anomaly detection. The benefits of the cloud-based architecture are enhanced scalability, improved system reliability, and optimized resource utilization. This allows for AI systems to handle growing volumes of data, enhance reliability by detecting and addressing issues proactively, and allocate resources efficiently, thereby reducing operational costs [227].

6. Human-Centric AI for Worker Safety

6.1. Wearable AI Devices for Health Monitoring

Due to the high stakes and sometimes fast-paced nature of assembling products on manual or semi-automatic assembly lines, stress, burnout, mental strain, and physical strain are often encountered by workers and laborers. An application of predictive maintenance is a wearable watch that detects the stress levels of people working on factory assembly lines. This is done by collecting the psychological data of workers, including heart rate variability, skin respiration rate, skin conductance, and electromyography. SVM is used to detect patterns in stress patterns, specifically skin conductance is processed through linear discriminant analysis (LDA) and classified using SVM. This technology is a tool that can enable stress management and contributes to worker well-being and productivity [159,228].

People with chronic illnesses who work in manufacturing industries can utilize wearable AI devices to monitor their health. Pre-existing chronic illnesses such as epilepsy can increase the risk of workplace accidents and injuries. The prediction and detection of epileptic seizures can be performed in real-time on a mobile device through specialized ANNs and DL models that classify electroencephalogram (EEG) data and group them into normal/abnormal EEG [229–231]. These models learn from individual patient data, synthetic data, EEG signals, video data, and wearable sensors. This technology is cardinal to Automated Seizure Diaries, which can be used for the diagnosis, treatment, and management of epilepsy and can aid in clinical trials [232,233]. Similarly, an experimental smartwatch called Embrace has been developed that measures sweat levels and sends an automated message and phone call to caregivers [233,234].

Smart sensors in wearable watches can also detect poor posture in employees while performing tasks such as pushing, pulling, and drilling. Through machine learning-based analysis and deep learning, improper and dangerous positions can be detected and alert the watch [73,233,235]. Other types of wearable devices used for health monitoring include eyewear, smart bands and e-patches [228].

Wearable devices should have the capability to connect to a personal mobile device to improve utility and ease of usage [235,236]. Biodata, alerts, device configuration, and the ability to connect to healthcare providers can be accessed through smartphones through the network in Figure 12. An alternate method to connect with healthcare workers can directly involve the wearable device and its sensors within it. Patient psychological data captured from the wearable device are linked to an IoT-Cloud database where data from a deep feed-forward neural network (DFFNN) are analyzed, classified, and sent to physicians without the patient being physically present. The healthcare provider can then track and analyze the data, notifying the patient about their condition and treatment without them being physically present [237].

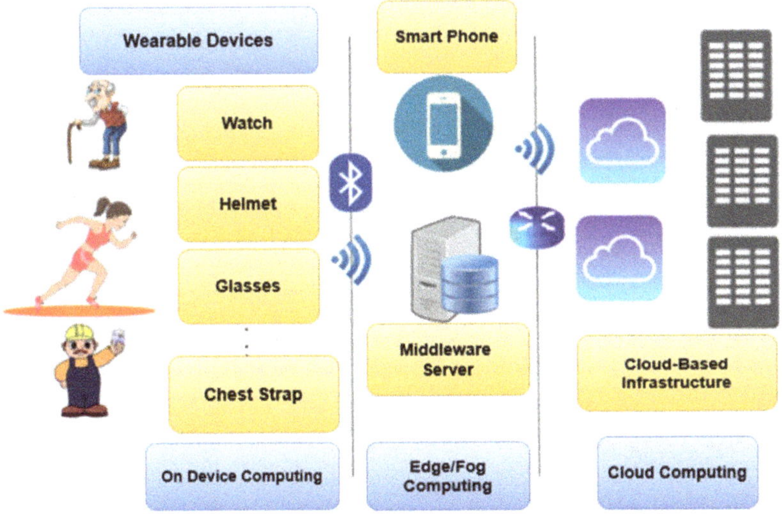

Figure 12. Wearable AI devices network to connect remotely to mobile devices [235].

6.2. Enhancing Safety Through Human–AI Collaboration

6.2.1. Training Using Different Methods

The types of manufacturing training include but are not limited to onboarding, safety, machine operation, management, developmental skills, and compliance training. Virtual Reality (VR), Augmented Reality (AR,) and Mixed Reality (MR) technologies are utilized to provide immersive training by simulating emergency and high-stakes situations in a controlled environment. The gamification of training modules through VR headset training creates an engaging and interactive environment that makes learning enjoyable. This form of learning often instills confidence and preparedness in employees, proving invaluable during real-life crises [238–240].

Manufacturing workers are also trained through IoT sensors on machinery that provide them with real-time insights into operational efficiency and performance. Low-level Key Performance Indicators (KPIs) are tools that monitor the overall performance of individual machines and workers. Forkan et al. investigated wearable MetaWear sensors, fine-grained KPIs, SVMs, and Raspberry Pi gateway devices that were used to detect the efficiency of workers and their knife utilization at a meat processing plant [241].

Mentoring and coaching are two methods of training that require supervision and effective intervention. Mentoring entails the transfer of skills through a sustained relationship from an individual with more experience in a particular area or field to a less experienced mentee. Alternatively, coaching is specifically targeted towards a certain task or skill and can utilize mentoring activities such as collaboration on projects and shadowing [242]. Coaching and mentoring can be facilitated through AI-driven chatbots, providing real-time assistance for onboarding employees by answering employee questions and providing feedback [243].

Apprenticeships are a combination of part time, formal education and experience at the workplace, thereby making it a method of efficient skill development. The benefits of apprenticeship include cognitive motivational effects, aid in integrating skills, increased youth employment rates, and better school-to-work transitions [244]. AI can facilitate apprenticeship and remote training through the creation of personalized training modules, tailored to individual learning speeds, strengths, and weaknesses. This method is superior to a one-size-fits-all training model since ML algorithms can analyze individual learning patterns, preferences, and performance, improving knowledge retention and making the training process more dynamic and effective [245].

6.2.2. LLM Models Used in Safety Standard Assessments

Large Language Models (LLMs) possess strong cognitive and reasoning capabilities, allowing them to process and analyze large volumes of complex data, essential for safety assessments. Their ability to understand and integrate multi-source sensor data ensures more accurate evaluations, improving the reliability of current safety standard assessments [246]. Additionally, LLMs have very impressive memory storages, allowing them to learn from historical experiences and apply past data to assess current situations more effectively [246]. Tools like ChatGPT excel in tasks that require creative thinking, such as brainstorming failure modes and identifying potential risk mitigation strategies [247].

LLMs are not strictly necessary for natural language processing (NLP), but they currently represent the most powerful approach for many NLP tasks. They can engage in human-like communications, as neural networks have demonstrated stronger abilities over humans in pattern recognition. These models can provide personalized and context-aware solutions while breaking down language barriers for smoother and more efficient communication [248]. Ensuring AI trustworthiness is very important for the adoption of

LLMs in manufacturing safety assessment. These models can improve decision-making by predicting potential risks and providing actionable information to the operators [249,250].

In manufacturing, LLMs can enhance and optimize quality control by integrating with existing safety assessment methods. Their ability to process large amounts of specialized manufacturing data, analyze historical patterns, and detect any irregularities allows for real-time monitoring and automated compliance inspection or checks [249,250]. Such an application is shown in Figure 13.

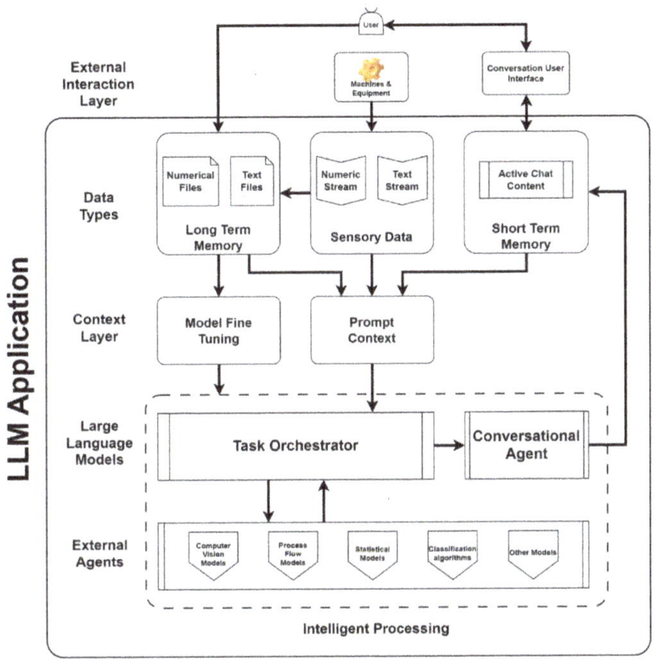

Figure 13. LLM integration in safety assessments, with interaction between data layers, contextual understanding, and external agents for accurate decision making [251].

6.2.3. Engineering Education

Human-centric AI is being included in engineering education to enhance worker safety in high-risk industries while bringing manufacturing industries in line with Industry 5.0, which emphasizes human–machine collaboration [252]. Some of the artificial intelligence (AI) tools being utilized to propel the change, from a traditional static educational approach to a more dynamic one, include the usage of Human Digital Twins (HDTs) and adaptive learning simulations. For instance, workers can now recreate real-life working conditions and avoid hazards like chemical additives by using VR and AR simulations using software like the Omniverse, which is already being utilized to improve a worker's danger recognition accuracy, train worker psychology, and their safety procedure implementations [253,254].

Moreover, there has been a rapid acceptance of these AI tools in today's manufacturing industry; for instance, German auto factories reported a 28% decrease in workplace accidents following the implementation of such AI-enhanced training simulations. However, the use of these tools in engineering education has come with several difficulties, including ethical questions about algorithmic biases and data privacy [255]. For instance, even though AI mapping and simulation tools, such as Honeywell's adaptive learning platform, help engineers become more proficient in hazard detection and prediction, there is a significant risk of bias perpetuation if these tools are not audited for fairness [256]. Similarly, when we examine wearable devices that track fatigue using EEG data, several problems regarding worker privacy occur. Furthermore, a 2024 ILO poll shows that nearly 58% of engineers now mistrust AI recommendations, making worker opposition to such abrupt shifts another significant obstacle [255]. Thus, to ensure hybrid human-AI oversight, autonomous hybrid safety ecosystems are essential to the future of worker education and occupational safety.

To implement this vision, industries must adopt the ISO 9241-210 standards [257,258], which outline that AI systems must be audited by an ethics review board aside from their commitment to investments in state-of-the-art AI infrastructure to enable real-time safety analytics while safeguarding worker privacy. Therefore, we may anticipate that workers in the future will have more secure and robust industrial environments if the aforementioned advancements and steps are implemented into the curriculum [259].

6.2.4. Quality Management and Control

Automated inspection and defect detection powered by AI has shown the potential to enhance quality management in manufacturing processes. AI-powered image analysis and processing techniques revolutionize quality control by improving the accuracy, efficiency, and comprehensiveness of inspections [260]. This technology utilizes machine learning, computer vision, and image processing algorithms to detect defects and anomalies with exceptional precision.

Predictive quality control powered by AI has gained huge traction as a means for quality control and management in manufacturing operations. Leveraging AI-driven advanced frameworks has led to significant improvements in defect detection for quality inspection and supported personalized road testing within the automotive manufacturing industry [261]. This approach contributes to production efficiency while also reducing costs and reducing time to manufacture, which adds some ease to the worker's tasks. The overall manufacturing process is also regulated with human supervision. The application of ML-integrated techniques have been used for measuring accountability based on the customer complaints with an accuracy of 64% [262].

Process optimization powered by AI has shown potential to enhance quality management in manufacturing environments. Real-time monitoring and process optimization frameworks driven by AI can revolutionize manufacturing performance [263]. Using machine learning algorithms, sensor data, and IoT connectivity, these systems enable continuous monitoring of production parameters, facilitating early fault prognosis and minimizing disruptions. The dynamic optimization in the manufacturing process that AI brings through real-time monitoring using data, adaptive control, and automated decision making provides ways to improve resource utilization, increasing efficiency as well as quality [263]. Furthermore, there are examples from different fields where AI contributes to the operational efficiency and workflow optimization in radiology departments in hospitals, which shows its versatility and applicability across various industries [264].

6.2.5. Human–AI Trust

Distrust towards AI systems in the manufacturing industry often arises from three main concerns: surveillance and manipulation, threats to autonomy and dignity, and unpredictability [265,266]. These issues are compounded by high-profile data breaches that expose the risks of AI misuse. Surveillance and manipulation concerns stem from AI systems that track sensitive worker data, such as performance metrics and health information, which can lead to fears of over-surveillance and unfair evaluations [267,268]. A notable example is the 2016 Facebook and Cambridge Analytica scandal [269], where personal data were misused without the consent of millions of Facebook users to simply harvest it for political targeting. In manufacturing, similar breaches could jeopardize worker privacy, undermining trust in AI systems used for performance monitoring or safety compliance. Another example is Samsung, where employees unintentionally leaked confidential information using ChatGPT to review internal documents [270]. These incidents highlight the critical need for robust data protection and transparency in AI usage.

Autonomy and dignity concerns arise when AI systems take over critical decision-making [271,272], particularly in high-risk environments like manufacturing. Workers may fear that automation could erode their autonomy, especially in tasks like safety monitoring or machine operation. This concern is further amplified by the potential for bias in AI systems [117,273], which could lead to discriminatory decisions. If not properly designed, AI systems may unintentionally favor certain groups, reinforcing existing inequalities. A notable example is Aon Consulting, which faced accusations of incorporating biases into its hiring assessment software, discriminating against individuals based on race and disability [274]. To mitigate these risks, regulatory frameworks such as the EU's Artificial Intelligence (AI) Act [275] ensure human oversight in AI decision-making, allowing for validation and intervention to prevent complete automation. Furthermore, methods to detect and reduce bias, such as test generation techniques and third-party audits [276–278], are essential to maintain fairness and trust in AI applications.

Unpredictability, or the lack of transparency in AI decision-making, also contributes to distrust [279]. When AI systems make critical decisions without clear explanations, workers may question the accuracy and reliability of those decisions. This is especially concerning in safety-critical scenarios, where the consequences of mistakes can be severe. Explainable AI (XAI) [280] becomes vital in such cases, as it ensures that AI systems provide transparent and understandable explanations for their decisions. Regulations like the AI Act [275] and ISO/IEC 27001c [281] emphasize the importance of explainability and robust security protocols to promote transparency. Providing clear reasoning for AI decisions helps workers verify that outcomes are based on sound logic, improving trust in the system. It also aids developers in detecting errors and refining the system, while policymakers can create appropriate regulations based on these insights.

In response to these concerns, regulatory frameworks like the General Data Protection Regulation (GDPR) [282], the AI Act [275], and ISO/IEC 27001c [281] have been developed to ensure data security, transparency, and accountability in AI systems. These regulations safeguard worker privacy, promote human oversight, and support explainability, creating a safer and more trustworthy environment in manufacturing. By addressing issues related to data protection, autonomy, and unpredictability, these frameworks help mitigate the risks of AI misuse and ultimately improve human trust in AI systems.

6.3. Benchmark Technologies of Human–AI Collaboration

To maximize safety outcomes, AI systems must complement human intuition, offering explainable feedback and fostering worker trust through transparency. Human–AI collaboration in predictive maintenance and worker safety allows for the integration of advanced technologies. These technologies combine sensor data, machine learning algorithms, and real-time monitoring systems to continuously and proactively monitor hazards, worker health, and predict equipment failures [263,264].

By merging human oversight with intelligent systems, industries can utilize a range of innovative technologies that support such collaboration in an industrial setting, as shown in Table 4, to mitigate risks, maintain operational efficiency, and improve overall workplace safety.

Table 4. Applications of predictive maintenance in worker safety.

Technology	Details of Technology	AI Model(s) Involved	Impact on Machinery/Workers	Specific Manufacturing Industry	Metrics Associated with AI Model	Ref.
Smartwatch	Collects physiological data of workers at the assembly line, including heart rate variability, skin respiration rate, skin conductance, and electromyography, and provides info on workers' mental and physical state	LR, DT, RF, KNN, SVM, XGBoost	Address mental health and stress affecting worker performance and productivity	Automatic and semi-automatic assembly lines	Matthew's Correlation Coefficient (MCC)	[160]
Computer vision	Utilizes digital images of gear teeth taken by endoscopic cameras. Images are then preprocessed and stored in the database.	*Techniques:* edge detection, Gray Level Co-Occurrence Matrix (GLCM), and DL with CNN	Surface wear on gear spurs	Gear systems	Classification accuracy rate (CAR), F1 score, recall	[182]
	A commercial camera with adjustable lenses is mounted. It is used to capture the wear of wire rope, serving as a prediction of the service life of components.	ANN, DNN	- Quality inspection and assurance of electrical components - Automate manual work to reduce manual labor, decrease costs and likelihood of error	Industrial manufacturing	N/A	[182]
ManuTrans	Uses sensor data to detect conditions on the manufacturing line and detect malfunctions	DL Comparison of Manutrans with: SVM, Autoregressive Integrated Moving Average (ARIMA), Long Short-Term Memory (LSTM)	Predicting exact moment of malfunction and severity in manufacturing lines	Pharmaceutical	Mean Squared Error (MSE)	[283]
Contrastive Predictive Coding (CPC)	Uses multi-sensor Internet of Things (IoT) infrastructure to measure different conditions of the production line (temperature, vibration, pressure) instead of training on vast datasets.	Self-supervised learning model	Anomaly detection and failure prediction	Hybrid R2R manufacturing	AUC-ROC, F1 score	[284]
Hierarchical Temporal Memory (HTM)	Mimics interactions between pyramidal neurons in the neocortex to efficiently learn over small datasets The learning model is weighted and constantly changing	DL	Early failure detection in roller-element bearings and 3D printers Detecting/classifying foreign materials on conveyor belts in cigarette manufacturing plants [11]	Additive manufacturing	Numenta Anomaly Benchmark (NAB)	[285,286]
Ensemble Learning Approach (ELA)	Concept of collaborative learning using multiple models. Each model is classified into 'boosting' or 'bagging'.	Cloud computing, Synthetic Minority Oversampling Technique (SMOTE), models 'ensembled': KNN, RF, decision jungle, XGBoost, LightGBM, reinforcement training	Quality management through forecasting defective products	Textile manufacturing	Two-dimension confused matrix. Used to estimate accuracy, precision, recall rate, F1 score, and MCC	[287]

7. Challenges and Ethical Considerations

7.1. Ethical Considerations

When implementing AI in manufacturing, several ethical conditions are essential to address, where transparency and accountability are the most critical. AI systems must be developed and deployed in ways that ensure fairness and minimize bias or unintended consequences [288]. The key challenges lie in its opaqueness, or simply the "black-box" nature of many AI models, which can lead to uncertain outcomes leading to concerns around traceability and responsibility [288].

Establishing clear frameworks for accountability in AI systems is important, particularly in manufacturing contexts where errors can have significant consequences. As AI becomes more prevalent in decision-making processes, determining responsibility for mistakes becomes increasingly complex. The explanation of AI systems can be extremely complex for individuals or regulators to interpret, which makes transparency alone insufficient for accountability [289]. To address this, comprehensive frameworks must be established that delineate responsibilities among various stakeholders, including AI developers, deployers, and end-users. These frameworks should incorporate regular audits, impact assessments, and clear mechanisms for redress when errors occur. For ensuring effective AI risk management and accountability, the adaptive risk governance frameworks must be capable of evolving in parallel with the technological improvements in the manufacturing industry [289].

Another important consideration is the impact on the workforce. AI implementation in manufacturing may lead to job displacement, requiring the reskilling of employees [290]. Small and medium-sized enterprises (SMEs) with financial constraints may face tremendous challenges, potentially exacerbating unemployment and inequality [291]. Furthermore, ensuring fairness in AI systems is critical, as they may unintentionally reinforce pre-existing biases embedded in their training data, leading to inequitable outcomes.

7.2. Data Privacy and Worker Consent

The current push for automation in manufacturing has intensified concerns over data privacy and worker consent. Technologies such as the Internet of Things (IoT) and Artificial Intelligence (AI) collect vast amounts of personal and operational data, often without clear boundaries on storage, access, or use [292]. These AI systems and smart technologies frequently collect data from workers, including their movements, decisions, and interactions [293]. However, workers often cannot give informed consent, as they may be unaware of the extent of data collection and how they are utilized [292–294].

An important issue is the ownership of collected and generated data. Typically, this ownership falls to the inventor or developer of the AI system, enabling employers to manage and utilize worker data without their input or control [295]. This raises ethical concerns about transparency, autonomy, and potential misuse of personal information in workplace decisions. Employers increasingly use invasive technologies, such as wearable tracking devices and digital reputation systems, to monitor performance and behavior. These systems collect large amounts of personal data, often without transparency [294].

While the EU General Data Protection Regulation (GDPR) recognizes these challenges, it does not provide sufficient protection in employment contexts, leaving workers vulnerable to exploitation [294]. Strengthening regulations and ethical frameworks is important in the advancement of the technology. Clearer policies on data ownership, bias mitigation, and worker consent will be essential in creating a better manufacturing environment. Differential privacy (DP) is an important tool used for protecting worker data privacy while allowing the industry to leverage the collection of valuable datasets. DP protects individual and organizational data by introducing controlled noise, preventing attackers from identifying specific workers or extracting sensitive information. In the Industrial Internet of Things (IIOT) and other smart manufacturing, large amounts of worker-related data, such as shift schedules, productivity metrics, and biometric readings, are collected to improve operations. Without strong and clear privacy measures, these data could be exploited causing privacy violations, discrimination, or unauthorized surveillance. DP fundamentally allows companies to analyze trends without compromising individual identities, such as anonymizing worker performance data, securing location data in logistics, and other sensitive employment information [296,297].

7.3. Reliability and Biases in AI Systems

Beyond privacy concerns, AI-driven monitoring systems can introduce biases into performance evaluations. For example, AI systems used for defect detection can disproportionately flag products handled by certain workers or teams based on biased training data. If a dataset contains more defects from a specific shift, the AI may systematically associate that shift with lower quality, leading to unfair performance evaluations and employment consequences [298]. Bias in AI-driven manufacturing can emerge stemming from unrepresentative datasets that overrepresent certain populations while underrepresenting others, such as those from low-income countries or ethnic minorities. Social biases, such as geographic, demographic, and economic factors, will all cause biases leading to inequitable outcomes [299].

Explainability in deep learning refers to a model's ability to provide clear and understandable explanations for its decisions, addressing the common issue of AI systems being 'black boxes'. This is important for building trust in the systems, as by improving explainability, these models improve reliability by being more consistent and dependable in their performance under changing environments in manufacturing [300].

Techniques such as visual explanations, feature-based methods, and uncertainty estimation allow operators to understand the factors influencing the model's predictions and output generation [300]. XAI methods, such as Class Activation Mapping (CAM) and saliency maps, provide the operators with insight into how the model makes decisions [301]. For instance, in the Ultrasonic Spot Welding (USW) quality prediction, visual explanations help engineers understand why a weld was classified as good or bad [300,301].

7.4. Grey Areas of Current Policies and Their Future

Furthermore, there is a noticeable trend and a prevalent issue in the growing global manufacturing sector, which is the lack of proper OSH implementation in certain countries or operations. As shown in Table 5, the agency that is required to build a thorough OSH culture is largely affected by the collective mindset and attitude toward safety in the region.

This leads to a gap in OSH similarities and organizational characteristics in countries that are culturally distant, which can be considered as a deficiency in worker safety. While most countries will demonstrate some degree of poor safety adherence, there are certain factors that heavily contribute to an unsafe culture.

These can be firm-specific, such as smaller operations allocating their fewer resources away from OSH implementation [68,302], or region/culture-wide norms such as lax regulatory frameworks, weak enforcement, or trends such as high temporary employment that tend towards poor worker safety [68,302]. In such scenarios, AI provides an easy solution to supplement OSH-deficient workplaces in the form of training, monitoring, and more. Integrating Artificial Intelligence (AI) into OSH policies can significantly enhance worker safety across various regions by addressing specific challenges and augmenting existing frameworks. Each country is familiar with their version of OSH, with some examples from notable manufacturing countries shown in Table 5.

Table 5. A sample of the occupational safety and health (OSH) organizations and codes followed by industry in each country.

Region	Country	Principal Organizations of OSH	Principal OSH Legislation	Year Enacted	Major Updates or Changes Made to the Standards	Pitfalls of these Standards for Worker Safety	Remarks	How AI Can Benefit These Policies?
Americas	Canada	Canadian Center for Occupational Health and Safety (CCOHS) [303]	Canada Labour Code [304]	1985	A 2024 amendment brought new employer requirements regarding employee treatment in termination and benefits. The Canada Labour Code is continuously updated each year [305].	• Some citizens feel the Canada Labour Code is not fully considerate of the work–life balance • Certain contractors are mislabeled as 'self-employed' [306].	CCOHS provides guidance and educational tools to use codes such as the Workplace Hazardous Materials Information System (WHMIS) [307].	• AI can analyze historical incident data to forecast potential hazards, enabling proactive interventions • AI-driven simulations can provide immersive safety training, enhancing hazard recognition and response • AI-powered workforce analytics can monitor workloads and detect patterns of overwork, enabling adaptive scheduling and alerts for excessive hours • Additionally, AI can flag discrepancies in employment classifications based on task analysis
	United States	Occupational Safety and Health Administration (OSHA) [29]	Occupational Safety and Health Act of 1970 [308]	1970	Several amendments have been implemented since 1970 [309]	• Establishing new standards is highly time and resource-consuming [310] • Resulting enforcement of outdated standards, such as perceptions of chemical safety or independent contractor representation [311]	OSHA serves as the principal enforcing agency to implement worker laws. The principal enforcement legislation is the Occupational Safety and Health Act of 1970 [308,312].	• AI can evaluate workstation setups to prevent musculoskeletal injuries via ergonomic assessments • Natural Language Processing (NLP) tools can scan and compare global safety standards, recommending timely updates and identifying gaps in current legislation • AI-powered incident analysis can provide real-time insights to support dynamic standard revisions

Table 5. *Cont.*

Region	Country	Principal Organizations of OSH	Principal OSH Legislation	Year Enacted	Major Updates or Changes Made to the Standards	Pitfalls of these Standards for Worker Safety	Remarks	How AI Can Benefit These Policies?
Europe	Switzerland	Federal Coordination Commission for Occupational Safety (EKAS) [68,313]	Arbeitsgesetz (ArG) (The Labour Act) [67]	1964	Switzerland has recently implemented many new policies regarding healthy work culture and overtime [314,315].	• Each sector has established specific legislation; some do not properly address OSH protocol [68].	The EKAS is defined as the central ruling organization. Each Swiss canton has its specific OSH organization managed by the EKAS. The Arbeitsgesetz (ArG) is the main Federal Labour Law, outlining worker practices and OSH [70,316].	• AI can ensure adherence to sector-specific OSH protocols by monitoring compliance in real-time, or in simple terms, introducing "Compliance Automation" • AI models can analyze data to identify potential safety hazards, facilitating timely interventions • Cross-sector AI compliance platforms can monitor operations across industries to identify and correct inconsistencies, offering unified dashboards for regulatory adherence across sectors
	Germany	Bundesanstalt für Arbeitsschutz und Arbeitsmedizin (BAuA) [69]	Arbeitsschutzgesetz (ArbSchG) [317]	1996	Several amendments have been implemented since 1996 [317].	• A culture of low OSH implementation is noticed in smaller-scale businesses [302].	The BAuA serves as the Federal Institute for OSH in Germany, which facilitates laws and regulations such as ArbSchG [318–320]	• AI-driven safety assistants (e.g., mobile apps) tailored for SMEs can automate hazard reporting, risk assessments, and compliance checks, reducing the burden on limited resources • AI tools can be incorporated to monitor and analyze worker behaviors to identify potential safety risks
Asia	Japan	Japan Organization of Occupational Health and Safety (JNIOSH) [321]	Japanese Industrial Safety and Health Act	1972	A 2006 amendment was established with an emphasis on addressing the issue of work–life balance [322]	• Strong work culture that trumps OSH compliance leads to high employee overwork [322,323]	While JNIOSH serves as the principal OSH organization within Japan, societies such as JSOH aim to promote education and research about worker health and safety. Their legislation is outlined by the Japanese Industrial Safety and Health Act [324].	• Wearable AI devices for workers can track fatigue indicators (heart rate, movement) and trigger alerts or interventions before exhaustion becomes a hazard, supporting enforcement of work–rest regulations

Table 5. *Cont.*

Region	Country	Principal Organizations of OSH	Principal OSH Legislation	Year Enacted	Major Updates or Changes Made to the Standards	Pitfalls of these Standards for Worker Safety	Remarks	How AI Can Benefit These Policies?
	China	State Administration of Work Safety (SAWS) [325]	Work Safety Law [326]	2002	The Worker Safety Law has undergone three amendments; in 2009, 2014, and 2021 [327].	• Many diseases are not included in policy, due to China's lack of occupational health. Should a worker suffer from an unlisted disease, it is not considered an occupational disease [328].	SAWS acts as the main OSH organization in China while facing the challenge of huge economic and social growth [329] China's Work Safety Law and Law on Prevention and Control of Occupational Diseases (2001) serve as the principal OSH legislation	• AI-driven health surveillance systems can detect early signs of unlisted occupational diseases by analyzing health trends and correlating them with workplace exposures, providing data to advocate for policy updates
	India	Ministry of Labour and Employment [330]	Factories Act [331]	1948	The Factories Amendment Bill in 2016 was amended to increase overtime hours [332]	• The Factories Act fails to address white-collar workers' overtime policy, leading to employee overwork [333] • Inadequate adoption of OSH legislation [334]	India's Factories Act is amongst many specialized workplace acts. In 2020, the Occupational Safety, Health, and Working Conditions Code was enacted to amend laws regarding OSH and workplace incidents [332,335]. These are all consolidated by the Ministry of Labour and Employment.	• Predictive AI tools can monitor overtime patterns and ergonomic risks for both industrial and office workers, highlighting unsafe trends and generating automated compliance reports to authorities

7.4.1. Open Areas for AI Governance

Currently, there are no universal frameworks for establishing ethical conduct when using AI that cater specifically to industrial requirements and standards. Data privacy laws hinder some of the functioning, as it varies from region to region, country to country, along with the algorithmic transparency and worker surveillance policies. This makes global framework development and standardization extremely complex, as different regions may interpret the generalized framework differently, which can lead to global ethical and legal mismatch. This can be implemented with the participation of neutral organizations like the United Nations, with a uniform and generalized framework that adapts to all regions and countries.

EU AI Act [275] and ISO 45001 [27] provide a much needed starting point for human-machine collaboration frameworks which can be applied for real-time hazard mitigation in industries. Some of the currently existing or newly created frameworks that relate to occupational safety and how they can be adapted for AI integration to improve worker safety are listed as follows:

- ***ISO 45001***: This international standard provides a management framework for occupational health and safety [27]. It can be extended to include AI-based risk detection, human-AI collaboration, and real-time safety feedback systems. AI governance can fit within existing clauses on continuous improvement, hazard identification, and performance evaluation.
- ***Europe AI Act (2024)***: This act, established in 2024, provides a risk-based regulatory framework for AI systems [275]. Classifies workplace AI safety systems as "High Risk", requiring conformity assessments, human oversight, and regular documentation.
- ***OECD AI Principles***: These international ethical guidelines, updated in 2024, emphasize accountability, robustness, and transparency in AI systems [336]. They are followed by OECD countries, the European Union, the United States, and the United Nations.
- ***NIST AI Risk Management Framework (USA)***: This voluntary toolkit from the National Institute of Standards and Technology (NIST) guides the implementation of trustworthy AI practices in manufacturing [337].
- ***ILO OSH Code of Practice***: This international framework informs workers about their rights [338]. It can be expanded to include worker rights in AI-monitored environments.

Based on this context and the literature, there are several areas where AI governance policies for manufacturing safety can be improved.

1. **Data Privacy and Surveillance Consent**: AI systems that process worker data, such as through computer vision or biometric tracking, must comply with the data protection laws of the country or regions, as applicable. Organizations should implement a privacy-by-design approach, offering opt-in consent and ensuring data anonymization within the system.
2. **AI Auditability and Traceability**: High-risk AI systems should maintain logs of decisions and actions for audit purposes to trace any inconsistencies. Model explainability must be ensured, particularly during incident investigations.
3. **Human-in-the-Loop Oversight**: AI systems used for critical tasks, such as predictive maintenance or real-time behavior monitoring, should require mandatory human involvement and oversight to avoid fully autonomous decision-making that could impact workers or lead to unwanted hazards.
4. **Governance Accessibility for SMEs**: National standardization agencies or industry consortiums should provide simplified AI governance frameworks for SMEs, along

with subsidized training and open-source compliance checklists, similar to the NIST AI Risk Management Toolkit.

7.5. Adoption and Integration Challenges

As shown in Figure 14, there are challenges in adopting advanced technologies, with a plethora of challenges that are generally common. Firstly, companies or businesses usually do not have adequate capital for the high initial investment required to integrate such advanced technologies, which leads to reluctance towards them. Especially for small and medium enterprises (SMEs), this discourages and induces hesitancy towards such shifts due to the underlying costs. Although AI deployment requires substantial investment, its long-term benefits in operational uptime, worker safety, and regulatory compliance can outweigh the costs [263,339].

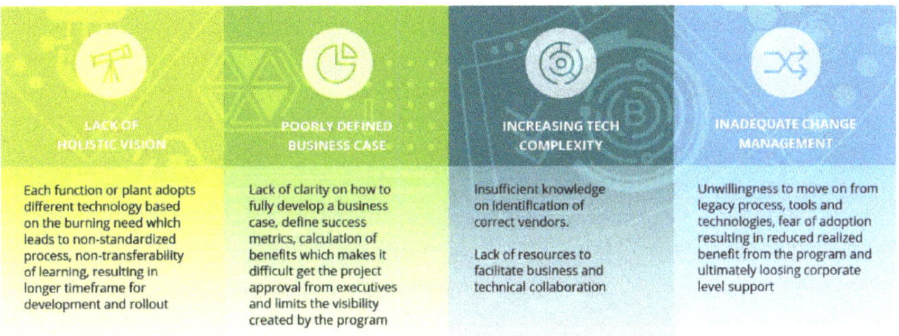

Figure 14. Challenges in the adoption of Predictive Maintenance (PdM). Credit to Deloitte [6].

Another critical challenge lies in the workforce's readiness to adapt, which is a quintessential factor, and without it, the adoption will be extremely challenging. There is hesitancy while working with new technologies due to fears of job displacement, lack of technical skills, or uncertainty about working alongside automated decision-making tools [340]. Without training and upskilling programs, the workers may struggle to effectively collaborate with AI, which can ultimately diminish the overall effectiveness of the shift towards it.

As discussed in Section 7.4, the gray areas of the policies and standards currently being implemented have not been adequately updated following AI integration or integration of advanced technologies where clear guidelines may have been established. As AI is a dynamic tool and there are still discoveries happening regarding its utility and the possibility to incorporate it, there is uncertainty in the regulatory guidelines and the standardizations [341].

Furthermore, data quality and infrastructure pose a major challenge for the implementation of PdM. PdM systems rely heavily on accurate, high-frequency data streams from machinery and operational processes. In many industrial environments, especially in older facilities with inconsistent data collection, outdated equipment and fragmented IT systems may limit the ability to deploy AI-based maintenance strategies at their full capacity [342,343].

7.6. Challenges and Solutions for SMEs

Small and medium-sized enterprises (SMEs) are the backbone of global manufacturing, especially in developing economies. AI can improve safety in these enterprises, but financial constraints often limit its use. The costs of AI, including sensors, software licenses, and infrastructure upgrades, can be too high for SMEs. In addition, SMEs often do not have dedicated experts such as data scientists, AI engineers, or IT security personnel, which makes AI integration harder and more expensive. SMEs also have limited historical data, which makes it difficult to train predictive models or computer vision systems. Furthermore, it can be challenging to retrofit existing machinery and safety processes with AI. SMEs face complex regulatory requirements, and adding AI governance could overwhelm them. A lack of information or awareness may lead to further problems. As discussed in Section 7.5, workers may also resist AI due to fears of job loss or unwillingness to learn, which can hinder adoption [344–346].

There are simple solutions to make AI more accessible to SMEs. One solution is to use easily accessible AI packages such as OpenCV, TensorFlow, and PyCaret for safety monitoring. These tools reduce technical barriers by offering pre-trained models, eliminating licensing costs, and benefiting from community-led updates [347]. Another solution is to use modular AI kits that are easy to deploy with pre-configured AI dashboards that require no coding. For example, vibration monitoring systems for predictive maintenance can be quickly implemented. Governments can help by offering subsidies, grants, or tax incentives for AI pilot programs focused on safety or by creating industry consortia to support such initiatives. As mentioned in Section 7.4.1, governments and national standardization bodies can also provide a simplified toolkit for SMEs. SMEs can also partner with universities, startups, or research centers to access AI consultancy, shared infrastructure, or digital literacy and employee training [347].

8. Future Directions in AI-Driven Worker Safety

As Industry 4.0 continues to evolve, the widespread adoption of smart factories remains a key principle, with AI and associated technologies driving the transformation of manufacturing processes. While current systems leverage "smart technology" such as sensors, actuators, and autonomous systems [348], there are several critical areas for further research and development.

One such area is the expansion of AI-driven Smart Factories for Manufacturing (ADSFM). Although these systems currently exhibit capabilities like self-perception, operational optimization, and dynamic reconfiguration, further exploration is needed to enhance their intelligent decision-making capabilities. Advancements in predictive maintenance, environmental health and safety (EHS), adaptive manufacturing, and digital twins can significantly improve the scalability and robustness of AI applications. Moreover, while AI applications such as computer vision, natural language processing, and intelligent robotics are already integrated, future work should focus on developing more reliable AI models that can handle diverse and dynamic manufacturing environments [349].

Automatic Identification (Auto-ID) and sensor fusion technologies, which play a fundamental role in smart factories, also present opportunities for future innovation. Research into improving sensor integration and data fusion for real-time, high-accuracy decision-making will be critical for enhancing control and optimizing factory efficiency. Additionally, expanding the capabilities of RFID and integrating newer sensor technologies for real-time data collection could pave the way for smarter, more adaptive manufacturing systems [350].

Blockchain technology, while already providing benefits in Additive Manufacturing (AM), has immense potential for further applications, particularly in ensuring data privacy and intellectual property (IP) protection [351]. Research into blockchain's role in enhancing transparency, security, and traceability within manufacturing workflows will be essential.

Furthermore, exploring how blockchain can facilitate monetization of AM processes and enable more efficient business models will open new avenues for the industry [352].

Despite these advancements, the integration of AI and blockchain technologies in manufacturing must be guided by strong global standards and regulatory frameworks. As the technology continues to mature, significant concerns regarding ethical use, data privacy, and misuse remain [353]. Future research should prioritize the development of standardized guidelines for the deployment of AI in smart factories, with an emphasis on transparency, fairness, and accountability. Additionally, exploring AI explainability and autonomous decision-making frameworks will be crucial for building trust among workers and stakeholders in automated environments.

In summary, while the potential of AI and blockchain in Industry 4.0 is vast, continued research, development, and regulatory efforts are essential to ensure that these technologies are deployed safely, ethically, and effectively across manufacturing sectors. Future work should focus on addressing these challenges while optimizing the capabilities of AI to enhance productivity, safety, and sustainability in smart manufacturing environments.

9. Conclusions

This study rigorously evaluates the AI models that have been used, the manufacturing industries that have benefited, the case studies, the policies, and their gray areas, and how AI can compensate for the inadequacies of the policies and standards that are currently implemented. Through the analysis, it is evident that AI has become a critical enabler in advancing occupational safety, hazard detection, and predictive maintenance across diverse manufacturing sectors. Through the integration of ML, computer vision, and real-time monitoring systems, AI offers to enhance operational efficiency and worker safety while also allowing us to address the gaps in regulatory compliance and possibilities to compensate for the weak implementation of occupational safety and health policies and standards.

However, AI serves as a powerful tool to bridge policy shortcomings and optimize safety practices. The current laws do not incorporate AI-related clauses that regulate its usage and avert any adversarial usage within the manufacturing industry. This study highlights the need for harmonizing AI integration with existing safety standards, fostering worker trust through transparent human–AI collaboration and ensuring continuous model improvement through feedback from real industrial scenarios.

Despite the current pitfalls and the nascent stage of technologies, with the current pace of discoveries and research, there is a huge opportunity for AI to contribute to worker safety. However, greater emphasis must be placed on developing adaptive, context-aware AI systems that align with the specific cultural, regulatory, and operational nuances of different manufacturing environments. Current global standardization efforts and interdisciplinary collaboration will be vital to maximizing the full potential of AI-driven safety systems, ensuring that both technological and human factors are balanced in safeguarding the future of industrial workspaces.

Author Contributions: Conceptualization, U.C.; methodology, M.K. and U.C.; software, K.A. (Kalana Abeywardena) and U.C.; validation, K.A. (Kalana Abeywardena), U.C. and K.B.; formal analysis, M.K., C.Z., S.M., H.K., K.A. (Kimi An), K.A. (Kalana Abeywardena), U.C. and K.B.; investigation, M.K., C.Z., S.M., H.K., K.A. (Kimi An), K.A. (Kalana Abeywardena) and U.C.; resources, K.A. (Kalana Abeywardena) and U.C.; data curation, M.K., C.Z., S.M., H.K. and K.A. (Kimi An); writing—original draft preparation, M.K., C.Z., S.M., H.K., K.A. (Kimi An), K.A. (Kalana Abeywardena) and U.C.; writing—review and editing, K.A. (Kalana Abeywardena), U.C. and K.B.; visualization, U.C.; supervision, K.B.; project administration, U.C. and K.B.; funding acquisition, U.C. and K.B. All authors have read and agreed to the published version of the manuscript.

Funding: This research received no external funding.

Data Availability Statement: There were no new data generated; all the relevant data are cited in the manuscript.

Conflicts of Interest: The authors declare no conflicts of interest.

References

1. Murtaza, A.A.; Saher, A.; Zafar, M.H.; Moosavi, S.K.R.; Aftab, M.F.; Sanfilippo, F. Paradigm shift for predictive maintenance and condition monitoring from Industry 4.0 to Industry 5.0: A systematic review, challenges and case study. *Results Eng.* **2024**, *24*, 102935. [CrossRef]
2. Li, B.h.; Hou, B.c.; Yu, W.t.; Lu, X.b.; Yang, C.w. Applications of artificial intelligence in intelligent manufacturing: A review. *Front. Inf. Technol. Electron. Eng.* **2017**, *18*, 86–96. [CrossRef]
3. Acemoglu, D.; Restrepo, P. Artificial intelligence, automation, and work. In *The Economics of Artificial Intelligence: An Agenda*; University of Chicago Press: Chicago, IL, USA, 2018; pp. 197–236.
4. Kusiak, A. Data mining: Manufacturing and service applications. *Int. J. Prod. Res.* **2006**, *44*, 4175–4191. [CrossRef]
5. Gupta, P.; Krishna, C.; Rajesh, R.; Ananthakrishnan, A.; Vishnuvardhan, A.; Patel, S.S.; Kapruan, C.; Brahmbhatt, S.; Kataray, T.; Narayanan, D.; et al. Industrial internet of things in intelligent manufacturing: A review, approaches, opportunities, open challenges, and future directions. *Int. J. Interact. Des. Manuf. (IJIDEM)* **2022**, 1–23. [CrossRef]
6. Predictive Maintenance Solutions. Available online: https://www2.deloitte.com/us/en/pages/operations/articles/predictive-maintenance-and-the-smart-factory.htm (accessed on 22 January 2025).
7. Revolutionizing OEM Manufacturing with AI. Available online: https://www.oemmagazine.org/engineering/automation/article/22883168/revolutionizing-oem-manufacturing-with-ai (accessed on 22 January 2025).
8. Pires, F.; Cachada, A.; Barbosa, J.; Moreira, A.P.; Leitão, P. Digital twin in industry 4.0: Technologies, applications and challenges. In Proceedings of the 2019 IEEE 17th international conference on industrial informatics (INDIN), Helsinki-Espoo, Finland, 22–25 July 2019; Volume 1, pp. 721–726.
9. Fu, Y.; Downey, A.R.; Yuan, L.; Zhang, T.; Pratt, A.; Balogun, Y. Machine learning algorithms for defect detection in metal laser-based additive manufacturing: A review. *J. Manuf. Processes* **2022**, *75*, 693–710. [CrossRef]
10. Acemoglu, D.; Restrepo, P. Tasks, automation, and the rise in US wage inequality. *Econometrica* **2022**, *90*, 1973–2016. [CrossRef]
11. Das, A.; Panda, S.; Datta, S.; Naskar, S.; Misra, P.; Chattopadhyay, T. AI based safety system for employees of manufacturing industries in developing countries. *arXiv* **2018**, arXiv:1811.12185.
12. Vernim, S.; Bauer, H.; Rauch, E.; Ziegler, M.T.; Umbrello, S. A value sensitive design approach for designing AI-based worker assistance systems in manufacturing. *Procedia Comput. Sci.* **2022**, *200*, 505–516. [CrossRef]
13. Khandelwal, P.; Khandelwal, A.; Agarwal, S.; Thomas, D.; Xavier, N.; Raghuraman, A. Using computer vision to enhance safety of workforce in manufacturing in a post covid world. *arXiv* **2020**, arXiv:2005.05287.
14. Petit, N.; De Cooman, J. Models of Law and Regulation for AI. In *The Routledge Social Science Handbook of AI*; Routledge: London, UK, 2021; pp. 199–221.
15. Hoffmann-Riem, W. Artificial intelligence as a challenge for law and regulation. In *Regulating Artificial Intelligence*; Springer: Cham, Switzerland, 2020; pp. 1–29.
16. Schuett, J. Defining the scope of AI regulations. *Law Innov. Technol.* **2023**, *15*, 60–82. [CrossRef]
17. Chohan, J.S.; Bilga, P.S. Occupational health hazards in small scale steel manufacturing industries: A case study. *Int. J. Manuf. Technol. Manag.* **2011**, *24*, 182–192. [CrossRef]
18. Reinhold, K.; Tint, P. Hazard profile in manufacturing: Determination of risk levels towards enhancing the workplace safety. *J. Environ. Eng. Landsc. Manag.* **2009**, *17*, 69–80. [CrossRef]
19. Nenonen, S. Fatal workplace accidents in outsourced operations in the manufacturing industry. *Saf. Sci.* **2011**, *49*, 1394–1403. [CrossRef]
20. Pacana, A. Comparative occupational risk assessment in a CNC machine tool operator position. *Prod. Eng. Arch.* **2017**, *16*, 28–31 [CrossRef]
21. Willquist, P.; Törner, M. Identifying and analysing hazards in manufacturing industry—A review of selected methods and development of a framework for method applicability. *Int. J. Ind. Ergon.* **2003**, *32*, 165–180. [CrossRef]
22. Dobrzyńska, E.; Kondej, D.; Kowalska, J.; Szewczyńska, M. State of the art in additive manufacturing and its possible chemical and particle hazards. *Indoor Air* **2021**, *31*, 1733–1758. [CrossRef]

23. Amirah, N.A.; Him, N.F.N.; Rashid, A.; Rasheed, R.; Zaliha, T.N.; Afthanorhan, A. Fostering a safety culture in manufacturing through safety behavior: A structural equation modelling approach. *J. Saf. Sustain.* **2024**, *1*, 108–116. [CrossRef]
24. Darabont, D.C.; Antonov, A.E.; Bejinariu, C. Key elements on implementing an occupational health and safety management system using ISO 45001 standard. *MATEC Web Conf.* **2017**, *121*, 11007. [CrossRef]
25. International Labour Organization. Safety and Health in the Non-Ferrous Metals Industries. Available online: https://www.ilo.org/resource/other/safety-and-health-non-ferrous-metals-industries (accessed on 18 January 2025).
26. International Labour Organization. Safety and Health in Shipbuilding and Ship Repair. Available online: https://www.ilo.org/resource/other/safety-and-health-shipbuilding-and-ship-repair-revised-edition (accessed on 15 January 2025).
27. International Organization for Standardization. *Occupational Health and Safety Management Systems: Requirements with Guidance for Use*; ISO: Geneva, Switzerland, 2018.
28. Robins, T.G.; Klitzman, S. Hazard communication in a large US manufacturing firm: The ecology of health education in the workplace. *Health Educ. Q.* **1988**, *15*, 451–472. [CrossRef]
29. Kim, Y.; Park, J.; Park, M. Creating a culture of prevention in occupational safety and health practice. *Saf. Health Work* **2016**, *7*, 89–96. [CrossRef]
30. Herron, C.; Braiden, P.M. A methodology for developing sustainable quantifiable productivity improvement in manufacturing companies. *Int. J. Prod. Econ.* **2006**, *104*, 143–153. [CrossRef]
31. Gupta, S.; Jain, S.K. A literature review of lean manufacturing. *Int. J. Manag. Sci. Eng. Manag.* **2013**, *8*, 241–249. [CrossRef]
32. National Health Service, England. Lean Six Sigma: Some Basic Concepts. Available online: https://www.england.nhs.uk/improvement-hub/wp-content/uploads/sites/44/2017/11/Lean-Six-Sigma-Some-Basic-Concepts.pdf (accessed on 20 January 2025).
33. U.S. Environmental Protection Agency. Lean Thinking and Methods: Kaizen. Available online: https://www.epa.gov/sustainability/lean-thinking-and-methods-kaizen (accessed on 15 January 2025).
34. Cheng, T.; Podolsky, S. *Just-In-Time Manufacturing: An Introduction*; Springer Science & Business Media: Berlin/Heidelberg, Germany, 1996.
35. American Society for Quality. Six Sigma. Available online: https://asq.org/quality-resources/six-sigma (accessed on 19 January 2025).
36. Rajurkar, K.; Hadidi, H.; Pariti, J.; Reddy, G. Review of sustainability issues in non-traditional machining processes. *Procedia Manuf.* **2017**, *7*, 714–720. [CrossRef]
37. Ellington, L.E.; Becerra, R.; Mallma, G.; da Fieno, J.T.; Nair, D.; Onchiri, F.; Nielsen, K.R. Knowledge acquisition and retention after a high flow training programme in Peru: A quasi-experimental single group pre–post design. *BMJ Open* **2020**, *10*, e035125. [CrossRef]
38. Çetintepe, S. Occupational health assessment and risks in automotive and wood manufacturing industries: A retrospective study. *Ann. Med. Res.* **2024**, *31*, 141–146.
39. Wang, L.; Dai, W.; Sun, D.; Zhao, Y. Risk evaluation for a manufacturing process based on a directed weighted network. *Entropy* **2020**, *22*, 699. [CrossRef] [PubMed]
40. Tabor, J. Occupational Hazard Prevention in Manufacturing Systems. *Appl. Mech. Mater.* **2015**, *718*, 227–232. [CrossRef]
41. Ikumapayi, O.; Omietimi, B.; Onu, P.; Ogedengbe, T.; Oluwafemi, J.; Afolalu, S.; Akinlabi, E. Non-Traditional Machining Techniques in Manufacturing Industries–An Overview. *E3S Web Conf.* **2023**, *430*, 01213. [CrossRef]
42. Elsheikh, A.; Ali, A.B.; Saba, A.; Faqeha, H.; Alsaati, A.A.; Maghfuri, A.M.; Abd-Elaziem, W.; El Ashmawy, A.A.; Ma, N. A review on sustainable machining: Technological advancements, health and safety considerations, and related environmental impacts. *Results Eng.* **2024**, *24*, 103042. [CrossRef]
43. Onawumi, A.; Olojede, M.; Agboola, O.; Ikubanni, P.; Odunola, A. Investigation into combined occupational hazards among automobile repairs workmen in Nigeria. *Int. J. Hum. Factors Ergon.* **2022**, *9*, 143–162. [CrossRef]
44. Du, D.l.; Qiu, J.; Zhao, H.y. Risk assessment study of manufacturing green supply chain based on grey theory. In Proceedings of the Proceedings of International Conference on Information Systems for Crisis Response and Management (ISCRAM), Washington, DC, USA, 25–27 November 2011; pp. 234–240.
45. Dyreborg, J.; Lipscomb, H.J.; Nielsen, K.; Törner, M.; Rasmussen, K.; Frydendall, K.B.; Bay, H.; Gensby, U.; Bengtsen, E.; Guldenmund, F.; et al. Safety interventions for the prevention of accidents at work: A systematic review. *Campbell Syst. Rev.* **2022**, *18*, e1234. [CrossRef] [PubMed]
46. Fatemi, F.; Dehdashti, A.; Jannati, M. Implementation of chemical health, safety, and environmental risk assessment in laboratories: A case-series study. *Front. Public Health* **2022**, *10*, 898826. [CrossRef]

47. California Department of Public Health. Introduction to Toxic Substances. Available online: https://www.cdph.ca.gov/Programs/CCDPHP/DEODC/OHB/HESIS/CDPH%20Document%20Library/introtoxsubstances.pdf (accessed on 25 January 2025).
48. Rout, B.; Sikdar, B. Hazard identification, risk assessment, and control measures as an effective tool of occupational health assessment of hazardous process in an iron ore pelletizing industry. *Indian J. Occup. Environ. Med.* **2017**, *21*, 56–76. [CrossRef] [PubMed]
49. Themann, C.L.; Masterson, E.A. Occupational noise exposure: A review of its effects, epidemiology, and impact with recommendations for reducing its burden. *J. Acoust. Soc. Am.* **2019**, *146*, 3879–3905. [CrossRef]
50. Chidambaram, V.; Gopalsamy, M.M.; Zindani, D.; Chandan, G.K.; Aslam, M.; Kanchan, B.K. Worker Exposure to Hand-Arm Vibration in Fettling Shop: An Experimental Study in Foundry. *Eng. Proc.* **2024**, *61*, 47. [CrossRef]
51. Weir, E.; Lander, L. Hand–arm vibration syndrome. *Cmaj* **2005**, *172*, 1001–1002. [CrossRef]
52. Baudin, C.; Vacquier, B.; Thin, G.; Chenene, L.; Guersen, J.; Partarrieu, I.; Louet, M.; Ducou Le Pointe, H.; Mora, S.; Verdun-Esquer, C.; et al. Occupational exposure to ionizing radiation in medical staff: Trends during the 2009–2019 period in a multicentric study. *Eur. Radiol.* **2023**, *33*, 5675–5684. [CrossRef] [PubMed]
53. Tustin, A.W. Evaluation of occupational exposure limits for heat stress in outdoor workers—United States, 2011–2016. *MMWR. Morb. Mortal. Wkly. Rep.* **2018**, *67*, 733–737. [CrossRef]
54. Dubik, K.; Akasika, C.; Naawie, N.; Dery, F.; Abunkudugu, F.A. Impact of Safety-Maintenance Practices on the Overall Performance of an Industry. *World J. Eng. Technol.* **2024**, *12*, 759–774. [CrossRef]
55. Park, D.U.; Zoh, K.E.; Jeong, E.K.; Koh, D.H.; Lee, K.H.; Lee, N.; Ha, K. Assessment of occupational health risks for maintenance work in fabrication facilities: Brief review and recommendations. *Saf. Health Work* **2024**, *15*, 87–95. [CrossRef]
56. Tabor, J. Use of safety analysis in industrial maintenance management. *Qual. Prod. Improv.-QPI* **2021**, *3*, 162–172.
57. Costella, M.F.; Dalcanton, F.; Cardinal, S.M.; Vilbert, S.S.; Pelegrini, G.A. Maintenance, occupational health and safety: A systematic review of the literature. *Gestão Produção* **2020**, *27*, e3922. [CrossRef]
58. Yan, T.; Ji, F.; Bi, M.; Wang, H.; Cui, X.; Liu, B.; Niu, D.; Li, L.; Lan, T.; Xie, T.; et al. Occupational stress and associated risk factors among 13,867 industrial workers in China. *Front. Public Health* **2022**, *10*, 945902. [CrossRef] [PubMed]
59. Hadler, N.M. Repetitive upper-extremity motions in the workplace are not hazardous. *J. Hand Surg.* **1997**, *22*, 19–29. [CrossRef]
60. Barthelme, J.; Sauter, M.; Mueller, C.; Liebers, F. Association between working in awkward postures, in particular overhead work, and pain in the shoulder region in the context of the 2018 BIBB/BAuA Employment Survey. *Bmc Musculoskelet. Disord.* **2021**, *22*, 1–12. [CrossRef]
61. Lohne, F.K.; Xu, K.; Fimland, M.S.; Palarea-Albaladejo, J.; Redzovic, S. Association between musculoskeletal pain and exposures to awkward postures during work: A compositional analysis approach. *Ann. Work Expo. Health* **2024**, *68*, 522–534. [CrossRef]
62. Bláfoss, R.; Skovlund, S.V.; Skals, S.; Sundstrup, E.; López-Bueno, R.; Calatayud, J.; Andersen, L.L. Duration and intensity of occupational lifting and risk of long-term sickness absence: Prospective cohort study with register follow-up among 45 000 workers. *Scand. J. Work Environ. Health* **2023**, *49*, 283. [CrossRef]
63. Al-Otaibi, S.T. Prevention of occupational back pain. *J. Fam. Community Med.* **2015**, *22*, 73–77. [CrossRef]
64. Chen, B.; Wang, L.; Li, B.; Liu, W. Work stress, mental health, and employee performance. *Front. Psychol.* **2022**, *13*, 1006580. [CrossRef]
65. Noblet, A.; LaMontagne, A.D. The role of workplace health promotion in addressing job stress. *Health Promot. Int.* **2006**, *21*, 346–353. [CrossRef]
66. Catapano, P.; Cipolla, S.; Sampogna, G.; Perris, F.; Luciano, M.; Catapano, F.; Fiorillo, A. Organizational and individual interventions for managing work-related stress in healthcare professionals: A systematic review. *Medicina* **2023**, *59*, 1866. [CrossRef] [PubMed]
67. Government of Switzerland. Swiss Labour Laws and Occupational Safety Regulations. Available online: https://www.fedlex.admin.ch/eli/cc/1966/57_57_57/de (accessed on 15 January 2025).
68. European Agency for Safety and Health at Work. OSH System at National Level—Switzerland. Available online: https://oshwiki.osha.europa.eu/en/themes/osh-system-national-level-switzerland (accessed on 15 January 2025).
69. Federal Ministry of Labour and Social Affairs. Occupational Safety and Health. Available online: https://www.bmas.de/EN/Labour/Occupational-Safety-and-Health/occupational-safety-and-health.html (accessed on 15 January 2025).
70. Federal Coordination Commission for Occupational Safety. EKAS: Occupational safety and health in Switzerland. Available online: https://www.ekas.ch/index-en.php?frameset=1 (accessed on 15 January 2025).
71. Çınar, Z.M.; Abdussalam Nuhu, A.; Zeeshan, Q.; Korhan, O.; Asmael, M.; Safaei, B. Machine learning in predictive maintenance towards sustainable smart manufacturing in industry 4.0. *Sustainability* **2020**, *12*, 8211. [CrossRef]

72. Rashid, A.B.; Kausik, A.K. AI revolutionizing industries worldwide: A comprehensive overview of its diverse applications. *Hybrid Adv.* **2024**, *7*, 100277. [CrossRef]
73. Rahman, M.M.; Hossain, A.; Sikder, M.A. Machine Learning Applications in Industry Safety: Analysis and Prediction of Industrial Accidents. In Proceedings of the 2024 International Conference on Smart Systems for applications in Electrical Sciences (ICSSES), Tumakuru, India, 3–4 May 2024; pp. 1–6.
74. Bokonda, P.L.; Ouazzani-Touhami, K.; Souissi, N. Predictive analysis using machine learning: Review of trends and methods. In Proceedings of the 2020 International Symposium on Advanced Electrical and Communication Technologies (ISAECT), Marrakech, Morocco, 25–27 November 2020; pp. 1–6.
75. Vukicevic, A.M.; Petrovic, M.N.; Knezevic, N.M.; Jovanovic, K.M. Deep learning-based recognition of unsafe acts in manufacturing industry. *IEEE Access* **2023**, *11*, 103406–103418. [CrossRef]
76. Hou, C.K.J.; Behdinan, K. Neural Networks with Input Dimensionality Reduction for Efficient Temperature Distribution Prediction in a Warm Stamping Process. *J. Appl. Comput. Mech.* **2022**, *8*, 1431–1444.
77. Hou, C.K.J.; Behdinan, K. Neural networks with dimensionality reduction for predicting temperature change due to plastic deformation in a cold rolling simulation. *AI EDAM* **2023**, *37*, e1. [CrossRef]
78. Hou, C.K.J.; Behdinan, K. Neural networks with dimensionality reduction for efficient springback prediction in deep drawing of multi-material cylindrical cups. *J. Exp. Theor. Artif. Intell.* **2025**, *37*, 111–130. [CrossRef]
79. Kazemi, Z.; Steeves, C.A. Machine learning for characterizing uncertain elastic properties of fused filament fabricated materials for topology optimization applications. *arXiv* **2024**, arXiv:2408.05850.
80. Kazemi, Z.; Steeves, C.A. Uncertainty quantification in material properties of additively manufactured materials for application in topology optimization. In Proceedings of the ASME International Mechanical Engineering Congress and Exposition, Columbus, OH, USA, 30 October–3 November 2022; Volume 86656, p. V003T04A009.
81. Sachdeva, I.; Ramesh, S.; Chadha, U.; Punugoti, H.; Selvaraj, S.K. Computational AI models in VAT photopolymerization: A review, current trends, open issues, and future opportunities. *Neural Comput. Appl.* **2022**, *34*, 17207–17229. [CrossRef]
82. Kakhki, F.D.; Freeman, S.A.; Mosher, G.A. Applied machine learning in agro-manufacturing occupational Incidents. *Procedia Manuf.* **2020**, *48*, 24–30. [CrossRef]
83. Ni, L.; Chen, F.; Ran, R.; Li, X.; Jin, N.; Zhang, H.; Peng, B. A deep learning-based model for predicting abnormal liver function in workers in the automotive manufacturing industry: A cross-sectional survey in Chongqing, China. *Int. J. Environ. Res. Public Health* **2022**, *19*, 14300. [CrossRef]
84. Shin, S.J.; Woo, J.; Rachuri, S. Predictive analytics model for power consumption in manufacturing. *Procedia Cirp* **2014**, *15*, 153–158. [CrossRef]
85. Lechevalier, D.; Narayanan, A.; Rachuri, S. Towards a domain-specific framework for predictive analytics in manufacturing. In Proceedings of the 2014 IEEE International Conference on Big Data (Big Data), Washington, DC, USA, 27–30 October 2014; pp. 987–995.
86. Leon-Medina, J.X.; Camacho, J.; Gutierrez-Osorio, C.; Salomón, J.E.; Rueda, B.; Vargas, W.; Sofrony, J.; Restrepo-Calle, F.; Pedraza, C.; Tibaduiza, D. Temperature prediction using multivariate time series deep learning in the lining of an electric arc furnace for ferronickel production. *Sensors* **2021**, *21*, 6894. [CrossRef] [PubMed]
87. Bouabdallaoui, Y.; Lafhaj, Z.; Yim, P.; Ducoulombier, L.; Bennadji, B. Predictive maintenance in building facilities: A machine learning-based approach. *Sensors* **2021**, *21*, 1044. [CrossRef]
88. Mawson, V.J. Coupling Simulation with Machine Learning for the Development of a Proactive HVAC System in the Manufacturing Sector. Ph.D. Thesis, University of Strathclyde, Glasgow, UK, 2020.
89. Massaro, A. Multi-level decision support system in production and safety management. *Knowledge* **2022**, *2*, 682–701. [CrossRef]
90. Relich, M. Predictive and prescriptive analytics in identifying opportunities for improving sustainable manufacturing. *Sustainability* **2023**, *15*, 7667. [CrossRef]
91. Belkin, M.; Hsu, D.; Ma, S.; Mandal, S. Reconciling modern machine-learning practice and the classical bias–variance trade-off. *Proc. Natl. Acad. Sci. USA* **2019**, *116*, 15849–15854. [CrossRef]
92. Aliferis, C.; Simon, G. Overfitting, underfitting and general model overconfidence and under-performance pitfalls and best practices in machine learning and AI. In *Artificial Intelligence and Machine Learning in Health Care and Medical Sciences: Best Practices and Pitfalls*; Springer: Cham, Switzerland, 2024; pp. 477–524.
93. Montesinos López, O.A.; Montesinos López, A.; Crossa, J. Overfitting, model tuning, and evaluation of prediction performance. In *Multivariate Statistical Machine Learning Methods for Genomic Prediction*; Springer: Cham, Switzerland, 2022; pp. 109–139.
94. Luo, X.; Li, X.; Goh, Y.M.; Song, X.; Liu, Q. Application of machine learning technology for occupational accident severity prediction in the case of construction collapse accidents. *Saf. Sci.* **2023**, *163*, 106138. [CrossRef]

95. Gudivada, V.; Apon, A.; Ding, J. Data quality considerations for big data and machine learning: Going beyond data cleaning and transformations. *Int. J. Adv. Softw.* **2017**, *10*, 1–20.
96. Martin, K. Predatory predictions and the ethics of predictive analytics. *J. Assoc. Inf. Sci. Technol.* **2023**, *74*, 531–545. [CrossRef]
97. Lu, J.; Liu, A.; Dong, F.; Gu, F.; Gama, J.; Zhang, G. Learning under concept drift: A review. *IEEE Trans. Knowl. Data Eng.* **2018**, *31*, 2346–2363. [CrossRef]
98. Žliobaitė, I.; Pechenizkiy, M.; Gama, J. An overview of concept drift applications. In *Big Data Analysis: New Algorithms for a New Society*; Springer: Cham, Switzerland, 2016; pp. 91–114.
99. Chen, Z.; Liu, B. *Lifelong Machine Learning*; Morgan & Claypool Publishers: San Rafael, CA, USA, 2018.
100. Liu, Z.; Xie, K.; Li, L.; Chen, Y. A paradigm of safety management in Industry 4.0. *Syst. Res. Behav. Sci.* **2020**, *37*, 632–645. [CrossRef]
101. Al-Bayati, A.J.; Rener, A.T.; Listello, M.P.; Mohamed, M. PPE non-compliance among construction workers: An assessment of contributing factors utilizing fuzzy theory. *J. Saf. Res.* **2023**, *85*, 242–253. [CrossRef] [PubMed]
102. Podgorski, D.; Majchrzycka, K.; Dąbrowska, A.; Gralewicz, G.; Okrasa, M. Towards a conceptual framework of OSH risk management in smart working environments based on smart PPE, ambient intelligence and the Internet of Things technologies. *Int. J. Occup. Saf. Ergon.* **2017**, *23*, 1–20. [CrossRef]
103. Yousif, I.; Samaha, J.; Ryu, J.; Harik, R. Safety 4.0: Harnessing computer vision for advanced industrial protection. *Manuf. Lett.* **2024**, *41*, 1342–1356. [CrossRef]
104. Yousif, I.; Burns, L.; El Kalach, F.; Harik, R. Leveraging computer vision towards high-efficiency autonomous industrial facilities. *J. Intell. Manuf.* **2024**, 1–26. [CrossRef]
105. Töreyin, B.U.; Dedeoğlu, Y.; Güdükbay, U.; Cetin, A.E. Computer vision based method for real-time fire and flame detection. *Pattern Recognit. Lett.* **2006**, *27*, 49–58. [CrossRef]
106. Geetha, S.; Abhishek, C.; Akshayanat, C. Machine vision based fire detection techniques: A survey. *Fire Technol.* **2021**, *57*, 591–623. [CrossRef]
107. Jeelani, I.; Asadi, K.; Ramshankar, H.; Han, K.; Albert, A. Real-time vision-based worker localization & hazard detection for construction. *Autom. Constr.* **2021**, *121*, 103448.
108. Fang, W.; Ding, L.; Zhong, B.; Love, P.E.; Luo, H. Automated detection of workers and heavy equipment on construction sites: A convolutional neural network approach. *Adv. Eng. Inform.* **2018**, *37*, 139–149. [CrossRef]
109. Jiang, D.; Li, G.; Tan, C.; Huang, L.; Sun, Y.; Kong, J. Semantic segmentation for multiscale target based on object recognition using the improved Faster-RCNN model. *Future Gener. Comput. Syst.* **2021**, *123*, 94–104. [CrossRef]
110. He, H.; Xu, H.; Zhang, Y.; Gao, K.; Li, H.; Ma, L.; Li, J. Mask R-CNN based automated identification and extraction of oil well sites. *Int. J. Appl. Earth Obs. Geoinf.* **2022**, *112*, 102875. [CrossRef]
111. Terven, J.; Córdova-Esparza, D.M.; Romero-González, J.A. A comprehensive review of yolo architectures in computer vision: From yolov1 to yolov8 and yolo-nas. *Mach. Learn. Knowl. Extr.* **2023**, *5*, 1680–1716. [CrossRef]
112. Liu, W.; Anguelov, D.; Erhan, D.; Szegedy, C.; Reed, S.; Fu, C.Y.; Berg, A.C. Ssd: Single shot multibox detector. In Proceedings of the Computer Vision–ECCV 2016: 14th European Conference, Amsterdam, The Netherlands, 11–14 October 2016; Proceedings, Part I 14; Springer: Cham, Switzerland, 2016; pp. 21–37.
113. Wang, M.; Wong, P.; Luo, H.; Kumar, S.; Delhi, V.; Cheng, J. Predicting safety hazards among construction workers and equipment using computer vision and deep learning techniques. In Proceedings of the 36th ISARC, Proceedings of the International Symposium on Automation and Robotics in Construction, Banff, Canada, 21–24 May 2019; Volume 36, pp. 399–406.
114. Lohani, D.; Crispim-Junior, C.; Barthélemy, Q.; Bertrand, S.; Robinault, L.; Tougne Rodet, L. Perimeter intrusion detection by video surveillance: A survey. *Sensors* **2022**, *22*, 3601. [CrossRef] [PubMed]
115. Tahir, G.A. Ethical Challenges in Computer Vision: Ensuring Privacy and Mitigating Bias in Publicly Available Datasets. *arXiv* **2024**, arXiv:2409.10533.
116. Xiang, A. Being "Seen" Versus "Mis-Seen": Tensions Between Privacy and Fairness in Computer Vision. *Harv. Jl Tech.* **2022**, *36*, 1.
117. Ragot, M.; Martin, N.; Cojean, S. Ai-generated vs. human artworks. a perception bias towards artificial intelligence? In Proceedings of the Extended Abstracts of the 2020 CHI Conference on Human Factors in Computing Systems, Honolulu, HI, USA, 25–30 April 2020; pp. 1–10.
118. Saini, D.K.; Ahir, D.; Ganatra, A. Techniques and challenges in building intelligent systems: Anomaly detection in camera surveillance. In *Proceedings of the First International Conference on Information and Communication Technology for Intelligent Systems: Volume 2*; Springer: Cham, Switzerland, 2016; pp. 11–21.
119. Zhu, L.; Menon, M.; Santillo, M.; Linkowski, G. Occlusion handling for industrial robots. In Proceedings of the 2020 IEEE/RSJ International Conference on Intelligent Robots and Systems (IROS), Las Vegas, NV, USA, 25–29 October 2020; pp. 10663–10668.

120. Youssef, S.M.; Hamza, M.A.; Fayed, A.F. Detection and tracking of multiple moving objects with occlusion in smart video surveillance systems. In Proceedings of the 2010 5th IEEE International Conference Intelligent Systems, London, UK, 7–9 July 2010; pp. 120–125.
121. Yang, T.; Chen, F.; Kimber, D.; Vaughan, J. Robust people detection and tracking in a multi-camera indoor visual surveillance system. In Proceedings of the 2007 IEEE International Conference on Multimedia and Expo, Beijing, China, 2–5 July 2007; pp. 675–678.
122. Kumar, C.; Singh, S. Security standards for real time video surveillance and moving object tracking challenges, limitations, and future: A case study. *Multimed. Tools Appl.* **2024**, *83*, 30113–30144. [CrossRef]
123. Moorthy, R.; Upadhya, V.; Holla, V.V.; Shetty, S.S.; Tantry, V.V. Challenges encountered in building a fast and efficient surveillance system: An overview. In Proceedings of the 2020 Fourth International Conference on I-SMAC (IoT in Social Mobile, Analytics and Cloud) (I-SMAC), Palladam, India, 7–9 October 2020; pp. 731–737.
124. Ricketts, J.; Barry, D.; Guo, W.; Pelham, J. A scoping literature review of natural language processing application to safety occurrence reports. *Safety* **2023**, *9*, 22. [CrossRef]
125. Zhou, Y.; She, J.; Huang, Y.; Li, L.; Zhang, L.; Zhang, J. A design for safety (DFS) semantic framework development based on natural language processing (NLP) for automated compliance checking using BIM: The case of China. *Buildings* **2022**, *12*, 780. [CrossRef]
126. Baker, H.; Smith, S.; Masterton, G.; Hewlett, B. Data-led learning: Using natural language processing (NLP) and machine learning to learn from construction site safety failures. In *Proceedings of the 36th Annual ARCOM Conference: Association of Researchers in Construction Management*; ARCOM: Leeds, UK, 2020; pp. 356–365.
127. Brundage, M.P.; Sexton, T.; Hodkiewicz, M.; Dima, A.; Lukens, S. Technical language processing: Unlocking maintenance knowledge. *Manuf. Lett.* **2021**, *27*, 42–46. [CrossRef]
128. Xu, H.; Liu, Y.; Shu, C.M.; Bai, M.; Motalifu, M.; He, Z.; Wu, S.; Zhou, P.; Li, B. Cause analysis of hot work accidents based on text mining and deep learning. *J. Loss Prev. Process Ind.* **2022**, *76*, 104747. [CrossRef]
129. Jahin, M.A.; Talapatra, S. A natural language processing-based classification and mode-based ranking of musculoskeletal disorder risk factors. *Decis. Anal. J.* **2024**, *11*, 100464. [CrossRef]
130. May, M.C.; Neidhöfer, J.; Körner, T.; Schäfer, L.; Lanza, G. Applying natural language processing in manufacturing. *Procedia CIRP* **2022**, *115*, 184–189. [CrossRef]
131. Ray, P.P. ChatGPT: A comprehensive review on background, applications, key challenges, bias, ethics, limitations and future scope. *Internet Things Cyber-Phys. Syst.* **2023**, *3*, 121–154. [CrossRef]
132. Gallegos, I.O.; Rossi, R.A.; Barrow, J.; Tanjim, M.M.; Kim, S.; Dernoncourt, F.; Yu, T.; Zhang, R.; Ahmed, N.K. Bias and fairness in large language models: A survey. *Comput. Linguist.* **2024**, *50*, 1097–1179. [CrossRef]
133. Bender, E.M.; Friedman, B. Data statements for natural language processing: Toward mitigating system bias and enabling better science. *Trans. Assoc. Comput. Linguist.* **2018**, *6*, 587–604. [CrossRef]
134. Ford's Cobots: Innovating Automotive Manufacturing Efficiency. Available online: https://profwurzer.com/fords-use-of-collaborative-robots (accessed on 10 April 2025).
135. Taylor, A. How Tesla and Ford Use Robotics to Revolutionize Manufacturing Efficiency. Available online: https://www.cleverence.com/articles/business-blogs/how-tesla-and-ford-use-robotics-to-revolutionize-manufacturing-efficiency/ (accessed on 10 April 2025).
136. Bi, Z.M.; Luo, C.; Miao, Z.; Zhang, B.; Zhang, W.J.; Wang, L. Safety assurance mechanisms of collaborative robotic systems in manufacturing. *Robot. Comput.-Integr. Manuf.* **2021**, *67*, 102022. [CrossRef]
137. Pauliková, A.; Gyurák Babel'ová, Z.; Ubárová, M. Analysis of the impact of human–cobot collaborative manufacturing implementation on the occupational health and safety and the quality requirements. *Int. J. Environ. Res. Public Health* **2021**, *18*, 1927. [CrossRef]
138. Ford Motor Enters Digital Era with Its Advanced Manufacturing Center. Available online: https://connectedmanufacturing.wbresearch.com/blog/ford-motor-company-advanced-manufacturing-center-strategy (accessed on 10 April 2025).
139. Shelby, L. Accelerating Tesla's Safety Culture. 2020. Available online: https://www.tesla.com/en_ca/blog/accelerating-teslas-safety-culture (accessed on 10 April 2025).
140. Industrial Computer Vision: Turn Your Regular CCTV into Smart AI Cameras—SeeWise.AI. 2023. Available online: https://www.seewise.ai/blogs/ai-to-ensure-safety-compliance-in-manufacturing.html (accessed on 10 April 2025).
141. How Intenseye Uses AI for Safety | Wire Journal International. Available online: https://www.intenseye.com/interviews/intenseye-qa-on-the-use-of-ai-for-safety (accessed on 10 April 2025).
142. Swire Coca-Cola: 27% decrease in LDR with Intenseye. Available online: https://www.intenseye.com/case-studies/swire-coca-cola (accessed on 10 April 2025).

143. Intenseye—Transforming Workplace Safety with AI. Available online: https://www.intenseye.com/ (accessed on 10 April 2025).
144. Koçtaş: Managing Missing PPE and Work at Height Risks with AI. Available online: https://www.intenseye.com/case-studies/koctas-managing-missing-ppe-and-work-at-height-risks-with-ai (accessed on 10 April 2025).
145. Burgess, M. AI Enhances Safety and Security in Asia—Linfox. Available online: https://www.linfox.com/ai-enhances-safety-and-security-in-asia/ (accessed on 10 April 2025).
146. Nikolakis, N.; Alexopoulos, K.; Xanthakis, E.; Chryssolouris, G. The digital twin implementation for linking the virtual representation of human-based production tasks to their physical counterpart in the factory-floor. *Int. J. Comput. Integr. Manuf.* **2019**, *32*, 1–12. [CrossRef]
147. Hovanec, M.; Korba, P.; Vencel, M.; Al-Rabeei, S. Simulating a digital factory and improving production efficiency by using virtual reality technology. *Appl. Sci.* **2023**, *13*, 5118. [CrossRef]
148. Use Case: NVIDIA Computing Platforms for Humanoid Robots. Available online: https://www.nvidia.com/en-us/use-cases/humanoid-robots/ (accessed on 10 April 2025).
149. Yang, X.; Deines, E.; Lauer, C.; Aurich, J.C. A human-centered virtual factory. In Proceedings of the MSIE 2011, Harbin, China, 8–9 January 2011; pp. 1138–1142.
150. Wang, H.; Lv, L.; Li, X.; Li, H.; Leng, J.; Zhang, Y.; Thomson, V.; Liu, G.; Wen, X.; Sun, C.; et al. A safety management approach for Industry 5.0′ s human-centered manufacturing based on digital twin. *J. Manuf. Syst.* **2023**, *66*, 1–12. [CrossRef]
151. What Are AI Agents? Available online: https://www.nvidia.com/en-us/glossary/ai-agents/ (accessed on 10 April 2025).
152. Bernardi, M.L.; Cimitile, M.; Pecori, R. Automatic Job Safety Report Generation using RAG-based LLMs. In Proceedings of the 2024 International Joint Conference on Neural Networks (IJCNN), Yokohama, Japan, 30 June–5 July 2024; pp. 1–8.
153. Lee, J.; Ahn, S.; Kim, D.; Kim, D. Performance comparison of retrieval-augmented generation and fine-tuned large language models for construction safety management knowledge retrieval. *Autom. Constr.* **2024**, *168*, 105846. [CrossRef]
154. NVIDIA Omniverse. Available online: https://www.nvidia.com/en-us/omniverse/ (accessed on 10 April 2025).
155. Home | Continental. Available online: https://www.continental.com/en/ (accessed on 10 April 2025).
156. Simulating Factories of the Future With Hyper-Scale, Hyper-Reality Digital Twins | GTC 24 2024 | NVIDIA On-Demand. Available online: https://www.nvidia.com/en-us/on-demand/session/gtc24-s62623/ (accessed on 10 April 2025).
157. Wistron. Available online: https://www.wistron.com/en (accessed on 10 April 2025).
158. McKenna, J. What Is a Virtual Factory, and How They're Making Industrial Digitalization a Reality. Available online: https://blogs.nvidia.com/blog/virtual-factories-industrial-digitalization/ (accessed on 10 April 2025).
159. Zonta, T.; Da Costa, C.A.; da Rosa Righi, R.; de Lima, M.J.; Da Trindade, E.S.; Li, G.P. Predictive maintenance in the Industry 4.0: A systematic literature review. *Comput. Ind. Eng.* **2020**, *150*, 106889. [CrossRef]
160. Hijry, H.; Naqvi, S.M.R.; Javed, K.; Albalawi, O.H.; Olawoyin, R.; Varnier, C.; Zerhouni, N. Real Time Worker Stress Prediction in a Smart Factory Assembly Line. *IEEE Access* **2024**, *12*, 116238–116249. [CrossRef]
161. Henkes, A.; Herrmann, L.; Wessels, H.; Kollmannsberger, S. Generative adversarial networks enable outlier detection and property monitoring for additive manufacturing of complex structures. *Eng. Appl. Artif. Intell.* **2024**, *136*, 108993. [CrossRef]
162. Mumbelli, J.D.; Guarneri, G.A.; Lopes, Y.K.; Casanova, D.; Teixeira, M. An application of Generative Adversarial Networks to improve automatic inspection in automotive manufacturing. *Appl. Soft Comput.* **2023**, *136*, 110105. [CrossRef]
163. Goodfellow, I.; Pouget-Abadie, J.; Mirza, M.; Xu, B.; Warde-Farley, D.; Ozair, S.; Courville, A.; Bengio, Y. Generative adversarial networks. *Commun. ACM* **2020**, *63*, 139–144. [CrossRef]
164. Zhang, L.; Wu, X.; Skibniewski, M.J.; Zhong, J. Dynamic risk assessment in construction projects using Bayesian networks. In Proceedings of the 31st ISARC, Proceedings of the International Symposium on Automation and Robotics in Construction, Sydney, Australia, 9–11 July 2014; Volume 31, p. 1.
165. Punyamurthula, S.; Badurdeen, F. Assessing production line risk using bayesian Belief networks and system dynamics. *Procedia Manuf.* **2018**, *26*, 76–86. [CrossRef]
166. Pereira, J.; Fragoso, M.; Todorov, M. Risk assessment using bayesian belief networks and analytic hierarchy process applicable to jet engine high pressure turbine assembly. *IFAC-PapersOnLine* **2016**, *49*, 133–138. [CrossRef]
167. Kang, K.; Ryu, H. Predicting types of occupational accidents at construction sites in Korea using random forest model. *Saf. Sci.* **2019**, *120*, 226–236. [CrossRef]
168. Hosseinzadeh, A.; Chen, F.F.; Shahin, M.; Bouzary, H. A predictive maintenance approach in manufacturing systems via AI-based early failure detection. *Manuf. Lett.* **2023**, *35*, 1179–1186. [CrossRef]
169. Lalwani, P.; Ramasamy, G. Human activity recognition using a multi-branched CNN-BiLSTM-BiGRU model. *Appl. Soft Comput.* **2024**, *154*, 111344. [CrossRef]
170. Shoorkand, H.D.; Nourelfath, M.; Hajji, A. A hybrid CNN-LSTM model for joint optimization of production and imperfect predictive maintenance planning. *Reliab. Eng. Syst. Saf.* **2024**, *241*, 109707. [CrossRef]

171. Borré, A.; Seman, L.O.; Camponogara, E.; Stefenon, S.F.; Mariani, V.C.; Coelho, L.d.S. Machine fault detection using a hybrid CNN-LSTM attention-based model. *Sensors* **2023**, *23*, 4512. [CrossRef] [PubMed]
172. Sii, H.S.; Ruxton, T.; Wang, J. A fuzzy-logic-based approach to qualitative safety modelling for marine systems. *Reliab. Eng. Syst. Saf.* **2001**, *73*, 19–34. [CrossRef]
173. Sun, W.; Jin, Y.; Dai, K.; Guo, Z.; Ma, F. Flexible manipulator trajectory tracking based on an improved adaptive particle swarm optimization algorithm with fuzzy PD control. *Mech. Sci.* **2025**, *16*, 125–141. [CrossRef]
174. Achouch, M.; Dimitrova, M.; Ziane, K.; Sattarpanah Karganroudi, S.; Dhouib, R.; Ibrahim, H.; Adda, M. On predictive maintenance in industry 4.0: Overview, models, and challenges. *Appl. Sci.* **2022**, *12*, 8081. [CrossRef]
175. Cadence PCB Solutions. Types of Predictive Maintenance. *Cadence PCB Design & Analysis Blog*, 2023. Available online: https://resources.pcb.cadence.com/blog/2023-types-of-predictive-maintenance (accessed on 15 January 2025).
176. Poór, P.; Basl, J.; Zenisek, D. Predictive Maintenance 4.0 as next evolution step in industrial maintenance development. In Proceedings of the 2019 International Research Conference on Smart Computing and Systems Engineering (SCSE), Colombo, Sri Lanka, 28 March 2019; pp. 245–253.
177. Ayvaz, S.; Alpay, K. Predictive maintenance system for production lines in manufacturing: A machine learning approach using IoT data in real-time. *Expert Syst. Appl.* **2021**, *173*, 114598. [CrossRef]
178. Nacchia, M.; Fruggiero, F.; Lambiase, A.; Bruton, K. A systematic mapping of the advancing use of machine learning techniques for predictive maintenance in the manufacturing sector. *Appl. Sci.* **2021**, *11*, 2546. [CrossRef]
179. Behera, P.K.; Sahoo, B.S. Leverage of multiple predictive maintenance technologies in root cause failure analysis of critical machineries. *Procedia Eng.* **2016**, *144*, 351–359. [CrossRef]
180. Aljohani, A. Predictive analytics and machine learning for real-time supply chain risk mitigation and agility. *Sustainability* **2023**, *15*, 15088. [CrossRef]
181. Nexteng Enviro Pvt Ltd. AI/ML Based Predictive Maintenance. Available online: https://nextengineering.in/solutions/ai-ml-based-predictive-maintenance/ (accessed on 2 January 2025).
182. Baumung, W.; Baumung, V. Application of Machine Learning and Vision for real-time condition monitoring and acceleration of product development cycles. *Procedia Manuf.* **2020**, *52*, 61–66. [CrossRef]
183. Bourou, S.; Maniatis, A.; Kontopoulos, D.; Karkazis, P.A. Smart detection system of safety hazards in industry 5.0. *Telecom* **2023**, *5*, 1–20. [CrossRef]
184. Canadian Centre for Occupational Health and Safety. Hazard and Risk—Risk Assessment. Available online: https://www.ccohs.ca/oshanswers/hsprograms/hazard/risk_assessment.html (accessed on 29 December 2024).
185. Occupational Safety and Health Administration. Hazard Identification and Assessment. Available online: https://www.osha.gov/safety-management/hazard-identification (accessed on 12 January 2025).
186. Albery, S.; Borys, D.; Tepe, S. Advantages for risk assessment: Evaluating learnings from question sets inspired by the FRAM and the risk matrix in a manufacturing environment. *Saf. Sci.* **2016**, *89*, 180–189. [CrossRef]
187. Oregon State University, Environmental Health and Safety. Hazard Identification and Control. Available online: https://ehs.oregonstate.edu/sites/ehs.oregonstate.edu/files/pdf/si/hazard_identification_and_control_si.pdf (accessed on 29 December 2024).
188. Occupational Safety and Health Administration. Right to refuse dangerous work. Available online: https://www.osha.gov/workers/right-to-refuse (accessed on 21 January 2025).
189. SafeStart. How to Navigate Ambiguity—A Threat to Stop-Work Authority 2024. Available online: https://safestart.com/news/ambiguity-threat-to-stop-work-authority/ (accessed on 21 January 2025).
190. Business Development Bank of Canada. What Is Operational Efficiency? Available online: https://www.bdc.ca/en/articles-tools/operations/operational-efficiency/3-critical-factors-operational-efficiency (accessed on 23 January 2025).
191. Supply Chain Mechanic. Basics of smoothing the manufacturing plan. *SupplyChainMechanic.com*, 2010. Available online: https://supplychain-mechanic.com/?p=314 (accessed on 23 January 2025).
192. U.S. Chemical Safety and Hazard Investigation Board. Hayes Lemmerz Dust Explosions and Fire. Available online: https://www.csb.gov/hayes-lemmerz-dust-explosions-and-fire/ (accessed on 23 January 2025).
193. U.S. Chemical Safety and Hazard Investigation Board. Chevron Richmond Refinery Fire. Available online: https://www.csb.gov/chevron-richmond-refinery-fire/ (accessed on 23 January 2025).
194. Rai, R.; Tiwari, M.K.; Ivanov, D.; Dolgui, A. Machine learning in manufacturing and industry 4.0 applications. *Int. J. Prod. Res.* **2021**, *59*, 4773–4778. [CrossRef]
195. Zehnder Clean Air Solutions. How Dust is Formed in Manufacturing. Available online: https://www.zehnder-cleanairsolutions.com/en/blog/how-dust-is-formed-in-manufacturing (accessed on 23 January 2025).

196. Canadian Centre for Occupational Health and Safety. Health and Safety Report—March 2016. Available online: https://www.ccohs.ca/newsletters/hsreport/issues/2016/03/ezine.html (accessed on 23 January 2025).
197. Canadian Centre for Occupational Health and Safety. Combustible Dust. Available online: https://www.ccohs.ca/oshanswers/chemicals/combustible_dust.html (accessed on 23 January 2025).
198. Dustcontrol. How Dust Harms Manufacturing and Productivity. Available online: https://dustcontrol.com/how-dust-harms-manufacturing-and-productivity/ (accessed on 23 January 2025).
199. Canadian Centre for Occupational Health and Safety. Workplace Housekeeping—Basic Guide. Available online: https://www.ccohs.ca/oshanswers/hsprograms/housekeeping/house.html (accessed on 20 January 2025).
200. Yaseen, A. Reducing industrial risk with AI and automation. *Int. J. Intell. Autom. Comput.* **2021**, *4*, 60–80.
201. Chinniah, Y.; Bourbonniere, R. Automation safety. *Prof. Saf.* **2006**, *51*, 26–33.
202. Negnevitsky, M.; Tomin, N.V.; Rehtanz, C. Preventing large-scale emergencies in modern power systems: AI approach. *J. Adv. Comput. Intell. Intell. Inform.* **2014**, *18*, 714–727. [CrossRef]
203. Hickok, M.; Maslej, N. A policy primer and roadmap on AI worker surveillance and productivity scoring tools. *AI Ethics* **2023**, *3*, 673–687. [CrossRef]
204. Rebbitt, D. When undetected deviation becomes an incident. *Can. Occup. Saf.* **2014**. Available online: https://www.thesafetymag.com/ca/news/opinion/when-undetected-deviation-becomes-an-incident/187069 (accessed on 24 January 2025).
205. Qi, R.; Zhang, J.; Spencer, K. A review on data-driven condition monitoring of industrial equipment. *Algorithms* **2022**, *16*, 9. [CrossRef]
206. Northwestern University McCormick School of Engineering. Improving the Occupational Health of Manufacturing Workers. Available online: https://www.mccormick.northwestern.edu/news/articles/2024/10/improving-the-occupational-health-of-manufacturing-workers/ (accessed on 24 January 2025).
207. Psico Smart. The Role of Artificial Intelligence in Enhancing Labor Fatigue Management Systems. Available online: https://psico-smart.com/en/blogs/blog-the-role-of-artificial-intelligence-in-enhancing-labor-fatigue-management-systems-170861 (accessed on 11 January 2025).
208. Hooda, R.; Joshi, V.; Shah, M. A comprehensive review of approaches to detect fatigue using machine learning techniques. *Chronic Dis. Transl. Med.* 2021, *in press*. [CrossRef] [PubMed]
209. Fiegler-Rudol, J.; Lau, K.; Mroczek, A.; Kasperczyk, J. Exploring Human–AI Dynamics in Enhancing Workplace Health and Safety: A Narrative Review. *Int. J. Environ. Res. Public Health* **2025**, *22*, 199. [CrossRef]
210. Chokor, A.; Naganathan, H.; Chong, W.K.; El Asmar, M. Analyzing Arizona OSHA injury reports using unsupervised machine learning. *Procedia Eng.* **2016**, *145*, 1588–1593. [CrossRef]
211. Khusainov, R.; Azzi, D.; Achumba, I.E.; Bersch, S.D. Real-time human ambulation, activity, and physiological monitoring: Taxonomy of issues, techniques, applications, challenges and limitations. *Sensors* **2013**, *13*, 12852–12902. [CrossRef]
212. International Labour Organization. Exposure to Hazardous Chemicals at Work and Resulting Health Impacts: A Global Review. 2021. Available online: https://www.ilo.org/sites/default/files/wcmsp5/groups/public/@ed_dialogue/@lab_admin/documents/publication/wcms_795460.pdf (accessed on 24 January 2025).
213. Khan, A.; Gupta, S.; Gupta, S.K. Multi-hazard disaster studies: Monitoring, detection, recovery, and management, based on emerging technologies and optimal techniques. *Int. J. Disaster Risk Reduct.* **2020**, *47*, 101642. [CrossRef]
214. Sivapriya, J.; Ramani, D.R.; Srivastava, R.P.; Kumar, K.; Nair, R.V. AI-Powered Anomaly and Threat Detection for Surveillance Footage Analysis. In Proceedings of the 2024 8th International Conference on Inventive Systems and Control (ICISC), Coimbatore, India, 29–30 July 2024; pp. 240–247.
215. Alghushairy, O.; Alsini, R.; Soule, T.; Ma, X. A review of local outlier factor algorithms for outlier detection in big data streams. *Big Data Cogn. Comput.* **2020**, *5*, 1. [CrossRef]
216. Liu, X.; Zhu, S.; Yang, F.; Liang, S. Research on unsupervised anomaly data detection method based on improved automatic encoder and Gaussian mixture model. *J. Cloud Comput.* **2022**, *11*, 58. [CrossRef]
217. Nassif, A.B.; Talib, M.A.; Nasir, Q.; Dakalbab, F.M. Machine learning for anomaly detection: A systematic review. *IEEE Access* **2021**, *9*, 78658–78700. [CrossRef]
218. Primartha, R.; Tama, B.A. Anomaly detection using random forest: A performance revisited. In Proceedings of the 2017 International conference on data and software engineering (ICoDSE), Palembang, Indonesia, 1–2 November 2017; pp. 1–6.
219. Ogbuke, N.J.; Yusuf, Y.Y.; Dharma, K.; Mercangoz, B.A. Big data supply chain analytics: Ethical, privacy and security challenges posed to business, industries and society. *Prod. Plan. Control* **2022**, *33*, 123–137. [CrossRef]
220. Phan, T.C.; Tran, H.C. Consideration of data security and privacy using machine learning techniques. *Int. J. Data Inform. Intell. Comput.* **2023**, *2*, 20–32.

221. Devineni, S.K.; Kathiriya, S.; Shende, A. Machine learning-powered anomaly detection: Enhancing data security and integrity. *J. Artif. Intell. Cloud Comput.* **2023**, *184*, 2–9. [CrossRef]
222. Elmrabit, N.; Zhou, F.; Li, F.; Zhou, H. Evaluation of machine learning algorithms for anomaly detection. In Proceedings of the 2020 International Conference on Cyber Security and Protection of Digital Services (Cyber Security), Dublin, Ireland, 15–19 June 2020; pp. 1–8.
223. Li, N.; Sun, M.; Bi, Z.; Su, Z.; Wang, C. A new methodology to support group decision-making for IoT-based emergency response systems. *Inf. Syst. Front.* **2014**, *16*, 953–977. [CrossRef]
224. Rodriguez, M.; Mohammed, M.; Mohammed, R.; Pasam, P.; Karanam, R.; Vennapusa, S.; Boinapalli, N. Oracle EBS and Digital Transformation: Aligning Technology with Business Goals. *Technol. Manag. Rev.* **2019**, *4*, 49–63.
225. Bazargan-Lari, M.R. Real-time response to contamination emergencies of urban water networks. *Iran. J. Sci. Technol. Trans. Civ. Eng.* **2018**, *42*, 73–83. [CrossRef]
226. Zhao, L. Business intelligence implementation in the framework of enhanced learning application. In Proceedings of the IOP Conference Series: Materials Science and Engineering, Krasnoyarsk, Russia, 4–6 April 2019; Volume 563, p. 032015.
227. Caiazzo, B.; Murino, T.; Petrillo, A.; Piccirillo, G.; Santini, S. An IoT-based and cloud-assisted AI-driven monitoring platform for smart manufacturing: Design architecture and experimental validation. *J. Manuf. Technol. Manag.* **2023**, *34*, 507–534. [CrossRef]
228. Nahavandi, D.; Alizadehsani, R.; Khosravi, A.; Acharya, U.R. Application of artificial intelligence in wearable devices: Opportunities and challenges. *Comput. Methods Programs Biomed.* **2022**, *213*, 106541. [CrossRef]
229. Kiral-Kornek, I.; Roy, S.; Nurse, E.; Mashford, B.; Karoly, P.; Carroll, T.; Payne, D.; Saha, S.; Baldassano, S.; O'Brien, T.; et al. Epileptic seizure prediction using big data and deep learning: Toward a mobile system. *EBioMedicine* **2018**, *27*, 103–111. [CrossRef]
230. Roy, S.; Kiral-Kornek, I.; Harrer, S. ChronoNet: A deep recurrent neural network for abnormal EEG identification. In Proceedings of the Artificial Intelligence in Medicine: 17th Conference on Artificial Intelligence in Medicine, AIME 2019, Poznan, Poland, 26–29 June 2019; Proceedings 17; Springer: Cham, Switzerland, 2019; pp. 47–56.
231. Ulate-Campos, A.; Coughlin, F.; Gaínza-Lein, M.; Fernández, I.S.; Pearl, P.; Loddenkemper, T. Automated seizure detection systems and their effectiveness for each type of seizure. *Seizure* **2016**, *40*, 88–101. [CrossRef]
232. Harrer, S.; Shah, P.; Antony, B.; Hu, J. Artificial intelligence for clinical trial design. *Trends Pharmacol. Sci.* **2019**, *40*, 577–591. [CrossRef]
233. Mirmomeni, M.; Fazio, T.; von Cavallar, S.; Harrer, S. From wearables to THINKables: Artificial intelligence-enabled sensors for health monitoring. In *Wearable Sensors*; Elsevier: London, UK, 2021; pp. 339–356.
234. Reynolds, M. A Medical-Grade Smartwatch Is Helping People Live with Epilepsy. *WIRED* 2018. Available online: https://www.wired.com/story/empatica-embrace-epilepsy-wearable-medical-device/ (accessed on 1 February 2025).
235. Sabry, F.; Eltaras, T.; Labda, W.; Alzoubi, K.; Malluhi, Q. Machine learning for healthcare wearable devices: The big picture. *J. Healthc. Eng.* **2022**, *2022*, 4653923. [CrossRef]
236. Mamdiwar, S.D.; R, A.; Shakruwala, Z.; Chadha, U.; Srinivasan, K.; Chang, C.-Y. Recent advances on IoT-assisted wearable sensor systems for healthcare monitoring. *Biosensors* **2021**, *11*, 372. [CrossRef] [PubMed]
237. Awotunde, J.B.; Ajagbe, S.A.; Florez, H. Internet of things with wearable devices and artificial intelligence for elderly uninterrupted healthcare monitoring systems. In Proceedings of the International Conference on Applied Informatics, Arequipa, Peru, 27–29 October 2022; pp. 278–291.
238. Ratushnyi, V. Determinants of Safe Evacuation of Large Production Factories. Available online: https://wsb.edu.pl/files//pages/634/security_forum_2_2023_doi_10_26410_sf_2_23_8.pdf (accessed on 18 April 2025).
239. Ruggiero, L. A review of augmented reality applications for building evacuation. *arXiv* **2018**, arXiv:1804.04186.
240. Doolani, S.; Wessels, C.; Kanal, V.; Sevastopoulos, C.; Jaiswal, A.; Nambiappan, H.; Makedon, F. A review of extended reality (xr) technologies for manufacturing training. *Technologies* **2020**, *8*, 77. [CrossRef]
241. Forkan, A.R.M.; Montori, F.; Georgakopoulos, D.; Jayaraman, P.P.; Yavari, A.; Morshed, A. An industrial IoT solution for evaluating workers' performance via activity recognition. In Proceedings of the 2019 IEEE 39th International Conference on Distributed Computing Systems (ICDCS), Dallas, TX, USA, 7–10 July 2019; pp. 1393–1403.
242. Manzi, A.; Hirschhorn, L.R.; Sherr, K.; Chirwa, C.; Baynes, C.; Awoonor-Williams, J.K. Mentorship and coaching to support strengthening healthcare systems: Lessons learned across the five Population Health Implementation and Training partnership projects in sub-Saharan Africa. *BMC Health Serv. Res.* **2017**, *17*, 5–16. [CrossRef] [PubMed]
243. Venusamy, K.; Rajagopal, N.K.; Yousoof, M. A study of human resources development through chatbots using artificial intelligence. In Proceedings of the 2020 3rd International Conference on Intelligent Sustainable Systems (ICISS), Thoothukudi, India, 3–5 December 2020; pp. 94–99.
244. Wolter, S.C.; Ryan, P. Apprenticeship. In *Handbook of the Economics of Education*; Elsevier: Amsterdam, The Netherlands, 2011; Volume 3, pp. 521–576.

245. Gorowara, N.; Prakash, A.; Correa, F.S.; Malik, V.; Mittal, R. AI Personalizing Training and Reskilling Employees for the Digital Age. In Proceedings of the 2024 International Conference on Emerging Smart Computing and Informatics (ESCI), Pune, India, 5–7 March 2024; pp. 1–6.
246. Sun, Y.; Ji, F. An Embodied Intelligence System for Coal Mine Safety Assessment Based on Multi-Level Large Language Models. *Sensors* **2025**, *25*, 488. [CrossRef] [PubMed]
247. Collier, Z.A.; Gruss, R.J.; Abrahams, A.S. How good are large language models at product risk assessment? *Risk Anal.* **2024**. [CrossRef]
248. Strasser, A. Pitfalls (and advantages) of sophisticated large language models. In *Ethics in Online AI-Based Systems*; Elsevier: London, UK, 2024; pp. 195–210.
249. Moreira, G.; Pleffken, D.; Santos, W.; Cerqueira, C.; Gotelip, M. Using LLMs to Automate Means of Compliance Assignment in Aerospace Defense Systems. *TechRxiv* **2024**. [CrossRef]
250. Li, Y.; Zhao, H.; Jiang, H.; Pan, Y.; Liu, Z.; Wu, Z.; Shu, P.; Tian, J.; Yang, T.; Xu, S.; et al. Large language models for manufacturing. *arXiv* **2024**, arXiv:2410.21418.
251. Garcia, C.I.; DiBattista, M.A.; Letelier, T.A.; Halloran, H.D.; Camelio, J.A. Framework for LLM applications in manufacturing. *Manuf. Lett.* **2024**, *41*, 253–263. [CrossRef]
252. Jędrych, E.; Rzepka, A. *Organizational Development, Innovation, and Economy 5.0: Challenges in the Digital Era*; Taylor & Francis: Abingdon, UK, 2024.
253. Nguyen, T.; Nguyen, D.H.; Nguyen, Q.T.; Tran, K.D.; Tran, K.P. Human-Centered Edge AI and Wearable Technology for Workplace Health and Safety in Industry 5.0. In *Artificial Intelligence for Safety and Reliability Engineering: Methods, Applications, and Challenges*; Springer: Cham, Switzerland, 2024; pp. 171–183.
254. Shah, I.A.; Mishra, S. Artificial intelligence in advancing occupational health and safety: An encapsulation of developments. *J. Occup. Health* **2024**, *66*, uiad017. [CrossRef]
255. Molek, N. The future of labour: How ai, robotics, and automation are redefining EU labour markets. In *Analysis of the State of the Art on the Future of Human Workforce*; Fakultet za organizacijske študije v Novem mestu: Novo Mesto, Slovenia, 2024; pp. 21–48. Available online: https://transform-euproject.eu/wp-content/uploads/2024/12/Analysis-of-the-State-o-the-Art-on-the-future-of-human-workforce.pdf#page=22 (accessed on 18 April 2025).
256. Dehghani, F.; Dibaji, M.; Anzum, F.; Dey, L.; Basdemir, A.; Bayat, S.; Boucher, J.C.; Drew, S.; Eaton, S.E.; Frayne, R.; et al. Trustworthy and Responsible AI for Human-Centric Autonomous Decision-Making Systems. *arXiv* **2024**, arXiv:2408.15550.
257. International Organization for Standardization. *Ergonomics of Human-System Interaction: Part 210: Human-Centred Design for Interactive Systems*; ISO: Geneva, Switzerland, 2010.
258. IEEE. Ethical Considerations of Autonomous Intelligent Systems (AIS). 2023. Available online: https://standards.ieee.org/wp-content/uploads/2023/08/ieee-ethical-considerations-ais-policy.pdf (accessed on 8 February 2025).
259. Trivedi, P.; Alqahtani, F.M. The advancement of artificial intelligence (AI) in occupational health and safety (OHS) across high-risk industries. *J. Infrastruct. Policy Dev.* **2024**, *8*, 6889. [CrossRef]
260. Sundaram, S.; Zeid, A. Artificial intelligence-based smart quality inspection for manufacturing. *Micromachines* **2023**, *14*, 570. [CrossRef] [PubMed]
261. Rydzi, S.; Zahradnikova, B.; Sutova, Z.; Ravas, M.; Hornacek, D.; Tanuska, P. A predictive quality inspection framework for the manufacturing process in the context of industry 4.0. *Sensors* **2024**, *24*, 5644. [CrossRef]
262. Silva, A.C.; Machado, J.; Sampaio, P. Predictive quality model for customer defects. *Tqm J.* **2024**, *36*, 155–174. [CrossRef]
263. Chadha, U.; Abraham, A.; Anilkumar, K.; Kuriyakkattil, V.; Singh, H.; Bane, S.; Chadha, A.; Armstrong, S.; Patterson, A. Synergizing Lean Six Sigma Framework Using Artificial Intelligence, Internet of Things, and Blockchain for Sustainable Manufacturing Excellence. *TechRxiv* **2024**. [CrossRef]
264. Letourneau-Guillon, L.; Camirand, D.; Guilbert, F.; Forghani, R. Artificial intelligence applications for workflow, process optimization and predictive analytics. *Neuroimaging Clin. N. Am.* **2020**, *30*, e1–e15. [CrossRef]
265. Afroogh, S.; Akbari, A.; Malone, E.; Kargar, M.; Alambeigi, H. Trust in AI: Progress, challenges, and future directions. *Humanit. Soc. Sci. Commun.* **2024**, *11*, 1–30. [CrossRef]
266. Bedué, P.; Fritzsche, A. Can we trust AI? An empirical investigation of trust requirements and guide to successful AI adoption. *J. Enterp. Inf. Manag.* **2022**, *35*, 530–549. [CrossRef]
267. Mehri, V.A.; Ilie, D.; Tutschku, K. Privacy and DRM requirements for collaborative development of AI applications. In Proceedings of the 13th International Conference on Availability, Reliability and Security, Hamburg, Germany, 27–30 August 2018; pp. 1–8.
268. Hintze, M. Science and privacy: Data protection laws and their impact on research. *Wash. J. Law Tech. Arts* **2018**, *14*, 103.

269. Wilson, R. Cambridge analytica, Facebook, and Influence Operations: A case study and anticipatory ethical analysis. In *Proceedings of the European Conference on Cyber Warfare and Security*; Academic Conferences International: London, UK, 2019; pp. 587–595.
270. John Lopez. Samsung Employees Use ChatGPT at Work, Unknowingly Leak Critical Source Codes. 2023. Available online: https://www.techtimes.com/articles/289996/20230404/samsung-employees-used-chatgpt-work-unknowingly-leaked-critical-source-codes.htm (accessed on 9 April 2025).
271. Tschopp, M. Is Artificial Intelligence Worth the Risk? 2019. Available online: https://www.scip.ch/en/?labs.20190718 (accessed on 9 April 2025).
272. Daugherty, P.; Wilson, H. *Human+ Machine: Reimagining Work in the Age of AI*; Harv. Bus. Rev. Press: Boston, MA, USA, 2018.
273. Brown, A. Biased algorithms learn from biased data: 3 kinds of biases found in AI datasets. *Forbes* **2020**, *7*. Available online: https://www.forbes.com/sites/cognitiveworld/2020/02/07/biased-algorithms/ (accessed on 18 April 2025).
274. Freedman, R. ACLU Asks FTC to Probe Aon AI Employment Assessment Tools for Bias. 2024. Available online: https://www.legaldive.com/news/aclu-ftc-complaint-aon-ai-employment-assessment-tools-biased/717662/ (accessed on 9 April 2025).
275. Madiega, T. *Artificial Intelligence Act*; European Parliament; European Parliamentary Research Service: Bruxelles, Brussel, 2021.
276. Srivastava, B.; Rossi, F. Towards composable bias rating of AI services. In Proceedings of the 2018 AAAI/ACM Conference on AI, Ethics, and Society, New Orleans, LA, USA, 2–3 February 2018; pp. 284–289.
277. Agarwal, A.; Lohia, P.; Nagar, S.; Dey, K.; Saha, D. Automated test generation to detect individual discrimination in AI models. *arXiv* **2018**, arXiv:1809.03260.
278. Celis, L.E.; Deshpande, A.; Kathuria, T.; Vishnoi, N.K. How to be fair and diverse? *arXiv* **2016**, arXiv:1610.07183.
279. Mueller, S.T.; Hoffman, R.R.; Clancey, W.; Emrey, A.; Klein, G. Explanation in human-AI systems: A literature meta-review, synopsis of key ideas and publications, and bibliography for explainable AI. *arXiv* **2019**, arXiv:1902.01876.
280. Wang, D.; Yang, Q.; Abdul, A.; Lim, B.Y. Designing theory-driven user-centric explainable AI. In Proceedings of the 2019 CHI Conference on Human Factors in Computing Systems, Glasgow, UK, 4–9 May 2019; pp. 1–15.
281. Disterer, G. ISO/IEC 27000, 27001 and 27002 for information security management. *J. Inf. Secur.* **2013**, *4*. [CrossRef]
282. Regulation, P. General data protection regulation. *Intouch* **2018**, *25*, 1–5.
283. Kavasidis, I.; Lallas, E.; Gerogiannis, V.C.; Charitou, T.; Karageorgos, A. Predictive maintenance in pharmaceutical manufacturing lines using deep transformers. *Procedia Comput. Sci.* **2023**, *220*, 576–583. [CrossRef]
284. Maguluri, L.P.; Suganthi, D.; Dhote, G.M.; Kapila, D.; Jadhav, M.M.; Neelima, S. AI-enhanced predictive maintenance in hybrid roll-to-roll manufacturing integrating multi-sensor data and self-supervised learning. *Int. J. Adv. Manuf. Technol.* 2024, *in press*. [CrossRef]
285. Malawade, A.V.; Costa, N.D.; Muthirayan, D.; Khargonekar, P.P.; Al Faruque, M.A. Neuroscience-inspired algorithms for the predictive maintenance of manufacturing systems. *IEEE Trans. Ind. Inform.* **2021**, *17*, 7980–7990. [CrossRef]
286. Rodriguez-Cobo, L.; Garcia-Allende, P.B.; Cobo, A.; Lopez-Higuera, J.M.; Conde, O.M. Raw material classification by means of hyperspectral imaging and hierarchical temporal memories. *IEEE Sens. J.* **2012**, *12*, 2767–2775. [CrossRef]
287. Hung, Y.H. Developing an improved ensemble learning approach for predictive maintenance in the textile manufacturing process. *Sensors* **2022**, *22*, 9065. [CrossRef] [PubMed]
288. Oyekunle, D.O.T.; Boohene, D.; Preston, D. Ethical Considerations in AI-Powered Work Environments: A Literature Review and Theoretical Framework for Ensuring Human Dignity and Fairness. *Int. J. Sci. Res. Manag. (IJSRM)* **2024**, *12*, 6166–6178. [CrossRef]
289. Cheong, B.C. Transparency and accountability in AI systems: Safeguarding wellbeing in the age of algorithmic decision-making. *Front. Hum. Dyn.* **2024**, *6*, 1421273. [CrossRef]
290. Naidu, G.; Narkhede, U.; Shahare, P.; Wankhede, P.; Deshkar, R.; Gudadhe, A. Ethical Considerations in AI-Driven Supply Chain Management: A Review of Emerging Trends and Challenges. In Proceedings of the 2024 2nd DMIHER International Conference on Artificial Intelligence in Healthcare, Education and Industry (IDICAIEI), Wardha, India, 29–30 November 2024; pp. 1–7.
291. Dey, P.K.; Chowdhury, S.; Abadie, A.; Vann Yaroson, E.; Sarkar, S. Artificial intelligence-driven supply chain resilience in Vietnamese manufacturing small-and medium-sized enterprises. *Int. J. Prod. Res.* **2024**, *62*, 5417–5456. [CrossRef]
292. Zeyu Tang, C.Z.; Zeng, Y. Research on data security in industry 4.0 manufacturing industry against the background of privacy protection challenges. *Int. J. Comput. Integr. Manuf.* 2024, *in press*. [CrossRef]
293. Trakadas, P.; Simoens, P.; Gkonis, P.; Sarakis, L.; Angelopoulos, A.; Ramallo-González, A.P.; Skarmeta, A.; Trochoutsos, C.; Calvo, D.; Pariente, T.; et al. An Artificial Intelligence-Based Collaboration Approach in Industrial IoT Manufacturing: Key Concepts, Architectural Extensions and Potential Applications. *Sensors* **2020**, *20*, 5480. [CrossRef]
294. Todolí-Signes, A. Algorithms, artificial intelligence and automated decisions concerning workers and the risks of discrimination: The necessary collective governance of data protection. *Transf. Eur. Rev. Labour Res.* **2019**, *25*, 465–481. [CrossRef]

295. Bisoyi, A. Ownership, liability, patentability, and creativity issues in artificial intelligence. *Inf. Secur. J. A Glob. Perspect.* **2022**, *31*, 377–386. [CrossRef]
296. Jiang, B.; Li, J.; Yue, G.; Song, H. Differential privacy for industrial internet of things: Opportunities, applications, and challenges. *IEEE Internet Things J.* **2021**, *8*, 10430–10451. [CrossRef]
297. Gärtner, S.; Oberle, M. Local Differential Privacy In Smart Manufacturing: Application Scenario, Mechanisms and Tools. In *Proceedings of the Conference on Production Systems and Logistics (CPSL 2022)*; Institutionelles Repositorium der Leibniz Universität Hannover: Hannover, Germany, 2022. [CrossRef]
298. Praça, J.A.P.R. Responsible AI-How Data-Based Decision Making Can Increase Bias in Smart Manufacturing. Master's Thesis, Universidade do Porto, Porto, Portugal, 2022.
299. Nwebonyi, N.; McKay, F. Exploring bias risks in artificial intelligence and targeted medicines manufacturing. *Bmc Med. Ethics* **2024**, *25*, 113. [CrossRef]
300. ŞAHiN, E.; Arslan, N.N.; Özdemir, D. Unlocking the black box: An in-depth review on interpretability, explainability, and reliability in deep learning. *Neural Comput. Appl.* **2024**, 1–107. [CrossRef]
301. Goldman, C.V.; Baltaxe, M.; Chakraborty, D.; Arinez, J.; Diaz, C.E. Interpreting learning models in manufacturing processes: Towards explainable AI methods to improve trust in classifier predictions. *J. Ind. Inf. Integr.* **2023**, *33*, 100439. [CrossRef]
302. Hoge, A.; Ehmann, A.T.; Rieger, M.A.; Siegel, A. Caring for workers' health: Do German employers follow a comprehensive approach similar to the total worker health concept? Results of a survey in an economically powerful region in Germany. *Int. J. Environ. Res. Public Health* **2019**, *16*, 726. [CrossRef]
303. CCOHS: Standards—ccohs.ca. Available online: https://www.ccohs.ca/topics/legislation/standards (accessed on 12 January 2025).
304. Canada Labour Code—laws-lois.justice.gc.ca. Available online: https://laws-lois.justice.gc.ca/eng/ACTS/L-2/index.html (accessed on 12 January 2025).
305. Costly goodbyes: Increased Termination Notice Entitlements Under the Canada Labour Code as of February 2024—blg.com. Available online: https://www.blg.com/en/insights/2023/09/canada-labour-code-changes-2024 (accessed on 12 January 2025).
306. Employment, Canada, S.D. What We Heard: Modernizing Federal Labour Standards—Canada.ca—canada.ca. Available online: https://www.canada.ca/en/employment-social-development/corporate/portfolio/labour/programs/labour-standards/reports/what-we-heard-modernizing-federal.html#s2 (accessed on 12 January 2025).
307. CCOHS—National OHS Centre—whmis.org. Available online: https://whmis.org/jurisdictions/ccohs.html (accessed on 12 January 2025).
308. Occupational Safety and Health Administration. About OSHA. Available online: https://www.osha.gov/aboutosha (accessed on 15 January 2025).
309. Occupational Safety and Health Administration. Historical Notes. Available online: https://www.osha.gov/laws-regs/oshact/historicalnotes (accessed on 15 January 2025).
310. Patel, R.H.; Davis, D.D. Occupational Safety and Health Administration Rules That Affect Healthcare. Available online: https://www.ncbi.nlm.nih.gov/books/NBK559327/ (accessed on 15 January 2025).
311. Rekus, J. OSHA: The Good, the Bad and the Ugly. Available online: https://www.ehstoday.com/archive/article/21914339/osha-the-good-the-bad-and-the-ugly (accessed on 15 January 2025).
312. U.S. Department of Labor. OSH Act of 1970. 2025. Available online: https://www.osha.gov/laws-regs/oshact/completeoshact (accessed on 15 January 2025).
313. International Safety Consultants. Federal Coordination Commission for Occupational Safety (EKAS). Available online: https://www.internationalsafetyconsultants.com/federal-coordination-commission-for-occupational-safety-ekas/ (accessed on 15 January 2025).
314. Lerch, A.; Ledergerber, S. Switzerland: Working and Rest Time: As of July 2023, Working and Rest Time Regulations will Be Made More Flexible for Certain Industries. Available online: https://leglobal.law/2023/05/23/switzerland-working-and-rest-time-as-of-july-2023-working-and-rest-time-regulations-will-be-made-more-flexible-for-certain-industries/ (accessed on 15 January 2025).
315. Vischer. Overview of the Most Recent Forms of Leave Under Swiss Employment Law. Available online: https://www.vischer.com/en/knowledge/blog/overview-of-the-most-recent-forms-of-leave-under-employment-law/ (accessed on 15 January 2025).
316. CIPD. Employment Law in Switzerland. Available online: https://www.hr-inform.co.uk/employment_law/employment-law-in-switzerland (accessed on 15 January 2025).
317. International Labour Organization. Switzerland: Employment Act. Available online: https://natlex.ilo.org/dyn/natlex2/r/natlex/fe/details?p3_isn=44880&cs=1zAJHGehp5UV8FIF-5XQ69KkgsW0RjWh561xehC-rWMX1DSb5FGnk6_gKIuoAnNV5ijlbVPP4c4E3hbK1SGpZaA (accessed on 15 January 2025).

318. Federal Ministry of Labour and Social Affairs. Occupational Safety and Health Act (Arbeitsschutzgesetz) (English Version). Available online: https://www.gesetze-im-internet.de/englisch_arbschg/englisch_arbschg.pdf (accessed on 16 January 2025).
319. European Agency for Safety and Health at Work. OSH System at National Level—Germany. Available online: https://oshwiki.osha.europa.eu/en/themes/osh-system-national-level-germany (accessed on 16 January 2025).
320. Federal Institute for Occupational Safety and Health. Technical Rules and Guidelines for Occupational Safety in Germany. Available online: https://www.baua.de/EN/Service/Technical-rules (accessed on 12 January 2025).
321. Japan Society for Occupational Health (SANEI). Overview of the Japan Society for Occupational Health. Available online: https://www.sanei.or.jp/english/about/overview/index.html (accessed on 13 January 2025).
322. Higashi, T. Trends and Issues of Occupational Safety and Health in Japan—Health and Safety Measures in Response to Globalization, Diversification of Employment Formats, and Declining Birth Rates. 2009. Available online: https://www.bollettinoadapt.it/old/files/document/6697HIGASHI_EU_03_20.pdf (accessed on 5 January 2025).
323. Hori, M. The trend and issues of occupational safety and health in Japan. *Procedia Eng.* **2012**, *43*, 610–614. [CrossRef]
324. Ojima, J. Features of the Japanese Industrial Safety and Health Act: Some key points regarding the organization of safety and health management. *Ind. Health* **2020**, *58*, 479–486. [CrossRef] [PubMed]
325. State Council of the People's Republic of China. State Administration of Work Safety. Available online: https://english.www.gov.cn/state_council/2014/09/09/content_281474986284037.htm (accessed on 15 January 2025).
326. China Labour Bulletin. Work Safety. Available online: https://clb.org.hk/en/content/work-safety (accessed on 15 January 2025).
327. LawInfoChina. Labor Law of the People's Republic of China. Available online: https://www.lawinfochina.com/display.aspx?id=35733&lib=law&EncodingName=big5 (accessed on 15 January 2025).
328. International Labour Organization. Comments and Suggestions on Amending the Law of the People's Republic of China on the Prevention and Control of Occupational Diseases. Available online: https://www.ilo.org/publications/comments-and-suggestions-amending-law-people%E2%80%99s-republic-china-prevention (accessed on 15 January 2025).
329. Zhou, Z. Understanding the administrative regulation on occupational health and trend in China. *J. Occup. Health* **2018**, *60*, 126–131. [CrossRef]
330. International Labour Organization. Ccupational Safety and Health: Global Trends and Challenges. Available online: https://www.ilo.org/media/287301/download (accessed on 15 January 2025).
331. Ministry of Labour and Employment, Government of India. Factories Act, 1948. Available online: https://labour.gov.in/sites/default/files/factories_act_1948.pdf (accessed on 15 January 2025).
332. PRS Legislative Research. Factories (Amendment) Bill. 2016. Available online: https://prsindia.org/files/bills_acts/bills_parliament/2016/%20Bill%20Text%20.pdf (accessed on 15 January 2025).
333. Sobhan, S. India's Archaic Labor Laws Allow Firms to Exploit Workers. *DW News*, 9 March 2024. Available online: https://www.dw.com/en/indias-archaic-labor-laws-allow-firms-to-exploit-workers/a-70121341#:~:text=Major%20international%20companies%2C%20including%20big%20tech%20firms%2C%20are,12-14%20hours%20a%20day%20Image%3A%20PUNIT%20PARANJPE%2FAFP%2FGetty%20Images (accessed on 15 January 2025).
334. Pingle, S. Occupational safety and health in India: Now and the future. *Ind. Health* **2012**, *50*, 167–171. [CrossRef]
335. Ministry of Labour and Employment, Government of India. Occupational Safety, Health, and Working Conditions Code, 2020 (Gazette Notification). Available online: https://labour.gov.in/sites/default/files/osh_gazette_0.pdf (accessed on 15 January 2025).
336. AI Principles Overview—oecd.ai. Available online: https://oecd.ai/en/ai-principles (accessed on 10 April 2025).
337. NIST AI. *Artificial Intelligence Risk Management Framework: Generative Artificial Intelligence Profile;* NIST Trustworthy and Responsible AI: Gaithersburg, MD, USA, 2024.
338. Elias, J. International labour standards, codes of conduct and gender issues: A review of recent debates and controversies. *Non-St. Actors Int'l L.* **2003**, *3*, 283. [CrossRef]
339. Kinkel, S.; Baumgartner, M.; Cherubini, E. Prerequisites for the adoption of AI technologies in manufacturing–Evidence from a worldwide sample of manufacturing companies. *Technovation* **2022**, *110*, 102375. [CrossRef]
340. Bianco, M. Overcoming the Social Barriers of AI Adoption. Ph.D. Thesis, Eindhoven University of Technology Eindhoven, Eindhoven, The Netherlands, 2021.
341. Rodríguez-Espíndola, O.; Chowdhury, S.; Dey, P.K.; Albores, P.; Emrouznejad, A. Analysis of the adoption of emergent technologies for risk management in the era of digital manufacturing. *Technol. Forecast. Soc. Change* **2022**, *178*, 121562. [CrossRef]
342. Skoumpopoulou, D.; Toliyat, S.M.H.; Ojra, A.; Shokri, A.; Hu, S. Challenges of achieving digital transformation in manufacturing firms: The case of predictive maintenance and spare part inventory management. *J. Manuf. Technol. Manag.* **2025**, *36*, 159–178. [CrossRef]

343. Elahi, M.; Afolaranmi, S.O.; Martinez Lastra, J.L.; Perez Garcia, J.A. A comprehensive literature review of the applications of AI techniques through the lifecycle of industrial equipment. *Discov. Artif. Intell.* **2023**, *3*, 43. [CrossRef]
344. Lu, X.; Wijayaratna, K.; Huang, Y.; Qiu, A. AI-enabled opportunities and transformation challenges for SMEs in the post-pandemic era: A review and research agenda. *Front. Public Health* **2022**, *10*, 885067. [CrossRef] [PubMed]
345. Chaudhuri, R.; Chatterjee, S.; Vrontis, D.; Chaudhuri, S. Innovation in SMEs, AI dynamism, and sustainability: The current situation and way forward. *Sustainability* **2022**, *14*, 12760. [CrossRef]
346. Oldemeyer, L.; Jede, A.; Teuteberg, F. Investigation of artificial intelligence in SMEs: A systematic review of the state of the art and the main implementation challenges. *Manag. Rev. Q.* **2024**, 1–43. [CrossRef]
347. Bettoni, A.; Matteri, D.; Montini, E.; Gładysz, B.; Carpanzano, E. An AI adoption model for SMEs: A conceptual framework. *IFAC-PapersOnLine* **2021**, *54*, 702–708. [CrossRef]
348. Lasi, H.; Fettke, P.; Kemper, H.G.; Feld, T.; Hoffmann, M. Industry 4.0. *Bus. Inf. Syst. Eng.* **2014**, *6*, 239–242. [CrossRef]
349. Wan, J.; Li, X.; Dai, H.N.; Kusiak, A.; Martinez-Garcia, M.; Li, D. Artificial-intelligence-driven customized manufacturing factory: Key technologies, applications, and challenges. *Proc. IEEE* **2020**, *109*, 377–398. [CrossRef]
350. Lucke, D.; Constantinescu, C.; Westkämper, E. Smart factory—A step towards the next generation of manufacturing. In Proceedings of the Manufacturing Systems and Technologies for the New Frontier: The 41 st CIRP Conference on Manufacturing Systems, Tokyo, Japan, 26–28 May 2008; Springer: London, UK, 2008; pp. 115–118.
351. Abeyratne, S.A.; Monfared, R.P. Blockchain ready manufacturing supply chain using distributed ledger. *Int. J. Res. Eng. Technol.* **2016**, *5*, 1–10.
352. Ghimire, T.; Joshi, A.; Sen, S.; Kapruan, C.; Chadha, U.; Selvaraj, S.K. Blockchain in additive manufacturing processes: Recent trends & its future possibilities. *Mater. Today: Proc.* **2022**, *50*, 2170–2180.
353. Ali, O.; Jaradat, A.; Kulakli, A.; Abuhalimeh, A. A comparative study: Blockchain technology utilization benefits, challenges and functionalities. *IEEE Access* **2021**, *9*, 12730–12749. [CrossRef]

processes

MDPI

Article

Scaling Up a Heater System for Devulcanization of Off-Spec Latex Waste: A Two-Phase Feasibility Study

Dalila Alias [1,2,*], Suganti Ramarad [2], Lik Yin Ng [1], Viknesh Andiappan [3], Jason B. C. Low [4], Fook Peng Leng [4], Jia Jia Leam [4] and Denny K. S. Ng [1,*]

1 Faculty of Engineering and Technology, Sunway University, Petaling Jaya 47500, Selangor, Malaysia
2 School of Engineering and Physical Sciences, Heriot-Watt University Malaysia, Presint 5, Putrajaya 62200, Malaysia
3 Research Centre for Sustainable Technologies, Faculty of Engineering, Computing and Science, Swinburne University of Technology Sarawak, Jalan Simpang Tiga, Kuching 93350, Sarawak, Malaysia
4 Karex Industries Sdn. Bhd., Pontian 82000, Johor, Malaysia
* Correspondence: dalilaa@sunway.edu.my (D.A.); dennyng@sunway.edu.my (D.K.S.N.)

Abstract

Although rubber waste devulcanization has been widely studied, its industrial-scale implementation remains limited due to challenges in process scalability. This study examines the feasibility of devulcanizing off-spec latex waste through a two-phase approach involving laboratory and pilot-scale trials. The latex waste was sourced from off-spec condom products composed of natural rubber latex. Laboratory-scale experiments were initially conducted to establish process parameters and generate baseline data, including gel content before and after the devulcanization process. Thermogravimetric analysis (TGA), gel permeation chromatography (GPC), and dynamic mechanical analysis (DMA) were employed. The laboratory findings have been used to design and operate the subsequent pilot-scale devulcanization process, using a retrofitted waste rubber machine. Samples from the pilot trials underwent the same analytical tests to assess consistency and process performance at scale. Results from the pilot scale experiments suggest that comparable levels of devulcanization were achieved, with gel contents of 52.5% and 55.2% achieved at the laboratory scale and pilot scale. GPC analysis confirmed a uniform distribution, with an increase in the number average molecular weight, indicating the scission of crosslinks in the sample. GPC analysis also revealed a decrease in dispersity index (Đ) value of 2.27 in lab scale conditions and 1.76 for pilot scale conditions, suggesting a more uniform molecular weight distribution and improved devulcanization efficiency, which enhances the possibility of recycling. The successful translation from lab-scale to the pilot setup highlights the process's potential for industrial rubber recycling using retrofitted equipment.

Keywords: rubber devulcanization; thermo-mechanical devulcanization; natural rubber latex recycling; laboratory- and pilot-scale process; process scalability and optimization; sustainable polymer processing

1. Introduction

In the 12th Malaysia Plan Kick-off conference, a circular economy plan for municipal and industrial waste was introduced [1]. This circular economy plan involves the recovery of waste materials, their recycling, and transformation to upcycle them, before reintroducing the recovered waste to the market as new raw materials. A study conducted by

McKinsey shows that a circular economy could yield a net economic gain of EUR 1.8 trillion annually by 2030 [2], suggesting that embracing sustainable practices can lead to significant financial benefits while mitigating environmental impact.

Natural rubber (NR) is a polymer consisting mainly of cis-1,4-polyisoprene and possesses desirable characteristics such as elasticity [3] and durability, making it suitable for various manufacturing applications. Malaysia is one of the key producers of NR in the world, along with Thailand, Indonesia, Cote d'Ivoire, Vietnam, China, India, Cambodia, Laos, and Myanmar. Table 1 shows the production of rubber by each country in the year 2024 [4].

Table 1. Production of natural rubbers in 2024.

Country	Production of Natural Rubber in 2024 (Million Tons)
Thailand	5.02
Indonesia	2
Côte d'Ivoire	1.8
Vietnam	1.3
China	0.88
India	0.88
Cambodia	0.53
Malaysia	0.39
Laos	0.35
Myanmar	0.33

In 2024, Malaysia's production of rubber is 409.4 thousand tons, a 6% increment compared to the year 2023, while the consumption of rubber throughout the same period also increased from 539.2 thousand tons to 571.5 thousand tons in the year of 2023 [4]. The top ten exports of Malaysia's rubber products are gloves, new pneumatic tires, tubing and piping, wires and electrical conductors, footwear, latex thread, and condoms, as shown in Figure 1.

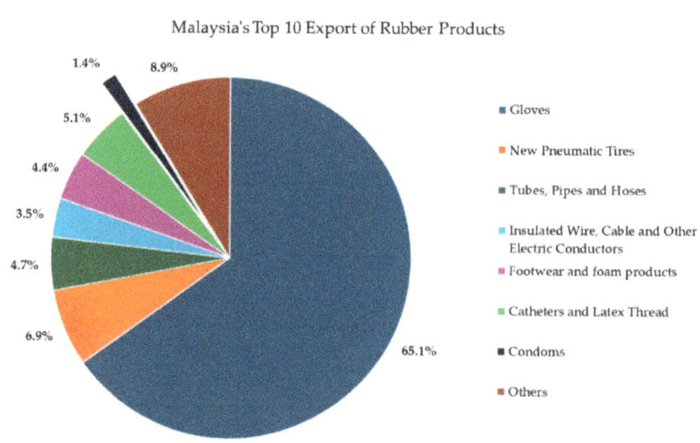

Figure 1. Malaysia's rubber export in 2024.

Among these rubber exports, Malaysia is recognized as one of the world's leading producers of condoms. In 2024, Malaysia exported condoms worth MYR 341 million [4], strengthening its position as one of the world's leading condom producers [3]. The primary component found in condoms is NR latex, which undergoes a vulcanization process to enhance the strength and resilience of the rubber. However, condom processing generates off-spec latex waste that is non-degradable, which poses an environmental concern.

Condom processing involves dipping, leaching, and stripping, followed by drying, quality control testing, and packaging [5], as shown by the general process flow in Figure 2. To produce strong and durable condoms, the vulcanization process is required. Vulcanization is the process of adding sulfur or sulfur-containing compounds into a mixture, with or without heat. In the quality control testing steps, physical and mechanical testing activities such as the burst test, tensile strength test, and leakage test were conducted. The condoms that adhere to the strict standards will be sent for packaging and labeling before being shipped off to the consumers, while the products that fail the quality control tests will be sent to the rubber recycling plant.

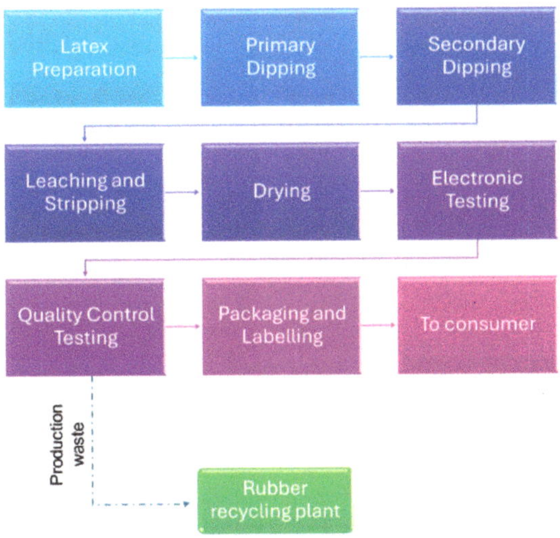

Figure 2. Condom processing flow.

While the vulcanization process has played a crucial role in ensuring the production of strong and reliable condoms, it also presents environmental challenges when dealing with off-spec latex waste such as defective products, trimming scraps, and expired inventory [6]. Note that the off-spec latex waste is vulcanized rubber, a strong 3D rubber network containing long polyisoprene molecular chains (C_5H_8) and crosslinked by sulfur (S), as illustrated in Figure 3.

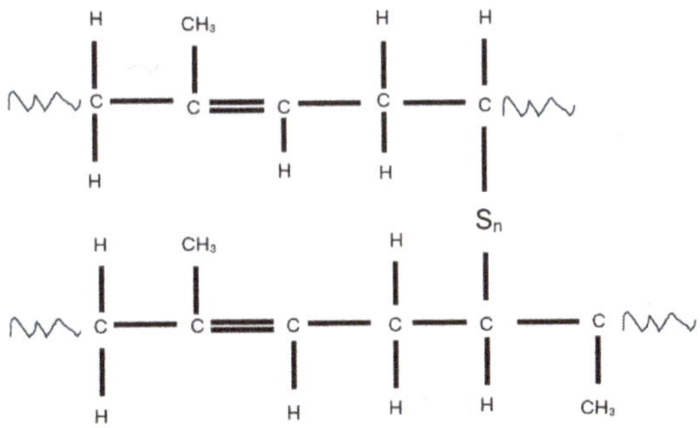

Figure 3. Chemical structure of vulcanized rubber.

Due to its vulcanized properties, the off-spec latex waste is resistant to natural degradation, making its disposal a significant environmental concern. As one of the world's leading exporters, Malaysia generates high production volumes, which in turn result in substantial amounts of production waste, highlighting the critical need for sustainable waste disposal strategies.

Other than concerns about disposal methods, there is also an increasing need to reclaim raw materials from waste, driven by the shift towards a circular economy. In the past years, efforts have been made to reclaim raw material from rubber waste using methods such as pyrolysis, devulcanization, and mechanical recycling. The difference between these techniques is presented in Table 2.

Table 2. Rubber reclaiming processes.

Rubber Reclaiming Process	Details	Advantages	Disadvantages	Reference
Pyrolysis	Convert rubber waste into smaller molecular mass products under high temperatures and a low-oxygen environment	Reduce landfill storage, obtain high-quality products of pyrolysis oil, carbon black, gas for electricity generation, and metal cord for the metallurgical industry.	High energy usage	[7,8]
Devulcanization	Breaking the sulfur crosslinks in vulcanized rubber	Process using heat, chemical agents, and mechanical shears to produce devulcanized rubber, which retains its natural properties of rubber	Process may break the main C-C polymer in addition to breaking the sulfur crosslinks, which will weaken the mechanical properties of devulcanized rubber	[9,10]
Mechanical recycling	Using mechanical processes such as grinding and shredding to produce smaller particles of rubber	Cost-effective components such as steel wires, textile fibers recovery, versatility of end products	Degradation of mechanical properties and inconsistent product quality	[11–14]

In Malaysia, there are several well-known tire recycling plants, such as Evergreen Corporate [15], that employ pyrolysis to generate synthetic diesel, carbon black, syngas, and recover steel wires from the used tires. A significant research gap can be seen in the recycling of off-spec latex waste from condom production. Unlike tire recycling, which has established recovery processes, the recycling of off-spec condom production waste remains largely unexplored due to its small volume and vulcanized structure.

To address this challenge, the devulcanization process has been identified as a potential recycling pathway for the treatment of off-spec latex waste originating from the condom manufacturing processes. Devulcanization, the process of breaking down vulcanized rubber into its constituent components [6], offers a promising solution for the recycling of off-spec latex waste from condom production. These recycled materials could then be utilized in the production of items like rubber components for automotive parts, footwear, and other consumer goods. In rubber product formulation, incorporating devulcanized rubber allows manufacturers to reduce the proportion of virgin NR required, while maintaining desirable product quality. A review by Xiao [16] reported that blends of devulcanized rubber with NR exhibit higher mechanical properties than blends of NR with non-devulcanized rubber. This improvement was attributed to the success of the devulcanization process, where crosslinks were broken while the main chains were preserved, producing compatible surface chemistry, which leads to the optimized blending of NR and devulcanized rubber, and leads to the improved mechanical properties of the final products.

Devulcanization can be achieved through various methods, including mechanical, thermal, chemical, ultrasonication, and microwave-assisted processes, and the combination of these methods. The devulcanization processes are summarized in Table 3.

Table 3. Devulcanization process.

Devulcanization Process	Details	Reference
Mechanical	• Using mechanical force such as shear • Often combined with heat • Use machinery such as a twin-screw extruder to apply energy to break the sulfur crosslinks • Solvent-free process • However, heat generated during the mechanical shear process can break the main carbon–carbon chain and reduce the mechanical properties of the devulcanized rubber	[17]
Thermal	• Heat is applied to overcome the activation energy of sulfur–sulfur and carbon–sulfur bonds • Processes can selectively break the crosslinks and avoid degradation of the main carbon–carbon chain.	[18]
Chemical	• Usage of chemicals to break carbon–sulfur or sulfur–sulfur bonds selectively • Sulfides such as tetramethylthiuram disulfide (TMTD), amine-based compounds, ionic liquids, or organic solvents can be used as devulcanizing agents • Processes can be limited by the toxicity of the chemical of choice	[19,20]
Ultrasonication	• Using high-frequency sound waves to cleave sulfur–sulfur or carbon–sulfur bonds • Higher amplitude can result in higher devulcanization	[21]
Microwave-assisted	• Molecular motion by microwave irradiations to raise the temperature that leads to scission of sulfur–sulfur and carbon–sulfur bonds • Works better with polar rubbers to absorb microwaves • Batch process • Faces challenges in scaling up	[22]
Combination of methods	• Combination of two or more methods to achieve devulcanization • Offers flexibility to process operators	[23–26]

Despite notable progress in devulcanization research, the majority of these practices remain confined to laboratory-scale experiments, with only limited adoption at the industrial level. Currently, the only known industrial-scale devulcanization plants are SSH Recycling in Scotland [27] and Tyromer [28] in Canada, which focus on tire recycling using the devulcanization process. While the use of devulcanization for treating the off-spec latex waste from condom production in laboratory-scale studies [29,30] has demonstrated its capability to break the crosslinked structure, its potential for scale-up and practical implementation requires further investigation through a technical feasibility study. To the best of our knowledge, there are no known large-scale facilities dedicated to processing off-spec latex waste. The lack of industrial implementation for off-spec latex waste highlights a significant gap in the field.

The objective of this study is to establish a thermo-mechanical devulcanization process for off-spec latex waste sourced from condom production waste at a laboratory scale and to modify the existing heaters used in waste rubber plants for pilot-scale devulcanization purposes. With the modification of a waste rubber machine, this study aims to identify optimal operating parameters while preserving the quality of devulcanized rubber for potential reuse in various applications.

2. Materials and Methods

2.1. Materials

Latex waste was sourced from off-spec condoms provided by Karex Industries Sdn. Bhd. (Pontian, Johor, Malaysia).

2.2. Methods

The thermo-mechanical devulcanization process of off-spec latex waste was carried out at the laboratory scale and pilot plant scale. The mechanical shear was applied using an overhead stirrer at the laboratory scale and an adjustable double-layer speed stirrer at the pilot plant scale. The thermal effect was regulated through controlled temperature adjustment via an oil bath and jacketed heater at the lab scale and pilot plant scale, respectively. The devulcanization process was free from any chemical agents or additives and took place in a closed vessel.

To mitigate odor emission and ensure laboratory safety, the devulcanization process was conducted in an open environment with adequate ventilation throughout the heating stage.

2.2.1. Lab-Scale Testing

To establish the devulcanization process at a lab scale, an experimental setup comprising a metal can, an oil bath with a temperature probe, an overhead stirrer, and two sets of retort stands with clamps, as depicted in Figure 4, was prepared to simulate the waste rubber tank used at the recycling plant.

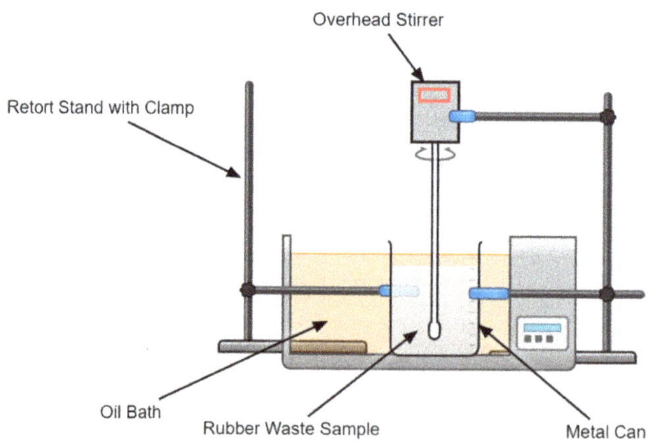

Figure 4. Laboratory-scale experimental setup for rubber devulcanization process.

Devulcanization of Rubber in a Lab-Scale Setup

The samples were prepared by exposing the off-spec latex waste to various temperatures and times. The shredded samples were loaded into a metal can, which was immersed in an oil bath. The temperature of the oil bath can be regulated to achieve a temperature set point. To analyze the influence of temperature and time on the devulcanization of rubber, temperature and time were varied according to Table 4. The overhead stirrer was set at a constant speed, 120 RPM. Residence times were adjusted according to temperature: shorter at high temperatures to avoid melting and excessive degradation, and longer at low temperatures to ensure adequate thermal activation for devulcanization.

Table 4. Operating condition for lab-scale devulcanization setup.

Sample	Temperature (°C)	Time (Minutes)
1	100	30
2	150	30
3	200	30
4	250	30
5	150	10
6	150	20
7	200	10
8	200	20

Scaled-Up Configuration of the Waste Rubber Machine for Rubber Devulcanization

The rubber devulcanization process was carried out in an existing waste rubber machine available in the rubber recycling plant. The waste rubber machine comprised two mixing chambers, with the first mixing chamber equipped with an electrical jacketed heater, as illustrated in Figure 5. The bottom stirrer in the first mixing chamber was designed to transfer the rubber from the first chamber to the second chamber.

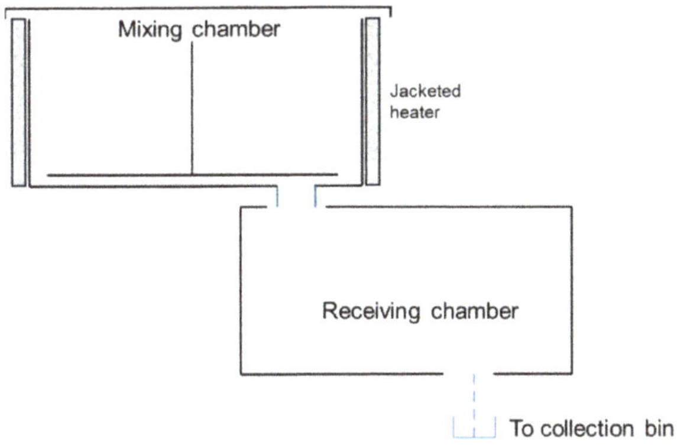

Figure 5. Existing rubber waste machine.

2.2.2. Devulcanization of Rubber in Modified-Waste Rubber Machine

The devulcanization process was performed in the modified waste rubber heater equipped with temperature sensors and an inverter motor. A constant weight of off-spec latex waste was used in each run. To evaluate the performance of the modified heater in the devulcanization process, the operating temperature and speed of the stirrer were varied. The off-spec latex waste was kept in ambient conditions before treatment. In each run, after the samples have reached the designated mixing time of 15 min, the waste rubber will be transferred into the second chamber for cooling before being discharged into a hopper bin. The operating conditions of the devulcanization process in the modified rubber machine are summarized in Table 5.

Characterization

a. Gel content analysis

Rubber samples were characterized by boiling toluene extraction as per the setup shown in Figure 5, adapted from ASTM D3616 [31], and the gel content value for each sample was determined according to Equation (1). Gel content value refers to the insoluble part of the rubber samples, which was extracted in a boiling toluene extraction process. A measure of 300 g of rubber was placed in a pouch made of 100-mesh-sized stainless-steel wire mesh and completely immersed in boiling toluene for 24 h. After the extraction, the samples were thoroughly washed using fresh toluene and dried for 48 h at 70 °C to eliminate the adsorbed solvent. Each sample underwent weighing twice: before extraction and after drying.

Table 5. Operating conditions for devulcanization in modified rubber machine.

Run	T (°C)	Speed (RPM)
1	130	120
2	130	150
3	130	180
4	140	120
5	140	150
6	140	180
7	150	120
8	150	150
9	150	180

Figure 6 shows the experimental setup for gel content analysis.

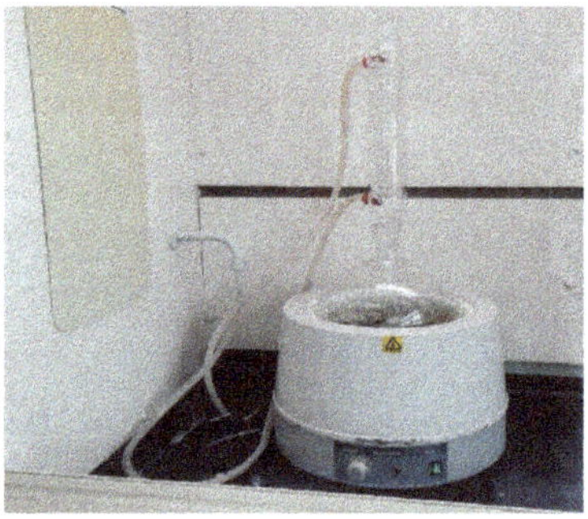

Figure 6. Laboratory setup for gel content analysis.

The gel and sol content values are determined using Equations (1) and (2), as below:

$$\text{Gel content } (\%) = \frac{(\text{Mass of the total dried sample})}{\text{Mass of the original sample}} \times 100 \tag{1}$$

$$\text{Sol content } (\%) = 100 - \text{Gel content} \tag{2}$$

b. Thermogravimetric analysis (TGA)

TGA tests were conducted using a thermogravimetric analyzer from Metler Toledo, Columbus, OH, USA. The analysis was performed at a 10 °C/minute heating rate under a N_2 atmosphere (50 mL/min). Approximately 8 to 10 mg of the samples was used in each test. The samples were heated from 25 to 600 °C. The temperature at the maximum decomposition rate (T_{max}) was determined from the TGA curve obtained.

c. Dynamic mechanical analysis (DMA)

DMA was carried out to determine the storage modulus (E′), loss tangent (tan δ), and associated glass transition temperature (T_g) as a function of temperature at a heating rate of 10 °C/min using a PerkinElmer Pyrus Diamond DMA from PerkinElmer Inc., Shelton, CT, USA. Test conditions included a specimen gauge length of 10 mm operated at a temperature range between −170 °C and 100 °C at 1 Hz frequency.

d. Gel permeation chromatography (GPC)

GPC was conducted using a Waters 1500 Series from Waters Corporation, Milford, MA, USA with a refractive index detector operating at 5 µm columns: Phenomenex (300 × 7.8 mm), Phenogel 5 µm 10^5 A (10K–1M, MW), Phenogel 5 µm 500 A (1K–15K, MW), and Phenogel 5 µm 50 A (100–3K, MW). The column temperature was set to 40 °C. Tetrahydrofuran (THF) was used as the mobile phase, with a flow rate of 1 mL/minute. The molecular weight distributions were calibrated using polystyrene (THF Dissolve) (Waters) standard solution.

e. Dispersity (Đ)

From the average molecular weight (M_w) and number average molecular weight (M_n) values obtained from the GPC analysis, the polydispersity index of the samples can be calculated using Equation (3).

$$Đ = \frac{M_w}{M_n} \quad (3)$$

3. Results and Discussion

3.1. Laboratory-Scale Devulcanization

To study the effect of operating temperature on the devulcanization process, the off-spec latex waste sample was exposed to different temperatures and times according to Table 4. At low temperatures, the physical structure of the latex waste was maintained, whilst at high temperatures, the off-spec waste was completely melted. Figure 7 shows that the thermal stability of the off-spec latex waste is temperature-dependent, with lower temperatures preserving its structural integrity, while at high temperatures, the sample undergoes a phase transition.

Figure 7. Off-spec latex sample after exposure to low temperature (100 °C) and high temperature (250 °C).

3.1.1. Effect of Temperature on Devulcanization Process

The devulcanization process aims to modify the chemical structure of vulcanized rubber, and this can be deduced by evaluating the gel content value. A devulcanization mechanism proposed in this work is shown in Figure 8. In general, a higher gel content value indicates a higher degree of a three-dimensional crosslinked network.

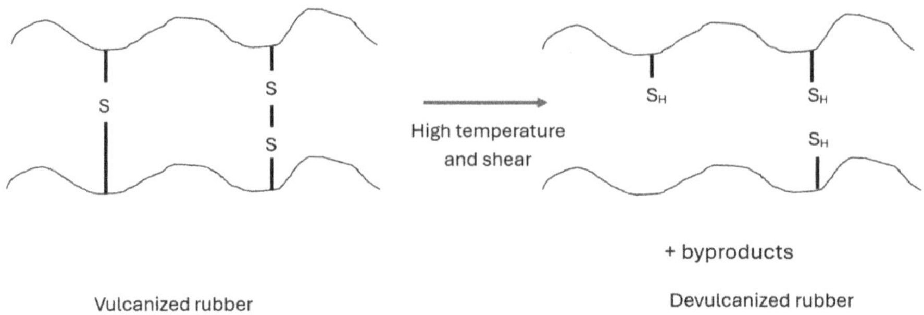

Figure 8. Proposed devulcanization mechanism.

Therefore, a reduction in gel content value indicates the breakdown of crosslinks and main-chain polymers during the process. To determine the fraction of crosslinked (insoluble or "gel") rubber and the soluble part of the rubber sample, gel content analysis is commonly used. The insoluble fraction or the gel fraction indicates the insoluble portion of the sample when the sample is exposed to a suitable solvent, such as toluene. To study the effect of operating temperature on the physical properties of the rubber waste, gel content analysis was carried out, and gel content percentage and soluble content percentage were calculated. The results were summarized in Table 6.

Table 6. Effect of temperature on gel content and T_{max} for devulcanization of off-spec latex waste at lab scale.

Temperature (°C)	Time (Minute)	Gel Content (%)	Sol Content (%)	Onset Degradation Temperature (°C)	T_{max} (°C)
25	0	84.09 ± 1.3	15.91 ± 1.3	345.0	377.17
100	30	79.36 ± 8.4	20.64 ± 8.41	339.11	375.33
150		58.43 ± 6.2	41.57 ± 6.2	340.31	377.33
200		0.91 ± 3.7	99.09 ± 3.7	339.53	377.17
250		0.37 ± 2.2	99.63 ± 2.2	339.10	376.33

As shown in Table 6, the control sample at 25 °C has a gel content value of 84.09%. After the devulcanization process, it is shown that exposing the off-spec latex waste to temperatures from 100 °C to 250 °C has reduced the gel content value to the range of 79.36 to 0.37%. The soluble content of the samples increases from 41.57% to 99.63% when the temperature increases from 150 °C to 200 °C. In general, the higher the final temperature of the sample, the smaller its gel content. As operating temperature increases, the gel content is reduced, thus proving that the rise in temperature is responsible for the occurrence of the devulcanization. The investigation of the thermal degradation of the prepared materials is important for assessing their potential applications when elevated temperatures are involved. However, as shown in the results, while operating at high temperatures, the gel content reduces. This is due to the fact that the thermal energy can scission the sulfur–sulfur bond, and carbon–sulfur bonds can also increase the possibility of undesirable reactions such as the scission of the polymer backbone, which can be reflected in an extremely low gel content value.

To further study the effect of temperature on the devulcanization process, the thermal stability of the samples was studied using thermogravimetry (TG) and derivative thermogravimetric analysis (DTG). From Table 6, it can be seen that the most stable sample at 30 min of devulcanization was the sample that had been exposed to operating temperatures of 150 °C and 200 °C, recording a T_{max} of 377.33 °C and 377.17 °C, respectively. As the temperature increased to 250 °C, the T_{max} shifted to 376.33 °C. The shift in the temperature at the maximum rate of decomposition indicates a reduction in thermal stability, likely due to chain scission and partial degradation of the rubber sample.

3.1.2. Effect of Time on Devulcanization Process

To study the effect of operating time on the rubber devulcanization process, the rubber wastes were exposed at various times, from 10 min to 30 min. From the previous results, due to the stability of rubber at 150 and 200 °C, these two operating temperatures were selected while the operating times varied, at 10, 20, and 30 min. The results of the percentage of devulcanization and T_{max} are shown in Table 7.

Table 7. Effect of time on gel content and T_{max} for devulcanization of off-spec latex waste at lab scale.

Temperature (°C)	Time (Minute)	Gel Content (%)	Sol Content (%)	T_{max} (°C)
150	10	87.36 ± 2.2	12.64 ± 2.2	376.83
	20	52.15 ± 2.2	47.85 ± 2.2	375.17
	30	58.43 ± 6.2	41.57 ± 6.2	377.33
200	10	26.59 ± 7.0	73.41 ± 7.0	376.50
	20	3.53 ± 1.4	96.47 ± 1.4	376.67
	30	0.91 ± 3.7	99.09 ± 3.7	377.17

From the results, it was shown that treatment time plays a significant role during the chemical devulcanization of rubber waste compounds. The results show that longer exposure times led to lower gel contents. The devulcanization process can thermally destabilize the rubber, thus promoting its degradation at a lower temperature. However, consideration of the optimum temperature does not solely depend on the degradation temperature data; the gel and sol contents must be considered. From data collected in this study, the optimum temperature and time are chosen as 150 °C at 20 min and 200 °C at 10 min.

The laboratory-scale test demonstrated that temperature and time control are essential not only for optimizing the devulcanization process but also for ensuring efficient material handling.

3.2. Pilot-Scale Devulcanization: Modification of Waste Rubber Tank

Initially, all off-spec latex waste was processed in a rubber waste machine without any control over critical parameters like operating temperature or stirring speed. The lack of process control has resulted in the inconsistent quality of recycled rubber and represented a missed opportunity to produce high-quality rubber with enhanced properties. To improve mixing efficiency and ensure uniform heat distribution, a two-layer impeller with adjustable speed was installed in the waste rubber machine. The purpose was twofold: to enhance the mixing of rubber waste within the tank and to prevent dead volume accumulation beneath the top stirrer [32]. To regulate the temperature in the rubber waste machine, four temperature sensors were placed at the top right, top left, bottom right, and bottom left of the machine. The design of the new stirrer and the position of the temperature sensors are presented in Figures 9 and 10.

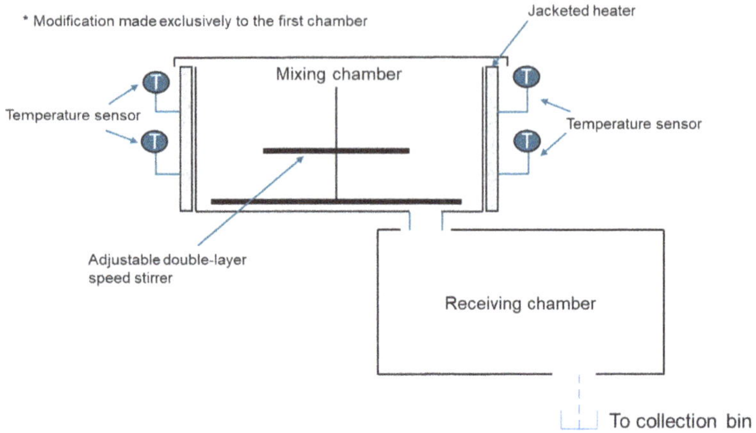

Figure 9. New stirrer design and thermocouple location for waste rubber machine.

Figure 10. New stirrer for the waste rubber machine.

3.2.1. Physical Appearance

After the modification of the waste rubber machine, the devulcanization process was carried out. Figure 11 presents the resulting devulcanized rubber after a thermo-mechanical devulcanization process.

Figure 11. Photos of the rubber samples before devulcanization (**left**) and after devulcanization process (**right**) in the modified tank.

During the devulcanization process, exposure to heat caused changes to the texture and appearance of the off-spec latex waste. Compared to a dry, non-sticky waste rubber sample, the devulcanized sample has a sticky, uncured rubber-like appearance. The stickiness and tackiness of the devulcanized rubber are due to the breakage of polymer crosslinks and the molecules being reverted to their original chains. However, it is crucial to closely monitor the process to prevent the samples from sticking to the outlet chute and obstructing the outlet. Following this initial observation, a more detailed understanding of the structural changes that occurred after the devulcanization process and their effect on the final properties of the devulcanized samples was required.

3.2.2. Gel and Sol Content Analysis

For this study, the waste rubber machine has been modified by adding a two-layer impeller and four temperature gauges. A comparison between gel and sol content between the vulcanized sample (control), the devulcanized sample (pre-modification of waste machine, 0 RPM), and the devulcanized sample at different impeller speeds was made and is summarized in Figure 12.

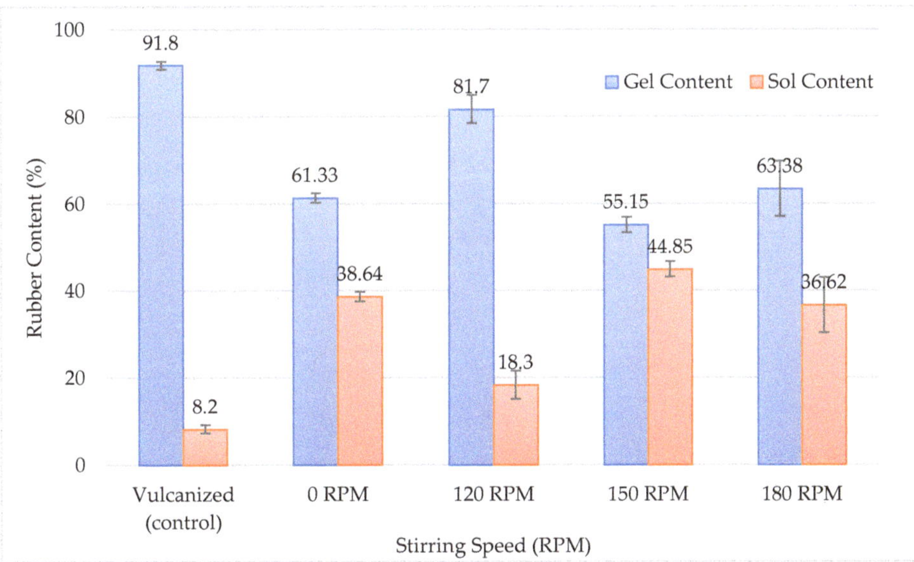

Figure 12. Gel and sol content of devulcanized sample at 150 °C.

In the modified waste rubber machine, the stirring speed was varied between 120, 150, and 180 RPM. Stirring of the material in the rubber tank helps to ensure uniform distribution of heat and prevent localized overheating and underheating during the devulcanization process. The stirring alone does not provide sufficient mechanical energy to break sulfur crosslinks; thus, the process has to be carried out at a certain temperature, which in this case is 150 °C. Close attention must be given to stirrer speed, as a low stirring speed will lead to non-uniform heat transfer, causing a temperature gradient between the rubber samples, and a high stirring speed will generate additional frictional heat, which may raise the temperature beyond the optimal range. An increase in gel content was recorded at 180 RPM, probably due to the increase in temperature inside the mixing tank and the possibility of the re-crosslinking of polymers at high temperature. At a speed of 150 RPM, it was observed that the gel content was the lowest, at 55.15%, making it a favorable set point for stirring speed.

The effect of operating temperature was also studied. From the laboratory-scale test, it was found that excessive heat leads to the complete melting of the rubber, which affects the chemical structure of the devulcanized product and makes handling difficult. Even though at the lab-scale the selected optimum temperature was 150 °C, with an operating time of 20 min, and 200 °C, with an operating time of 10 min, the temperature selected for operation at the pilot plant scale was limited to below 150 °C. This selection was made to maintain a controlled heating profile with sufficient time for stabilization and to prevent the rubber from completely melting inside the modified rubber tank and possibly clogging the discharged outlet. The operating temperature of the rubber waste machine varied between 130, 140, and 150 °C.

The percentages of soluble (sol) and insoluble (gel) content after devulcanization are shown in Figure 13.

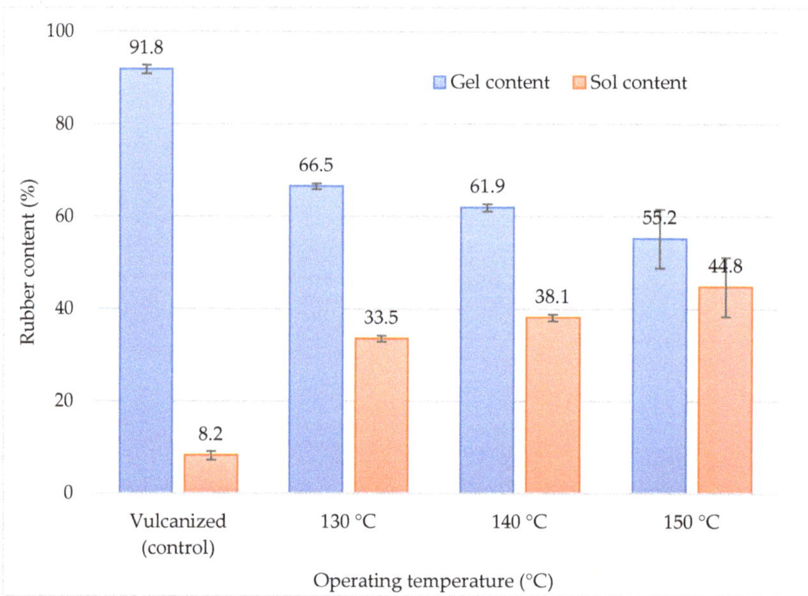

Figure 13. Gel and sol content value at three different operating temperatures at 150 RPM.

By keeping the rotation speed constant at 150 RPM and varying the operating temperature to 130, 140, and 150 °C, it is observed that as the temperature increases, the gel content value decreases. The gel content value decreased from 91.8% to the lowest gel content value of 55.15% at 150 °C.

Temperature is one of the important factors in the devulcanization process, acting as a dual influence: both a facilitator of carbon–sulfur (C-S) bond and sulfur–sulfur (S-S) bond crosslink cleavage and a potential contributor to rubber degradation due to carbon–carbon (C-C) bond scission. Appropriate temperature levels provide the energy required to break S-S and C-S bonds within the rubber's crosslinked network, enabling effective devulcanization. The energy needed to break the C-S and S-S bonds is 273 kJ/mol and 227 kJ/mol, respectively. However, careful selection is required to prevent main-chain degradation. The bond-dissociating energy of C-C and C = C is 348 kJ/mol and 621 kJ/mol, respectively [33]. The set-point temperature of 130 °C is inadequate to break a significant amount of carbon–sulfur and sulfur–sulfur bonds, and this was reflected by the high gel content value of the samples. At temperatures of 140 °C and 150 °C, more of the S-S and C-S bonds within the rubber network start to weaken, thus resulting in a lower gel content.

Based on observations during the operation, it was determined that a minimum of 15 min of mixing must be performed for the devulcanization of rubber to take place. During the devulcanization process, an increase in temperature is expected. However, the increase in temperature in the tank should be monitored, as it should not exceed 170 °C. Once the temperature of the waste rubber machine has reached 170 °C, the rubber must be directly discharged. This step must be carried out to prevent rubber from aggregating and forming clumps inside the tank and making it difficult for it to be discharged from the equipment. For this study, the optimum operating conditions have been identified as being a temperature of 150 °C at 150 RPM.

The reported gel (55.2%) and sol (44.8%) fractions obtained at 150 °C and 150 RPM represent the analytical distribution of crosslinked and solubilized segments within the same rubber matrix, rather than two physically separated phases. The devulcanized rubber remains as a single, cohesive piece, in which partial network cleavage occurs while a portion of the original crosslinks is retained. Such partial devulcanization is intentionally targeted to preserve mechanical integrity and prevent excessive chain scission. The resulting material can be directly reused or blended with virgin rubber in new formulations without the need for physical separation. Therefore, the process achieves effective material recycling and minimizes waste generation.

3.2.3. Thermal Analysis

Figure 14 presents the evaluation of devulcanized rubber by thermogravimetric analysis (TGA) from 25 °C to 600 °C. The thermogram shows that the first weight loss (14%) occurs below 325 °C, corresponding to a loss of water and other highly volatile materials such as oils, antioxidants, and accelerators.

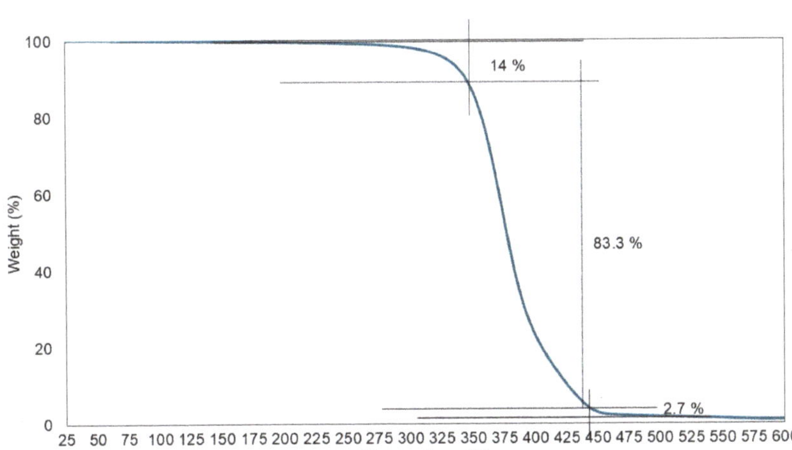

Figure 14. Thermogravimetric analysis (TGA) result of devulcanized rubber.

The second weight loss, at 325–475 °C, was due to degradation of the rubber chain, cis-1,4-polyisoprene. Since the devulcanized rubber is mainly natural rubber, only small amounts of residue remained after 475 °C, which was due to the presence of inorganic fillers such as zinc oxide, which is a common additive in the condom manufacturing process. Derivative thermogravimetry (DTG) peaks that were derived from TGA curves were analyzed to identify the T_{max} of the devulcanized sample, which is summarized in Table 8.

Table 8. Maximum degradation temperature of control sample, pre-modification sample, and pilot-scale devulcanized rubber sample.

Sample	Temperature	Speed (RPM)	T_{max} (°C)
Vulcanized (Control)	25	0	375.8
Pre-modification	150	0	376.8
Post-modification	150	150	374.3

From Table 8, it is clear that the vulcanized (control) sample has a T_{max} of 375.8 °C, and the pre-modification rubber sample has a T_{max} of 376.8 °C. The increase in the degradation temperature of the pre-modification devulcanized rubber is probably due to the uneven temperature distribution in the mixing chamber. Some regions in the mixing chamber have a high temperature profile, leading to partial re-crosslinking during the devulcanization process, which leads to a more thermally stable network structure. This re-crosslinking may result from the interaction of residual sulfur bonds between polymer chains, resulting in the improved thermal stability of the sample. However, excessive re-crosslinking could also reduce the flexibility of the sample, potentially affecting its processability and reusability in new applications.

At 150 °C and 150 RPM, T_{max} recorded a value of 374.3 °C, which is a lower value compared to the control sample. There was a leftward shift in the DTG peak for devulcanized rubber compared to the vulcanized (control) sample. The shift in the DTG curve indicates changes in the thermal decomposition behavior of the devulcanized rubber. This shift is probably due to the reduction in molecular mass, which is due to the scission of crosslinks and macromolecules (C-S, S-S, and C-C). In the devulcanization process, sulfur crosslinks and macromolecular chains were broken, resulting in fewer crosslinks in the rubber sample, thus making the polymer chains less stable. These results corroborate the gel content value, as both criteria are strongly related to structural changes induced during the devulcanization process. It is expected that a lower gel content value leads to an earlier DTG peak due to reduced thermal stability; however, a well-controlled devulcanization process can balance the gel content value and temperature at the maximum decomposition rate to maintain reusability in new rubber products.

3.2.4. Molecular Weight Distribution

To study the molecular mass weight and distribution of the devulcanized rubber sample, a gel permeation chromatography analysis (GPC) was conducted. It should be noted that the GPC analysis was performed on a soluble rubber sample. The number and weight average molecular weights (M_n and M_w) and the dispersity (Đ) are presented in Table 9.

Table 9. GPC results.

Sample	Mn	Mw	Đ
Control	6850	13,460	1.97
Lab scale: 150 °C, 20 min	7240	16,420	2.27
Pilot plant: 150 °C, 150 RPM	62,440	109,900	1.76

The results for the GPC analysis in Table 9 show a comparison of the control, lab-scale, and pilot plant samples. Due to the scission of sulfur bonds, the molecular weight (Mw) of the devulcanized samples increased to 16,418 and 109,907, at the lab scale and pilot plant scale, respectively. This is an indication that in lab- and pilot plant-scale samples, more macromolecular chains are freed from the complex 3D network structure of vulcanized rubber.

The dispersity (Đ) of the vulcanized sample (control sample) was 1.97, reflecting the initial molecular weight distribution of the crosslinked rubber. After the devulcanization process, an increase in molecular weight distribution occurred due to the partial breakdown of the rubber network, resulting in increase in the Đ to 2.27. At the pilot scale, the Đ decreased to 1.76, suggesting a more uniform molecular weight distribution and improved control over the devulcanization process at a larger scale.

The devulcanization process produces a higher molecular weight rubber as the sulfur bonds are being scissioned, but the targeted scission process has helped to control the distribution of the molecular weight distribution.

3.2.5. Dynamics Mechanical Properties of Devulcanized Rubber

Figure 15 illustrates the effect of the devulcanization process on loss tangent (tan δ) as a function of test temperature for the devulcanized rubber and control samples.

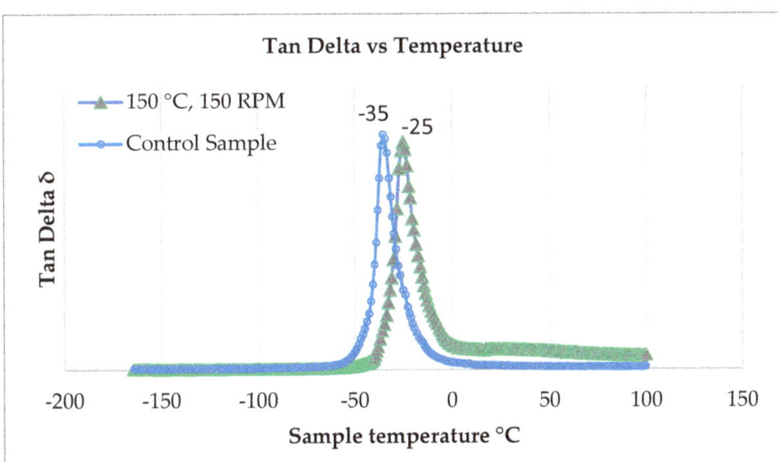

Figure 15. Effect of devulcanization on loss tangent as a function of temperature.

Tan δ reflects the dynamic behavior of the devulcanized rubber. The tan δ versus temperature curves showed damping peaks around temperatures between −60 and −10 °C, or else known as a transition region. The damping peaks of the curves represent the glass transition temperature (T_g). The glass transition temperature of the devulcanized sample is −25 °C, compared to that of the control sample, which is −35 °C. The glass transition temperature of the devulcanized sample is higher than the control sample, signifying that the devulcanized sample is less flexible at the molecular level compared to the control sample. This is due to the increased number average molecular weight and weight average molecular weight of the devulcanized sample. A higher Mn and Mw results in an increased free chain length, which enables active physical entanglements, hence increasing the Tg of the devulcanized sample. It was also observed that the width of both curves is similar, thus indicating that the devulcanization process in the modified heater does not alter the uniformity of the devulcanized rubber sample, as corroborated the Đ values from the GPC analysis in Table 9. This behavior shows promise for the reuse of devulcanized rubber as feed material in the new rubber processing applications.

The storage modulus vs. temperature curve is shown in Figure 16.

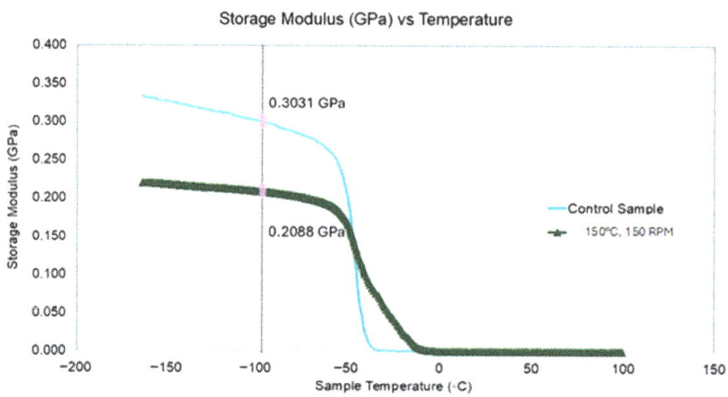

Figure 16. Storage modulus (GPa) vs. temperature for control sample and devulcanized rubber sample.

The storage modulus of the vulcanized (control) sample was higher than the devulcanized rubber samples. This may be due to the strong carbon–sulfur and sulfur–sulfur bonds available in the control sample. At −100 °C, the storage modulus for the control sample is 0.3031 GPa compared to the devulcanized sample, with 0.2088 GPa of storage modulus. The lower storage modulus of the devulcanized sample is deduced to be due to the reduced complexity of 3D network chains because of devulcanization. Hence, the decrease in storage modulus value and the shift in glass transition temperature recorded indicated that the devulcanization process has occurred.

4. Conclusions

In this work, a devulcanization process was carried out on off-spec latex waste originating from condoms. Laboratory-scale devulcanization was conducted, and it was found that the off-spec latex waste sample had been successfully devulcanized, with 58.4% gel content. A waste rubber machine was retrofitted with a two-layer impeller and a temperature gauge to demonstrate the devulcanization process at a pilot plant scale. The effect of stirring speed and operating temperature was investigated. Gel content analysis showed that the value decreased from 91.8% to 55.15% as the temperature increased, indicating that increasing the temperature can enhance the devulcanization process. However, the operating temperature must be limited to below 170 °C to prevent the rubber samples from agglomerating and the formation of clumps at the outlet chute. It was also observed that the devulcanization process at 150 RPM and 150 °C gave the most prominent result in terms of gel and sol contents. The results from TGA, GPC, and DMA indicated that devulcanization had occurred with minimal loss of properties, as evidenced by the comparable thermal stability of the devulcanized rubber at 374.3 °C; a more uniform molecular weight distribution, with Đ of 1.76; and a moderate shift in glass transition temperature from −35 °C to −25 °C, indicating that the molecular and thermal characteristics of the rubber were largely maintained. The thermo-mechanical devulcanization process in the retrofitted waste rubber machine shows good promise in making value-added rubber products from rejected condom waste. This work directly supports Sustainable Development Goal 9—Industry, Innovation and Infrastructure—by promoting sustainable rubber recycling technologies, and helps to create high-value rubber products for various industries.

Author Contributions: Conceptualization, S.R., D.K.S.N., L.Y.N. and V.A.; methodology, S.R., D.A. and D.K.S.N.; validation, S.R., D.A., D.K.S.N., J.B.C.L. and F.P.L.; formal analysis, D.A., F.P.L. and J.B.C.L.; resources, F.P.L., J.B.C.L., J.J.L. and D.K.S.N.; supervision, D.K.S.N.; project administration, S.R. and D.A.; funding acquisition, S.R., D.K.S.N. and J.J.L. All authors have read and agreed to the published version of the manuscript.

Funding: This research was funded by Karex Industries Sdn. Bhd and Heriot-Watt University Malaysia Sdn. Bhd., grant number P22Karex.

Data Availability Statement: The original contributions presented in this study are included in the article. Further inquiries can be directed to the corresponding authors.

Acknowledgments: For the support and insightful discussions provided by the project's industrial partner, Karex Industries Sdn. Bhd, the authors extend their gratitude. In particular, the authors would like to thank the staff and technicians from Karex Industries Sdn. Bhd., whose direct and indirect involvement was instrumental in the successful completion of this project.

Conflicts of Interest: The authors declare that there is no conflict of interest. This research was funded by Karex Industries Sdn Bhd., which collaborated with the authors in designing the study. The funder had no influence on data collection, data analysis, interpretation of results, or the decision to submit the manuscript for publication.

Abbreviations

The following abbreviations are used in this manuscript:

NR	Natural rubber
TGA	Thermogravimetric analysis
GPC	Gel permeation chromatography
DMA	Dynamic mechanical analysis
Đ	Dispersity
DTG	Derivative thermogravimetry

References

1. Umeswara, S.S. Towards a Circular Economy Waste Management in Malaysia. In Proceedings of the MEA 12MP Kick-Off Conference, Kuala Lumpur, Malaysia, 1–4 July 2019.
2. Growth Within: A Circular Economy Vision for a Competitive Europe. Available online: https://www.mckinsey.com/capabilities/sustainability/our-insights/growth-within-a-circular-economy-vision-for-a-competitive-europe (accessed on 20 December 2024).
3. Tono, T.; Nakazawa, Y.; Sato, K.; Hashimoto, M.; Aizawa, M.; Ikeda-Fukazawa, T. Effects of thermal history of silica composite polyisoprene rubber on structure of contact water. *Chem. Phys. Lett.* **2024**, *847*, 141373. [CrossRef]
4. Malaysian Rubber Council. World Rubber Production, Consumption and Trade. Available online: https://www.myrubbercouncil.com/industry/world_production.php (accessed on 26 February 2025).

5. Jawjit, W.; Pavasant, P.; Kroeze, C.; Tuffey, J. Evaluation of the potential environmental impacts of condom production in Thailand. *J. Integr. Environ. Sci.* **2021**, *18*, 89–114. [CrossRef]
6. Markl, E.; Lackner, M. Devulcanization technologies for recycling of tire-derived rubber: A review. *Materials* **2020**, *13*, 1246. [CrossRef]
7. Tang, X.; Chen, Z.; Liu, J. Dynamic pyrolysis behaviors, products, and mechanisms of waste rubber and polyurethane bicycle tires. *J. Hazard. Mater.* **2021**, *402*, 123516. [CrossRef]
8. Maga, D.; Aryan, V.; Blömer, J. A comparative life cycle assessment of tyre recycling using pyrolysis compared to conventional end-of-life pathways. *Resour. Conserv. Recycl.* **2023**, *199*, 107255. [CrossRef]
9. Brunella, V.; Aresti, V.; Romagnolli, U.; Muscato, B.; Girotto, M.; Rizzi, P.; Luda, M.P. Recycling of EPDM via Continuous Thermo-Mechanical Devulcanization with Co-Rotating Twin-Screw Extruder. *Polymers* **2022**, *14*, 4853. [CrossRef]
10. Simon, D.Á.; Bárány, T. Microwave Devulcanization of Ground Tire Rubber and Its Improved Utilization in Natural Rubber Compounds. *ACS Sustain. Chem. Eng.* **2023**, *11*, 1797–1808. [CrossRef]
11. Karunarathna, S.; Linforth, S.; Kashani, A.; Liu, X.; Ngo, T. Effect of recycled rubber aggregate size on fracture and other mechanical properties of structural concrete. *J. Clean. Prod.* **2021**, *314*, 128230. [CrossRef]
12. Ceretti, D.V.A.; Edeleva, M.; Cardon, L.; D'hooge, D.R. Molecular Pathways for Polymer Degradation during Conventional Processing, Additive Manufacturing, and Mechanical Recycling. *Molecules* **2023**, *28*, 2344. [CrossRef]
13. Candau, N.; Leblanc, R.; Maspoch, M.L. A comparison of the mechanical behaviour of natural rubber-based blends using waste rubber particles obtained by cryogrinding and high-shear mixing. *Express Polym. Lett.* **2023**, *17*, 1135–1153. [CrossRef]
14. Muzyka, R.; Mumtaz, H.; Sobek, S.; Werle, S.; Adamek, J.; Semitekolos, D.; Charitidis, C.A.; Tiriakidou, T.; Sajdak, M. Solvolysis and oxidative liquefaction of the end-of-life composite wastes as an element of the circular economy assumptions. *J. Clean. Prod.* **2024**, *478*, 143916. [CrossRef]
15. Evergreen Corporate Sdn Bhd. Evergreen Corporate Making the World a Better Place. Available online: https://www.evergreencorporate.com/ (accessed on 6 March 2025).
16. Xiao, Z.; Pramanik, A.; Basak, A.K.; Prakash, C.; Shankar, S. Material recovery and recycling of waste tyres—A review. *Clean. Mater.* **2022**, *3*, 100115. [CrossRef]
17. Guo, L.; Wang, C.; Lv, D.; Ren, D.; Zhai, T.; Sun, C.; Liu, H. Rubber reclamation with high bond-breaking selectivity using a low-temperature mechano-chemical devulcanization method. *J. Clean. Prod.* **2021**, *279*, 123266. [CrossRef]
18. Pirityi, D.Z.; Bárány, T.; Pölöskei, K. Recycling of EPDM rubber via thermomechanical devulcanization: Batch and continuous operations. *Polym. Degrad. Stab.* **2024**, *230*, 111014. [CrossRef]
19. Gschwind, L.; Jordan, C.S.; Vennemann, N. Devulcanization of ethylene-propylene-diene monomer rubber waste. Effect of diphenyl disulfide derivate as devulcanizing agent on vulcanization, and devulcanization process. *J. Appl. Polym. Sci.* **2022**, *139*, 52141. [CrossRef]
20. Cataldo, F. Thermochemistry of Sulfur-Based Vulcanization and of Devulcanized and Recycled Natural Rubber Compounds. *Int. J. Mol. Sci.* **2023**, *24*, 32623. [CrossRef]
21. Shabani, M.; Jamshidi, M. Recycling NR/SBR waste using probe sonication as a new devulcanizing method; study on influencing parameters. *RSC Adv.* **2022**, *12*, 26264–26276. [CrossRef]
22. Ghamarpoor, R.; Jamshidi, M. Synergistic effect of microwave assisted devulcanization of waste NBR rubber and using superhydrophobic/superoleophilic silica nanoparticles on oil-water separation. *Alex. Eng. J.* **2023**, *69*, 67–84. [CrossRef]
23. Li, J.; Jiang, N.; Sheng, Z.; Qin, L.; Li, C.; Wang, Y.; Liu, Z.; Yang, J.; Li, J. Devulcanization of ground tire rubber using a plasma-assisted fluidized-bed system. *Appl. Surf. Sci.* **2025**, *684*, 161784. [CrossRef]
24. Ghowsi, M.A.; Jamshidi, M. Recycling waste nitrile rubber (NBR) and improving mechanical properties of Re-vulcanized rubber by an efficient chemo-mechanical devulcanization. *Adv. Ind. Eng. Polym. Res.* **2023**, *6*, 255–264. [CrossRef]
25. Costamagna, M.; Brunella, V.; Luda, M.P.; Romagnolli, U.; Muscato, B.; Girotto, M.; Baricco, M.; Rizzi, P. Environmental assessment of rubber recycling through an innovative thermo-mechanical devulcanization process using a co-rotating twin-screw extruder. *J. Clean. Prod.* **2022**, *348*, 131352. [CrossRef]
26. Colom, X.; Cañavate, J.; Formela, K.; Shadman, A.; Saeb, M.R. Assessment of the devulcanization process of EPDM waste from roofing systems by combined thermomechanical/microwave procedures. *Polym. Degrad. Stab.* **2021**, *183*, 109450. [CrossRef]
27. SSH Recycling. Available online: https://sshrecyclingltd.org.uk/our-process/ (accessed on 17 March 2025).
28. Tyromer Inc. Technology Rubber Devulcanization. Available online: https://tyromer.com/technology/ (accessed on 17 March 2025).
29. Asaro, L.; Gratton, M.; Poirot, N.; Seghar, S.; Hocine, N.A. Devulcanization of natural rubber industry waste in supercritical carbon dioxide combined with diphenyl disulfide. *Waste Manag.* **2020**, *118*, 647–654. [CrossRef]

30. Saiwari, S.; Nobnop, S.; Bueraheng, Y.; Thitithammawong, A.; Hayeemasae, N.; Salaeh, S. Segregated MWCNT Structure Formation in Conductive Rubber Nanocomposites by Circular Recycling of Rubber Waste. *ACS Appl. Polym. Mater.* **2022**, *4*, 7463–7475. [CrossRef]
31. *ASTM D3616-95(2024)*; Standard Test Method for Rubber—Determination of Gel, Swelling Index, and Dilute Solution Viscosity. ASTM International: West Conshohocken, PA, USA, 2024.
32. Ge, M.; Zheng, G. Fluid–Solid Mixing Transfer Mechanism and Flow Patterns of the Double-Layered Impeller Stirring Tank by the CFD-DEM Method. *Energies* **2024**, *17*, 1513. [CrossRef]
33. Walvekar, R.; Afiq, Z.M.; Ramarad, S.; Khalid, S. Devulcanization of Waste Tire Rubber Using Amine Based Solvents and Ultrasonic Energy. *MATEC Web Conf.* **2018**, *152*, 01007. [CrossRef]

processes

MDPI

Article

A Two-Level Facility Layout Design Method with the Consideration of High-Risk Facilities in Chemical Industries

Guanxin Xu, Siyu Xu and Yufei Wang *

State Key Laboratory of Heavy Oil Processing, China University of Petroleum, Beijing 102249, China; 2022310241@student.cup.edu.cn (G.X.); 2018213117@student.cup.edu.cn (S.X.)
* Correspondence: wangyufei@cup.edu.cn

Abstract

Understanding facility layout design in chemical industries requires multi-disciplinary knowledge and experience. The recent work mainly focuses on improving safety and calculating the efficiency of the design. However, in chemical industries, facilities are always located in frames, so both facility layout and frame layout should be considered in the design, as well as safety. Such a situation has not been well studied. In this work, facilities are divided into several frames and then placed in a fixed area. The risk resources located in the frames and out of the frames are both contained, and the safety distances are compliant with relative regulations. Optimization and some heuristic rules are applied to obtain the layout of each frame and the whole plant. Moreover, fire embankments are considered to achieve a more realistic and reasonable final layout. As a result, compared with the initial one, the actual and potential safety factors and the reasonable degree of the factory layout are both improved. The total costs are reduced by 7.38×10^4 \$$\cdot$a^{-1}. Through these steps, the effectiveness of the proposed approach is proven.

Keywords: facility layout problem; safety factor; genetic algorithm; surplus rectangle fill algorithm

1. Introduction

Understanding facility layout problems (FLPs) requires multi-disciplinary knowledge and experience. Experienced designers can offer a feasible industry layout but cannot optimize the layout. With the help of a computer-aided systematic method, an optimized layout design with lower costs can be obtained. The recent research on computer-aided layout optimization mainly focuses on improving the safety of the layout and calculating its efficiency.

For the safety aspect, several improved approaches for quantitative risk assessments (QRAs) combined with uncertain assumptions [1] or dynamic and accident simulations [2] were proposed. The previous works about layout optimization integrated with safety factors can be roughly classified into two categories. One is using the loss incurred by the accidents to measure the risk level [3]. Alves et al. [4] mitigated the hazard by minimizing consequences to nearby residential areas in the event of accidents. They used the Monte Carlo method to estimate the superposition of accident effect areas onto population polygons. According to the case evaluation, their method could obtain feasible layouts and effectively reduce risk levels. In addition, Caputo et al. [5] presented an optimization method based on the Genetic Algorithm (GA), and evaluated safety issues by calculating the expected annual loss due to the secondary unit damage, which is caused by the primary accidents happening in nearby process units. A case study was described to show the effectiveness of their approach. Although the safety level can be improved through these methods, it is often incomplete since all aspects of the loss, such as compensation for casualties and costs of environmental pollution treatment, cannot be completely covered. In the other type of work, facilities were studied through risk assessment to obtain minimum safety distances, which were accounted for in layout optimization [6]. For instance, Ahumada et al. [7] obtained the shortest isolation distances between units on the basis of loss of human life and structural damage. However, in addition to the results calculated by the mathematical models, the requirements of security-related administrative regulations for the minimum safety distance must be complied with in real design works. In addition, the former works failed to add common protection facilities into the layout, such as a fire embankment.

For the research on the algorithm used in layout design, various algorithms like Coral Reefs Optimization (CRO) [8], Variable Neighborhood Search (VNS) and Simulated Annealing (SA) [9] are examined in recent studies. Among all of them, GA is an algorithm based on the mechanisms of natural selection, which is regarded as one of the most adopted and effective algorithms in FLPs [10]. Mohamadi et al. [11] proposed a new approach including a bi-level GA to solve FLPs in an open field. They compared the proposed approach with four approaches from the literature to evaluate efficiency and performance. The quality of the method is verified by improving the best solution obtained in previous studies and obtaining the optimal solution. Besbes et al. [12] used an approach by combining GA with A * Search Algorithm to establish a 3D layout model. They studied the effectiveness of the combined algorithm, comparing it with other optimizing methods. The results show that although their method had a lower convergence speed, it could obtain a better layout. In addition, different metaheuristics can also be combined for hybrid algorithms, in order to obtain better calculating performance [13].

In the previous works, several factors were taken into account, such as material transportation cost [14], cable cost [15], production time [16], total distance of logistics and space occupation [17]. Moreover, multi-floor structures were proposed [18]. A model considering inner frame structures could reduce the interconnection cost including land, floor, pipeline construction and material handling [19]. However, in the actual layout design work, the facility layout is generally carried out in a certain position with a fixed length and width. In other words, the land cost is determined before designing the layout. So, what needs to be carried out is the increase in the utilization rate of fixed space, rather than a reduction of land costs.

In real chemical industries, facilities are arranged in several frames, and frames, high facilities and high-risk facilities are arranged in the plant area. However, by studying previous works, it is found that such two-level facilities layout design has not been well studied. In addition, safety aspects have not been well incorporated in two-level facility layout designs. In this work, GA and the Surplus Rectangle Fill Algorithm (SRFA) are combined into a two-level facility layout design method. The layouts of the facilities and frames are optimized with the minimum annual total cost as the objective function. The safety distances of risk resources are compliant with relative regulations and the corresponding costs are added to optimization objectives. In addition, fire embankments are considered in the industrial plant. To improve the practicality of the model, the area of the whole plant is fixed, and the total land utilization rate should be as high as possible.

2. Mathematical Model and Algorithm

2.1. Problem Statement

In this work, the size of the whole plant is fixed. Some basic data, such as sizes, numbers and safety distances of the facilities and flow information, are known. Frame layout and plant layout are the two specific levels of layouts. Through modeling and optimization, a complete plant layout can be acquired by considering multiple aspects. The information obtained should include the following:

- Coordinates of each facility in the frame region.
- Relative locations of frames and special facilities, including towers, reactors and compressor rooms.
- Various costs and internal layout of each frame and the whole layout.

FLPs usually contain plenty of variables and constraints. Thus, several assumptions are required to simplify the model and reduce the calculation time.

- The facilities on the same floor are arranged on a two-dimensional plane.
- Plants, frames and all the facilities are placed orthogonally.
- High facilities, including towers and reactors, are regular circles; the other facilities, plants and frames are regular rectangles.
- Compressors are placed in a rectangular area to form a compressor room, regarded as an original facility.
- Special facilities, including high facilities and the compressor room, are placed apart from the frames in the plant layout.
- The materials across two frames are defined as cross-frame connections (CFCs). The exit points of CFCs are set as the midpoints of sides of rectangle frames.
- If several risk facilities are clustered together in the plant, fire embankment(s) can be constructed between the risk source clustering area and other facilities. The corresponding safety distances can be reduced.
- If a single-floor layout is incapable of containing all the facilities, multi-floor structures can be considered in the frame.
- If multiple floors are applied, vertical materials and floor requirements for special facilities should be concentrated on.

2.2. Constraints

Frame layout deals with multiple facilities possessing various shapes and functions. Therefore, more constraints are necessary to be considered generally. To achieve a reasonable and effective frame layout, a couple of constraints are set.

Facility orientation constraints are set up to determine the length and width of each facility within the frame area. Facilities can be placed in different directions, while the direction and the original size can determine the size of the layout, which is defined as "visual size". In addition, the term "length" is defined as the size parallel to the x-axis and "width" is that in the y direction. A particular relationship between the visual size and original size of a facility can be developed. A binary variable d is defined to illustrate this relationship. It is stipulated that when $d_i = 0$, the facility i is placed vertically, that is, the original length equals its visual width. On the contrary, when $d_i = 0$, it is arranged horizontally. Then, the relationship between the visual and actual size of the facility can be described as follows:

$$l_i = (1 - d_i)l_{o,i} + d_i w_{o,i} \quad (1)$$

$$w_i = (1 - d_i)w_{o,i} + d_i l_{o,i} \quad (2)$$

where l_i and w_i are the visual length and width of facility i (m); $l_{o,i}$ and $w_{o,i}$ are the original length and width of facility i (m).

Non-overlapping constraints are applied to avoid overlapping regions of distinct facilities. These constraints are related to the length, width and the coordinates of the midpoint of each facility. Facility i and j are forced to accommodate one of the following inequality constraints to prevent a cross area:

$$\left|x_i - x_j\right| \geq \frac{1}{2}l_j + \frac{1}{2}l_i \tag{3}$$

$$\left|y_i - y_j\right| \geq \frac{1}{2}w_j + \frac{1}{2}w_i \tag{4}$$

where x_i and y_i are the center coordinates of facility i; x_j and y_j are that of facility j. Inequalities (3) and (4) represent the situations that facility i is on the right and left, top and bottom sides of facility j, respectively.

Boundary constraints demand that all the facilities cannot exceed the frame bounds. The following four constraints must be satisfied simultaneously to limit the location of a facility:

$$x_i - \frac{1}{2}l_i \geq 0 \tag{5}$$

$$x_i + \frac{1}{2}l_i \geq L \tag{6}$$

$$y_i - \frac{1}{2}w_i \geq 0 \tag{7}$$

$$y_i + \frac{1}{2}w_i \geq W \tag{8}$$

where L and W are the length and width of the frame (m).

According to the certain area of the plant, the sizes of the frames and special facilities are indispensable to meet boundary constraints:

$$X_i - \frac{1}{2}L_i \geq 0 \tag{9}$$

$$X_i + \frac{1}{2}L_i \leq L_p \tag{10}$$

$$Y_i - \frac{1}{2}W_i \geq 0 \tag{11}$$

$$Y_i + \frac{1}{2}W_i \leq W_p \tag{12}$$

where X_i and Y_i are the center coordinates of frame i; L_i and W_i are the length and width mentioned before (m). L_p and W_p are the fixed length and width of the plant (m).

Pump constraints are established to restrict all the pumps of a frame within a rectangular area for centralized management. In each frame, pumps are arranged in an orderly manner as a whole. In addition, the pump area is defined as the smallest rectangle enclosing all the pumps, even if it is not completely filled by them, in order to prevent irregular shapes in the model. The pump area is regarded as a facility to be optimized together with others. Since the relative positions of the pumps within the pump area are determined, the actual coordinates in the frame can be described as follows:

$$x_{pi} = x_{pri} + x_{pa} \tag{13}$$

$$y_{pi} = y_{pri} + y_{pa} \tag{14}$$

where x_{pi} and y_{pi} are the center coordinates of pump i in the frame; x_{pri} and y_{pri} are that in the pump area. x_{pa} and y_{pa} are the lower left coordinates of the whole pump area.

Floor constraints for specific kinds of facilities are requisite when multi-floors are applied in the frame. For instance, pumps must be placed on the first floor to prevent cavitation. Air coolers are required to be assigned on the top floor to guarantee a cooling effect. A couple of parallel heat exchangers or air coolers with similar features should be set as a whole.

2.3. Objective Function

The total cost generally serves as the measure of layout performance [20], generally containing land cost (*LC*), floor cost (*FC*), pipeline construction cost (*PCC*) and material handling cost (*MHC*). In this work, the fixed area of the whole plant means fixed *LC*. So the total annual cost (*TAC*, $), containing the corresponding *FC*, *PCC* and *MHC*, is regarded as the objective function in each frame:

$$TAC = FC + PCC + MHC \tag{15}$$

FC ($) is associated with the frame area:

$$FC = Af \cdot UFC \cdot S \tag{16}$$

$$Af = \frac{I(1+I)^T}{(1+I)^T - 1} \tag{17}$$

$$S = \max(x_i + l_i) \cdot \max(y_i + w_i) \tag{18}$$

where *UFC* is the unit floor cost ($\$\cdot m^{-2}$); *S* is the frame area (m^2); Af is the annualized factor; *I* is the interest rate; and *T* is life cycle of the industrial park (y).

PCC ($) is related to the pipeline investment:

$$PCC = Af \cdot \sum_{i=1}^{n} (UPC_i \cdot MD_i) \tag{19}$$

$$MD = |x_i - x_j| + |y_i - y_j| \tag{20}$$

where *UPC* is the unit pipeline cost ($\$\cdot m^{-1}$), which is obtained by the method proposed by Stijepovic and Linke [21]; *MD* is the Manhattan distance (m) between connected facilities.

$$UPC = 0.82wt_{pipe} + 185D_{out}^{0.48} + 6.8 + 295D_{out} \tag{21}$$

where wt_{pipe} is the unit quality of pipelines (kg/m); D_{out} is the outer diameter of pipelines (m). They are determined by Equations (22)–(24) [21]:

$$wt_{pipe} = 644.3D_{inner}^2 + 72.5D_{inner} + 0.4611 \tag{22}$$

$$D_{out} = 1.052D_{inner} + 0.005251 \tag{23}$$

$$D_{inner} = \sqrt{\frac{4Q}{\pi\rho u}} \tag{24}$$

where D_{inner} is the inner diameter of pipelines (m), Q is the mass flow rate of the medium in the pipelines (kg·s^{-1}), ρ is the density of the materials (kg·m^{-3}) and u is the flow rate (m·s^{-1}).

MHC ($) is related to the power consumed by the pumps for materials transporting, which can be calculated as follows:

$$MHC = C_E \cdot H \cdot \sum P \tag{25}$$

where C_E is the cost required for the pumps to consume per unit of operating power ($·kW^{-1}·h^{-1}), H is the annual operating hours (h) and P is the operating power of the pumps consumed by transporting materials (kW), which can be calculated as follows:

$$P = \frac{Q \cdot h_f}{\eta} \tag{26}$$

$$h_f = \lambda \frac{MD \cdot u^2}{2D_{inner}} + \alpha \cdot gz \tag{27}$$

where h_f is the energy used during the material handling process for overcoming resistance and gravity (J·kg^{-1}), η is the pump operation efficiency, λ is the friction coefficient, g is the gravity constant (m·s^{-2}), z is the length of vertical material transportation (m) and α is a binary variable, which is introduced to stipulate whether there is a connection transporting materials vertically from bottom to top.

2.4. Optimization Algorithm

Since complex FLPs involve multiple variables and constraints, it is difficult to solve them manually. Therefore, algorithms are often utilized in optimization progress. In this work, a composite algorithm combining GA and SRFA is used. A simple introduction is proposed for the two algorithms and their combined approach.

Compared with other metaheuristics, GA has evident advantages in dealing with FLPs. Due to the strong randomness of solutions generating, GA possesses significant strength in global searching [22]. There are numerous individuals having one-to-one correspondence with facilities, respectively. Just like genes on chromosomes determine the performance of individuals, GA sets a number of variables to determine the number, orientation and placement order of each facility, the length of each frame and the facility number in each floor (if there is a multi-floor structure). Variables are arbitrarily valued within the predetermined range to reflect the randomness of the optimization. The desired results with minimized TAC can be obtained through GA.

Although GA is suitable to solve FLPs according to its features, it has disadvantages in acquiring a final precise layout. It is difficult to consider the sizes and safety distances of facilities while using GA, causing hardly generating efficient and tight layouts. Therefore, an additional algorithm is required to solve the problems. SRFA [23] is capable of dividing a rectangular plate into several parts. This feature is in line with the concept of arranging rectangular facilities within a rectangular area. SRFA creates a residual rectangle data set to collect available space for the arrangement of facilities, which is utilized to obtain the final layout along with the coordinates of the facilities.

Although SRFA proves to be appropriate for arranging rectangular facilities in a given region, it cannot independently solve FLPs since it is unable to minimize TAC through iteration. In addition, it requires initial and essential conditions like facility placement order and orientation. These can be available through GA. Thus, a hybrid algorithm combined with GA and SRFA can achieve optimization of the frame layout. For the optimization of frame layouts, GA randomly generates feasible solutions, including facility placement order, orientation and frame bottom length, based on the given information such as facility numbers and sizes. With all the data above, SRFA designs the precise layout from the residual rectangle data set to achieve corresponding facility coordinates, and returns them to GA as input parameters, in order to minimize TAC with flow information. The iteration continues until the termination condition is satisfied. Then, the layout with the smallest fitness value is selected as the output result. In addition, for the optimization of plant layouts, the initial information entered into GA is about the frames and CFCs. Figure 1 shows the workflow of the hybrid algorithm of frame layout optimization. The algorithm is carried out in MATLAB 2019b, and the relative operators in GA are the default in the toolbox.

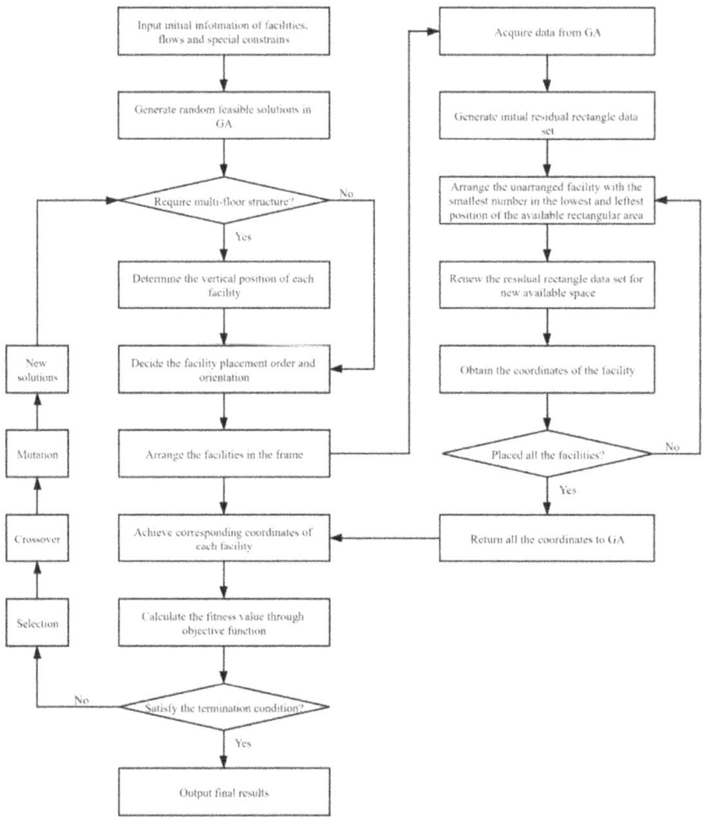

Figure 1. Workflow of the hybrid algorithm.

3. Optimization Approach

In this work, the implementation of frame layouts for facilities and plant layouts for frames are the two levels of layout design. Before dividing the facilities into a couple of frames, the safety distances of the high-risk facilities are considered. A quantitative approach is used with the objective of minimizing the number of cross-frame material connections during the segmentation of facilities. The initial size and internal layout of each frame are obtained through the first-step optimization. Then, the frames are arranged together with the special facilities (towers, reactors and compressor room in this work) in the specified area according to the combination of optimization and designer decisions. According to the layout optimization of frames in the previous step, the exit positions of the CFCs are determined on the boundaries of the frames. The second-step optimization in each frame is carried out to obtain an improved layout of the plant with consideration of CFCs. Moreover, fire embankment(s) can be added between risk resources and other facilities in the plant to reduce safety distances. Therefore, the third-step optimization of the plant layout is required. The above steps can be described in detail as follows.

- Facilities are classified depending on their features. Longer fire distances specified in the regulations are necessary to be added to the sizes of the hazard resources. Parallel heat exchangers should be arranged as a whole. Pumps and compressors are required to be placed in an orderly manner within the pump area and compressor room, respectively. The special facilities, like reactors, towers and compressor rooms, should be separated from the frames. Attention should be paid to the relevant floor constraints associated with particular facilities such as pumps and air coolers if there is a multi-floor structure in the frame.
- According to the flow information, material connections are categorized into internal ones within the frame and cross-frame ones.
- The facilities, except the special ones, are divided into several frames. The number of frames is determined according to the actual situation. The classification principle is to find out the cutting positions with the least total number of CFCs. This approach will be elaborated in detail in the following statements.
- Each frame is optimized utilizing the hybrid algorithm, while only internal connections (ICs) are considered. Then, the initial sizes and layouts of frames are obtained. Multi-floor frame(s) may be required to prevent exceeding the specified area of the plant.
- Hybrid algorithms are applied to arrange the frames together with the special facilities within the specified space, as the same way of arranging facilities within the frame. The positions can be altered according to designer decisions.
- After arranging the frames and special facilities, an initial plant layout is obtained. In addition, the exits of CFCs are determined at some midpoints of the sides of the frames.
- Each frame is re-optimized by additionally considering CFCs, and the sizes and structures of frames are updated.
- The whole plant layout is updated with the updated frames and changeless exit positions of CFCs.
- Fire embankment(s) are constructed between risk resources and other facilities to reduce safety distances. Sizes and structures of a part of frames probably convert during manual adjustment.
- The whole plant layout is optimized again with the adjusted frames.

The designer's decisions, added to the model during the process of arranging frames and special facilities, can save unnecessary space by ensuring enough safety distances. Frames and special facilities are arranged by computational optimization theoretically. However, the calculation results may lead to reduced space utilization. For instance, when a facility with high risk is located near the edge of the plant, the safety distance on the side close to the edge is unnecessary. The manual adjustment of the example is shown in Figure 2. Therefore, designer decisions are required to prevent such problems.

A quantitative method is proposed to sort facilities into several frames. CFCs usually mean longer material transportation distances. Therefore, the minimization of the number of them while sorting can theoretically reduce the costs of the final plant layout.

To minimize the number of CFCs, some work needs to be carried out. All the facilities, except special ones, are arranged in a single-floor area. The number of frames depends on the result obtained. The width of each frame is described as follows:

$$\frac{H}{n}i - h \leq cp_i \leq \frac{H}{n}i + h \tag{28}$$

$$W_i = \begin{cases} cp_i, i = 1 \\ cp_i - cp_{i-1}, 1 < i < n \\ H - cp_{i-1}, i = n \end{cases} \tag{29}$$

where H is the width of the whole single-floor area (m), m is the number of frames, cp_i is the coordinates in the direction of y-axis of cutting point i and h is the range that cp_i can fluctuate. W_i is the width if frame i.

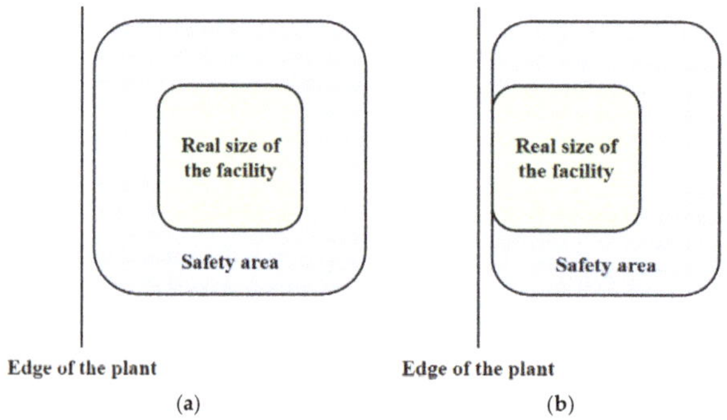

Figure 2. Manual adjustment of the position of facility: (**a**) original position and safety area; (**b**) modified position and safety area.

After determining all the cp_i, the frame number i each facility belongs to can be obtained. If the two facilities connected by one flow are placed in different frames, the flow connection is regarded as a cross-frame one. The hybrid algorithm is adopted in the objective function to minimize the total number of CFCs (TN):

$$TN = \sum_{i=1, j=1, i \neq j}^{n} N_{ij} \tag{30}$$

In the single-floor layout in Figure 3, facilities are simplified as center points (in red). The cutting lines (in green) are set to be perpendicular to the y-axis and move within a certain range to divide all the facilities into several frames. The width of each frame can be slightly different in order to minimize the number of CFCs. All the flows are drawn in the form of right-angle lines. The connections judged as a cross-frame one by the approach are drawn in black and the internal ones are drawn in blue. Obviously, when more than two frames are required, if the length of a connection exceeds the width of a single frame, it will be cut by multiple cutting lines. Therefore, if the total number of cutting points is treated as the objective, a suboptimal result may be obtained due to the repeated calculation of the number of CFCs. The calculation method proposed above can avoid this issue.

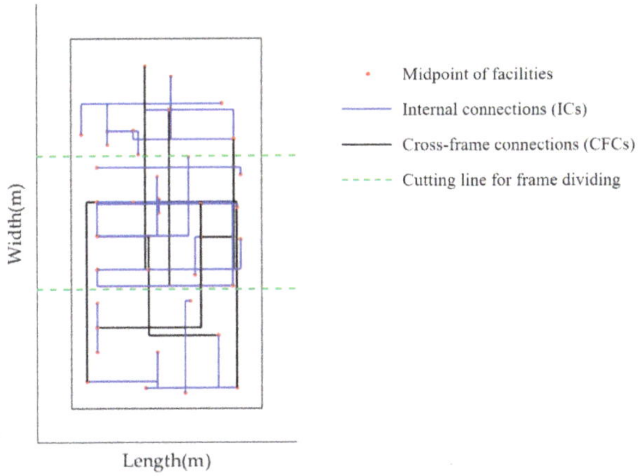

Figure 3. Schematic diagram for choosing cutting points.

4. Case Study and Result Analysis

To verify the effectiveness of the proposed method, a 70 m $\times$ 70 m-area plant with 141 facilities and 247 material connections is designed and optimized. In this case, numerous rectangular facilities, with different visual and original lengths, and widths are placed in a fixed rectangular space. Facilities possess various functions and specific placement constraints. Risk resources arranged in frames and separated from them are contained. Parallel placements and centralized arrangements are involved. Multi-floor structure frame(s) and fire embankment(s) are utilized. Therefore, if the case is verified as successfully coped with, other types of block layouts can be solved through the proposed method with slight modifications according to the situation.

In this case, there are ten special facilities, which are riser reactor (RR), settler–regenerator (SR), fractionating tower (FT), stripping tower (ST), absorption–desorption tower (ADT), stabilization tower (STA), reabsorption tower (RT) and three compressors. The high-risk compressors are arranged together in a compressor room for ease of management and fire prevention. The basic data on the flow information, the sizes and categories of the facilities have been acquired. The remaining 131 facilities, containing 5 risk resource facilities, 82 heat exchangers and air coolers, 6 vessels and 38 pumps, should be divided into several frames and then optimized with the objective of minimizing the number of CFCs.

Before the optimization, the number of frames required should be defined. There is no doubt that the fewer frames there are, the fewer CFCs. The respective optimized results when 3, 4, 5 and 6 frames are required are shown in Table 1.

Table 1. Information on the total number of CFCs of different arrangement plans.

Number of Frames	Total Number of CFCs
3	31
4	35
5	49
6	70

The results of 3 and 4 frames are similar, and far fewer than that of 5 and 6 frames. To receive the minimal number of CFCs, the layout should be divided into three frames. However, numerous facilities being divided into one frame leads to the excessive width of each frame if the number of frames is only three. With the consideration of safety factors, fire and emergency accesses should be constructed both inside and outside the frames. The increase in the length of the passages, caused by the longer frame widths, will extend escape time. More outer safe passages can be arranged with smaller frames. Thus, the allocation plan of four frames and three cutting points is selected.

Heat exchangers, including air coolers, are placed in the same direction for neatness. A couple of parallel ones can be regarded as a whole to simplify the diagram. The bottom length is set to be approximately 30 m. The optimization results of the facility arrangement and the frame segmentation are shown in Figure 4a,b, as well as the ICs and CFCs. The calculation process takes around 600 s.

The calculation results show that the length and width of the single-floor layout are 32.97 m and 124 m. The width of each frame is approximately 32 m. Thus, the locations of the three cutting points are about 32 m, 64 m and 96 m in the y-direction, respectively. Four frames are named A to D from the bottom to the top. The optimization result is shown in Tables 2 and 3.

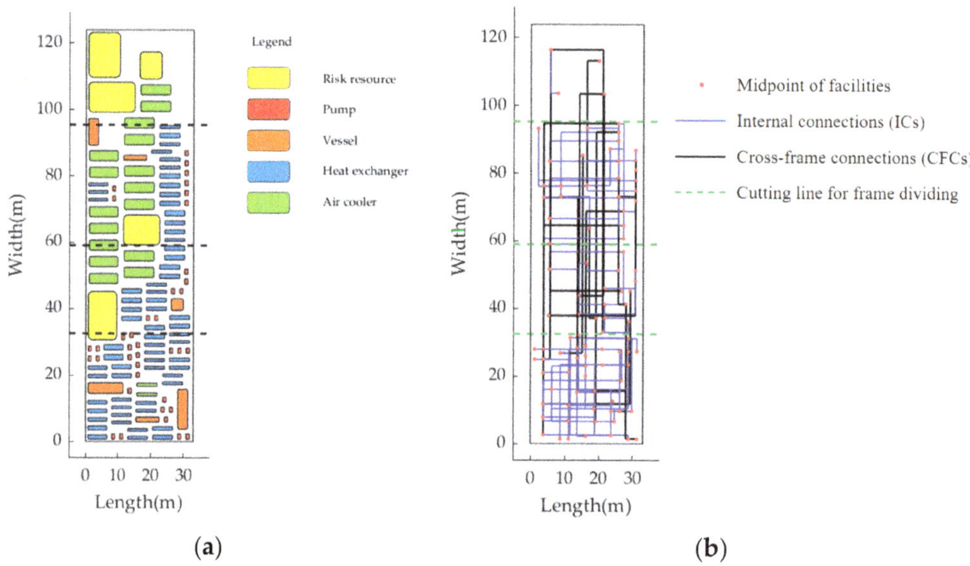

Figure 4. Facility separation of the single-floor layout: (**a**) the result of the facility arrangement and frame segmentation; (**b**) ICs and CFCs in the layout.

Table 2. Information about CFCs between each frame.

	A	B	C	D	Total Number of CFCs
A	-	6	8	1	
B	-	-	8	9	35
C	-	-	-	3	

Table 3. Information about cutting point positions and widths of each frame.

Frame	Precise Extents in Widths (m)		Width of Each Facility (m)
	Lower Point (m)	Upper Point (m)	
A	0	32.55	41
B	32.55	59.10	14
C	59.10	95.45	18
D	95.45	124.00	5

Obviously, facilities are arranged more closely in frames A and C. According to the result shown in Figure 4, more than half of the pumps are arranged in frame A; risk resources are mostly placed in frame D; and parallel air coolers are mainly contained in frame C. Thus, it can be figured out that facilities with similar functions tend to be centrally arranged in the same frame. This situation is in line with the actual layouts.

According to the layout of an actual plant, special facilities (reactors, towers and compressors) should be located outside the frames. Therefore, the connections for material transport between the special facilities and frames are also regarded as cross-frame ones. Table 4 presents the number of facilities, ICs and relative CFCs of each frame.

Table 4. Information on facilities and corresponding connections of each frame.

Frame	Number of Facilities Inside	Number of ICs	Number of Connections Between this Frame and Special Facilities
A	44	69	41
B	18	23	14
C	28	40	18
D	4	1	5

Through the above steps, four frames are determined. However, this step is only used to assign facilities to minimize the number of CFCs, without consideration of costs. The sizes and layouts should be then optimized to minimize *TAC*. In this step, only ICs are considered because the material existing points of CFCs are not determined since the relative locations of frames are not fixed.

All the frames are firstly optimized to be single-floor, in order to maximize the utilization rate of the certain plant area. The optimized layouts are listed in Table 5.

Table 5. Initial size of single-floor layout of each frame.

Frame	Length (m)	Width (m)	Area (m^2)	Total Area (m^2)
A	23.07	46.70	1077.40	3724.00
B	26.11	31.60	825.11	
C	29.80	41.00	1222.00	
D	33.12	18.10	599.48	

When the special facilities are added to the whole plant, the safe distances of all the frames and special facilities should be considered. As a result, the total area required exceeds the fixed space. So, multi-floor frame(s) are required. Frame D cannot be transformed because of the very few number of facilities contained in the frame. The information on optimized multi-layer frames A, B and C are shown in Table 6.

Table 6. Initial size of multi-floor layout of each frame.

Frame	Length (m)	Width (m)	Area (m^2)	Reduced Area (m^2)
A	30.83	19.10	588.86	488.54
B	31.11	15.60	485.27	339.85
C	26.09	25.00	652.19	569.82

After comparing three optimized multi-layer frames, it is found that if frame A or B is transformed into a multi-floor one alone, the total area required still cannot meet the land restrictions. Thus, frame C is selected to be a double-floor layout to ensure the feasibility of the whole plant due to the most reduced area. The selected results of each frame are shown in Table 7.

Table 7. Selected initial size of each frame.

Frame	Length (m)	Width (m)	Area (m^2)	Total Area (m^2)
A	23.07	46.70	1077.40	3154.18
B	26.11	31.60	825.11	
C	26.09	25.00	652.19	
D	33.12	18.10	599.48	

Through the layout optimization based on economy, the sizes and internal structures have been adjusted for the convenience of the further arrangement of the plant layout. It is necessary to confirm the exit positions of CFCs at the edge of the frame. All the positions are set at the midpoint of any side of the frames for simplification. The whole plant layout of the first optimization is presented in Figure 5.

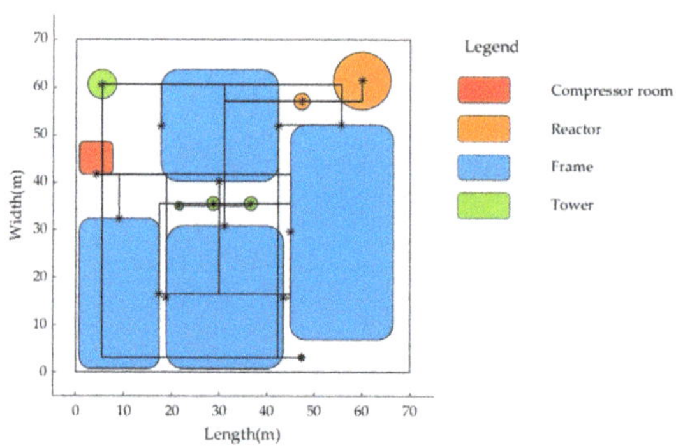

Figure 5. Plant layout with CFCs and exit positions of the first optimization.

In Figure 5, an initial plant layout is proposed. However, there are still several problems in the layout obtained by theoretical calculation. CFCs (connected with "*") are not taken into account. In addition, the location of ST is far away from other towers, which also goes against the experience of an actual design process. Thus, modifications of the initial frames are required to achieve an optimal layout. CFCs are added to the objective function in this step to minimize the total cost in each frame, and the modified results are shown in Table 8.

Table 8. Modified size of each frame.

Frame	Length (m)	Width (m)	Area (m^2)	Total Area (m^2)
A	23.58	46.60	1098.88	3211.89
B	26.61	32.00	851.64	
C	25.27	25.80	652.08	
D	19.10	31.90	609.29	

It is determined that the shapes and sizes of the four frames basically remain the same, which means the positions of material exit points can be applied to the modified plant layout. The result is shown in Figure 6.

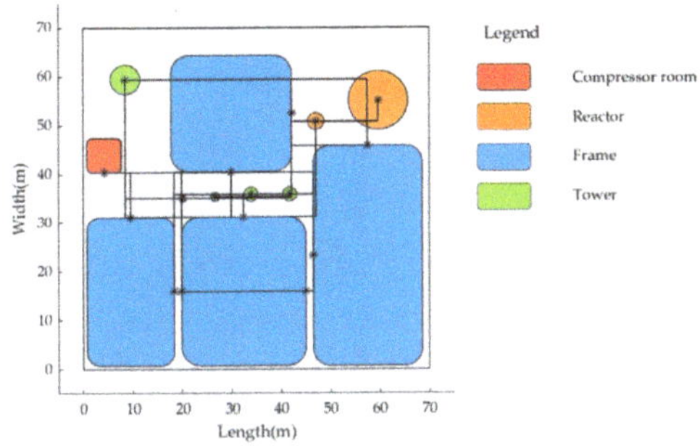

Figure 6. Plant layout with CFCs and exit positions of the second optimization.

In the modified plant layout with the consideration of CFCs, the positions of towers are also more reasonable. As mentioned earlier, risk resource facilities are mostly placed in frame D. According to the regulations, fire embankments can be constructed to decrease the pipeline length. In frame D, manual modification is added to obtain a structure including fire embankments. Figure 7 shows the modification approach of the structure.

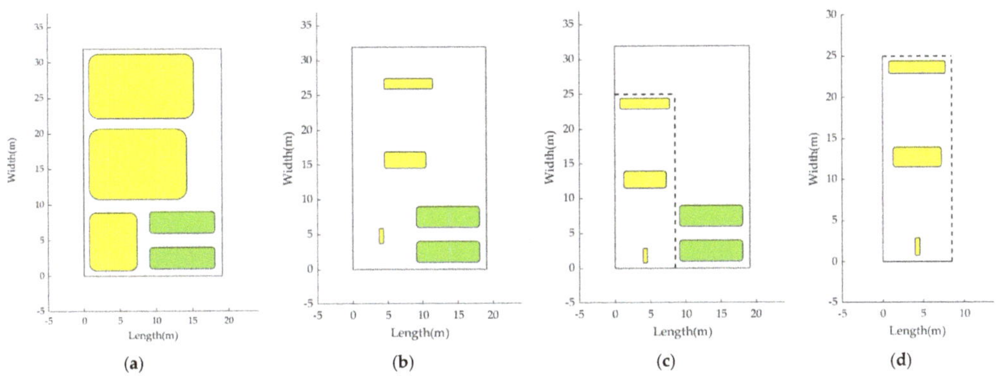

Figure 7. Adjustment process of frame D: (**a**–**d**) step 1 to step 4.

Figure 7a shows the original layout of the facilities in frame D with safety distances, and (b) shows the net sizes of them. Adjusted positions and added fire embankments are shown in (c). The structure of frame D is changed and the area is saved. To obtain a more reasonable and neater plant layout, the air cooler in frame D is moved to frame B, in order to minimize the increased number of CFCs. New frame D is obtained with the method above, and new frame B is achieved through optimization with consideration of new facilities and CFCs to minimize TAC. Figure 8 shows the plant layout with the original structures of frames A and C and adjusted B and D, and the sizes of adjusted frames B and D are shown in Table 9.

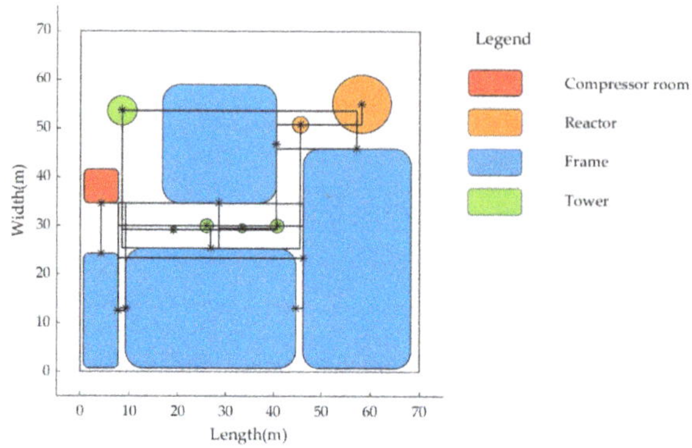

Figure 8. Plant layout with CFCs and exit positions of the third optimization.

Table 9. Sizes of the adjusted frames.

Frame	Length (m)	Width (m)	Area (m^2)
B	26.12	36.90	963.93
D	8.50	25.00	212.50

With the addition of fire embankments, the plant layout is more in line with the actual situation. The number of CFCs passing through the frames is further reduced to 9. This means lower hidden costs of the actual layout.

The respective *TAC* of initial, modified and adjusted plant layouts should be compared to prove the effectiveness of the optimization and the improvement of fire production. However, the contents of *PCC* and *MHC* in the objective function are different between the initial plant layout and the modified and adjusted one. Therefore, the *PCC* and MHC of cross-frame connections are calculated according to flow information, exit position and facility position, and are added to the costs of the initial results. Then, the comparison is made between the three layouts. Information on the three layouts is listed in Tables 10–12 and Figure 9. It should be noted that the area utilization in Tables 10–12 is calculated by the total area of all the frames. The area of all special facilities is constant in the three layouts. So, the comparison of the total area of all frames can reflect the real utilization rate.

Table 10. Results of the first optimization.

	IC		CFC
PCC (10^4 $\$\cdot a^{-1}$)	2.24		1.37
MHC (10^5 $\$\cdot a^{-1}$)	2.11		1.73
LC (10^4 $\$\cdot a^{-1}$)	0		6.69
FC (10^3 $\$\cdot a^{-1}$)	5.35		0
TAC (10^5 $\$\cdot a^{-1}$)	2.39		2.53
Total cost (10^5 $\$\cdot a^{-1}$)		4.92	
Area (m^2)	3154.18		4900.00
Area utilization (%)		64.37	

Table 11. Results of the second optimization.

	IC		CFC
PCC (10^4 \$·a^{-1})	2.71		1.43
MHC (10^5 \$·a^{-1})	1.94		1.42
LC (10^4 \$·a^{-1})	0		6.69
FC (10^3 \$·a^{-1})	5.34		0
TAC (10^5 \$·a^{-1})	2.26		2.23
Total cost (10^5 \$·a^{-1})		4.50	
Area (m^2)	3211.89		4900.00
Area utilization (%)		65.55	

Table 12. Results of the third optimization.

	IC		CFC
PCC (10^4 \$·a^{-1})	2.62		1.36
MHC (10^5 \$·a^{-1})	1.58		1.49
LC (10^4 \$·a^{-1})	0		6.69
FC (10^3 \$·a^{-1})	5.34		0
TAC (10^5 \$·a^{-1})	1.89		2.29
Total cost (10^5 \$·a^{-1})		4.18	
Area (m^2)	2927.39		4900.00
Area utilization (%)		59.74	

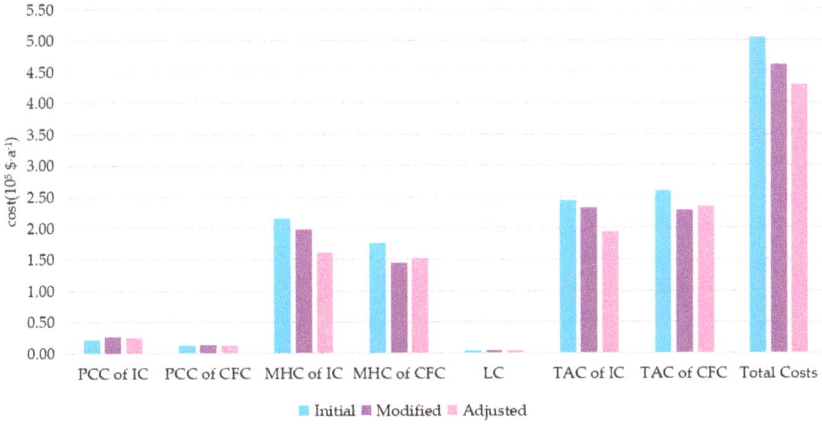

Figure 9. Cost comparison of three layouts.

According to the tables and figures, the land utilization rate of the second layout is the highest of the three, while that of the third one drops to the lowest due to the addition of fire embankments and the reduction in safety distances. The vacated land can be used to build fire protection facilities to improve the safety of the whole layout further. It seems to be contrary to the original idea of reducing costs by maintaining the highest possible land utilization. However, in fact, the total cost of the third layout is 4.18×10^5 \$·a^{-1}, which is 14.99% lower than that of the first one (4.92×10^5 \$·a^{-1}) and 6.93% lower than the second one (4.50×10^5 \$·a^{-1}), mainly due to the reduction in MHC of IC. This is because the flow rates of adjusted ICs are faster, or their medium temperatures are higher. Thus, when the lengths of them are reduced, the resulting reduction in *MHC* is more obvious. In addition, in the three layouts, a tendency of CFCs to be concentrated in a certain area is demonstrated. This is consistent with the actual layout design. The centralized arrangement of pipelines also facilitates subsequent design works, such as erecting pipe racks and constructing safe passages outside frames.

As a conclusion, the actual and potential safety factors and reasonable degree of the factory layout are both improved, by increasing the fire distance and adding fire embankments. The total costs of the plant are also effectively reduced through the combination of optimization algorithm and manual adjustment. As a result, compared with the initial layout, the total costs are reduced by 540,086.70 ¥/a after modification and adjustment.

5. Conclusions

In this work, efforts are made to arrange facilities in a two-level layout and increase the safety level of a plant layout through several steps. By using the proposed method, facilities can be well arranged in a frame and frames can then be well arranged in a plant. The risk of the whole plant is reduced, mainly by increasing safety distances of the risk facilities and adding safety protection facilities according to the structures and layout. This method added more options for the safety design of FLPs, facilitating multifaceted coordinated optimization. Compared with the previous work, the rationality of the layout has been improved. In the case study, the total costs of the layout after modification and adjustment are reduced as the objective function. After modification and adjustment, the *TAC* of the third layout is 4.18×10^5 \$·a^{-1}, which is 7.38×10^4 \$·a^{-1} lower than that of the first one and 3.12×10^4 \$·a^{-1} lower than that of the second one. According to the lowest area utilization rate (59.74%) of the three layouts, more fire protection equipment can be built in the third one for the further improvement of safety. In addition, the third layout is more reasonable and realistic, due to the addition of fire embankments and the tendency towards centralized arrangement of pipelines. Thus, the effectiveness and rationality of the proposed method can be proven.

As mentioned in the introduction, FLPs are complex and multi-branched. This work emphasizes the safety factor of two-level layout designs but ignores some other realistic problems, such as the 3D structures of the frames and facilities and the space occupied by the pipelines. Therefore, one of the future directions is to combine as many practical factors as possible. In addition, manual operation is still required in this work. For the convenience of the application in the industry in the future, complete automation of the solution is requisite.

Author Contributions: Conceptualization, Y.W.; methodology, Y.W.; software, G.X.; validation, G.X.; formal analysis, G.X.; investigation, G.X. and S.X.; data curation, S.X.; writing—original draft preparation, G.X.; writing—review and editing, Y.W.; visualization, G.X.; supervision, Y.W.; project administration, Y.W.; funding acquisition, Y.W. All authors have read and agreed to the published version of the manuscript.

Funding: Financial support from the National Natural Science Foundation of China under Grant (No. 22022816 and 22078358) are gratefully acknowledged.

Data Availability Statement: The data presented in this study are available on request from the corresponding author due to privacy.

Conflicts of Interest: The authors declare no conflicts of interest.

References

1. Ko, C.; Lee, H.; Kim, K.; Lee, W.B. Quantitative risk assessment integrated with dynamic process simulation for reactor section in heavy oil desulfurization process. *J. Loss Prev. Process Ind.* **2020**, *66*, 104158. [CrossRef]

2. Berner, C.L.; Flage, R. Creating risk management strategies based on uncertain assumptions and aspects from assumption-based planning. *Reliab. Eng. Syst. Saf.* **2017**, *167*, 10–19. [CrossRef]
3. Wang, R.; Wang, Y.; Gundersen, T.; Wu, Y.; Feng, X.; Liu, M. A layout design method for an industrial park based on a novel arrangement algorithm—Consideration of pipe network and multiple hazard sources. *Chem. Eng. Sci.* **2020**, *227*, 115929. [CrossRef]
4. Alves, D.T.S.; de Medeiros, J.L.; Araújo, O.d.Q.F. Optimal determination of chemical plant layout via minimization of risk to general public using Monte Carlo and Simulated Annealing techniques. *J. Loss Prev. Process Ind.* **2016**, *41*, 202–214. [CrossRef]
5. Caputo, A.C.; Pelagagge, P.M.; Palumbo, M.; Salini, P. Safety-based process plant layout using genetic algorithm. *J. Loss Prev. Process Ind.* **2015**, *34*, 139–150. [CrossRef]
6. Athar, M.; Shariff, A.M.; Buang, A.; Nazir, S.; Hermansyah, H.; See, T.L. Process equipment common attributes for inherently safer process design at preliminary design stage. *Process Saf. Environ. Prot.* **2019**, *128*, 14–29. [CrossRef]
7. Brunoro Ahumada, C.; Quddus, N.; Mannan, M.S. A method for facility layout optimisation including stochastic risk assessment. *Process Saf. Environ. Prot.* **2018**, *117*, 616–628. [CrossRef]
8. Garcia-Hernandez, L.; Garcia-Hernandez, J.A.; Salas-Morera, L.; Carmona-Muñoz, C.; Alghamdi, N.S.; de Oliveira, J.V.; Salcedo-Sanz, S. Addressing Unequal Area Facility Layout Problems with the Coral Reef Optimization algorithm with Substrate Layers. *Eng. Appl. Artif. Intell.* **2020**, *93*, 103697. [CrossRef]
9. Palubeckis, G. An Approach Integrating Simulated Annealing and Variable Neighborhood Search for the Bidirectional Loop Layout Problem. *Mathematics* **2020**, *9*, 5. [CrossRef]
10. Pierreval, H.; Caux, C.; Paris, J.L.; Viguier, F. Evolutionary approaches to the design and organization of manufacturing systems. *Comput. Ind. Eng.* **2003**, *44*, 339–364. [CrossRef]
11. Mohamadi, A.; Ebrahimnejad, S.; Soltani, R.; Khalilzadeh, M. An integrated approach based on a bi-level genetic algorithm and a combined zone-lp for the facility layout problem. *S. Afr. J. Ind. Eng.* **2019**, *30*, 87–101. [CrossRef]
12. Besbes, M.; Zolghadri, M.; Costa Affonso, R.; Masmoudi, F.; Haddar, M. 3D facility layout problem. *J. Intell. Manuf.* **2020**, *32*, 1065–1090. [CrossRef]
13. Tayal, A.; Solanki, A.; Singh, S.P. Integrated frame work for identifying sustainable manufacturing layouts based on big data, machine learning, meta-heuristic and data envelopment analysis. *Sustain. Cities Soc.* **2020**, *62*, 102383. [CrossRef]
14. Barbosa-Povoa, A.P.; Mateus, R.; Novais, A.Q. Optimal design and layout of industrial facilities: A simultaneous approach. *Ind. Eng. Chem. Res.* **2002**, *41*, 3601–3609. [CrossRef]
15. Jung, S.; Ng, D.; Laird, C.D.; Mannan, M.S. A new approach for facility siting using mapping risks on a plant grid area and optimization. *J. Loss Prev. Process Ind.* **2010**, *23*, 824–830. [CrossRef]
16. Ebrahimi, A.; Kia, R.; Komijan, A.R. Solving a mathematical model integrating unequal-area facilities layout and part scheduling in a cellular manufacturing system by a genetic algorithm. *SpringerPlus* **2016**, *5*, 1254. [CrossRef]
17. Zhao, Y.L.; Lu, J.S.; Yi, W.C. A new cellular manufacturing layout: Multi-floor linear cellular manufacturing layout. *Int. J. Adv. Robot. Syst.* **2020**, *17*, 1729881420925300. [CrossRef]
18. Ejeh, J.O.; Liu, S.; Papageorgiou, L.G. Optimal layout of multi-floor process plants using MILP. *Comput. Chem. Eng.* **2019**, *131*, 106573. [CrossRef]
19. Xu, S.; Wang, Y.; Feng, X. Plant Layout Optimization for Chemical Industry Considering Inner Frame Structure Design. *Sustainability* **2020**, *12*, 2476. [CrossRef]
20. Zha, S.S.; Guo, Y.; Huang, S.H.; Wu, Q.; Tang, P.Z. A hybrid optimization approach for unequal-sized dynamic facility layout problems under fuzzy random demands. *Proceeding Inst. Mech. Eng. Part B-J. Eng. Manuf.* **2019**, *234*, 382–399. [CrossRef]
21. Stijepovic, M.Z.; Linke, P. Optimal waste heat recovery and reuse in industrial zones. *Energy* **2011**, *36*, 4019–4031.
22. Gómez, A.; Fernández, Q.I.; De la Fuente García, D.; García, P.J. Using genetic algorithms to resolve layout problems in facilities where there are aisles. *Int. J. Prod. Econ.* **2003**, *84*, 271–282. [CrossRef]
23. Tao, X.; Wang, H.; Li, Z. Optimal Solution of Rectangular Part Layout Based on Rectangle-Filling Algorithm. *China Mech. Eng.* **2003**, *14*, 1104–1107.

processes

MDPI

Article

Esterification of Kenaf Core Fiber as a Potential Adsorbent for Oil Removal from Palm Oil Mill Effluent (POME)

Nor Halaliza Alias [1,2,*], Luqman Chuah Abdullah [1,*], Thomas Choong Shean Yaw [1], Siti Nurul Ain Md Jamil [3], Teo Ming Ting [4], Ahmad Jaril Asis [5], Chuan Li Lee [6] and Abel Adekanmi Adeyi [7]

1 Department of Chemical and Environmental Engineering, Faculty of Engineering, Universiti Putra Malaysia, Serdang 43400, Selangor, Malaysia; csthomas@upm.edu.my
2 School of Chemical Engineering, College of Engineering, Universiti Teknologi MARA, Shah Alam 40450, Selangor, Malaysia
3 Department of Chemistry, Faculty of Science, Universiti Putra Malaysia, Serdang 43400, Selangor, Malaysia; ctnurulain@upm.edu.my
4 Radiation Processing Technology Division, Malaysian Nuclear Agency, Kajang 43000, Selangor, Malaysia; tmting@nm.gov.my
5 Sime Darby Research Sdn. Bhd., Lot 42700, Pulau Carey, Banting 42960, Selangor, Malaysia
6 Faculty of Biotechnology and Biomolecular Sciences, Universiti Putra Malaysia, Serdang 43400, Selangor, Malaysia; chuanli@upm.edu.my
7 Department of Chemical and Petroleum Engineering, College of Engineering, Afe Babalola University Ado-Ekiti (ABUAD), PMB 5454, Ado-Ekiti 360211, Ekiti State, Nigeria; abeladeyi@abuad.edu.ng
* Correspondence: norhalaliza@uitm.edu.my (N.H.A.); chuah@upm.edu.my (L.C.A.)

Abstract

Palm oil mill effluent (POME) is a major contributor to industrial oily wastewater in Malaysia, demanding effective treatment solutions. This study explores the potential of esterified kenaf core (EKC) fiber as an oil adsorbent for oil removal from POME, optimized using a full central composite design (CCD) within the response surface methodology (RSM) framework. The optimum conditions achieved 76% oil removal efficiency, with a 1:0.5 ratio of mercerized kenaf core to stearic acid (MKC:SA), 15 wt% of catalyst, and 1 h reflux time during the esterification process. The regression model exhibited strong predictive capability, with a significant quadratic correlation and an R^2 value of 0.94. The Fourier transform infrared (FTIR) spectroscopy revealed the existence of ester functional groups characterized by significant hydrophobicity and a decrease in hydroxyl groups, indicating the chemical changes of EKC. Moreover, the scanning electron microscopy (SEM) research demonstrated structural alterations in EKC, including heightened surface roughness, fibrillation, and pore development, which improved oil adhesion relative to raw kenaf core (RKC). These findings indicate that EKC provides an effective, environmentally sustainable solution for managing oil wastewater issues in the palm oil sector, facilitating enhanced ecological sustainability and resource management.

Keywords: palm oil mill effluent (POME); oily wastewater; oil removal; adsorption; kenaf core; esterification

1. Introduction

The management of oily wastewater has emerged as a crucial area of investigation owing to its intricate composition and considerable environmental consequences. This wastewater, typically generated by industrial processes, encompasses a variety of pollutants, including oils and grease [1], petroleum derivatives such as gasoline, kerosene, and diesel [2], and elevated levels of dissolved organic and inorganic compounds [1,3]. Its dangerous characteristics present a significant risk to ecosystems, human health, and natural resources. In 2020, the global discharge of oily wastewater worldwide has been expected to reach 54 billion m^3 [4]. These contaminants are recognized for their capacity to disturb aquatic ecosystems, taint potable and groundwater supplies, diminish agricultural yields, impede animal development, and degrade environmental quality [1,5–9]. The dire effects highlight the necessity for creative and effective treatment techniques.

In Malaysia, palm oil mill effluent (POME) is a leading source of industrial oily wastewater. POME is a colloidal suspension consisting of 95–96% water, 0.6–0.7% oil and grease, and 4–5% total solids [6,7]. It has high biological oxygen demand (BOD) and chemical oxygen demand (COD) with the value of 50,000 mg/L and 25,000 mg/L, respectively, coupled with significant oil and grease content (4000–8000 mg/L), acidic pH (4–5), and discharged temperature of 80 to 90 °C. The Malaysian Department of Environment (DOE) mandates a maximum allowable limit of 50 mg/L for oil and grease in discharged POME. Meeting these standards requires advanced and effective treatment solutions to address the environmental challenges posed by POME [9–13].

Various treatment methods have been explored for POME, including anaerobic digestion (ponding system) [14], anaerobic sludge fixed film reactor [15], integrated anaerobic-aerobic bioreactor [16], anaerobic sludge blanket reactor [17], and co-digestion in a solar-assisted bioreactor [18]. Conventional treatments such as the ponding system have low operating costs and are more convenient. However, these anaerobic digestion treatments require a long retention time and a large area to treat the POME and produce an unpleasant odor [19,20]. The long treatment period is mostly due to the slow microbe growing rates [21]. Other existing treatments for POME have been addressed by [22,23], which include micro and ultrafiltration, coagulation, flotation, membrane reactor, electrocoagulation, and Fenton technology. These technologies tend to be inefficient, costly, time-consuming, operationally complex, or require highly skilled operators [24]. The oil residual from POME can also be removed using the solvent extraction method [25], sedimentation–flocculation–coagulation methods [26], or physiochemical treatment [27].

However, as emphasized by [28], coagulation alone is often ineffective for the complete removal of pollutants, leaving residual impurities in the effluent. Moreover, the presence of organic compounds in natural coagulants that do not participate in the coagulation process can result in dissolved organic matter remaining in the treated water, which may promote bacterial regrowth and lead to the formation of harmful disinfection by-products. This highlights the limitations of coagulation as a standalone treatment and underscores the need for additional treatment stages [29]. In this context, the adsorption process stands out as one of the most efficient and extensively used methods of wastewater treatment, efficiently targeting both organic and inorganic contaminants [7,9], and is particularly good at recovering residual oil in water [30]. This is due to its simplicity, cost-effectiveness, and minimal environmental impact. Unlike other methods, adsorption systems produce minimum sludge and allow for the regeneration and reuse of adsorbents [23,31,32]. However, the selection of a suitable adsorbent remains critical, as different sorbents might have distinct adsorption mechanisms and thus affect the efficiency of the adsorption in treating oily wastewater [7].

Natural sorbents, such as chitosan flakes and powder [9], rubber powder [13], sago bark [33], chitosan in empty fruit bunch (EFB) filter [34], natural zeolites [35], and raw kenaf core fiber [24], have been investigated for their ability to remove the oil from POME. While these materials are affordable and environmentally friendly, they often exhibit low hydrophobicity and low buoyancy, which limit their oil adsorption capacity [36,37]. To overcome these limitations, the surface of the natural sorbent was modified by changing the chemical structure, where the functional groups attached to natural fibers are changed to more hydrophobic groups to increase the oil adsorption capacity and its hydrophobicity properties [36–41]. Recently, modified adsorbents were developed for the oil removal from POME. These include the modified empty fruit bunch (EFB) through acetylation [42], alginate and mangrove composite beads coated with chitosan (AMCBCC) [43], modified oil palm leave (OPL) and modified oil palm frond (OPF) using lauric acid [44], esterified sago bark (ESB) [33], and a magnetic composite prepared from palm shell-based carbon [45].

The utilization of fatty acid in increasing the hydrophobicity of the natural sorbent through an esterification reaction to produce an oil adsorbent has been reported. This includes modified plantain peel fiber (PPF) using stearic acid [46], esterified sago bark (ESB) using stearic acid [33], oleic and palmitic acid-treated bark (OTB and PTB) [41], modified banana trunk fibers using oleic acid [47], and sawdust grafted with oleic acid [48]. Fatty acid is used in esterification reactions to convert the hydroxyl group in the natural fiber into an ester group, which has high hydrophobic characteristics and hence increases the oil adsorption capacity [36].

Kenaf (*Hibiscus cannabinus* L.), a fast-growing fiber plant, is recognized as a sustainable resource due to its adaptability to various climates, rapid growth, and minimal crop rotation needs. It thrives under different environmental and soil conditions, making it suitable for industrial applications. The core fiber of kenaf, which makes up about 65% of its stem, has shown significant potential as a cost-effective adsorbent for wastewater treatment [49–51]. With a price of approximately RM500 per tonne, kenaf offers a cost-effective, environmentally friendly alternative that is widely available worldwide [52]. According to [53], the ground kenaf core is able to remove the oil from water, with an adsorption capacity of 3.95 g oil/g. To date, the use of kenaf has been extensively reported for several applications, such as paper, pulp and other fiber products, jeans, indoor panels and other interior components for high-end cars (BMW and Mercedes Benz), composite materials for buildings or the construction industry, brighter bulletproof vest, erosion control in the geotextiles industry and soil remediation for hydrocarbonate infestations [50], human and animal feed [54], filler or a natural anti-degradant in producing thermoplastic elastomer [55], and adsorption [56,57].

Notwithstanding its numerous applications, the utilization of kenaf as an oil adsorbent for the treatment of oily wastewater remains predominantly unexamined. In contrast to non-renewable resources like coal, which is frequently utilized in energy generation and industrial applications, and valued at roughly $115 per ton in January 2025 [58], kenaf offers a cost-effective and renewable alternative for wastewater treatment. Its cost-effectiveness, coupled with its worldwide accessibility, renders it a suitable option for sustainable industrial practices. Chemical modifications to kenaf may enhance its properties, rendering it an effective and environmentally sustainable adsorbent. Creating kenaf-derived adsorbents for industrial wastewater treatment can diminish dependence on limited resources such as coal, lessen environmental repercussions, and promote sustainable industrial methodologies.

This study focuses on the modification of raw kenaf core (RKC) fiber through mercerization and esterification with stearic acid to produce esterified kenaf core (EKC) fiber as an efficient oil adsorbent for POME treatment. This reaction replaces the hydroxyl groups in MKC with ester groups, which exhibit strong hydrophobic characteristics. The hydroxyl group, being a reactive functional group, is highly susceptible to chemical modification. Stearic acid, a long-chain fatty acid with 18 carbon atoms, was selected due to its significant hydrophobicity and lipophilic properties [59], both of which are critical in improving the oil adsorption efficiency of EKC. Optimization of the production process was conducted using response surface methodology (RSM) with a central composite design (CCD), examining parameters such as the ratio of MKC to SA, the percentage of the catalyst (calcium oxide, CaO), and reflux time. The modified EKC is expected to exhibit superior hydrophobicity and oil adsorption capacity, offering a cost-effective and environmentally sustainable solution for POME treatment. By advancing the development of natural adsorbents, this study addresses the pressing need for innovative and efficient wastewater management technologies in the palm oil industry.

2. Materials and Methods

2.1. Physical Treatment of POME

The palm oil mill effluent (POME) was collected from the sludge pit area at the Sime Darby KKS Labu Mill, Negeri Sembilan, Malaysia, at a temperature of 90 °C. The raw effluent (POME) was filtered using a muslin cloth strainer to remove millimeter-sized solid particles. The filtered POME was subsequently analyzed for pH, total solid, and oil and grease content. To preserve its properties, the sample was stored at 4 °C prior to the pretreatment.

2.2. Adsorbent Preparation

2.2.1. Raw Kenaf Core Fiber Preparation

The kenaf stalks used in the present study were obtained from Lembaga Kenaf dan Tembakau Negara (LKTN), Kelantan, Malaysia. Initially, the entire stalks of raw kenaf core were chipped using a chipper mill to reduce them to a smaller size. The chipped raw kenaf core (RKC) was then thoroughly dried in an industrial oven set to 60 °C for 7 days to ensure that all moisture was removed. Following this, the chipped RKC was ground, washed with distilled water, and dried in the oven at 60 °C for 24 h. The material was sieved to a particle size of 2 mm using a stainless steel sieve shaker and kept in a desiccator until further use.

2.2.2. Mercerization of Raw Kenaf Core (RKC) Fiber Preparation

A total of 10 g of raw kenaf core (RKC) fiber was immersed in a 20 wt% sodium hydroxide (NaOH) solution, prepared by dissolving 50 g NaOH in 250 mL of distilled water, for a duration of 24 h. Following the soaking process, the mercerized kenaf core (MKC) fibers were thoroughly rinsed with distilled water to remove any residual alkali. The washed fibers were then dried in an oven at 60 °C for 24 h to achieve consistent dryness before being utilized in the subsequent esterification reaction.

2.3. Single Factor Experiment

A single-factor experiment was conducted to determine the optimal range of oil removal from palm oil mill effluent (POME) for subsequent optimization in the esterification reaction. The range selected was based on a previous study conducted by [33]. Three independent parameters were studied within their respective ranges: the ratio of MKC to stearic acid (1:0.5, 1:1, and 1:2), percentages of catalyst (5 wt%, 10 wt%, and 15 wt%), and reflux time (1, 4.5, and 8 h). Each parameter was maintained at a constant level, while the effect of one factor was investigated by varying its level individually. The impact of each factor on the percentage of oil removal from POME was analyzed accordingly.

2.4. Design of Experiment (DOE) and Statistical Analysis

The response surface methodology (RSM) was applied using a statistical design experiment based on the central composite design (CCD). The analysis of the experimental data was conducted using Design-Expert Version 7.0.0 software [60]. In accordance with the result from the preliminary single-factor experiment, the independent variables with their corresponding levels were then applied to the CCD, with an alpha (α) value of ±1. The independent variables involved in EKC optimization were the MKC:SA ratio (factor A), percentages of catalyst (factor B), and refluxing time (factor C). The response in this study was the oil removal efficiency of EKC from POME. Each factor was studied at two levels, which were the low level (-1) and the high level (+1). The central values (zero level) chosen for the experimental design were 1:1 (factor A), 10% (factor B), and 4.5 h (factor C). Table 1 shows the analyzed factors and their respective notations and levels. The value ranges of MKC:SA, catalyst percentage, and refluxing time were chosen based on single-factor experiments.

Table 1. Parameters of oil removal efficiency of EKC from POME and the respective treatment levels.

Notation	Factor	Unit	Actual Value		
			Low Level (-1)	**High Level (+1)**	**Mean (0)**
A	MKC:SA ratio	-	0.5	1.5	1.0
B	Percentage of catalyst	%	5	15	10
C	Refluxing time	hours	1	8	4.5

The experimental response, which refers to the percentage of oil removal from POME attained at distinct combinations of the factors' levels, was then used to initiate an empirical regression model relating to the studied factors based on the second-order polynomial model as shown in Equation (1):

$$y = \beta_o \sum_{i=1}^{k} \beta_i x_i + \sum_{i=1}^{k} \beta_{ii} x^2{}_i + \sum_{i=1}^{k} \sum_{j=i+1}^{k-1} \beta_{ij} x_i x_j \quad (1)$$

where y is the response variable, which is the oil removal from POME (%); β_o, β_i, β_{ii}, and β_{ij} are the regression coefficients for the intercept, linear, quadratic, and interaction, respectively; x_i and x_j are the independent variables (factors) influencing the response values; and k is the number of studied variables. The analysis of variance (ANOVA) provided a statistical analysis, where it revealed the correlation between all the factors involved, the significant of each factor tested, and the accuracy of the developed model. Meanwhile, the adequacy of the fitted model was assessed through the residual analysis. All statistical evaluations considered a 5% significance level, where the p-values were required to be less than 0.05.

2.5. Esterification Reaction

The mercerized kenaf core (MKC) (4 g) was placed in a 500 mL three-neck round-bottom flask containing 100 mL of ethyl acetate as the reaction medium. Stearic acid was introduced according to the specified MKC-to-stearic acid ratios of 1:0.5, 1:1, and 1:1.5. Calcium oxide (CaO) was used as the catalyst and added at 5 wt%, 10 wt%, and 15 wt%, respectively, relative to the mass of MKC. The esterification reaction was conducted under reflux conditions for durations of 1 h, 4.5 h, and 8 h, depending on the experimental design. Upon completion of the reaction, the mixture was filtered to separate the modified fibers. The product was subsequently washed with ethanol and then acetone to remove unreacted materials and impurities. The esterified kenaf core (EKC) was dried in an oven at 40 °C for 3 h. After drying, the EKC was cooled to ambient temperature and stored in a desiccator to prevent moisture absorption prior to further analysis. The esterification apparatus utilized in this study is depicted in Figure 1.

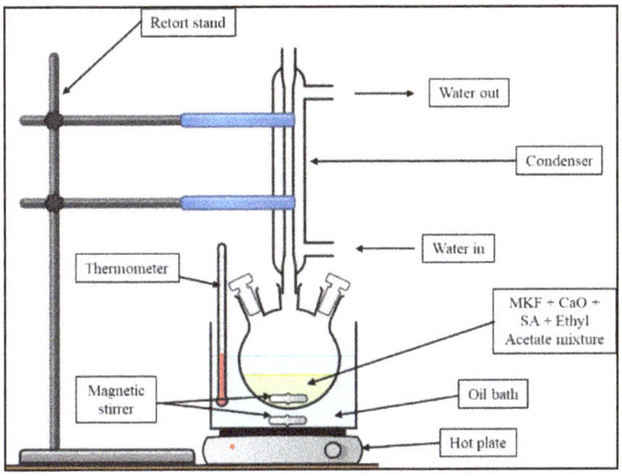

Figure 1. Setup of the esterification reaction.

2.6. *Application of Esterified Kenaf Core on POME*

The oil removal efficiency of the esterified kenaf core (EKC) was evaluated using palm oil mill effluent (POME). A sample of 0.08 g of EKC was mixed with 20 mL of POME and stirred at 130 rpm for 30 min at ambient temperature using a magnetic stirrer. Following the reaction, the mixture was filtered, and the filtrate was collected for oil and grease analysis. All experiments were performed in triplicate to ensure the reliability and consistency of the results.

2.7. *Determination of Oil and Grease (O&G) Content in POME*

The oil content in POME was determined using a hexane extraction method, adhering to the guidelines outlined in the Standard Methods for the Examination of Water and Wastewater, Section 5520B [61]. Approximately 20 mL of the filtrate (treated POME) was transferred into a separatory funnel. A 1:1 hydrochloric acid (HCl) to water solution was prepared, and 0.5 mL of this solution was added to the filtrate to adjust the pH $\leq$ 2 before introducing 3 mL of hexane. The mixture was vigorously shaken for 2 min and allowed to settle for 10 min to form two distinct layers.

The organic and oil-containing layer was separated by gravity filtration, with the procedure repeated in triplicate for accuracy. The collected filtrate was dried over 2 g of anhydrous sodium sulfate (Na_2SO_4). After drying, the Na_2SO_4 was separated from the oil using Whatman No. 40 filter paper. Residual hexane in the oil sample was removed using a rotary evaporator (Heidolph Laborota 4000 Efficient) operated for 15 min to ensure complete solvent removal. The oil sample was subsequently dried in an oven at 103 °C for 10 min, cooled to room temperature, and weighed. The weight of the oil sample was recorded as the oil and grease (O&G) content, which was calculated using Equation (2):

$$O\&G\left(\frac{mg}{L}\right) = \left[\left(\frac{\left(W_f - W_i\right)g \times 1000\frac{mg}{g}}{V(L)}\right)\right] \quad (2)$$

where W_f and W_i are the final and initial weight of the flask, respectively (g); and V is the volume of POME (L).

Meanwhile, the oil removal efficiency can be determined from Equation (3):

$$Oil\ removal\ efficiency(\%) = \left[\frac{C_O - C}{C_O}\right] \times 100\% \quad (3)$$

where Co and C are the initial oil and grease concentration (mg/L) in POME and the oil and grease concentration (mg/L) in the filtrate, respectively [33,45].

2.8. Preliminary Adsorption Analysis

The adsorption capacity of RKC and EKC for oil removal from POME was investigated at varied initial pH values (2–8) as a typical case study. The batch experiment was performed using 0.4 g of RKC and EKC, respectively, which were mixed with 100 mL of POME. The mixture was stirred under constant speed at 200 rpm at 30 °C (room temperature) for a contact time of 30 min. The oil removal from POME was analyzed according to the procedure described in Section 2.7. The experiments were performed in triplicate, and the average values were recorded.

2.9. Characterization

2.9.1. Attenuated Total Reflectance—Fourier Transform Infrared Spectroscopy (ATR-FTIR)

ATR-FTIR analysis was performed using a Perkin-Elmer Spectrum 100 model (Perkin-Elmer (M) Sdn. Bhd., Petaling Jaya, Malaysia). The technique employs infrared light to scan samples and detect characteristic chemical bonds and functional groups [62]. Both raw kenaf core (RKC) and esterified kenaf core (EKC) fibers were analyzed to assess whether the intended modifications were successfully achieved, particularly by observing changes in their functional groups.

2.9.2. Scanning Electron Microscopy (SEM)

In this study, the surface morphology of raw kenaf core (RKC) and esterified kenaf core (EKC) fiber was observed using an S-3400N SEM (Hitachi, Hi-Tech Instrument Sdn. Bhd., Puchong, Malaysia) at magnifications of 100× and 600×, respectively [63].

3. Results and Discussions

3.1. Reaction Mechanism of Esterification Reaction

Figure 2 shows the reaction mechanism between the hydroxyl group from RKC, CaO, and SA in the esterification reaction. The purpose of using CaO was to deprotonate the abundant hydroxyl group in cellulose [33]. The alkoxide ion produced after deprotonation reacted with SA, which finally produced an ester group that has the ability to adsorb the oil from POME due to its high hydrophobicity compared to MKC and RKC.

3.2. Single Factor Experiment

The analysis of the individual effect of each independent factor associated with the oil removal efficiency was examined using a single-factor experiment. Furthermore, it provides a practical range for consecutive optimization studies. The range covered in this study included the ratio of MKC to SA (MKC:SA), the percentages of catalyst, and the reflux time of the esterification reaction.

3.2.1. Effect of the Ratio of MKC to Stearic Acid (MKC:SA)

Figure 3a shows the effect of the ratio of MKC:SA ranging from 1:0.5, 1:1, and 1:2. The amount of catalyst percentage and reflux time were fixed at 5 wt% of MKC and 1 h, respectively. As the amount of SA increased, oil removal efficiency decreased noticeably.

At a ratio of 1:0.5, the efficiency was measured at 75.3%, but at 1:2, it dropped to 42.4%. Due to an excess of SA in the reaction, the ratio should not be greater than 1:1. This is evidenced by the existence of a white solid SA residue that partially coated the surface of EKC following the 1:2 ratio experiment (Figure 4). Furthermore, the higher ratio (mass) of SA to MKC would increase the demand for ethanol to remove excess SA from the EKC surface. This excessive use of ethanol renders the washing process economically inefficient, especially with the elevated amount of SA. Therefore, it can be concluded that the lowest amount of SA, with a ratio of 1:0.5 (MKC:SA), was adequate to esterify the MKC in this single-factor study, as it gave the highest oil removal efficiency (75.3%).

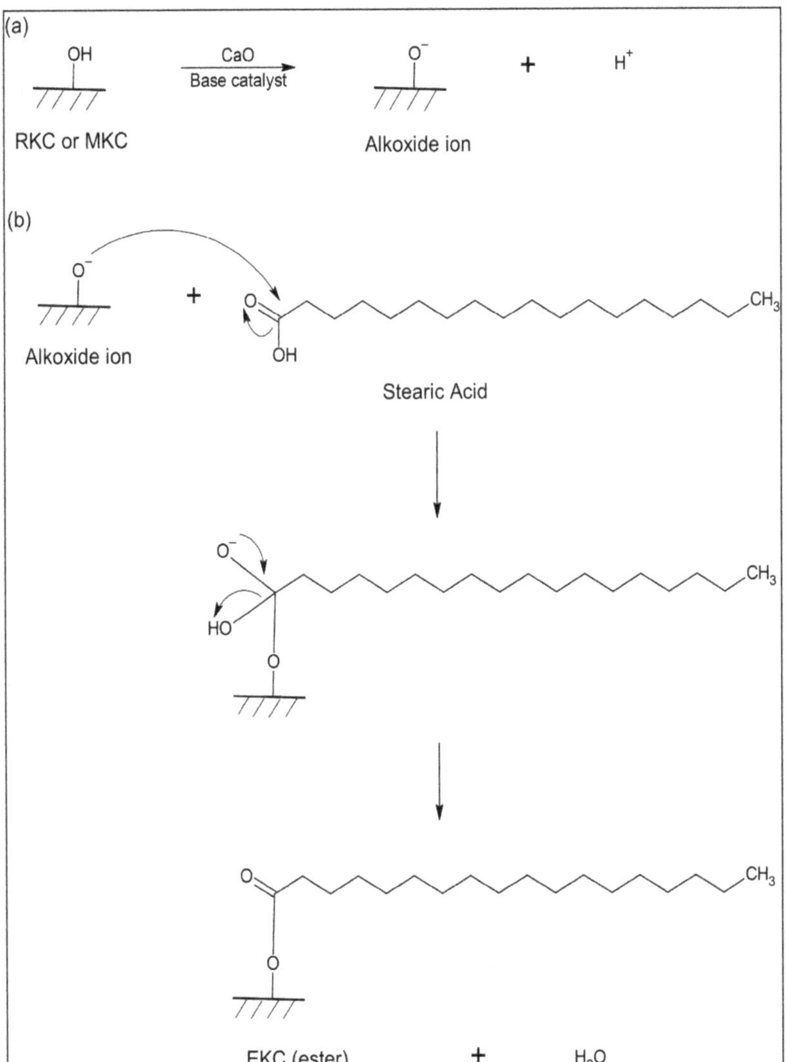

Figure 2. Proposed reaction mechanism of esterification starting from (**a**) the deprotonation of RKC or MKC, followed by (**b**) the esterification reaction with stearic acid to produce an ester (EKC).

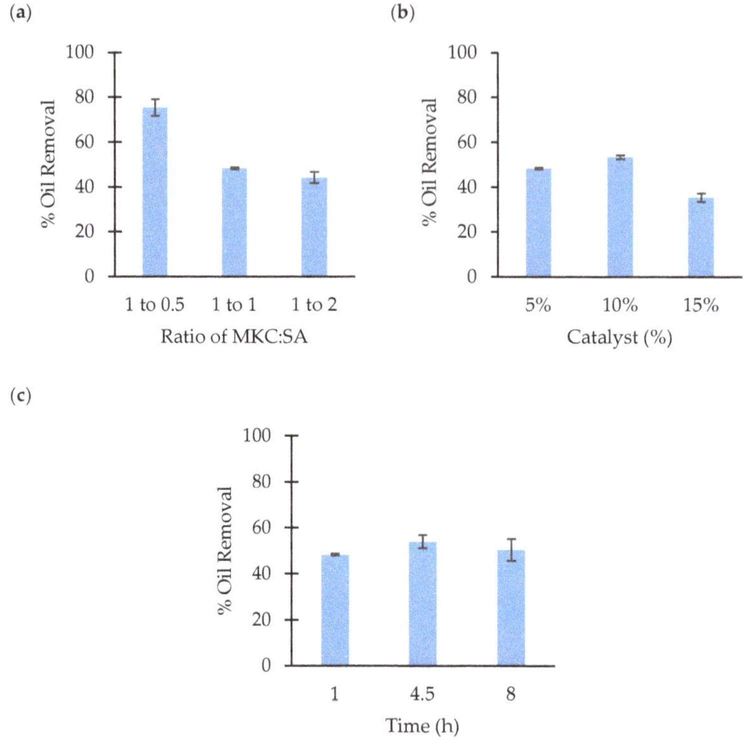

Figure 3. Single factor experiment for (**a**) the effect of the ratio of MKC:SA, (**b**) the effect of the catalyst, and (**c**) the effect of refluxing time on oil removal efficiency from POME through the esterification reaction.

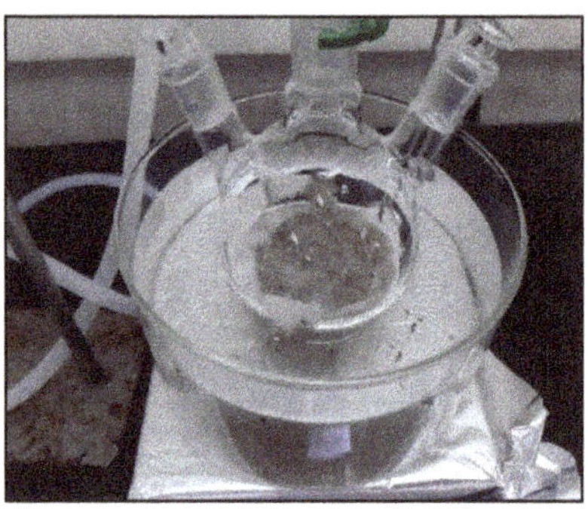

Figure 4. White solid SA residue partially coating the surface of EKC at an MKC:SA ratio of 1:2 (reaction condition: 5 wt% of CaO and 1 h of reflux time).

3.2.2. Effect of Catalyst Percentages

The effect of catalyst percentages ranged from 5 wt%, 10 wt%, and 15 wt% of MKC. The ratio of MKC:SA was fixed at 1:1, with a reflux time of 1 h. Referring to Figure 3b, it is apparent that as the catalyst percentage increased from 5 wt% to 10 wt% of MKC, there was a corresponding rise in the oil removal efficiency. Oil removal efficiencies at these catalyst percentages were 48.3% and 53.5%, respectively. This observation can be attributed to the higher catalyst percentage added into the reaction, which induces more hydroxyl groups in the MKC structure to deprotonate [64]. Consequently, this results in more SA attaching to the MKC surface, ultimately leading to an increase in the oil removal efficiency. However, the graph showed a significant decrease in oil removal efficiency from POME when using EKC, which fell to 35.4% as the percentage of catalyst increased to 15 wt% of MKC. This was probably because the addition of 15 wt% was not suitable, as this amount might be too immoderate for this experiment setup, which used a 1:1 ratio of MKC:SA and a 1 h reflux time. The addition of excess catalyst caused a saponification reaction to occur due to the existence of the calcium hydroxide ($Ca(OH)_2$) compound in CaO, which has basic properties. Furthermore, the saponification reaction made the solution too viscous, making it difficult for CaO to dissolve when present in excess [65,66].

It was also noticeable that the solvent (ethyl acetate) used in this reaction became cloudy (Figure 5) when 15 wt% (CaO amount) was used compared to 5 wt% due to the excess CaO amount at a 1 h reflux time. The reaction between CaO and water (a byproduct of the esterification reaction) produced $Ca(OH)_2$, which caused the formation of a cloudy solvent [65,67]. Another possible reason for the cloudy solvent was the unstable nature of calcium oxide, which tends to leach into the reaction medium [68]. Figure 3b reveals that a 10 wt% catalyst amount achieved the highest oil removal efficiency from POME using EKC, reaching 53.5% in this single-factor study.

Figure 5. Appearance of cloudy solvent after using 15 wt% of catalyst (CaO) in the esterification reaction (reaction condition: MKC:SA ratio of 1:1 and 1 h of reflux time).

3.2.3. Effect of Refluxing Time

Figure 3c depicts the effect of reflux time in the esterification reaction. The range studied was between 1 h, 4.5 h, and 8 h. The time was set to be limited to 8 h of operation only to simulate the normal operation in the chemical industry [69]. In the effect of reflux time, the ratio of MKC:SA and percentage of catalyst were set at 1:1 and 5 wt% of MKC, respectively. The graph shows that the oil removal efficiency increased as the reflux time increased from 1 h (48.3%) to 4.5 h (54.0%), whereby a longer refluxing time exposed the MKC surface and ultimately increased the number of ester sites. Thus, high oil removal efficiency was achieved with a longer refluxing time. Nevertheless, Figure 3c shows the deterioration of oil removal efficiency from POME to 50.4% at 8 h (after long hours of operation). This was attributed to the formation of a white solid (residual SA, which appeared after 4.5 h of reflux time) on the EKC surface, where the layer is fully coated with the white solid. The increase in reflux time beyond the optimum might cause the esterification reaction to proceed in the backward direction since the reaction is reversible [70]. This led to an increase in the acid value (SA) used in this experiment. Based on Figure 3c, the reflux time of 4.5 h yielded the maximum oil removal efficiency, at 54.0% in this single-factor study.

3.3. Optimization of Esterified Kenaf Core (EKC) Using RSM

3.3.1. Fitting the Model and Analysis of Variance (ANOVA)

A central composite design (CCD) was performed by measuring the esterification factors, namely, the ratio of MKC:SA, the percentage of catalyst, and reflux time. A design matrix comprising 18 single-block experiments, conducted in a randomized order with six repetitions at the center point, was produced. The main objective of this research was to obtain optimized oil removal efficiency (Response). Therefore, the actual esterification factors and the average response are presented in Table 2.

Table 2. Design matrix based on CCD with experimental response data of oil removal efficiency from POME through the esterification reaction.

Run	Factor A	Factor B	Factor C	Average Response	Predicted Value	Residual Error (%)
	A: Ratio of MKC:SA	B: Percentage of Catalyst (%)	C: Refluxing Time (h)	Y: Oil Removal Efficiency (%)		
1	1	10	4.5	60.5	64.25	5.84
2	0.5	15	8	71.5	71.56	0.08
3	1.5	5	8	71.8	71.60	0.28
4	0.5	5	8	72.2	71.85	0.48
5	0.5	15	1	76.0	76.21	0.28
6	1	10	4.5	64.2	64.25	0.08
7	1	10	4.5	63.5	64.25	1.17
8	1	10	4.5	66.8	64.25	3.82
9	1.5	15	8	75.3	75.21	0.12
10	0.5	5	1	74.2	74.30	0.14
11	1	10	1	74.7	74.10	0.80
12	1.5	15	1	69.6	69.95	0.50
13	1	10	8	74.9	75.50	0.79
14	1	10	4.5	65.9	64.25	2.50
15	1	15	4.5	73.3	72.78	0.71
16	1	10	4.5	64.6	64.25	0.54
17	1	5	4.5	69.5	70.02	0.74
18	1.5	5	1	64.2	64.15	0.08

The data obtained from the experimental value (Response) were fitted to a second-order polynomial equation. The empirical quadratic model for the oil removal efficiency, Y (%), with coded factors, is shown in Equation (4):

$$Y(\%) = +64.25 - 1.63A + 1.38B + 0.70C + 0.98AB + 2.48AC - 0.55BC - 10.10A^2 + 7.15B^2 + 10.55C^2 \quad (4)$$

Equation (4) presents the regression model, which involves all terms, including the linear terms (A, B, C), the interaction terms (AB, AC, BC), the quadratic terms (A^2, B^2, C^2), and their respective regression coefficient values. The positive and negative signs in the linear terms demonstrate increasing and decreasing factor levels, respectively [71]. Analysis of variance (ANOVA) serves to analyze the adequacy and quality of the fitted model. Additionally, ANOVA helps distinguish the variation of change in the combination of factor levels (treatments) from the variation due to random errors, allowing for the measurement of the generated responses.

Table 3 shows the ANOVA summary for the esterification reaction, representing the significance of the model terms, including linear, interaction, and quadratic terms. According to [72], the analysis is also important to validate the accuracy, credibility of the model, and goodness-of-fit of the quadratic model (Equation (4)) obtained from the experimental results. The second-order polynomial model can be confirmed as either significant or not significant using ANOVA analysis.

Table 3. Analysis of variance (ANOVA) for the esterification reaction.

Source	Sum of Squares	Mean Square	F Value	p-Value (Prob > F)	
Model	376.89	41.88	13.09	0.0007 *	Significant
A-MKC:SA	21.13	21.13	6.6	0.0332 *	
B-% Catalyst	19.04	19.04	5.95	0.0406 *	
C-Time	4.9	4.9	1.53	0.251	
AB	7.61	7.61	2.38	0.1617	
AC	49.01	49.01	15.32	0.0045 *	
BC	2.42	2.42	0.76	0.4098	
A^2	78.98	78.98	24.68	0.0011 *	
B^2	76.68	76.68	23.97	0.0012 *	
C^2	166.95	166.95	52.18	<0.0001 **	
Residual	25.6	3.2			
Lack of Fit (LOF)	1.62	0.54	0.11	0.9489	Not Significant
Pure Error	23.97	4.79			
Corrected Total	402.49				
Analysis of The Regression Equation					
Standard Deviation (SD)	1.79	R^2	0.9364		
Mean	69.59	Adjusted R^2	0.8649		
Correction of Variation (CV) in %	2.57	Predicted R^2	0.8654		
PRESS	54.17	Adequate Precision	9.046		

* $p < 0.05$, significant. ** $p < 0.0001$, highly significant.

Based on Table 3, the ANOVA results indicated that the quadratic model was statistically significant ($p < 0.05$), suggesting that the model was the best fit for all experimental data collected on oil removal efficiency from POME using EKC. Among the model terms, the linear terms involving the ratio of MKC:SA (A) and the percentage of catalyst (B), the interaction terms of the ratio of MKC:SA and reflux time (AC), and the quadratic terms of the ratio of MKC:SA (A^2), percentage of catalyst (B^2), and reflux time (C^2) were observed as statistically significant, with p-values less than 0.05. On the other hand, the remaining terms, namely the reflux time (C), interaction terms of the ratio of MKC:SA and the percentage of catalyst (AB), and the percentage of catalyst and reflux time (BC) were identified as insignificant since the p-values were greater than 0.05. According to Table 3, the most significant factor was the quadratic term of reflux time (C^2), with the lowest p-value nearly approaching zero (<0.0001). This suggests that the quadratic factor of C^2 can highly influence oil removal efficiency. The strong nonlinearity (curvature) of reflux time (C) indicates a divergence in the oil removal response from POME as reflux time (C) increased. It shows that further increases in reflux time in the esterification may reduce oil removal efficiency from POME. This was due to the residual SA appearing in white solid form on the EKC surface (fully coated) when the reflux time increased from 4.5 h to 8 h (as discussed in Section 3.2.3). Thus, more volume of ethanol is needed in the EKC washing process, which leads to an uneconomical effect. Improper washing of EKC can block its pores with SA and CaO, reducing its ability to adsorb oil from POME.

The fitness of the model was also evaluated through the lack of fit (LOF) analysis. A significant model should have a non-significant LOF, which proposes that the residual error (the difference between the observed response and the predicted response) has not exceeded the pure error produced from randomized replicated design points [73,74]. In this study, the p-value obtained for the LOF analysis was 0.9489 (p-value > 0.05), which indicated the insignificance of LOF. There was a 94.89% chance that a lack of fit F-value this large could occur due to noise. The F-value for LOF (0.11) also showed insignificance, comparable to the pure error of the experiment. Therefore, the second-order (quadratic model) suggested could be used to predict response values (oil removal efficiency.)

The coefficient of determination (R^2), also known as R-squared, is another critical measurement used to evaluate the fitness of the model. Basically, the R-squared value indicates how much of the data variation around the mean value (variance) of the observed responses can be accounted for by the model. Based on Table 3, it was observed that the R-squared value obtained in this study was 0.9364, suggesting that almost 93.64% of the variation in the oil removal efficiency response can be described by the independent variables, while only 6% of the variation cannot be described by the model [75]. Thus, the R-squared projected in this study showed the adequacy of the quadratic polynomial model, as a high R-squared value implies that the experimental values are well aligned with the predicted values.

The adequate precision measures the signal-to-noise ratio, where the range of the predicted values at the design point is compared to the average prediction variance. Based on this study, the adequate precision value obtained was 9.046. In order to have an adequate model, a ratio greater than 4 is desirable since this indicates the ability of the model to distinguish the origin of data variations. Thus, it signifies that the model can be used to navigate the design space. Besides, further verification of the model fitness could be done using the CV% (coefficient of variation) value. The CV% describes the degree of dispersion of the data points around the mean value. A lower value of CV% (preferably less than 10%) recommends a good reproducibility of the model [73]. The CV% value observed in this study was 2.57%, which indicates that the experimental value had high precision and excellent reliability [76,77] since it showed a low variation in the mean value [78].

The diagnostic plots of the predicted responses versus actual responses for oil removal efficiency from POME are presented in Figure 6a. Generally, a diagnostic plot provides a visual representation of data to evaluate the goodness-of-fit of model regression. The discrepancy between the predicted responses and actual responses can be envisaged through the fitted plots in Figure 6a. The data points were randomly dispersed along and close to the straight line, except for one point (blue) that deviated from the linear line with a residual of −3.75 (5.84% error). However, the plots were consistent, signifying that the quadratic model was well fitted to obtain oil removal efficiency due to the high value of R-squared (0.9364).

Based on the normal probability plot in Figure 6b, the residuals were disseminated along the straight line, except for two points at the bottom left (blue and red) that deviated from the linear line with residuals of −3.75 (5.84% error) and −0.6 (0.79% error), respectively. Nevertheless, the normal probability plot of the residuals envisaged was approximately linear, supporting the condition that the error terms are normally distributed [79]. As reported by [74], residual data should be independent of each other and follow a normal distribution with constant variance. Thus, it was confirmed that the plot followed the normal or linear distribution. Meanwhile, the plot of residual versus predicted responses portrays the suitability of the selected model and the correlation of the data. All the diagnostic plots (Figure 6a,b) confirmed that the ANOVA assumptions were satisfied based on the residual analysis above. Thus, the regression analysis produced the estimation of unbiased coefficient estimates with the least variance, suggesting the adequacy of the model in calculating oil removal efficiency.

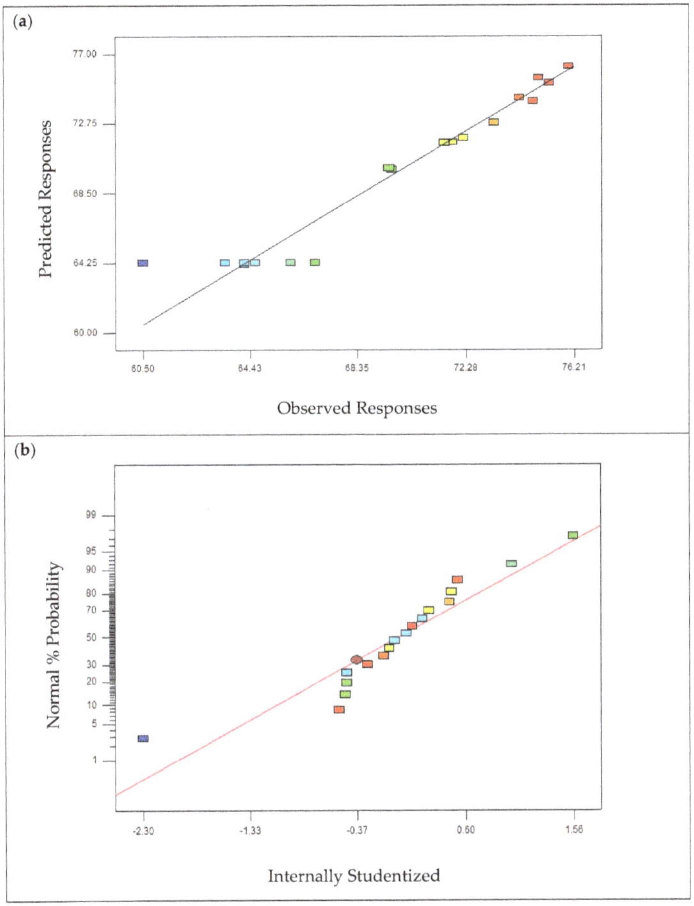

Figure 6. Diagnostic plots of (**a**) predicted responses vs. observed responses, and (**b**) the normal probability plot for the esterification reaction. The different color in the figure represents the data point plotted based on the Experiment Run. No.

3.3.2. Effect of Esterification Factors on Percentage Oil Removal from POME in Response Surface Plots Analysis

Figure 7a presents the surface and contour plots of oil removal efficiency as a function of the MKC:SA ratio (A) and catalyst percentage (B) at a constant reflux time (C) of 4.5 h. The maximum oil removal efficiency of EKC was 73.3%, which corresponds to a 1:1 ratio of MKC:SA and a 15 wt% catalyst amount added to the esterification reaction. At the lowest percentage of catalyst (5 wt%) with 1:1 of MKC:SA ratio, only about 69.5% oil removal was achieved. This shows that the oil removal efficiency of EKC was higher at 15 wt% than at 5 wt% catalyst. As reported in Section 3.2.2, the increase in catalyst amount allowed it to deprotonate more hydroxyl groups in the MKC structure, which eventually created more SA attached to the MKC surface. In addition, a lower factor of MKC:SA ratio and a higher percentage of catalyst were required to facilitate the removal of oil from POME, as the combination of these terms (AB) has been proven to reduce the efficiency of the EKC.

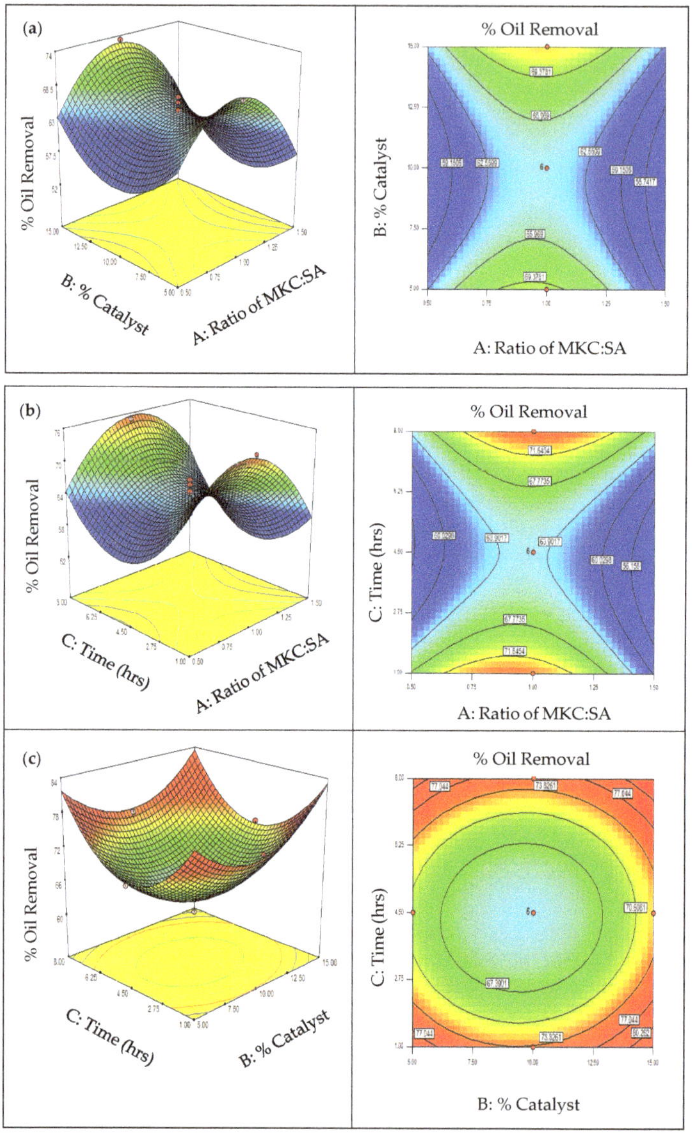

Figure 7. Response surface plots of the interaction effects of esterification factors on oil removal efficiency from POME: (**a**) ratio of RKC:SA and percentage of catalyst (AB), (**b**) ratio of RKC:SA and time (AC), and (**c**) percentage of catalyst and time (BC).

Meanwhile, Figure 7b illustrates the response surface plots of oil removal efficiency as a function of the MKC:SA ratio (A) and reflux time (C) at a fixed catalyst percentage (B) of 10 wt%. The shape of the surface plot obtained followed a trend similar to Figure 7a. The maximum oil removal efficiency of EKC was 74.9% at a 1:1 ratio of MKC:SA and 8 h reflux time (maximum duration). On the other hand, when the reflux time was reduced to 1 h (minimum duration), there was a slight difference in oil removal efficiency (74.7%) with the same ratio of MKC:SA (1:1). It was depicted that the increase in reflux time had a small significant effect on the removal of oil from POME, as the difference was only 0.2% when using 10 wt% catalyst in the esterification reaction. The increase in refluxing time has increased the fatty acid ester [80,81]. Based on this study, the increase in reflux time has increased the oil removal efficiency of EKC if the percentage of catalyst used is constant at 10 wt%. Thus, the combined effect of a lower ratio of MKC:SA (1:1) and a higher refluxing time (8 h) resulted in maximum oil removal efficiency using EKC.

Lastly, Figure 7c illustrates the response surface plots of oil removal efficiency as a function of catalyst percentage (B) and reflux time (C) at a constant ratio of MKC:SA (1:1). Based on the plot, at a 1 h reflux time, the oil removal efficiency of EKC using 5 wt% and 15 wt% of catalyst amount were 79% and 83%, respectively. However, as the reflux time increased to 8 h of operation, there was a slight increase in the oil removal efficiency of EKC. These corresponded to 82% and 83% of oil removal efficiency, recorded at the percentage catalyst of 5 wt% and 15 wt%, respectively. It was observed that the increase in catalyst percentage in the esterification reaction has increased the removal of oil from POME at both reflux times (1 h and 8 h). Nevertheless, it was noticeable that the higher reflux time gave a better oil removal efficiency from POME. The values of oil removal efficiency obtained were approximately close to each other, thus it can be concluded that 1 h is sufficient to conduct the reflux or esterification reaction due to economic factors. In fact, at 6 h reflux time and 15 wt% catalyst, a substantial amount of white solid SA residual covered half of the EKC surface. It was observed that the reaction at the highest reflux time (8 h) and the highest consumption of catalyst percentages (15 wt%) contributed the most enormous amount of white solid residual of SA and CaO, which fully covered the EKC surface. This was not observed under other reaction conditions in this study.

3.3.3. Optimization and Model Validation

The optimum condition for the oil removal efficiency of EKC was determined using numerical optimization, where the desirability value was used as an indicator. Ordinarily, the most optimum condition was suggested when the desirability value is equal to 1.0, which represents a completely desirable value or ideal response [82]. The response (oil removal efficiency) is usually calculated based on the independent factors at different combination levels set in CCD, according to the optimization goal of oil removal efficiency. Based on Table 4, the optimum conditions for maximum value of oil removal efficiency were predicted at the ratio of MKC to SA of 1:0.88 using 5.22% catalyst and a reflux time of 1.08 h. The predicted responses were attributed to 78.97% oil removal efficiency for EKC, corresponding to a desirability of 1.0. Thus, this implies that the predicted points were approximately acceptable compared to the observed response.

Meanwhile, the precision of the regression model displayed in the earlier section and the credibility of the optimum condition could be verified through the experimental work (run in triplicates), at the proposed optimum operating condition. As shown in Table 4, the experimental or observed response obtained was 72.4%, which fell between the 95% prediction intervals, namely within 72.85% (95% PI low) and 85.10% (95% PI high). Hence, the result obtained showed the suitability and validation of the quadratic polynomial model, and the regression model could be verified.

Table 4. Numerical optimization and model validation for oil removal efficiency from POME.

Factors	Name	Optimization Goal	Lower Limit	Upper Limit	Prediction Point
A	KC:SA ratio	Is in range	0.5	1.5	0.88
B	Percentage of catalyst (%)	Is in range	5	15	5.22
C	Refluxing time (hours)	Is in range	1	8	1.08
Response					
Y	Oil removal (%)	Maximum	60.5	76.0	78.97
Predicted response	Desirability	95% PI low	95% PI high	Observed response	
78.97	1.0	72.85	85.10	72.4	

3.3.4. Optimum Condition of Oil Removal Efficiency Using EKC as an Adsorbent to Remove Oil from POME

The experimental design for oil removal efficiency has been developed as described in Section 3.2, where the observed and predicted responses of the oil removal efficiency were tabulated in Table 2. The purpose of this study was to determine the optimum percentage of oil removal efficiency (Response) of EKC by conducting 18 experimental runs (all in triplicates) through the CCD analysis in RSM. The optimum condition to produce EKC with a maximum percentage of oil removal efficiency (76%) was a 1:0.5 ratio of MKC:SA, 15 wt% catalyst, and 1 h of reflux time. The maximum percentage of oil removal efficiency obtained for EKC was higher compared to RKC tested at the same dosage of 0.4 g, which achieved only 51% of oil removal efficiency. This shows that the mercerization and esterification reaction successfully increased the hydrophobicity and oil removal efficiency of RKC.

3.4. Case Study of Oil Adsorption Capacity

There are various factors controlling the oil adsorption process and the oil removal efficiency of an adsorbent. In a batch oil adsorption system, one of the key components that is remarkable in analyzing oil removal is the influence of solution pH. This is because the change in pH has an effect on the characteristics of the adsorbent surface, binding sites, and emulsion breaking [31,35,83]. The pH of the solution has a high impact on the emulsion stability by controlling the degree of ionization of the residual oil [35,84]. According to [64], different adsorbents would react differently toward the pH change during oil removal.

This study examined the influence of pH on the oil removal efficiency from POME for both RKC and EKC at pH levels of 2, 4.8, 7, and 8, as illustrated in Figure 8. This assessment highlights the impact of variations in pH on the adsorption efficacy of the materials. The analysis revealed that EKC consistently demonstrated higher oil removal efficiency than RKC. Both materials exhibited similar trends, with significantly higher removal efficiencies observed under acidic and basic conditions compared to neutral pH (7). However, EKC showed greater stability across the tested pH range due to the presence of ester groups, which enhanced its adsorption capacity for oil. The highest oil removal efficiencies for RKC and EKC were achieved at pH 2, with values of 76.8% and 91.4%, respectively. This improvement under strongly acidic conditions is attributed to the destabilization of oil droplets, leading to their increased size and easier adsorption onto the material surface.

This phenomenon highlights the enhanced performance of both RKC and EKC under acidic conditions [9,31,35].

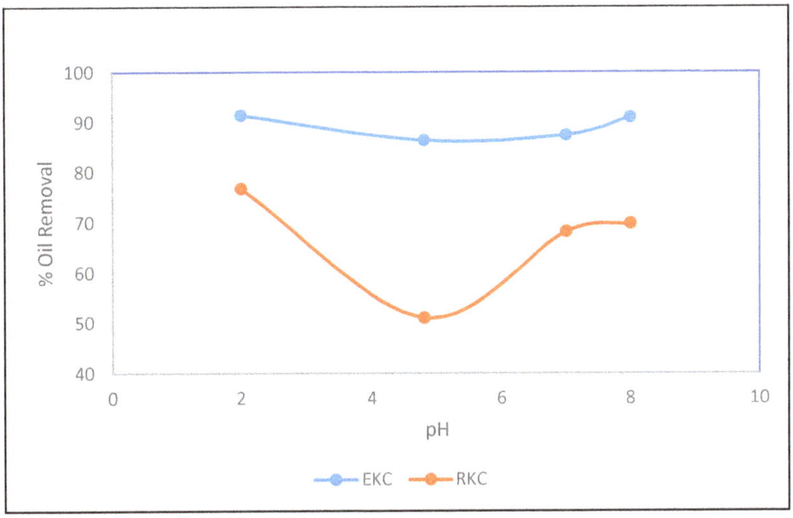

Figure 8. Effect of pH for RKC and EKC on oil removal from POME.

The lowest oil removal efficiency was observed at the original pH of POME (pH 4.8) for both RKC and EKC, with values of 51.0% and 86.4%, respectively. Beyond this point, the oil removal efficiency increased under strongly basic conditions, achieving 69.8% for RKC and 91.0% for EKC. However, the elevated removal efficiency at high pH levels was found to be inaccurate due to the saponification process. This process occurs when excessive NaOH is added to the POME solution, leading to the hydrolysis of residual oil into glycerol and fatty acid salts (soap), which are more soluble in water than in the organic solvent (n-hexane). This phenomenon, commonly observed with light and medium oils such as palm oil, can overstate the true adsorption efficiency of the materials.

While the maximum oil removal efficiency was attained under highly acidic circumstances (pH 2), it is noteworthy that the efficiency at the initial pH of POME (pH 4.8) was only slightly reduced (5%) for EKC. This finding highlights the importance of utilizing the original pH of POME for practical applications, as it obviates the necessity for substantial pH modifications while preserving high oil removal efficiency. Moreover, the enhanced efficacy of EKC as an oil adsorbent relative to RKC underscores the material's applicability in practical scenarios. These findings indicate the pivotal importance of pH in enhancing oil removal procedures and offer significant insights into the practical application of adsorption technology for POME treatment. Table 5 shows the comparison of percentage oil removal from POME using various adsorbents. The values of the percentage of oil removal for RKC and EKC are comparable with other types of adsorbents used in adsorbing the oil from POME.

Table 5. Comparison of oil removal percentages from POME using different adsorbents.

Adsorbent	Percentage of Oil Removal (%)	pH	Time (min)	Reference
RKC	76.8	2.0	30	This study
EKC	91.4	2.0	30	This study
Sago bark	48.2	2.0	30	[31]
Esterified sago bark	68.5	2.0	30	[31]
Natural zeolite	70.0	3.0	Not mentioned	[35]
Poly(dimethylsiloxane) (PDMS)	39.0	3.7	14	[83]
Carbon nanotubes (CNT's)—PDMS	64.0	3.7	10	[83]

3.5. Characterization

3.5.1. Attenuated Total Reflectance—Fourier Transform Infrared Spectroscopy (ATR-FTIR)

ATR-FTIR analysis was carried out to characterize the functional group that is present before and after the esterification reaction. Figure 9 presents the FTIR spectra for these samples, highlighting significant changes following esterification and subsequent oil adsorption. The FTIR spectra of RKC revealed key peaks at 3854 cm^{-1}, 3813 cm^{-1}, 3748 cm^{-1}, 3460 cm^{-1}, and 3025 cm^{-1}, corresponding to O-H stretching, C=C stretching of ketone (2326 cm^{-1}), and C=O stretching of ester (1747 cm^{-1}) in the cellulosic fiber. This indicates the presence of hydroxyl groups and native ester bonds inherent to RKC. Following esterification, the FTIR spectra of EKC showed a notable reduction in the O-H peak intensity compared to RKC, confirming the successful replacement of hydroxyl groups with ester groups [37,48,64]. This transformation is further validated by the shift in the C=O ester peak from 1747 cm^{-1} in RKC to 1735 cm^{-1} in EKC, indicating the formation of ester bonds. The presence of these ester groups is pivotal, as they impart hydrophobicity to EKC, facilitating oil adsorption from POME [24,64,85].

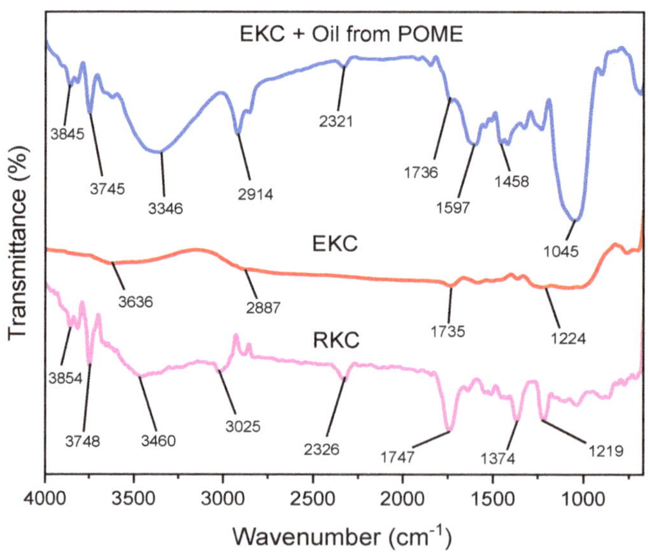

Figure 9. FTIR spectra of RKC, EKC, and EKC after oil adsorption from POME.

After oil adsorption, the FTIR spectra of EKC demonstrated several distinctive changes. The spectral range between 3845 and 3346 cm^{-1} exhibited peaks attributable to O-H stretching groups, which likely originated from hydroxyl groups attached to hydrocarbon chains within the oil mixture. A significant new peak at 2914 cm^{-1} was observed, confirming the presence of C-H stretching vibrations associated with aliphatic hydrocarbons, a major component of oils [84]. Notably, peaks corresponding to C=C stretching of aromatic hydrocarbons (1597 cm^{-1}) and C-H bending vibrations (1458 cm^{-1}) appeared in the EKC spectra post-adsorption, providing further evidence of oil attachment to the EKC surface [44,86]. Additionally, a prominent peak at 1045 cm^{-1} indicated C-O bending vibrations in primary alcohols, alongside the hydrogen-bonded O-H stretching at 3346 cm^{-1} [87]. The decrease in intensity of the ester peak (1736 cm^{-1}) in EKC after oil adsorption, compared to RKC and untreated EKC, suggests strong interactions between the ester groups and the oil molecules. This observation further supports the role of ester groups in capturing and binding oil to the EKC surface. FTIR analysis demonstrates the successful modification of RKC into EKC via esterification, enhancing its hydrophobic nature for efficient oil adsorption. The spectral changes observed after oil adsorption confirm the presence of aliphatic hydrocarbons, aromatic C=C bonds, and functional groups from the adsorbed oil, providing strong evidence of EKC's efficacy as an oil adsorbent.

3.5.2. Scanning Electron Microscopy (SEM)

The surface morphology of RKC and EKC was examined using scanning electron microscopy (SEM) to observe the structural changes resulting from mercerization and esterification treatments. Figure 10 presents the SEM micrographs of RKC at two different magnifications: 100× (Figure 10a) and 600× (Figure 10b). At 100× magnification, the surface of RKC exhibited relatively smooth and irregular textures, with only minor pores visible (shown in yellow arrow in Figure 10a). The smoothness of the surface is attributed to the natural waxes and oils present in the lignocellulosic fibers of RKC, which coat the surface [88]. When magnified to 600×, a more detailed view reveals the presence of rigid, open cylindrical pores with an irregular distribution (Figure 10b). Furthermore, impurities, marked with yellow arrows in Figure 10b, were detected on the RKC surface. These impurities likely originated from the salt (NaCl) used in the pulping process, where sodium ions (Na^+) from sodium hydroxide (NaOH) react with chloride ions (Cl^-) from hydrochloric acid (HCl) during acid precipitation [89].

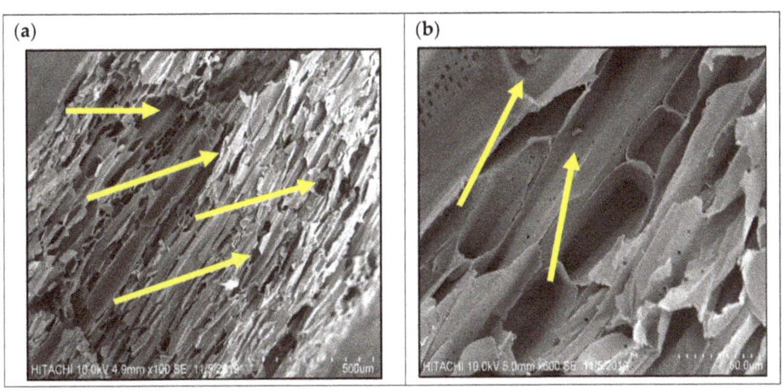

Figure 10. SEM micrograph of RKC at (**a**) 100× and (**b**) 600× magnifications.

Following mercerization and esterification, the surface morphology of EKC showed more pronounced changes. Figure 11 presents the SEM micrographs of EKC at 100× magnification. The surface exhibited increased porosity, with more pores developed compared to RKC (indicated by arrows in Figure 11a). Additionally, the cross-sectional image in Figure 11b revealed the formation of internal pores within the EKC structure. The mercerization process notably increased the surface roughness, contributing to a larger surface area [90,91]. These rough and porous characteristics are crucial for enhancing the oil adsorption capacity of EKC, making it more efficient at capturing oil from POME. The increased surface area provides more active sites for oil adsorption, improving interfacial bonding between the oil and the EKC surface [24,90,92,93].

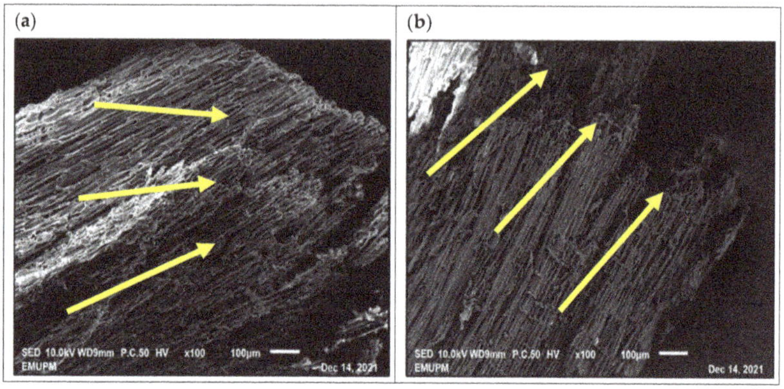

Figure 11. SEM micrograph of EKC using 100× magnification. (**a**) A rougher texture was noticeable on the EKC surface. (**b**) Cross-section image of EKC.

At 600× magnification, Figure 12 provides a closer view of EKC's surface after mercerization and esterification. More uniform pores are evident (indicated by arrows in Figure 12a), with the pores aligned in a honeycomb structure (Figure 12b). This formation of heterogeneous pores is attributed to the expansion of smaller interior pores in the fibers and fibrils, resulting from the alkaline treatment with NaOH during mercerization [94]. These open cylindrical pores and perforations offer excellent sites for oil adsorption, as they provide a larger surface area for oil to be trapped and retained on the EKC surface [95].

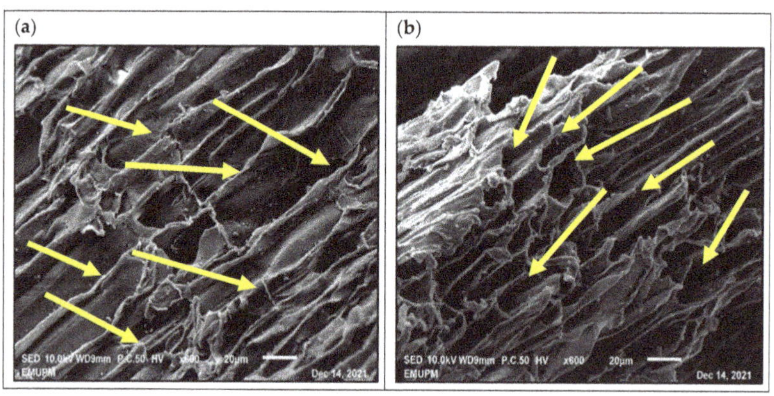

Figure 12. SEM micrograph of EKC using 600× magnification. (**a**) More uniform pores were visible on the EKC surface. (**b**) Pores were aligned in a heterogeneous pore group of honeycomb structure.

In addition to the pore formation, fibrillation on the EKC surface was observed, likely due to the mechanical disruption of fibers during the mercerization process [91,96].

The presence of CaO in the esterification reaction is also thought to contribute to fibrillation [33]. Furthermore, mercerization effectively cleaned the fibers, removed impurities, and enhanced the surface properties of EKC [88]. The SEM analysis reveals significant morphological transformations in EKC after mercerization and esterification. The development of a rougher texture, increased porosity, and the formation of a honeycomb pore structure significantly enhance the adsorption capacity of EKC for oil from POME. These structural changes, including the expansion of fibrils and removal of impurities, are critical for improving the interfacial interaction between the oil and the EKC surface, thereby increasing the overall oil removal efficiency.

4. Conclusions

This research reveals the effective modification of raw kenaf core (RKC) fiber via mercerization and esterification, enhancing its hydrophobic properties and oil extraction efficacy for palm oil mill effluent (POME) remediation. The mercerization process revealed more hydroxyl groups on the mercerized kenaf core (MKC) fiber surface, which then reacted with stearic acid (SA) during esterification to produce ester groups, thereby markedly enhancing its surface hydrophobicity. A regression model with an R^2 value of 0.94 determined the optimum parameters for oil removal as a 1:0.5 MKC:SA ratio, 15 wt% catalyst, and a reflux duration of 1 h, resulting in a 76% oil removal efficiency, in contrast to 51% with untreated RKC. The case study revealed that the highest oil adsorption capacity was obtained in acidic conditions for both RKC and EKC, compared to alkali and neutral conditions. However, EKC exhibited a higher adsorption capacity compared to RKC, demonstrating that EKC has the potential to become a good oil adsorbent. FTIR research affirmed the effective production of esters and the reduction of hydroxyl groups, while SEM imaging demonstrated enhanced surface roughness, fibrillation, and porosity in the modified cellulose, leading to improved oil adsorption. The findings indicate that EKC is an effective and economical adsorbent for oil extraction from POME, with the potential for future optimization via batch studies and kinetic modeling to assess its full industrial applicability.

Author Contributions: Conceptualization, N.H.A., L.C.A. and S.N.A.M.J.; methodology, N.H.A.; software, N.H.A.; validation, N.H.A. and L.C.A.; formal analysis, N.H.A.; investigation, N.H.A.; writing—original draft preparation, N.H.A.; writing—review and editing, N.H.A., L.C.A., C.L.L. and A.A.A.; supervision, L.C.A., T.C.S.Y., S.N.A.M.J., T.M.T. and A.J.A.; funding acquisition, L.C.A. and A.J.A. All authors have read and agreed to the published version of the manuscript.

Funding: This research received no external funding.

Data Availability Statement: The original contributions presented in this study are included in the article. Further inquiries can be directed to the corresponding authors.

Acknowledgments: The authors would like to thank the Department of Chemical and Environmental Engineering, Faculty of Engineering, Universiti Putra Malaysia, Serdang, Selangor, Malaysia, for providing research amenities for this project and Sime Darby Research Sdn. Bhd. and School of Chemical Engineering, College of Engineering, Universiti Teknologi MARA, Shah Alam, Selangor, Malaysia, for financial support.

Conflicts of Interest: Ahmad Jaril Asis was employed by Sime Darby Research Sdn. Bhd.; the remaining authors declare that the research was conducted in the absence of any commercial or financial relationships that could be construed as a potential conflict of interest.

References

1. Adetunji, A.I.; Olaniran, A.O. Treatment of Industrial Oily Wastewater by Advanced Technologies: A Review. *Appl. Water Sci.* **2021**, *11*, 1–19. [CrossRef]
2. Jamaly, S.; Giwa, A.; Hasan, S.W. Recent Improvements in Oily Wastewater Treatment: Progress, Challenges, and Future Opportunities. *J. Environ. Sci* **2015**, *37*, 15–30. [CrossRef] [PubMed]
3. Wei, Y.; Jin, Y.; Zhang, W. Treatment of High-Concentration Wastewater from an Oil and Gas Field via a Paired Sequencing Batch and Ceramic Membrane Reactor. *Int. J. Environ. Res. Public. Health* **2020**, *17*, 1953. [CrossRef] [PubMed]
4. Wang, Y.; Zhang, Y.; Liang, L.; Tu, F.; Li, Z.; Tang, X.; Dai, L.; Li, L. Research Progress on Membrane Separation Technology for Oily Wastewater Treatment. *Toxics* **2024**, *12*, 794. [CrossRef] [PubMed]
5. Abuhasel, K.; Kchaou, M.; Alquraish, M.; Munusamy, Y.; Jeng, Y.T. Oilywastewater Treatment: Overview of Conventional and Modern Methods, Challenges, and Future Opportunities. *Water* **2021**, *13*, 980. [CrossRef]
6. de Medeiros, A.D.M.; da Silva Junior, C.J.G.; de Amorim, J.D.P.; Durval, I.J.B.; de Santana Costa, A.F.; Sarubbo, L.A. Oily Wastewater Treatment: Methods, Challenges, and Trends. *Processes* **2022**, *10*, 743. [CrossRef]
7. Haan, T.Y.; Isma Nordin, P.M.; Ahmad Juanda, N.'I.; Mohd Shafi, M.A.; Krishnan, P. A Review on Adsorption Process for the Treatment of Oily Wastewater. *Adv. Environ. Eng. Res.* **2023**, *4*, 1–21. [CrossRef]
8. Ramli, A.N.; Mohd Ghazi, R. Removal of Oil and Grease in Wastewater Using Palm Kernel Shell Activated Carbon. In *IOP Conference Series: Earth and Environmental Science, Proceedings of the 2nd International Conference on Tropical Resources and Sustainable SciencesPengkalan Chepa, Malaysia, 10–11 August 2020*; IOP Publishing Ltd.: Bristol, UK, 2020; Volume 549.
9. Ahmad, A.L.; Sumathi, S.; Hameed, B.H. Adsorption of Residue Oil from Palm Oil Mill Effluent Using Powder and Flake Chitosan: Equilibrium and Kinetic Studies. *Water Res.* **2005**, *39*, 2483–2494. [CrossRef]
10. Madaki, Y.S.; Seng, L. Palm Oil Mill Effluent (POME) from Malaysia Palm Oil Mills: Waste or Resource. *Int. J. Sci. Environ.* **2013**, *2*, 1138–1155.
11. Liew, R.K.; Chai, C.; Yek, P.N.Y.; Phang, X.Y.; Chong, M.Y.; Nam, W.L.; Su, M.H.; Lam, W.H.; Ma, N.L.; Lam, S.S. Innovative Production of Highly Porous Carbon for Industrial Effluent Remediation via Microwave Vacuum Pyrolysis plus Sodium-Potassium Hydroxide Mixture Activation. *J. Clean. Prod.* **2019**, *208*, 1436–1445. [CrossRef]
12. Ma, A.N. Environmental Management for the Palm Oil Industry. *Palm Oil Dev.* **2000**, *30*, 1–10.
13. Ahmad, A.L.; Bhatia, S.; Ibrahim, N.; Sumathi, S. Adsorption of Residual Oil from Palm Oil Mill Effluent Using Rubber Powder. *Braz. J. Chem. Eng.* **2005**, *22*, 371–379. [CrossRef]
14. Igwe, J.C.; Onyegbado, C.C. A Review of Palm Oil Mill Effluent (Pome) Water Treatment. *Glob. J. Environ. Res.* **2007**, *1*, 54–62.
15. Zinatizadeh, A.A.L.; Salamatinia, B.; Mohamed, A.R.; Isa Hasnain, M. Palm Oil Mill Effluent Digestion in an Up-Flow Anaerobic Sludge Fixed Film Bioreactor. *Int. J. Environ. Res.* **2007**, *1*, 264–271.
16. Chan, Y.J.; Chong, M.F.; Law, C.L. An Integrated Anaerobic-Aerobic Bioreactor (IAAB) for the Treatment of Palm Oil Mill Effluent (POME): Start-up and Steady State Performance. *Process Biochem.* **2012**, *47*, 485–495. [CrossRef]
17. Khemkhao, M.; Nuntakumjorn, B.; Techkarnjanaruk, S.; Phalakornkule, C. UASB Performance and Microbial Adaptation during a Transition from Mesophilic to Thermophilic Treatment of Palm Oil Mill Effluent. *J. Environ. Manag.* **2012**, *103*, 74–82. [CrossRef] [PubMed]
18. Zaied, B.K.; Nasrullah, M.; Islam Siddique, M.N.; Zularisam, A.W.; Singh, L.; Krishnan, S. Enhanced Bioenergy Production from Palm Oil Mill Effluent by Co-Digestion in Solar Assisted Bioreactor: Effects of Hydrogen Peroxide Pretreatment. *J. Environ. Chem. Eng.* **2020**, *8*, 1–29. [CrossRef]
19. Lee, M.D.; Lee, P.P. Performance of Chitosan as Natural Coagulant in Oil Palm Mill Effluent Treatment. In *Promising Techniques for Wastewater Treatment and Water Quality Assessment*; IntechOpen: London, UK, 2020; pp. 1–21.
20. Wu, T.Y.; Mohammad, A.W.; Md Jahim, J.; Anuar, N. Pollution Control Technologies for the Treatment of Palm Oil Mill Effluent (POME) through End-of-Pipe Processes. *J. Environ. Manag.* **2010**, *91*, 1467–1490. [CrossRef] [PubMed]
21. Jadhav, P.; Muhammad, N.; Bhuyar, P.; Krishnan, S.; Abd Razak, A.S.; Zularisam, A.W.; Nasrullah, M. A Review on the Impact of Conductive Nanoparticles (CNPs) in Anaerobic Digestion: Applications and Limitations. *Environ. Technol. Innov.* **2021**, *23*, 1–16. [CrossRef]
22. Saad, M.S.; Joe, N.G.; Shuib, H.A.; Hakim Wirzal, M.D.; Adi Putra, Z.; Khan, M.R.; Busquets, R. Techno-Economic Analysis of an Integrated Electrocoagulation-Membrane System in Treatment of Palm Oil Mill Effluent. *J. King Saud. Univ. Sci.* **2022**, *34*, 1–10. [CrossRef]
23. Adeleke, O.A.; Ismail, N.; Fadilat, A.; Apandi, N.; Latiff, A.A.A.; Saphira, M.R.; Daud, Z.; Ab Aziz, N.A.; Al-Gheethi, A.; Kumar, V.; et al. Principles and Mechanism of Adsorption for the Effective Treatment of Palm Oil Mill Effluent for Water Reuse. In *Nanotechnology in Water and Wastewater Treatment: Theory and Applications*; Elsevier: Amsterdam, The Netherlands, 2019; pp. 1–33. ISBN 9780128139028.

24. Alias, N.H.; Abdullah, L.C.; Choong, T.S.Y.; Md Jamil, S.N.A.; Asis, A.J.; Rusli, N.H. Oil Removal From Palm Oil Mill Effluent (POME) by Using Kenaf Core (*Hibiscus cannabinus*). *Adv. Eng. Res.* **2020**, *200*, 158–163.
25. Ahmad, A.L.; Sithamparam, K.; Ismail, S. Extraction of Residual Oil from Palm Oil Mill Effluent (POME) Using Organic Solvent. *ASEAN J. Sci. Technol. Dev.* **2003**, *20*, 385–394. [CrossRef]
26. Ahmad, A.L.; Bhatia, S.; Ibrahim, N.; Ismail, S. Coagulation-Sedimentation-Extraction Pretreatment Methods for the Removal of Suspended Solids and Residual Oil from Palm Oil Mill Effluent (POME). *IIUM Eng. J.* **2002**, *3*, 25–33.
27. Laohaprapanon, T.; Prasertsan, P.; Kittikun, A.H. Physical and Chemical Separation of Oil and Suspended Solids from Palm Oil Mill Effluent. *Asian J. Energy Environ.* **2005**, *6*, 39–55.
28. Ang, W.L.; Mohammad, A.W. State of the Art and Sustainability of Natural Coagulants in Water and Wastewater Treatment. *J. Clean. Prod.* **2020**, *262*, 1–18. [CrossRef]
29. Sánchez-Martín, J.; Ghebremichael, K.; Beltrán-Heredia, J. Comparison of Single-Step and Two-Step Purified Coagulants from Moringa Oleifera Seed for Turbidity and DOC Removal. *Bioresour. Technol.* **2010**, *101*, 6259–6261. [CrossRef]
30. Pintor, A.M.A.; Vilar, V.J.P.; Botelho, C.M.S.; Boaventura, R.A.R. Oil and Grease Removal from Wastewaters: Sorption Treatment as an Alternative to State-of-the-Art Technologies. A Critical Review. *Chem. Eng. J.* **2016**, *297*, 229–255. [CrossRef]
31. Wahi, R.; Chuah Abdullah, L.; Nourouzi Mobarekeh, M.; Ngaini, Z.; Choong Shean Yaw, T. Utilization of Esterified Sago Bark Fibre Waste for Removal of Oil from Palm Oil Mill Effluent. *J. Environ. Chem. Eng.* **2017**, *5*, 170–177. [CrossRef]
32. Stawiński, W.; Węgrzyn, A.; Dańko, T.; Freitas, O.; Figueiredo, S.; Chmielarz, L. Acid-Base Treated Vermiculite as High Performance Adsorbent: Insights into the Mechanism of Cationic Dyes Adsorption, Regeneration, Recyclability and Stability Studies. *Chemosphere* **2017**, *173*, 107–115. [CrossRef] [PubMed]
33. Wahi, R.; Abdullah, L.C.; Ngaini, Z.; Nourouzi, M.M.; Choong, T.S.Y. Esterification of M. Sagu Bark as an Adsorbent for Removal of Emulsified Oil. *J. Environ. Chem. Eng.* **2014**, *2*, 324–331. [CrossRef]
34. Mohd Nawawi, M.G.; Othman, N.; Sadikin, A.N.; Khalid, S.S.; Yusof, N. Development of Empty Fruit Bunch Filter with Addition of Chitosan for Pre-Treatment of Palm Oil Mill Effluent. *J. Chem. Nat. Resour. Eng.* **2008**, *2*, 8–16.
35. Shavandi, M.A.; Haddadian, Z.; Ismail, M.H.S.; Abdullah, N.; Abidin, Z.Z. Removal of Residual Oils from Palm Oil Mill Effluent by Adsorption on Natural Zeolite. *Water Air Soil. Pollut.* **2012**, *223*, 4017–4027. [CrossRef]
36. Viju, S.; Thilagavathi, G.; Vignesh, B.; Brindha, R. Oil Sorption Behavior of Acetylated Nettle Fiber. *J. Text. Inst.* **2019**, *110*, 1415–1423. [CrossRef]
37. Asadpour, R.; Sapari, N.; Isa, M.H.; Orji, K.U. Investigation of Modified Mangrove Bark on the Sorption of Oil in Water. *Appl. Mech. Mater.* **2014**, *567*, 74–79. [CrossRef]
38. Asadpour, R.; Yavari, S.; Kamyab, H.; Ashokkumar, V.; Chelliapan, S.; Yuzir, A. Study of Oil Sorption Behaviour of Esterified Oil Palm Empty Fruit Bunch (OPEFB) Fibre and Its Kinetics and Isotherm Studies. *Environ. Technol. Innov.* **2021**, *22*, 1–12. [CrossRef]
39. Zamparas, M.; Tzivras, D.; Dracopoulos, V.; Ioannides, T. Application of Sorbents for Oil Spill Cleanup Focusing on Natural-Based Modified Materials: A Review. *Molecules* **2020**, *25*, 4522. [CrossRef]
40. Teli, M.D.; Valia, S.P. Grafting of Butyl Acrylate on to Banana Fibers for Improved Oil Absorption. *J. Nat. Fibers* **2016**, *13*, 470–476. [CrossRef]
41. Asadpour, R.; Sapari, N.; Isa, M.H.; Orji, K.U. Enhancing the Hydrophobicity of Mangrove Bark by Esterification for Oil Adsorption. *Water Sci. Technol.* **2014**, *70*, 1220–1228. [CrossRef]
42. Jahi, N.; Othaman, R.; Lazim, A.M.; Ramli, S. Empty Fruit Bunch Cellulose Based Sorbent for Oil Sorption in Palm Oil Mill Effluent. *Sains Malays.* **2020**, *49*, 2323–2333. [CrossRef]
43. Jawad, R.J.; Ismail, M.H.S.; Siajam, S.I. Adsorption of Heavy Metals and Residual Oil from Palm Oil Mill Effluent Using a Novel Adsorbent of Alginate and Mangrove Composite Beads Coated with Chitosan in a Packed Bed Column. *IIUM Eng. J.* **2018**, *19*, 1–14. [CrossRef]
44. Jahi, N.; Ling, E.S.; Othaman, R.; Ramli, S. Modification of Oil Palm Plantation Wastes as Oil Adsorbent for Palm Oil Mill Effluent (POME). *Malays. J. Anal. Sci.* **2015**, *19*, 31–40.
45. Ngarmkam, W.; Sirisathitkul, C.; Phalakornkule, C. Magnetic Composite Prepared from Palm Shell-Based Carbon and Application for Recovery of Residual Oil from POME. *J. Environ. Manag.* **2011**, *92*, 472–479. [CrossRef] [PubMed]
46. Ekwueme, B.N.; Ezema, C.A.; Asadu, C.O.; Onu, C.E.; Onah, T.O.; Ike, I.S.; Orga, A.C. Isotherm Modelling and Optimization of Oil Layer Removal from Surface Water by Organic Acid Activated Plantain Peels Fiber. *Arab. J. Chem.* **2023**, *16*, 1–19. [CrossRef]
47. Sathasivam, K.; Mas Haris, M.R.H. Adsorption Kinetics and Capacity of Fatty Acid-Modified Banana Trunk Fibers for Oil in Water. *Water Air Soil. Pollut.* **2010**, *213*, 413–423. [CrossRef]
48. Banerjee, S.S.; Joshi, M.V.; Jayaram, R.V. Treatment of Oil Spill by Sorption Technique Using Fatty Acid Grafted Sawdust. *Chemosphere* **2006**, *64*, 1026–1031. [CrossRef] [PubMed]
49. Crane, J.C. Kenaf Fiber-Plant Rival of Jute 1. *Econ. Bot.* **1947**, *1*, 334–350. [CrossRef]

50. Saba, N.; Jawaid, M.; Hakeem, K.R.; Paridah, M.T.; Khalina, A.; Alothman, O.Y. Potential of Bioenergy Production from Industrial Kenaf (*Hibiscus cannabinus* L.) Based on Malaysian Perspective. *Renew. Sustain. Energy Rev.* **2015**, *42*, 446–459. [CrossRef]
51. Wilson, F.D.; Menzel, M.Y. Kenaf (*Hibiscus cannabinus*), Roselle (*Hibiscus sabdariffa*). *Econ. Bot.* **1964**, *18*, 80–91. [CrossRef]
52. Anynomous Kenaf: GOVT to Expand Kenaf Cultivation Nationwide. Available online: https://www.bernama.com/en/news.php?id=1868782 (accessed on 23 January 2025).
53. Tan, J.Y.; Low, S.Y.; Ban, Z.H.; Siwayanan, P. A Review on Oil Spill Clean-up Using Bio-Sorbent Materials with Special Emphasis on Utilization of Kenaf Core Fibres. *Bioresources* **2021**, *16*, 8394–8416. [CrossRef]
54. Kujoana, T.C.; Weeks, W.J.; Van der Westhuizen, M.M.; Mabelebele, M.; Sebola, N.A. Potential Significance of Kenaf (*Hibiscus Cannabinus* L.) to Global Food and Feed Industries. *Cogent Food Agric.* **2023**, *9*, 1–16. [CrossRef]
55. Hayeemasae, N.; Viet, C.X.; Masa, A.; Shuib, R.K.; Ismail, H.; Surya, I. The Use of Kenaf Fibre as a Natural Anti-Degradant in Recycled High-Density Polyethylene and Natural Rubber-Based Thermoplastic Elastomers. *Polymers* **2023**, *15*, 1–16. [CrossRef] [PubMed]
56. Sajab, M.S.; Chia, C.H.; Zakaria, S.; Mohd Jani, S.; Khiew, P.S.; Chiu, W.S. Removal of Copper(II) Ions from Aqueous Solution Using Alkali-Treated Kenaf Core Fibres. *Adsorpt. Sci. Technol.* **2010**, *28*, 377–386. [CrossRef]
57. Sajab, M.S.; Chia, C.H.; Zakaria, S.; Jani, S.M.; Ayob, M.K.; Chee, K.L.; Khiew, P.S.; Chiu, W.S. Citric Acid Modified Kenaf Core Fibres for Removal of Methylene Blue from Aqueous Solution. *Bioresour. Technol.* **2011**, *102*, 7237–7243. [CrossRef] [PubMed]
58. Anynomous Coal. Available online: https://tradingeconomics.com/commodity/coal (accessed on 23 January 2025).
59. Quora, Inc. Why Are Fatty Acids Hydrophobic? Available online: https://www.quora.com/Why-are-fatty-acids-hydrophobic (accessed on 16 July 2021).
60. Stat-Ease, Inc. Design-Expert V22.0: Anova Output. Available online: https://www.statease.com/docs/latest/contents/analysis/anova-output/ (accessed on 5 December 2022).
61. Clesceri, L.S.; Greenbreg, A.E.; Eaton, A.D. *Standard Methods for the Examination of Water and Wastewater*, 20th ed.; APHA, AWWA, WEF: Washington, CD, USA, 1999.
62. RTI Laboratories FTIR Analysis | RTI Laboratories. Available online: https://rtilab.com/techniques/ftir-analysis/ (accessed on 23 May 2018).
63. Laboratory Testing Inc. Scanning Electron Microscopy, SEM Analysis. Available online: https://www.labtesting.com/services/materials-testing/metallurgical-testing/sem-analysis/ (accessed on 26 May 2018).
64. Wahi, R. Application of Esterified Sago Bark for Removal of Oil from Palm Oil Mill Effluent. Ph.D. Thesis, Universiti Putra Malaysia, Serdang, Selangor, Malaysia, 2014.
65. Satriadi, H.; Khaibar, A.; Almakhi, M.M. Biodiesel Production by Using Heterogeneous Catalyst from Fly Ash and Limestone. In Proceedings of the 2017 International Conference on Sustainable Energy Engineering and Application (ICSEEA), Jakarta, Indonesia, 23–24 October 2017; pp. 41–44.
66. Hadiyanto, H.; Lestari, S.P.; Abdullah, A.; Widayat, W.; Sutanto, H. The Development of Fly Ash-Supported CaO Derived from Mollusk Shell of Anadara Granosa and Paphia Undulata as Heterogeneous CaO Catalyst in Biodiesel Synthesis. *Int. J. Energy Environ. Eng.* **2016**, *7*, 297–305. [CrossRef]
67. Basumatary, S.F.; Brahma, S.; Hoque, M.; Das, B.K.; Selvaraj, M.; Brahma, S.; Basumatary, S. Advances in CaO-Based Catalysts for Sustainable Biodiesel Synthesis. *Green. Energy Resour.* **2023**, *1*, 1–31. [CrossRef]
68. Wong, Y.C.; Tan, Y.P.; Taufiq-Yap, Y.H.; Ramli, I. Effect of Calcination Temperatures of CaO/Nb_2O_5 Mixed Oxides Catalysts on Biodiesel Production (Kesan Suhu Pengkalsinan Oksida Campuran CaO/Nb_2O_5 Ke Atas Penghasilan Biodiesel). *Sains Malaysiana.* **2014**, *43*, 783–790.
69. Abd Aziz, A.S. *The Basics of Chemical Plant Operations*, 1st ed.; Azharin Shah bin Abd Aziz: Kemaman, Malaysia, 2023; ISBN 978-629-99043-0-4.
70. Saha, R.; Goud, V.V. Ultrasound Assisted Transesterification of High Free Fatty Acids Karanja Oil Using Heterogeneous Base Catalysts. *Biomass Convers. Biorefin* **2015**, *5*, 195–207. [CrossRef]
71. Abu Bakar, N.N.; Latip, J.; Sapian, S.; Mohd Khalid, R. Optimization of Roselle (*Hibiscus sabdariffa* Linn.) Anthocyanin Extraction Parameter by Response Surface Modeling and Potential of Roselle Agro-Waste as Alternative Sources of Anthocyanin. *Sains Malays.* **2023**, *52*, 3147–3162. [CrossRef]
72. Madhumita, M.; Guha, P.; Nag, A. Extraction of Betel Leaves (*Piper betle* L.) Essential Oil and Its Bio-Actives Identification: Process Optimization, GC-MS Analysis and Anti-Microbial Activity. *Ind. Crops Prod.* **2019**, *138*, 1–12. [CrossRef]
73. Samadi, M.; Zainal Abidin, Z.; Yoshida, H.; Yunus, R.; Awang Biak, D.R.; Lee, C.H.; Lok, E.H. Subcritical Water Extraction of Essential Oil from Aquilaria Malaccensis Leaves. *Sep. Sci. Technol.* **2020**, *55*, 2779–2798. [CrossRef]
74. Anderson, M.J.; Whitcomb, P.J. *Anderson_Whitcomb 2017 RSM-Optimizing Processes Using. Response Surface Methods for Design of Experiments*, 2nd ed.; CRC Press, Taylor & Francis Group: Boca Raton, FL, USA, 2017; ISBN 978-1-4987-4598-7.

75. Frost, J. Statistics by Jim: How To Interpret R-Squared in Regression Analysis. Available online: https://statisticsbyjim.com/regression/interpret-r-squared-regression/ (accessed on 11 March 2023).
76. Zhang, L.; Cheng, Z.; Guo, X.; Jiang, X.; Liu, R. Process Optimization, Kinetics and Equilibrium of Orange G and Acid Orange 7 Adsorptions onto Chitosan/Surfactant. *J. Mol. Liq.* **2014**, *197*, 353–367. [CrossRef]
77. Yu, P.; Chao, X. Statistics-Based Optimization of the Extraction Process of Kelp Polysaccharide and Its Activities. *Carbohydr. Polym.* **2013**, *91*, 356–362. [CrossRef] [PubMed]
78. Prakash Maran, J.; Manikandan, S. Response Surface Modeling and Optimization of Process Parameters for Aqueous Extraction of Pigments from Prickly Pear (Opuntia Ficus-Indica) Fruit. *Dye. Pigment.* **2012**, *95*, 465–472. [CrossRef]
79. The Pennsylvania State University 4.6-Normally Probability Plot of Residuals. Available online: https://online.stat.psu.edu/stat501/lesson/4/4.6 (accessed on 16 March 2023).
80. Burmana, A.D.; Tambun, R.; Haryanto, B.; Alexander, V. Effect of Reaction Time on Biodiesel Production from Palm Fatty Acid Distillate by Using PTSA as a Catalyst. In *IOP Conference Series: Materials Science and Engineering, Proceedings of the 2nd International Conference on Industrial and Manufacturing Engineering (ICI&ME 2020), Medan, Indonesia, 3–4 September 2020*; IOP Publishing Ltd.: Bristol, UK, 2020; Volume 1003.
81. Tang, C.Z.; Tao, H.X.; Zhan, X.Q.; Xie, X.A. Mechanism of Esters Formation during Cellulose Liquefaction in Sub-and Supercritical Ethanol. *Biresources* **2014**, *9*, 4946–4957. [CrossRef]
82. Vera Candioti, L.; De Zan, M.M.; Cámara, M.S.; Goicoechea, H.C. Experimental Design and Multiple Response Optimization. Using the Desirability Function in Analytical Methods Development. *Talanta* **2014**, *124*, 123–138. [CrossRef] [PubMed]
83. Salman, N.A.S.; Mohamed Saheed, M.S.; Saqib, N.U.; Abdul Latip, A.F.; Adnan, R. Polydimethylsiloxane-Carbon Nanotubes-Based Hydrophobic Sponge for Treatment of Palm Oil Mill Rffluents with Enhanced Reusability. *J. Environ. Chem. Eng.* **2023**, *11*, 1–15. [CrossRef]
84. Jaber, W.S.; Alwared, A.I. Removal of Oil Emulsion from Aqueous Solution by Using Ricinus Communis Leaves as Adsorbent. *Springer Nat. Appl. Sci.* **2019**, *1*, 1–12. [CrossRef]
85. Vitz, E.; Moore, J.W.; Shorb, J.; Prat-Resina, X.; Wendorff, T.; Hahn, A. 20.5: Nonpolar Lipids. Available online: https://chem.libretexts.org/Bookshelves/General_Chemistry/Book:_ChemPRIME_(Moore_et_al.)/20:_Molecules_in_Living_Systems/20.05:_Nonpolar_Lipids (accessed on 25 April 2023).
86. Sidik, S.M.; Jalil, A.A.; Triwahyono, S.; Adam, S.H.; Satar, M.A.H.; Hameed, B.H. Modified Oil Palm Leaves Adsorbent with Enhanced Hydrophobicity for Crude Oil Removal. *Chem. Eng. J.* **2012**, *203*, 9–18. [CrossRef]
87. Xiaobing, L.I.; Chunjuan, Z.; Jiongtian, L. Adsorption of Oil from Waste Water by Coal: Characteristics and Mechanism. *Min. Sci. Technol.* **2010**, *20*, 778–781. [CrossRef]
88. Chukwuma, O.E. Effect of Alkali Treatment on Kenaf Fibre Quality. *Sci. Arena Publ. Spec. J. Agric. Sci.* **2017**, *3*, 12–21.
89. Syed Hashim, S.N.A.; Zakaria, S.; Chia, C.H.; Pua, F.L.; Syed Jaafar, S.N. Chemical and Thermal Properties of Purified Kenaf Core and Oil Palm Empty Fruit Bunch Lignin. *Sains Malays.* **2016**, *45*, 1649–1653.
90. Wong, C.; McGowan, T.; Bajwa, S.G.; Bajwa, D.S. Impact of Fiber Treatment on the Oil Absorption Characteristics of Plant Fibers. *Bioresources* **2016**, *11*, 6452–6463. [CrossRef]
91. Ilyas, R.A.; Sapuan, S.M.; Ishak, M.R.; Zainudin, E.S. Effect of Delignification on the Physical, Thermal, Chemical, and Structural Properties of Sugar Palm Fibre. *Bioresources* **2017**, *12*, 8734–8754. [CrossRef]
92. Wahi, R.; Abdullah, L.C.; Choong, T.S.Y.; Ngaini, Z.; Nourouzi, M.M. Oil Removal from Aqueous State by Natural Fibrous Sorbent: An Overview. *Sep. Purif. Technol.* **2013**, *113*, 51–63. [CrossRef]
93. Abbas, A.G.N.; Abdul Aziz, F.N.A.; Abdan, K.; Mohd Nasir, N.A.; Norizan, M.N. Kenaf Fibre Reinforced Cementitious Composites. *Fibers* **2022**, *10*, 3. [CrossRef]
94. Koistinen, A.; Wang, H.; Hiltunen, E.; Vuorinen, T.; Maloney, T. Refinability of Mercerized Softwood Kraft Pulp. *Cellulose* **2024**, *31*, 6471–6484. [CrossRef]
95. Zaini, N.; Kamarudin, K.S.N. Adsorption of Carbon Dioxide on Monoethanolamine (MEA)-Impregnated Kenaf Core Fiber by Pressure Swing Adsorption System (PSA). *J. Teknol.* **2014**, *68*, 2180–3722. [CrossRef]
96. Mwaikambo, L.Y.; Ansell, M.P. Chemical Modification of Hemp, Sisal, Jute, and Kapok Fibers by Alkalization. *J. Appl. Polym. Sci.* **2002**, *84*, 2222–2234. [CrossRef]

processes

MDPI

Article

Trends in Sustainable Inventory Management Practices in Industry 4.0

Silvia Carpitella [1] and Joaquín Izquierdo [2,*]

[1] Deptartment of Manufacturing Systems Engineering and Management, California State University Northridge, 18111 Nordhoff St., Northridge, CA 91330, USA; silvia.carpitella@csun.edu

[2] Fluing-Institute for Multidisciplinary Mathematics, Universitat Politècnica de València, Cno. de Vera s/n, 46022 Valencia, Spain

* Correspondence: jizquier@upv.es

Abstract

This study examines 52 recently published papers on sustainable inventory management in Industry 4.0, intending to bridge theory and practice through a comprehensive literature review. By analyzing the latest advancements discussed over the past two years, covering 2024 and 2025, we identify key trends shaping the field and highlight existing gaps that may require further exploration. Focusing on this time frame is particularly relevant because it reflects how companies have recently started using artificial intelligence more practically to support sustainability goals. During these years, AI has been applied to improve how inventory is tracked, how demand is predicted, and how resources are managed to reduce waste. These tools are making supply chains more efficient while helping organizations to lower their environmental impact. In this regard, our work aims to provide a deeper understanding of how sustainable strategies are evolving in response to technological innovations, offering insights for researchers and practitioners seeking to enhance efficiency and environmental responsibility in modern supply chains.

Keywords: sustainability; inventory management; Industry 4.0; review

1. Introduction

With the advancement of Industry 4.0, sustainable inventory management has become increasingly crucial in improving supply chain efficiency and resilience [1]. As industries transition to more environmentally responsible operations, the integration of such digital technologies as automation, real-time data systems, and smart analytics is reshaping traditional inventory management practices. Apart from considerably streamlining operations, these tools are contributing to increasing the importance of logistics performance in global markets, influenced by shifting customer expectations, stricter environmental regulations, and intensified international competition [2]. Sustainable inventory management today goes beyond improving efficiency as it plays a direct role in achieving broader environmental goals. Specifically, it contributes to reducing material waste [3], minimizing carbon emissions across the supply chain [4], and optimizing resource utilization through more accurate forecasting and planning [5]. The strategic adoption of sustainable inventory practices is essential for ensuring long-term economic growth while mitigating environmental impacts. Analyzing how these practices have evolved, particularly through the use of digital and automated tools, offers essential insight into how industries are aligning sustainability with competitiveness in the modern industrial landscape [6].

For this reason, the main goal of this paper is to conduct a comprehensive review of relevant and updated research on sustainable inventory management in the context of Industry 4.0. This objective will be pursued through the following three key areas of focus.

- This study includes a structured search for the recent literature, specifically journal articles, conference papers, and book chapters, on sustainable inventory management and Industry 4.0 technologies from the last two calendar years (2024 and 2025).
- The identified articles are thoroughly screened, categorized, and reviewed to synthesize key managerial insights, relevant techniques, and emerging trends related to sustainable practices in inventory management in Industry 4.0.
- Research gaps and promising future developments are formalized by discussing areas where innovative approaches can contribute to improving sustainability in inventory management in the evolving digital transformation landscape.

The structure of this paper is as follows. Section 2 presents a comprehensive literature review, including an overview of sustainable inventory management in Industry 4.0, a detailed description of the methodology used to conduct this review, and a categorization of studies based on their thematic focus. Section 3 synthesizes the results derived from this research with an emphasis on sustainability aspects that can be crucial in management practices. Finally, Section 4 summarizes the main findings of this study and proposes potential directions for future research in sustainable inventory management within Industry 4.0.

2. Literature Review

2.1. *Sustainable Inventory Management in Industry 4.0*

As industries strive to balance operational efficiency with environmental responsibility, sustainability is a key organizational priority. As observed in [7], achieving a balance between operational goals and sustainability requires a fundamental shift in how supply chain practices are designed. This involves aligning economic, social, and environmental objectives. In the era of Industry 4.0, where digital advancements are reshaping production and logistics, companies are rethinking inventory management practices to not only reduce carrying costs but also to minimize waste, use resources more efficiently, and reduce environmental impact [8]. Recent studies reflect a shared understanding that sustainable inventory management plays a critical role in long-term profitability [9], helps organizations comply with environmental regulations [3,10], and supports growing customer expectations for responsible and transparent practices [11]. This subsection provides a conceptual overview of these developments, drawing from the current literature to show how digital tools enable companies to balance sustainability with operational performance better.

One of the fundamental principles of sustainable inventory management is minimizing excess stock while avoiding shortages. In this regard, Nobil et al. [12] consider the impact of shortages in raw materials on production decisions, integrating environmental concerns by factoring in carbon emissions and the potential for carbon trading. Singh et al. [13] present an optimal control-based inventory model that minimizes excess stock while preventing shortages in an imperfect production system, aiming to balance supply, minimize losses, and ensure efficient stock management. As discussed in [14], overstocking leads to higher storage costs, increased material waste, and unnecessary resource consumption, while understocking disrupts production and affects customer satisfaction. To address these challenges, businesses are adopting data-driven demand forecasting, automated tracking systems, and real-time monitoring to improve inventory accuracy and reduce inefficiencies [15,16]. Another key aspect of sustainable inventory management refers to the integration of environmentally friendly practices into supply chain operations [17,18], for example, by implementing energy-efficient storage solutions [19] and prioritizing suppliers with sustainable practices [20]. In such a direction, many companies are also exploring circular economy models, where returned or unused products are refurbished, reused, or repurposed instead of being discarded [21]. Reducing transportation-related emissions is another important goal. To such an aim, localized supply chains may help minimize long-distance transportation, improving supply chain resilience while mitigating low-magnitude disruptions, as underlined by McDougall and Davis [22]. In this context, collaboration among such stakeholders as manufacturers, suppliers, and regulatory bodies remains crucial for advancing sustainability efforts in inventory management [23].

Despite the discussed benefits, challenges remain in transitioning to sustainable inventory management, especially given the pressures retailers face in meeting customer expectations [24]. Notably, as discussed by Agyemang [25], initial costs for implementing green technologies and redesigning supply chain processes can be significant. Additionally, balancing sustainability with cost efficiency requires careful planning and investment in long-term solutions. Considering all of these aspects, the future of sustainable inventory management will likely focus on continuous improvement and innovation, for which embracing emerging technologies and refining logistics strategies will be instrumental to meeting sustainability goals while maintaining competitive advantages in the current global market scenario [26].

Broadly speaking, sustainable inventory management involves a careful balance between reducing excess stock and avoiding shortages while also addressing broader environmental goals. Current practices reflect a growing focus on minimizing waste, improving forecasting accuracy, and integrating eco-friendly solutions across supply chains. At the same time, businesses face ongoing challenges, including the cost of transitioning to green technologies and the need to align sustainability with operational efficiency. Looking ahead, continued innovation, collaboration, and the adoption of practical strategies will be essential for companies seeking to strengthen their environmental performance and competitiveness in a changing global market.

2.2. *Methodological Approach Used for the Review*

We will conduct a structured literature search and review on sustainable inventory management practices in Industry 4.0 based on the systematic approach proposed by Muka et al. [27]. The first step was to investigate whether the topic had already been reviewed, and we were able to check that several reviews have been developed since 2024 that are relevant to the subject of the present work. However, none of these works specifically focus on a broad analysis of sustainable inventory management in Industry 4.0. This motivates us in our the present review to propose a more comprehensive and unique categorization of papers while incorporating the most recent contributions in the field. To ensure a thorough and diverse collection of relevant studies, a literature search was carried out across four major scientific databases, namely Scopus, IEEE Xplore, PubMed, and Google Scholar, each of them offering distinct advantages in terms of indexing peer-reviewed journals, conference proceedings, and multidisciplinary research. Figure 1 outlines the publication items by the database. A total of 59 papers have been collected, all within the time range of 2024 to the present. These papers have been categorized by type (journal papers, conference papers, and book chapters). The bar chart illustrates their distribution across the databases, while the pie chart provides an overview.

An advanced search was performed in each database in March 2025 by selecting fields and entering the following keywords: "*Sustainability*" AND "*Inventory Management*" AND "*Industry 4.0*". The added date range refers to items published from 2024 to the present. In IEEE Xplore and PubMed, the search was conducted using the "all field" category, meaning that the selected keywords were applied across the entire content of each document, including titles, abstracts, keywords, and full text when available, ensuring a broad and inclusive search. For Scopus, the search was restricted to article titles, abstracts, and keywords to balance relevance and comprehensiveness. Scopus indexes many sources, and this refinement helps focus on the most pertinent studies. In the case of Google Scholar, a more extensive and less curated database, the results were reviewed only up to the third page to ensure that only the most relevant and widely cited papers were considered [28].

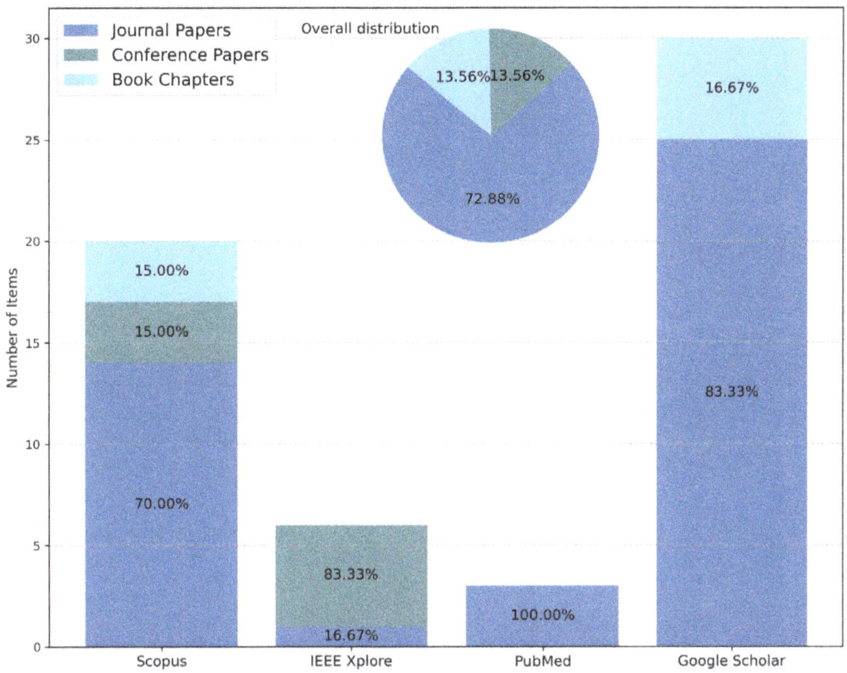

Figure 1. Publication distribution across databases.

After compiling all the references and abstracts into a single document, we conducted a thorough screening process to eliminate duplicates, reducing the total number of items for analysis from 59 to a smaller, refined set comprising 52 items. Next, we grouped these filtered papers based on their publishers and performed an in-depth analysis to determine their core subject matter. Each paper was assigned to one of six distinct topic categories (TCs) during the classification process. This process was conducted by carefully examining each paper's abstract, objectives, methodologies, and main contributions. We employed a qualitative content analysis approach to identify the dominant theme of each work, considering both explicit keywords and the underlying research aims. Full texts were consulted to clarify the primary focus along with the abstracts. The six topic categories were derived based on recurring patterns across the literature, ensuring that each category reflects a coherent thematic area. This method allowed for consistency in classification while remaining flexible enough to accommodate interdisciplinary studies that overlap across themes. The distribution of the analyzed papers, both by publisher and their assignment to TC, is illustrated in Figure 2. This figure provides an overview of how the papers are grouped based on their source and how they have been categorized according to their primary subject matter. Since some papers address multiple topics, we categorized each based on its most relevant primary focus while acknowledging its potential significance in other areas. This approach ensures that the classification remains structured and meaningful while preserving the awareness of interdisciplinary connections. The following sections provide a detailed description of each TC.

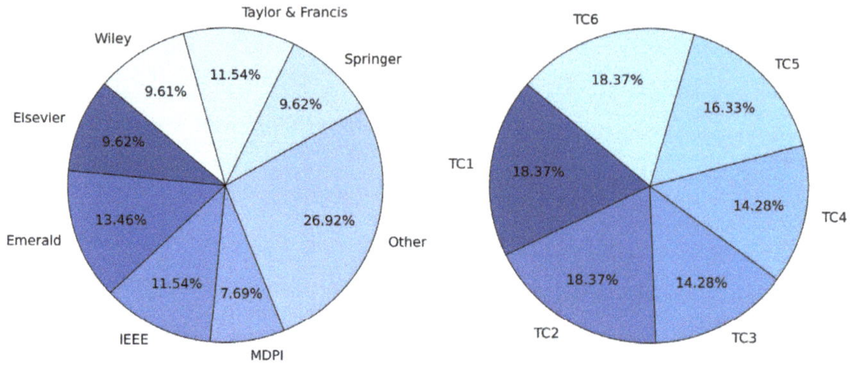

Figure 2. Distribution of papers by publisher and their assignment to topic categories.

- TC_1 Industry 4.0 and digital transformation. Papers in this category underline that Industry 4.0 brings digital tools like smart sensors, automation, and real-time data analytics to inventory management, improving efficiency and reducing waste. With predictive analytics, businesses can optimize stock levels, prevent shortages, and reduce excess inventory, leading to a more sustainable supply chain.
- TC_2 Sustainability and circular economy. Papers in this category discuss sustainable inventory management focusing on minimizing waste, reusing materials, and reducing environmental impact. Industry 4.0 technologies enable better tracking of product lifecycles, allowing businesses to extend product use, implement recycling strategies, and adopt circular economy practices that reduce resource consumption.
- TC_3 Artificial intelligence and machine learning applications. Papers in this category focus on AI and machine learning transforming inventory management by predicting demand, optimizing stock levels, and reducing overproduction. These technologies analyze patterns in supply and demand, adjusting inventory dynamically to prevent waste and improve efficiency, making supply chains more sustainable.
- TC_4 Supply chain and logistics management. Papers sorted into this category deal with the role of efficient supply chain and logistics operations for sustainable inventory management. Industry 4.0 enables automated inventory tracking, smart warehouses, and optimized transportation routes, reducing excess storage, cutting emissions, and ensuring a smoother, more sustainable flow of goods.
- TC_5 Sustainable manufacturing and lean production. Papers in this category are focused on lean production combined with Industry 4.0 technologies, helping manufacturers reduce waste and energy consumption. Smart monitoring systems and automated inventory controls ensure that materials are used efficiently, minimizing overproduction and improving sustainability without compromising productivity.

- TC_6 Blockchain, digital twins, and emerging technologies. Papers in this category refer to relevant technologies. Blockchain enhances transparency and security in supply chains, reducing fraud and ensuring ethical sourcing. Digital twins create virtual models of inventory systems, allowing businesses to test and optimize strategies before implementation. Emerging technologies like the Internet of Things (IoT) and 3D printing further improve significant sustainability aspects in inventory management.

2.3. *Examination of Papers in the Designated Topic Categories*

2.3.1. Industry 4.0 and Digital Transformation

Asrol [29] develops a systematic literature review on Industry 4.0 adoption in supply chain operations, using both qualitative and quantitative analyses, including Preferred Reporting Items for Systematic Reviews and Meta-Analyses (PRISMA) and bibliometric methods. The author identifies five key dimensions to assess Industry 4.0 readiness: technology and IT infrastructure, supply chain integration, manufacturing and inventory, leadership and human resources, and sustainability. While this study highlights the role of digital transformation maturity models in general supply chain operations, our review aims to take a more targeted approach to Industry 4.0 technologies, enhancing sustainable inventory management. Another interesting review has been developed by Valenzuela-Cobos et al. [30], exploring the adoption of Industry 4.0 in logistics management and its advancement in Latin America. This review mainly provides a regional perspective on the impact of Industry 4.0 on logistics flexibility and risk management by considering how optimizing material and information flow helps to enhance competitiveness.

In this context, Singh et al. [31] examine the relationship between Industry 4.0 and environmental sustainability, particularly in manufacturing contexts. The authors discuss the main environmental challenges posed by such advanced manufacturing technologies as Cyber-Physical Systems (CPSs), including pollution and increased resource consumption, exploring how integrating Industry 4.0 with sustainable development initiatives can influence economic and business changes while supporting Supply Chain Decarbonization (SCD) and achieving net-zero targets. Akhouri et al. [32] analyze the integration of Supply Chain Management (SCM) with Industry 4.0 technologies to enhance sustainability in Indian manufacturing. Several key SCM factors are the object of study, with a particular emphasis on collaboration, cost efficiency, and adaptive inventory management in a technologically evolving landscape. Gantayat et al. [33] observe that Industry 5.0 builds on Industry 4.0 by emphasizing sustainability, human-centricity, and human–machine collaboration to create more socially aligned industrial systems. Their research highlights the role of Large Language Models (LLMs) in achieving higher levels of innovation in sustainable industrial practices and human-centered processes.

Rehman et al. [34] investigate the impact of Industry 4.0 technologies with a specific focus on the international performance of Small and Medium Enterprises (SMEs), applying dynamic capabilities theory and institutional theory. Using data from 301 Malaysian exporting SMEs and structural equation modeling, the study finds that most Industry 4.0 technologies positively influence SMEs' global performance by enhancing their ability to adapt and innovate at an international scale. An exploratory case study of an Indian automobile supply chain is developed in [35], where the need for collaboration between supply chain entities and support from large manufacturers is considered the main driver of sustainability and Business Model Innovation (BMI) in the whole automotive sector. Ahmed et al. [36] focus on the medical device industry of emerging economies, using Pakistan as a case study. An integrated Multi-Criteria Decision-Making (MCDM) approach is proposed to evaluate and rank Industry 4.0 technologies based on their effectiveness in enhancing the performance of marketing strategies. The study identifies Big Data, Virtual Reality (VR) and Augmented Reality (AR), IoT, and machine learning as the most impactful technologies within the reference context. Bakeries have been another sector recently analyzed in the literature [37] in order to evaluate the impact of digitalization to optimize sustainable inventory management and product quality while driving profitability.

2.3.2. Sustainability and Circular Economy

Kumar et al. [38] provide a categorized review of 121 papers on Industry 4.0 and Smart Manufacturing (SM) published between 2011 and 2022. This research identifies China, the USA, India, and Germany as leading countries in Industry 4.0 research, underlining that conceptual models are currently predominant while case studies and mathematical modeling are gaining popularity. This review also explores the integration of Lean Six Sigma with Industry 4.0 for efficiency improvements but notes gaps in understanding long-term impacts and applications beyond manufacturing. Concerning Lean Six Sigma and Industry 4.0 integration, a holistic study is proposed in [39]. This paper is motivated by the disconnection between quality management and circular economy principles, aiming to bridge the gap in assessing Industry 4.0 production value based on sustainability considerations. Specifically, an integrated approach using Lean Six Sigma and the Delphi technique is proposed to map key quality elements against sustainability factors through Adapted Interpretive Structural Modeling (AISM), facilitating a shift toward sustainable manufacturing ecosystems. Sasso et al. [40] also focus on the relation between lean management and circular economy to address complex market competition challenges. They suggest that lean practices promote a smooth transition towards circular economy by optimizing resource use and supporting recycling and remanufacturing, aided by Just in Time (JIT) and Value Stream Mapping (VSM) methodologies.

Considering this perspective, Marak et al. [41] explore the interconnections between Industry 4.0, the circular economy, and sustainable performance, addressing a research gap primarily focused on bivariate relationships among these concepts. In detail, this study presents a theoretical framework to clarify this complex interlinkage by examining the drivers that facilitate the implementation of Industry 4.0 technologies, circular economy strategies, and sustainability practices while identifying practical challenges that firms commonly face in integrating these elements to achieve sustainability goals. Saleh and AlShafeey have explored another interesting line of research in the direction of this integration [42], leveraging the use of sentiment analysis and association rule-mining techniques on 6759 abstracts from the Scopus database. In this regard, sentiment analysis reveals a predominantly positive perception of Industry 4.0's impact across three sustainability dimensions: Environmental, Economic, and Social (EES). Association rule mining demonstrates the strongest link between Industry 4.0 and economic sustainability, followed by social and environmental sustainability, highlighting benefits such as automation of hazardous tasks, job creation, and emissions reduction through real-time monitoring, among others. An important finding of this research is that EES aspects are frequently analyzed together in the Industry 4.0 context. For example, Amir et al. [43] discuss the impact of the transition from Industry 4.0 to Industry 5.0 on EES sustainability. The authors highlight the metaverse's role and emerging technologies such as blockchain, IoT, VR, and AR in transforming real-world interactions into virtual experiences, contributing to sustainability across various domains, including supply chain networks, e-commerce, smart manufacturing, inventory management, etc.

Relevant cases have been developed in various practical contexts. Singh et al. [44] apply a risk assessment framework based on MCDM to mitigate Supply Chain Risks

(SCRs) and enhance sustainability in the context of Additive Manufacturing Technology (AMT). This study reveals that AMT adoption significantly reduces risks related to lead time fluctuations, waste generation, supplier dependency, inventory management, and logistics, proving to be a strategic approach for improved supply chain sustainability and innovation in green technologies. A Brazilian case on food-tech startups integrating Industry 4.0 technologies to advance the circular economy is proposed in [45]. The study analyzes eight startups using questionnaires, interviews, and content analysis, revealing their role as key players in driving sustainable transformation while offering insights on leveraging technology in the food sector. Zaki [46] focuses on the hospitality sector by assessing circular economic practices in green hotels in Saudi Arabia and Egypt and analyzing specific practices that can contribute to increased hotel performance.

2.3.3. Artificial Intelligence and Machine Learning Applications

Artificial intelligence (AI) is becoming increasingly popular in inventory management practices. In this field, a study of 123 SMEs was led by Rana et al. [47] through various analytical methods, stressing the direct connection between implementing AI-based approaches and obtaining gains in terms of security increase and cost reduction in Supply Chain 4.0. A systematic review of the integration of AI in SCM during the transition from Industry 4.0 to Industry 6.0 is carried out in [48], underlining its role in enhancing human–AI collaboration and sustainability. Using the PRISMA framework, this study analyzes the literature from 2010 to 2023 and identifies key areas where AI has significantly contributed to supply chain management, including improved accuracy in demand forecasting, enhanced inventory optimization, and data-driven decision-making. Additionally, the study highlights critical challenges such as cybersecurity vulnerabilities, ethical concerns, and the need for upskilling the workforce to fully leverage AI capabilities. The findings of our review confirm and expand upon these insights, with a growing emphasis on the practical implementation of predictive analytics and AI-driven tools aimed at advancing sustainable and resilient logistics strategies. Focusing on the particular case of sustainable innovation in the U.S. fashion industry, Faisal et al. [49] identified a research gap in integrating AI into existing apparel manufacturing frameworks and examined AI's role in production streamlining and personalized consumer experiences by pointing out the importance of demand forecasting.

Traditional inventory models face significant challenges in accurate prediction due to growing demand uncertainty regarding the need to improve forecasting accuracy and optimize stock levels. Kumar et al. [50] discuss the critical role of inventory management in maintaining a smooth supply chain with a specific focus on backorder prediction. They explain how companies leverage machine learning techniques to predict backorders more precisely, intending to reduce financial pressure and disruptions in supply chain operations while positively impacting customer satisfaction. Similarly, Ogidan et al. [51] emphasize the role of machine learning in manufacturing and its impact on quality and productivity within the Fourth Industrial Revolution. The discussion includes manufacturing-related machine learning algorithms, implementation challenges, and future research prospects by reviewing practical applications on quality process control, predictive maintenance, inventory management, and supply chain optimization.

Paraschos et al. [52] integrate reinforcement learning with lean green manufacturing principles for optimizing decision-making in a multi-stage degrading manufacturing/remanufacturing system. The reinforcement learning-based approach integrates pull production, predictive maintenance, and circular economy strategies to minimize material consumption while extending product lifecycles, as validated through experimental analysis. In another study [53], the same authors explore the role of Industry 4.0 and smart manufacturing technologies in enhancing decision-making by collecting real-time data on machinery and production. The authors focus on process scheduling in manufacturing, emphasizing the importance of adapting operations to dynamic product demand and customer trends and proposing a reinforcement learning framework for optimized manufacturing and maintenance control.

2.3.4. Supply Chain and Logistics Management

Industry 4.0 technologies and sustainability practices in the logistics industry converge to shape the evolving framework of Sustainable Logistics 4.0, as discussed by Di Nardo et al. [54]. Through a comprehensive literature review and bibliometric network analysis of studies from 2013 to 2023 sourced from Scopus and Google Scholar, the authors formalize how digital interventions consistently contribute to the improvement of sustainability in logistics. As observed by Holloway [55], sustainable inventory management plays a vital role in modern supply chain and logistic management, balancing efficiency with environmental responsibility. While financial constraints and organizational resistance may pose challenges to technological innovation, strategic investments and collaboration can drive meaningful change in terms of resilience while complying with existing regulations.

For example, Pinagapani et al. [56] discuss the design and implementation of a material-handling prototype robot and monitoring system within the framework of Industry 4.0. The authors highlight the role played by Automated Material Handling Systems (AMHSs) in improving efficiency in inventory storage and warehouses by reducing human intervention, also exploring automated queuing technology for continuous stacking and unstacking, with the ultimate goal of optimizing material placement based on weight capacity. Choudhary et al. [57] analyze the role of Radio Frequency Identification (RFID) and next-generation IoT technologies in enhancing supply chain flexibility while addressing security challenges and in the presence of regulatory constraints. This research discusses traditional inventory models such as Economic Order Quantity (EOQ) and Economic Production Quantity (EPQ) and how IoT-driven solutions improve real-time data transmission and enhance inventory mobility. Additionally, the impact of digitization, cloud computing, and drones is discussed concerning the future of logistics. Al-Okaily et al. [58] lead a literature review on the intersection of management practices and Industry 4.0 technologies in driving supply chain sustainability, synthesizing findings on how approaches like Total Quality Management (TQM) and JIT contribute to efficiency and waste reduction. At the same time, the study highlights how blockchain, IoT, and big data enhance transparency and decision-making, concluding that all of these elements collectively contribute to improving sustainability efforts by minimizing environmental impact. Understanding the context in which these technologies and management practices are implemented is crucial, as industry-specific challenges and market dynamics can significantly influence their effectiveness in driving sustainable supply chain transformation.

In relation to practical scenarios, Khan et al. [59] elaborate on the need for optimizing spare parts warehousing to support decarbonization efforts in the oil and gas sector, proposing a smart warehouse management system powered by Industry 4.0 technologies to enhance logistics efficiency and sustainability for a major Norwegian company. Kumar et al. [60] analyze the operational performance of the Indian Railway supply chain using Industry 4.0 technologies to enhance customer service and inventory management. The authors propose an MCDM-based supply chain model to assess service quality and identify inventory-related challenges, such as integration systems, financial behavior, and management perspectives, helping decision-makers design a procurement-driven supply chain strategy in the railway sector. Focusing on the specific case of the Indian supply chain landscape, Pandey et al. [61] identify and benchmark the following key sustainability enablers to build a Reliable Supply Chain (RSC): digitalization, decentralization, smart factory technologies, and data security. The authors use an integrated MCDM approach to analyze the influence of these enablers and their interrelationships, demonstrating the EES benefits of Industry 4.0 adoption. Another relevant case study refers to the horticulture sector. In this regard, Singh et al. [62] focus on automated production management through technologies that enable the real-time tracking of crop growth and yield predictions, essential for managing inventory efficiently. By integrating AI and machine learning, this study demonstrates how farmers can optimize harvesting schedules, mitigate overproduction or shortages, and synchronize their output with market demand while promoting a more sustainable supply chain in agricultural inventory management. Huerta-Soto et al. [63] analyze optimization strategies for the modernization of the dairy supply chain, showing that AI and machine learning are emerging as preferred tools for improving inventory planning and logistics concerning traditional mathematical modeling. However, while modernization offers efficiency gains, the study also highlights environmental concerns and economic challenges smaller farms face.

2.3.5. Sustainable Manufacturing and Lean Production

Sustainable manufacturing within Industry 4.0 has been extensively studied by Harikannan and Vinodh [64] through the review of 442 related articles published between 2009 and 2022. In their work, the authors develop a conceptual framework for manufacturing industries willing to adopt sustainable practices aimed at creating economic value through digital technologies. This framework is highly relevant in light of the current literature review, as many recent studies align with its principles by emphasizing the use of digital technologies to drive both sustainability and economic value in industry. One study [64] also discusses how to enhance production efficiency while addressing sustainability concerns in the manufacturing sector, which, due to its resource-intensive nature, significantly impacts EES aspects. The improvement of these EES aspects via sustainable production strategies is widely discussed by Ospino et al. [65], who led a bibliometric analysis of 68 articles from Scopus and Web of Science by using scientific software and the PRISMA method and investigating how Industry 4.0 technologies can contribute to Lean and Green production practices in response to climate change and resource depletion. EES dimensions in sustainable manufacturing are also the object of a work of research led by Ahmad et al. [66], who connect each of these dimensions with suitable Key Performance Indicators (KPIs), e.g., resource efficiency and pollutant emissions for environmental sustainability, remanufacturing productivity and technology investments for economic sustainability, and social fairness initiatives and community feedback for social sustainability. The same authors [67] develop another paper on sustainability across various sectors. In detail, they explore the role of smart sensing in aviation, energy grids, resource extraction, and urban development. These advancements directly impact inventory management by tracking resources in real-time and making supply chains more efficient. Specific applications include AI-driven mine planning, IoT-enabled forest monitoring, smart ocean communication networks, and precision agriculture.

Considering practical applications in global markets, Harikannan et al. [68] use Structural Equation Modeling (SEM) to analyze and interpret data from experts in India's automotive component manufacturing sector, indicating a strong connection between Industry 4.0 and sustainable manufacturing for EES performance. In a similar context, Alkaraan et al. [69] observe how companies transition to a Green Servitisation-Oriented

Business Model (GS-OBM) while improving Environmental, Social, and Governance (ESG) performance. Using results from the UK Innovation Survey between 2012 and 2021, the authors show that Industry 4.0 technologies help integrate green servitisation strategies into sustainable supply chains, with corporate governance further strengthening this connection. Raissi [70] deals with corporate social responsibility by using structural equation modeling on survey data from 175 companies in Saudi Arabia and showing that digitalization enhances environmental decision-making. The study highlights how Industry 4.0 helps companies to improve service quality and stay competitive while also pointing out the difficulties they face in using digital technology to support sustainability. Sasaki [71] focuses on the Southeast Asian wood industry, showing how Industry 4.0 technologies can enhance carbon storage in Harvested Wood Products (HWPs). The author reviews past supply chain practices and proposes policy recommendations to encourage early adoption of advanced technologies for optimizing forest utilization and management.

2.3.6. Blockchain, Digital Twins, and Emerging Technologies

Various works in the literature specifically focus on using emerging technology in the field of reference. Hariyani et al. [72] review the transformative role of blockchain technology in manufacturing management and industrial engineering by analyzing 480 peer-reviewed papers from the Scopus database. The results show that blockchain significantly improves sustainable supply chain management in terms of real-time traceability and smart contract automation, apart from providing interesting opportunities for integration with IoT to build smarter manufacturing systems. Another insightful review of 60 blockchain-based frameworks has been carried out in [73], exploring the use of this technology in food supply chains and highlighting key applications such as transparency, traceability, and security. Moreover, areas where blockchain's potential is likely underutilized have emerged, for example food donation, waste management, and supply chain financing. The review also elaborates on a roadmap for future research on the role played by blockchains for sustainable food supply chains. Kulshrestha et al. [74] develop a review on Spare Parts Management (SPM) in the Industry 4.0 era and the related contribution of emerging technologies. They implement a content-based analysis of 118 papers published between 1998 and 2022, categorizing SPM research into the following categories: inventory management, spare types, circularity based on 6Rs, performance indicators, and strategic and operational aspects. An interesting result from this research refers to highlighted research gaps at the intersection of SPM and Industry 4.0, emphasizing the need for further exploration to maximize technological benefits. Findings from the current literature confirm this need, showing increased efforts to close these gaps through practical applications of AI, IoT, and predictive analytics in spare parts inventory and circular strategies.

While global supply chain expansion and lean strategies have improved efficiency, they have also increased fragility and complexity. In this regard, a study on the impact of Industry 4.0 technologies on the resilience of manufacturing companies against the COVID-19 pandemic has been led by Madrid-Guijarro et al. [75]. This study makes use of Partial Least Squares Structural Equation Modeling (PLS-SEM) involving 304 manufacturing firms to confirm the positive influence of Industry 4.0 technologies on supply chain resilience to COVID-19. Furthermore, to address supply chain challenges in Industry 4.0, other useful technologies are gaining traction, for example, digital twins (DTs), which are virtual replicas of supply chain systems used to enhance their resilience and visibility, as stated in [76]. This work uses case studies, literature reviews, and qualitative methods to thoroughly evaluate DTs' benefits, even providing a conceptual framework to guide professionals in implementing this technology for stronger SCM. An important aspect concurring to achieve this goal is waste management optimization, as highlighted by Fatorachian and Pawar [77]. Also, in this case, it is demonstrated that the combined use of digital solutions considerably leads to waste reduction through real-time monitoring, predictive maintenance, and increased transparency, aligning supply chain operations with sustainability and Net Zero targets. For example, Gonzalez et al. [78] leverage fog computing for designing an inventory management system aimed at improving sustainability in supply chains while effectively addressing demand uncertainties through the integration of an attention mechanism. Rather than relying solely on cloud services, fog computing helps outperform conventional demand estimation methods based on historical data in different demand modeling scenarios. Another significant technology in this field has been explored by Alzahmi and Shamayleh [79], who examine 3D printing's impact on spare parts management. By enabling local production, 3D printing reduces inventory costs and shipping delays, supporting a shift from "make to stock" to "make to order" models. However, challenges like high costs and size limitations currently hinder widespread adoption. The study highlights the need for further research on the integration of 3D printing with Industry 4.0 technologies as well as the development of regulatory frameworks, which are essential aspects for achieving broader implementation. Andres et al. [80] discuss the transition from Industry 4.0 to Industry 5.0, which emphasizes smart and interconnected logistics operations with a specific focus on inventory management and warehousing. The authors analyze enabling Industry 5.0 technologies through real-world case studies, assessing their current applications and identifying areas for future exploration to drive the practical implementation of Logistics 5.0.

3. Formalization of Results

In the present section, we will formalize the insights gathered from the literature review by organizing them into structured tables, each dedicated to one of the previously identified topic categories. For each TC, we will summarize key aspects, including the methods and topics explored in the literature, the benefits associated with these approaches, the challenges researchers have encountered, and the future trends anticipated in that specific field. This structured approach will provide a clearer understanding of the current research state and highlight potential future research directions.

Table 1 reports key managerial insights for sustainable inventory management in Industry 4.0, emphasizing how digital transformation enhances supply chain efficiency, reduces costs, and supports environmental sustainability. Integrating IoT, CPSs, and data-driven decision-making enables more agile and resilient inventory management. However, the table also highlights critical challenges, including IT infrastructure readiness, regional disparities in technology adoption, and the high resource consumption of digital solutions. Future trends in this category suggest that Industry 5.0 will drive further innovation by combining sustainability with human-centric approaches. At the same time, AI, big data, and machine learning will play an increasingly vital role in optimizing supply chains. Managers should align digital strategies with environmental goals while promoting collaboration across supply chain entities to overcome adoption barriers.

Table 2 synthesizes insights on sustainable inventory management in Industry 4.0, highlighting how the integration of advanced digital technologies with circular economy principles is transforming supply chain operations. The findings emphasize that Lean Six Sigma methodologies combined with Industry 4.0 technologies contribute significantly to enhancing manufacturing efficiency, reducing waste, and promoting sustainable inventory management. The adoption of JIT and VSM further supports the transition to circular economy models by optimizing resource utilization and facilitating recycling efforts. Additionally, Industry 4.0 is driving sustainability by creating new job opportunities, lowering emissions, and increasing supply chain resilience through digital monitoring and predictive analytics. However, Table 2 also highlights several challenges that managers must address. The limited availability of case studies and mathematical modeling approaches in Industry 4.0 research restricts the ability to develop generalizable frameworks for sustainable inventory management. Moreover, the lack of integration between quality management systems and circular economy strategies creates gaps in the implementation of sustainability initiatives across different sectors. The complexity of aligning digital transformation efforts with circular economy objectives remains an issue, as many industries struggle with sector-specific barriers that hinder the full realization of Industry 4.0's sustainability potential. Future trends of research outlined in Table 2 suggest an increasing convergence between Lean Six Sigma and Industry 4.0, fostering more structured and efficient sustainability practices. The adoption of AI-driven sentiment analysis and association rule mining is expected to play a greater role in assessing the EES impacts of digital transformation. Additionally, the expansion of Industry 5.0 incorporating immersive technologies such as VR and AR may enhance sustainable inventory management through real-time visualization. AMTs will also play a growing role in mitigating SCRs, reducing lead time fluctuations while promoting green innovation across industries.

Table 1. Summary of TC_1 Industry 4.0 and digital transformation.

Areas	Key Findings
Methods and topics	- Systematic literature review on Industry 4.0 and supply chain operations [29]. - Local Industry 4.0 adoption focusing on logistic flexibility and risk management [30]. - IoT, CPSs and environmental sustainability in manufacturing contexts [31]. - Integration of SCM with Industry 4.0 technologies in Indian manufacturing [32]. - Industry 5.0 and human-centric industrial systems [33]. - Empirical study on Industry 4.0 and SMEs' international competitiveness [34]. - Exploratory study on sustainability and BMI in Indian automotive [35]. - MCDM approach for ranking Industry 4.0 technologies [36]. - Digitalization impact on sustainable inventory in the bakery industry [37].
Benefits	- Enhanced supply chain efficiency, inventory management, and logistics [30,32]. - Support for environmental sustainability and supply chain decarbonization [31]. - Improved collaboration and cost efficiency in SCM [32,35]. - Industry 5.0 advances sustainability and human–machine collaboration [33]. - Digitalization promotes business model innovation in market competitiveness [34,35].
Challenges	- IT infrastructure readiness is critical for successful Industry 4.0 adoption [29]. - Regional disparities in Industry 4.0 implementation [30]. - High resource consumption and environmental impact of CPSs [31]. - Collaboration barriers existing between supply chain entities [35]. - Sector-specific variability in digital transformation effectiveness [36,37].
Future trends	- Growth of Industry 5.0 with sustainability and human-centric innovations [33]. - Stronger alignment of Industry 4.0 with environmental sustainability goals [31,37]. - Policies supporting regional digitalization efforts and SMEs [30,34]. - Greater integration of AI, big data, and machine learning in supply chains [36]. - Expanding digitalization strategies across diverse industries [32,37].

To capitalize on these trends, managers should prioritize digitalization strategies that align with circular economy principles, integrate sustainability-focused quality management frameworks, and foster collaboration across supply chain networks to address implementation challenges. AI-driven decision-making, predictive analytics, and digital twin technologies can further enhance the effectiveness of sustainable inventory practices by proactively addressing sector-specific barriers.

Table 3 reports results from the literature on the application of AI and machine learning in sustainable inventory management within Industry 4.0. As it is possible to observe, AI-driven solutions are transforming supply chain operations by enhancing security, cost efficiency, and decision-making. Particularly, machine learning techniques improve demand forecasting, inventory optimization, and backorder prediction, reducing financial pressures and preventing supply disruptions. Reinforcement learning has also been integrated with lean green manufacturing systems, allowing for dynamic scheduling and real-time adaptability. These advances contribute to more resilient and efficient supply chains, supporting sustainability by minimizing waste and improving resource utilization. However, the adoption of AI-driven approaches presents several challenges that may limit its widespread implementation, such as, for example, cybersecurity risks, workforce skill gaps, and other technical barriers. Additionally, integrating AI into traditional manufacturing processes may pose compatibility issues, requiring significant investments in infrastructure and employee training. Reinforcement learning models, while promising, require extensive real-world validation before they can be reliably applied to industrial settings. In the future, AI-driven sustainability initiatives are expected to further support circular economy practices by optimizing resource flows and reducing waste. Future research will also explore AI–human collaboration in supply chain management.

Table 2. Summary of TC_2 sustainability and circular economy.

Areas	**Key Findings**
Methods and topics	- Review on Industry 4.0 and SM and related conceptual and mathematical models [38]. - Lean Six Sigma and Industry 4.0 integration through AISM and Delphi technique [39]. - Lean management and circular economy synergy via JIT and VSM [40]. - Theoretical framework on Industry 4.0, circular economy and sustainability [41]. - Sentiment analysis and association rule mining revealing with impact on EES [42]. - Analysis of Industry 5.0's impact on EES, blockchain, IoT, VR, and AR [43]. - MCDM-based risk assessment framework for mitigating SCRs within AMT [44]. - Case on food-tech startups integrating Industry 4.0 for circular economy [45]. - Circular economy practices in green hotels in Saudi Arabia and Egypt [46].
Benefits	- Manufacturing efficiency integrating Lean Six Sigma and Industry 4.0 [38,39]. - Circular economy transition eased by lean practices, resource use, and recycling [40]. - Industry 4.0 drives sustainability through job creation and lower emissions [42]. - Contribution to sustainability across supply chains and inventory management [43]. - AMT reduces lead time fluctuations, waste, and supplier dependency [44]. - Food-tech startups and hospitality leverage Industry 4.0 for sustainability [45,46].
Challenges	- Limited case studies and mathematical approaches in Industry 4.0 research [38]. - Lack of integration between quality management and circular economy [39]. - Unclear interconnection between Industry 4.0 and circular economy strategies [41]. - Sector-specific challenges in applying Industry 4.0 for sustainability [45,46].
Future trends	- Increasing adoption of Lean Six Sigma within Industry 4.0 [39]. - Stronger alignment of circular economy principles with Industry 4.0 [40,41]. - Wider use of sentiment analysis and AI techniques for Industry 4.0 impact [42]. - Expansion of Industry 5.0 with metaverse and immersive technologies [43]. - Greater role of AMT in reducing SCRs and promoting green innovation [44].

Table 3. Summary of TC_3 artificial intelligence and machine learning applications.

Areas	Key Findings
Methods and topics	- AI-based approaches in Supply Chain 4.0 improve security and cost efficiency [47]. - Systematic review on AI's role in SCM from Industry 4.0 to Industry 6.0 [48]. - Machine learning for backorder prediction and inventory optimization [50]. - Reinforcement learning integrated with lean green manufacturing systems [52].
Benefits	- AI enhances demand forecasting, inventory management, and decision-making [48]. - Machine learning reduces financial pressures and prevents supply disruptions [50]. - Reinforcement learning improves scheduling by adapting to real-time data [53]. - AI in fashion manufacturing enables personalized consumer experiences [49].
Challenges	- AI implementation faces cybersecurity risks and workforce skill gaps [48]. - Machine learning requires overcoming technical and implementation barriers [51]. - AI integration into traditional manufacturing may pose compatibility issues [49]. - Reinforcement learning models need extensive real-world validation [52].
Future trends	- Increased adoption of predictive maintenance and smart inventory management [51]. - AI and machine learning driving sustainability and circular economy practices [52]. - Further exploration of AI–human collaboration in supply chain optimization [48]. - Adaptive AI solutions tailored for complex manufacturing environments [53].

Table 4 collects the main insights on sustainable inventory management in Industry 4.0, with a focus on supply chain and logistics management. We can see that the integration of digital interventions such as AMHSs, RFID, IoT, and AI-driven predictive analytics is transforming inventory planning and logistics efficiency. In addition, smart factory technologies and AI applications in supply chain performance evaluation contribute to sustainability by optimizing resource allocation and reducing carbon footprints.

Table 4. Summary of TC_4 supply chain and logistics management.

Areas	Key Findings
Methods and topics	- Bibliometric network analysis on digital interventions in sustainable logistics [54]. - AMHSs optimizes warehouse efficiency through automated queuing [56]. - RFID and IoT enhance supply chain flexibility, security, and inventory mobility [57]. - MCDM applied to supply chain performance evaluation in the railway sector [60]. - AI and machine learning enable predictions for sustainable horticulture logistics [62].
Benefits	- Industry 4.0 enhances logistics efficiency, waste reduction, and transparency [58]. - AI-driven warehouse management supports decarbonization efforts in industry [59]. - Smart factory technologies improve supply chain reliability and sustainability [61]. - AI and machine learning improve dairy supply chain inventory planning [63]. - IoT solutions improve real-time data transmission and inventory management [57].
Challenges	- Financial constraints and organizational resistance hinder technology adoption [55]. - Industrial challenges and market dynamics influence technology effectiveness [58]. - Smaller farms face environmental challenges in supply chain modernization [63]. - Integration of IoT and blockchain requires overcoming regulatory constraints [57]. - Implementation and scalability issues for AI in agricultural supply chains [62].
Future trends	- Increased investment in AI-driven predictive analytics for sustainable logistics [54]. - Expansion of decentralized supply chain to enhance resilience and efficiency [61]. - Further exploration of AI-enabled automated inventory tracking [62]. - Evolution of blockchain and IoT integration for transparent and efficient SCM [58]. - Role of smart warehouse management in green supply chain initiatives [59].

These findings align with and build upon the literature review focused on the previous period 2013–2023 [54], which highlighted how digital interventions have consistently contributed to improving sustainability in logistics through enhanced forecasting, inventory control, and decision-making processes. Our review confirms this trend, with a growing emphasis on investment in AI-driven predictive analytics for sustainable logistics. However, challenges like financial constraints, regulatory barriers, and organizational resistance may exist with scalability issues, especially in agricultural supply chains, further complicating implementation. In the future, increased investment in AI-driven analytics and blockchain–IoT integration will enhance transparency and efficiency. Managers should balance technology adoption with workforce readiness and regulatory compliance to maximize Industry 4.0's benefits for sustainable inventory management.

Table 5 reports the key findings on sustainable inventory management within Industry 4.0 and lean production, emphasizing the role of smart technologies in improving resource efficiency and corporate sustainability. On the one hand, digitalization enhances transparency, supports green servitization strategies, and enables data-driven decision-making. On the other hand, the resource-intensive nature of manufacturing, specific market dynamics, and the existence of policy barriers represent complex challenges to widespread adoption. As future lines of research, AI-driven decision support systems integrated with smart sensing technologies will be critical for integrating sustainability into manufacturing operations, also taking into account digitalized corporate governance. Managers need to be equipped to face cost constraints and regulatory challenges while implementing Industry 4.0 innovations to achieve long-term sustainability goals.

Table 5. Summary of TC_5 sustainable manufacturing and lean production.

Areas	Key Findings
Methods and topics	- Review and elaboration of a conceptual framework for economic value creation [64]. - Bibliometric analysis using PRISMA on Lean and Green production [65]. - EES sustainability dimensions linked to suitable KPIs measuring their impacts [66]. - SEM applied to Industry 4.0 adoption in India's automotive manufacturing [68]. - ESG improvement through GS-OBM based on UK Innovation Survey data [69].
Benefits	- Smart technologies enhance efficiency while addressing sustainability concerns [64]. - AI and IoT monitoring improve resource tracking in several sectors [67]. - Industry 4.0 strengthens corporate sustainability strategies and service quality [70]. - Digitalization in forestry and wood industries supports carbon storage in HWPs [71]. - Industry 4.0 help integrate green servitisation strategies into supply chains [69].
Challenges	- Manufacturing's resource-intensive nature makes sustainability goals complex [64]. - Difficulties in digitalization for sustainability-driven decision-making [70]. - Market-specific dynamics influence the efficacy of sustainable manufacturing [68]. - Policy and cost barriers hinder the adoption of forest management technologies [71]. - ESG-focused business models require corporate governance adjustments [69].
Future trends	- Increasing reliance on AI-driven decision support for sustainable manufacturing [66]. - Smart sensing expansion in aviation, energy grids, sustainable urban planning [67]. - Integration of digitalization and sustainable initiatives in corporate governance [70]. - Development of policy incentives for early adoption of Industry 4.0 in forestry [71]. - Data-driven frameworks for EES sustainability in production [64].

Table 6 highlights how emerging technologies are shaping sustainable inventory management in Industry 4.0. Specifically, blockchain improves traceability and automation, helping businesses track products in real-time. DTs enhance supply chain resilience by providing better visibility and predictive capabilities, while 3D printing is making it easier for companies to shift from keeping large inventories to producing items on demand, reducing waste and storage costs. Despite these advantages, significant challenges include high implementation costs, regulatory uncertainties, and infrastructure limitations. In the future, blockchain is expected to play a bigger role in food supply chains, while DTs will continue to improve real-time decision-making. As 3D printing technology becomes more cost-effective, its adoption is likely to grow. Managers should explore ways to integrate these technologies into their operations, balancing sustainability and adaptability.

Table 6. Summary of TC_6 blockchain, digital twins, and emerging technologies.

Areas	Key Findings
Methods and topics	- Blockchain is reviewed with a focus on smart contracts and real-time traceability [72]. - DT is explored to enhance resilience and visibility of supply chain systems [76]. - A review of 3D printing in SPM highlights a shift toward localized production [79].
Benefits	- Blockchain and IoT improve supply chain sustainability and automation [72]. - Digital solutions help reduce waste and align with sustainability goals [77]. - Fog computing enhances inventory management through forecast accuracy [78]. - 3D printing eases a transition from "make to stock" to "make to order" models [79]. - Industry 4.0 contribute to resilience during crises like the COVID-19 pandemic [75].
Challenges	- Blockchain adoption still remains limited in underutilized management areas [73]. - Despite its benefits, 3D printing adoption faces cost barriers and size limitations [79]. - Infrastructure and regulatory gaps in integrating digital solutions [77].
Future trends	- Research on Industry 4.0 to 5.0 transition about smarter logistics/warehousing [80]. - Expansion of blockchain applications in food supply chains is expected [73]. - More integration of DT with supply chains will improve real-time adaptability [76]. - 3D printing in SPM will advance with cost efficiency and regulations [79].

4. Conclusions and Future Research

This study comprehensively reviews recent advancements in sustainable inventory management within Industry 4.0, examining 52 papers published in 2024 and 2025. Our analysis bridges theory and practice by identifying key trends, benefits, challenges, and future directions in how digital transformation is driving sustainability in supply chains, providing valuable guidance for researchers and practitioners aiming to improve efficiency and environmental responsibility. The identified papers have been categorized and discussed in relation to the following topic categories: Industry 4.0 and digital transformation (TC_1), sustainability and circular economy (TC_2), artificial intelligence and machine learning applications (TC_3), supply chain and logistics management (TC_4), sustainable manufacturing and lean production (TC_5), and blockchain, digital twins, and emerging technologies (TC_5). The findings have been formalized and reported accordingly.

Broadly speaking, it is possible to observe a consensus in the literature about the evidence that Industry 4.0 is redefining sustainable inventory management by integrating several technologies and AI-driven analytics to enhance efficiency, reduce costs, and support the development of sustainability strategies in diverse management areas. Circular economy principles, Lean Six Sigma, and digitalization are promoting effective waste reduction and resource optimization, reinforcing the shift toward more adaptive and resilient supply chains. Despite these benefits, barriers persist to the adoption of Industry 4.0 technologies in making inventory management practices more sustainable. For instance, infrastructure gaps, regional disparities, as well as the high resource consumption of digital solutions, hinder widespread implementation. Furthermore, on the one hand, AI applications have the potential to improve forecasting and decision-making considerably; on the other hand, they raise concerns over critical aspects such as cybersecurity, workforce readiness, and integration with traditional systems. The existence of complex regulatory constraints and financial limitations further complicate the transition to digitalized supply chains as well as sustainability initiatives for inventory management.

Future research will likely focus on bridging these gaps by advancing AI–human collaboration and expanding blockchain–IoT integration for a higher degree of transparency. The transition to Industry 5.0 is expected to emphasize human-centric sustainability as well as the use of immersive technologies together with improved AI-driven governance. To fully capitalize on these advancements, managers will need to align digital strategies with environmental goals and invest in workforce training while creating collaboration across supply chain networks for the long-term sustainability of inventory management practices.

Author Contributions: Conceptualization, methodology, software, validation, resources, data curation, and visualization, formal analysis and investigation, S.C. and J.I.; writing—original draft preparation, S.C.; writing—review and editing, J.I.; supervision, project administration, S.C. and J.I. All authors have read and agreed to the published version of the manuscript.

Funding: This research received no external funding.

Data Availability Statement: The raw data supporting the conclusions of this article will be made available by the authors on request.

Conflicts of Interest: The authors declare no conflicts of interest.

Abbreviations

The following abbreviations are used in this manuscript:

AI	Artificial Intelligence
AISM	Adapted Interpretive Structural Modeling
AMHS	Automated Material Handling System
AMT	Additive Manufacturing Technology
AR	Augmented Reality
BMI	Business Model Innovation;
CPS	Cyber-Physical System
DT	Digital Twin
EES	Environmental, Economic and Social
EOQ	Economic Order Quantity
EPQ	Economic Production Quantity
ESG	Environmental, Social, and Governance
GS-OBM	Green Servitisation-Oriented Business Model
HWP	Harvested Wood Product
IoT	Internet of Things;
JIT	Just in Time
KPI	Key Performance Indicator
LLM	Large Language Model
MCDM	Multi-Criteria Decision-Making
PLS-SEM	Partial Least Squares Structural Equation Modeling
RFID	Radio Frequency Identification
RSC	Reliable Supply Chain
SCD	Supply Chain Decarbonization
SCM	Supply Chain Management
SCR	Supply Chain Risk
SEM	Structural Equation Modeling
SM	Smart Manufacturing
SME	Small and Medium Enterprise

SPM	Spare Parts Management
TQM	Total Quality Management
VSM	Value Stream Mapping
VR	Virtual Reality

References

1. Huang, K.; Wang, K.; Lee, P.K.; Yeung, A.C. The impact of industry 4.0 on supply chain capability and supply chain resilience: A dynamic resource-based view. *Int. J. Prod. Econ.* **2023**, *262*, 108913. [CrossRef]
2. Sun, X.; Yu, H.; Solvang, W.D.; Wang, Y.; Wang, K. The application of Industry 4.0 technologies in sustainable logistics: A systematic literature review (2012–2020) to explore future research opportunities. *Environ. Sci. Pollut. Res.* **2022**, *29*, 9560–9591. [CrossRef] [PubMed]
3. Anand, G.; Vashisht, P.; Singh, S.P.; Mittal, M. Sustainable Inventory Control and Management. In *Sustainable Inventory Management: Perspectives from India*; Springer: Berlin, Germany, 2025; pp. 1–24.
4. Becerra, P.; Mula, J.; Sanchis, R. Sustainable inventory management in supply chains: Trends and further research. *Sustainability* **2022**, *14*, 2613. [CrossRef]
5. Villacis, M.Y.; Merlo, O.T.; Rivero, D.P.; Towfek, S. Optimizing Sustainable Inventory Management using An Improved Big Data Analytics Approach. *J. Intell. Syst. Internet Things* **2024**, *11*, 15. [CrossRef]
6. Ahmed, U.; Carpitella, S.; Certa, A.; Izquierdo, J. A feasible framework for maintenance digitalization. *Processes* **2023**, *11*, 558. [CrossRef]
7. Nweje, U.; Taiwo, M. Supply chain management: Balancing efficiency and environmental responsibility. *World J. Adv. Res. Rev.* **2025**, *25*, 1547–1564. [CrossRef]
8. Doborjginidze, G.; Petriashvili, L.; Inaishvili, M. Optimization of Inventory Management in the Supply Chain. *J. Commun. Comput.* **2021**, *16*, 1–5. [CrossRef]
9. Keswani, M.; Khedlekar, U. Optimizing pricing and promotions for sustained profitability in declining markets: A Green-Centric inventory model. *Data Sci. Financ. Econ.* **2024**, *4*, 83–131. [CrossRef]
10. Pattnaik, S.; Nayak, M.M.; Abbate, S.; Centobelli, P. Recent trends in sustainable inventory models: A literature review. *Sustainability* **2021**, *13*, 11756. [CrossRef]
11. Arunadevi, E.; Umamaheswari, S. Enhancing ameliorating items with sustainable inventory management strategies: A Weibull approach for time and price-dependent demand trends. *Ain Shams Eng. J.* **2024**, *15*, 103065. [CrossRef]
12. Nobil, E.; Cárdenas-Barrón, L.E.; Garza-Núñez, D.; Treviño-Garza, G.; Céspedes-Mota, A.; de Jesús Loera-Hernández, I.; Smith, N.R.; Nobil, A.H. Sustainability inventory management model with warm-up process and shortage. *Oper. Res. Perspect.* **2024**, *12*, 100297. [CrossRef]
13. Singh, R.; De, P.K.; Barman, A.; Narang, P. Optimal analysis of profit maximization production inventory system under an imperfect environment and shortage. *OPSEARCH* **2024**, 1–31. [CrossRef]
14. Ikpe, V.; Shamsuddoha, M. Functional model of supply chain waste reduction and control strategies for retailers—The USA retail industry. *Logistics* **2024**, *8*, 22. [CrossRef]
15. Nweje, U.; Taiwo, M. Leveraging Artificial Intelligence for predictive supply chain management, focus on how AI-driven tools are revolutionizing demand forecasting and inventory optimization. *Int. J. Sci. Res. Arch.* **2025**, *14*, 230–250. [CrossRef]
16. Tadayonrad, Y.; Ndiaye, A.B. A new key performance indicator model for demand forecasting in inventory management considering supply chain reliability and seasonality. *Supply Chain Anal.* **2023**, *3*, 100026. [CrossRef]
17. Carpitella, S. Overcoming barriers to digital transformation towards greener supply chains in automotive paint shop operations. *Sustainability* **2024**, *16*, 1948. [CrossRef]
18. Mzougui, I.; Carpitella, S.; Certa, A.; Izquierdo, J. A strategic approach to safeguard global supply chains against COVID-19 disruptions. *Math. Methods Appl. Sci.* **2025**, *48*, 7875–7889. [CrossRef]
19. Pattanayak, B.K.; Khader, H.A. Energy-Efficient Storage System Optimization and Recent Trends in Enhancing Energy Management and Access Microgrid: A Review. *J. Adv. Sci. Eng. Technol.* **2024**, *7*, 39–54.
20. Cinnirella, V.; Carpitella, S.; Coco, A.; Frangiamore, D.D.M.; de Geronimo, R.P. Sustainable suppliers evaluation in the waste management sector: The case of a leading Sicilian enterprise. *IFAC-PapersOnLine* **2022**, *55*, 66–71. [CrossRef]
21. Psarommatis, F.; May, G. A Cost–Benefit Model for Sustainable Product Reuse and Repurposing in Circular Remanufacturing. *Sustainability* **2025**, *17*, 245. [CrossRef]
22. McDougall, N.; Davis, A. The local supply chain during disruption: Establishing resilient networks for the future. *J. Clean. Prod.* **2024**, *462*, 142743. [CrossRef]

23. Chauhan, C.; Kaur, P.; Arrawatia, R.; Ractham, P.; Dhir, A. Supply chain collaboration and sustainable development goals (SDGs). Teamwork makes achieving SDGs dream work. *J. Bus. Res.* **2022**, *147*, 290–307. [CrossRef]
24. Farooq, A.; Abbey, A.B.N.; Onukwulu, E.C. Inventory optimization and sustainability in retail: A conceptual approach to data-driven resource management. *Int. J. Multidiscip. Res. Growth Eval.* **2024**, *5*, 1356–1363. [CrossRef]
25. Agyemang, M.; Zhu, Q.; Adzanyo, M.; Antarciuc, E.; Zhao, S. Evaluating barriers to green supply chain redesign and implementation of related practices in the West Africa cashew industry. *Resour. Conserv. Recycl.* **2018**, *136*, 209–222. [CrossRef]
26. Sharma, R.; Gupta, H. Harmonizing sustainability in industry 5.0 era: Transformative strategies for cleaner production and sustainable competitive advantage. *J. Clean. Prod.* **2024**, *445*, 141118. [CrossRef]
27. Muka, T.; Glisic, M.; Milic, J.; Verhoog, S.; Bohlius, J.; Bramer, W.; Chowdhury, R.; Franco, O.H. A 24-step guide on how to design, conduct, and successfully publish a systematic review and meta-analysis in medical research. *Eur. J. Epidemiol.* **2020**, *35*, 49–60. [CrossRef] [PubMed]
28. Carpitella, S.; Jia, X.; Narimani, R. A review on current trends on optimal EVCS location. In Proceedings of the International Workshop on Simulation for Energy, Sustainable Development & Environment, Tenerife, Spain, 18–20 September 2024; pp. 1–10.
29. Asrol, M. Industry 4.0 adoption in supply chain operations: A systematic literature review. *Int. J. Technol.* **2024**, *15*, 544–560. [CrossRef]
30. Valenzuela-Cobos, A.; Vera-Cabanilla, B.; Castillo-Heredia, L.; Valenzuela-Cobos, J. Industry 4.0 in logistics management in Latin America: A bibliometric review. *J. Ind. Eng. Manag.* **2025**, *18*, 115–129. [CrossRef]
31. Singh, R.; Filho, W.L.; Katragadda, R.; Khan, S. A Conceptual Study on Utilizing Technology for Attaining Net Zero Through Industry 4.0. In *Zero Carbon Industry, Eco-Innovation and Environmental Sustainability*; Springer: Berlin, Germany, 2025; pp. 53–69.
32. Akhouri, D.; Dastidar, N.G.; Tripathy, S.; Singhal, D.; Panda, D. An ISM Approach among the Key Factors of SCM for Industry 4.0 Towards Sustainability in Indian Manufacturing Sectors. In *Decision-Making Models and Applications in Manufacturing Environments*; Apple Academic Press: Palm Bay, FL, USA, 2024; pp. 445–462.
33. Gantayat, P.K.; Kaur, T.; Majhi, M.; Jena, U.K.; Das, S.; Soni. From Efficiency to Innovation: LLM 5.0 and the Future of Industry. In Proceedings of the 2025 International Conference on Multi-Agent Systems for Collaborative Intelligence (ICMSCI), Erode, India, 20–22 January 2025; pp. 551–558.
34. Rehman, S.U.; Jabeen, F.; Shahzad, K.; Riaz, A.; Bhatti, A. Industry 4.0 technologies and international performance of SMEs: Mediated-moderated perspectives. *Int. Entrep. Manag. J.* **2025**, *21*, 25. [CrossRef]
35. Krishnan, R.; Phan, P.Y.; Krishnan, S.N.; Agarwal, R.; Sohal, A. Industry 4.0-driven business model innovation for supply chain sustainability: An exploratory case study. *Bus. Strategy Environ.* **2025**, *34*, 276–295. [CrossRef]
36. Ahmed, I.; Feng, B.; Emmanuel Yeboah, K.; Feng, J.; Jumani, M.S.; Ali, S.A. Leveraging Industry 4.0 for marketing strategies in the medical device industry of emerging economies. *Sci. Rep.* **2024**, *14*, 27664. [CrossRef]
37. Melesse, T.Y.; Orrù, P.F. The Digital Revolution in the Bakery Sector: Innovations, Challenges, and Opportunities from Industry 4.0. *Foods* **2025**, *14*, 526. [CrossRef]
38. Kumar, P.; Bhamu, J.; Bhadu, J.; Saraswat, P. Exploring industry 4.0 technologies for sustainable development goals: A systematic literature review. *Int. J. Prod. Manag. Eng.* **2025**, *13*, 27–44. [CrossRef]
39. Shrivastava, A.; Mishra, R.P. Integrating Lean Six Sigma and Industry 4.0 for advancing the circular economy: A holistic research study. *Qual. Manag. J.* **2025**, *32*, 2–14. [CrossRef]
40. Sasso, R.A.; Filho, M.G.; Ganga, G.M.D. Synergizing lean management and circular economy: Pathways to sustainable manufacturing. *Corp. Soc. Responsib. Environ. Manag.* **2025**, *32*, 543–562. [CrossRef]
41. Marak, Z.R.; Sharma, A.; Uikey, A.A. Exploring the interlinkages between industry 4.0, circular economy, and sustainable performance. In *Implementing ESG Frameworks Through Capacity Building and Skill Development*; IGI Global Scientific Publishing: Hershey, PA, USA, 2025; pp. 301–328.
42. Saleh, M.A.S.; AlShafeey, M. Examining the synergies between industry 4.0 and sustainability dimensions using text mining, sentiment analysis, and association rules. *Sustain. Futur.* **2025**, *9*, 100423. [CrossRef]
43. Amir, M.; Perween, U.; Al-Amin, A.Q. Environmental, Economic, and Social Sustainability and the Virtual World. In *Green Metaverse for Greener Economies*; CRC Press: Boca Raton, FL, USA, 2024; pp. 126–141.
44. Singh, S.; Misra, S.C.; Singh, G. Greening the supply chain: Leveraging additive manufacturing for sustainable risk management. *Bus. Strategy Environ.* **2024**, *33*, 8233–8246. [CrossRef]
45. da Silva, T.H.H.; Sehnem, S. The utilization of Industry 4.0 technologies to enhance the circular economy through the engagement of stakeholders in Brazilian food technology companies. *Bus. Strategy Environ.* **2025**, *34*, 129–161. [CrossRef]
46. Zaki, K. Enabling hotel circularity via Industry 4.0 innovations for enhanced hotel performance: Insights from Saudi Arabia and Egypt. *J. Hosp. Tour. Insights* **2025**, *8*, 915–936. [CrossRef]

47. Rana, G.; Sharma, R.; Parashar, B. Application of Artificial Intelligence for the Success of Supply Chain Operations in the Age of Data Analytics. In Proceedings of the International Conference on Smart Devices (ICSD), Dehradun, India, 2–3 May 2024; pp. 1–6.
48. Samuels, A. Examining the integration of artificial intelligence in supply chain management from Industry 4.0 to 6.0: A systematic literature review. *Front. Artif. Intell.* **2025**, *7*, 1477044. [CrossRef]
49. Faisal, M.H.; Chowdhury, S.S.; Ahmed, M.F.; Rahman, Z. AI-Powered Business Management: Advancing Lean Manufacturing and Sustainable Innovation in the US Fashion Industry. *Innov. Eng. J.* **2025**, *2*, 1–11. [CrossRef]
50. Kumar, P.; Bhonde, V.N.; Rajak, S.; Sundararaman, M. Machine learning applications in inventory management. In *Industry 4.0 for Manufacturing Systems: Concepts, Technologies, and Applications*; CRC Press: Boca Raton, FL, USA, 2025; p. 137.
51. Ogidan, E.T.; Olawale, O.P.; Dimililer, K. Machine learning applications in industry 4.0: Opportunities and challenges. In *Handbook of Intelligent and Sustainable Manufacturing*; CRC Press: Boca Raton, FL, USA, 2025; pp. 284–304.
52. Paraschos, P.D.; Koulinas, G.K.; Koulouriotis, D.E. Reinforcement learning-based optimization for sustainable and lean production within the context of industry 4.0. *Algorithms* **2024**, *17*, 98. [CrossRef]
53. Paraschos, P.D.; Koulinas, G.K.; Koulouriotis, D.E. A reinforcement learning/ad-hoc planning and scheduling mechanism for flexible and sustainable manufacturing systems. *Flex. Serv. Manuf. J.* **2024**, *36*, 714–736. [CrossRef]
54. Di Nardo, M.; Gallab, M.; Murino, T.; Wu, J.; Pandey, S. Integrating Sustainability and Industry 4.0: A Framework for Sustainable Logistics 4.0. *Circ. Econ. Sustain.* **2025**, 1–39. [CrossRef]
55. Holloway, S. Inventory Management as a Key Driver of Sustainability in Supply Chains. *SSRN* **2025**. [CrossRef]
56. Pinagapani, A.K.; Abhishek, G.; Athipathi, S.; Nithin Karthik, R.V.; Pragatheesh, G. Automated material handling for inventory management system. *Acta IMEKO* **2024**, *13*, 1–8. [CrossRef]
57. Choudhary, V.; Patel, K.; Niaz, M.; Panwala, M.; Mehta, A.; Choudhary, K. Implementation of Next-Gen IoT to Facilitate Strategic Inventory Management System and Achieve Logistics Excellence. In Proceedings of the 2024 International Conference on Trends in Quantum Computing and Emerging Business Technologies, Pune, India, 22–23 March 2024; pp. 1–6.
58. Al-Okaily, M.; Younis, H.; Al-Okaily, A. The impact of management practices and industry 4.0 technologies on supply chain sustainability: A systematic review. *Heliyon* **2025**, *10*, e36421. [CrossRef]
59. Khan, N.; Solvang, W.D.; Yu, H.; Rolland, B.E. Towards the design of a smart warehouse management system for spare parts management in the oil and gas sector. *Front. Sustain.* **2024**, *5*, 1426089. [CrossRef]
60. Kumar, M.; Vipin; Agarwal, A. Evaluation of Indian Railway's Supply Chain Parameters for Improved Sustainability Under Industry 4.0 Transformation: An Application of AHP and TOPSIS. *Eng. Manag. J.* **2024**, 1–13. [CrossRef]
61. Pandey, A.K.; Pratap, S.; Dwivedi, A.; Khan, S.A. Industry 4.0 and supply chain sustainability: Benchmarking enablers to build reliable supply chain. *Benchmarking Int. J.* **2025**, *32*, 433–458. [CrossRef]
62. Singh, G.; Bhagat, N.; Thakur, S. Automated Production Management in Horticulture: An Industry 4.0 Perspective. In *Blockchain-Enabled Internet of Things Applications in Healthcare: Current Practices and Future Directions*; Bentham Science Publishers: Sharjah, United Arab Emirates, 2025; pp. 258–282.
63. Huerta-Soto, R.; Ramirez-Asis, E.; Tarazona-Jiménez, J.; Nivin-Vargas, L.; Norabuena-Figueroa, R.; Guzman-Avalos, M.; Reyes-Reyes, C. Predictable inventory management within dairy supply chain operations. *Int. J. Retail. Distrib. Manag.* **2025**, *53*, 1–17. [CrossRef]
64. Harikannan, N.; Vinodh, S. State of art review on sustainable manufacturing and Industry 4.0. *Bus. Strategy Environ.* **2025**, *34*, 872–913. [CrossRef]
65. Ospino, E.S.; Guardo, A.P.; Castellano, R.Z.; Fajardo-Moreno, W.S. Sustainable Lean & Green 4.0 production: A bibliometric analysis. In Proceedings of the Proceedings of the LACCEI International Multi-Conference for Engineering, Education and Technology 2024, San Jose, CA, USA, 17–19 July 2024.
66. Ahmad, I.; Shabbir, M.; Zulfiqar, A.; Khan, J.S.; Çalışkan, H.; Hong, H. Principles and frameworks of sustainable manufacturing. In *Smart Cities and Sustainable Manufacturing*; Elsevier: Amsterdam, The Netherlands, 2025; pp. 1–7.
67. Ahmad, I.; Zulfiqar, A.; Shabbir, M.; Khan, J.S.; Çalışkan, H.; Hong, H. Advanced sensors, monitoring, and control systems for environmental sustainability. In *Smart Cities and Sustainable Manufacturing*; Elsevier: Amsterdam, The Netherlands, 2025; pp. 47–55.
68. Harikannan, N.; Vinodh, S.; Antony, J. Analysis of the relationship among Industry 4.0 technologies, sustainable manufacturing practices and organizational sustainable performance using structural equation modelling. *TQM J.* **2025**, *37*, 42–72. [CrossRef]
69. Alkaraan, F.; Elmarzouky, M.; de Sousa Jabbour, A.B.L.; Jabbour, C.J.C.; Gulko, N. Maximising sustainable performance: Integrating servitisation innovation into green sustainable supply chain management under the influence of governance and Industry 4.0. *J. Bus. Res.* **2025**, *186*, 115029. [CrossRef]
70. Raissi, N.; Hakeem, A.; Haidar, H.M. Sustainable Industry 4.0 and corporate social responsibility challenges in environmental decision-making effectiveness. *J. Asia Bus. Stud.* **2025**. [CrossRef]

71. Sasaki, N. Carbon Storage in Harvested Products along the Wood Supply Chain through Industry 4.0 Technologies: Challenges, Opportunities, and Policy Implications. *Energy Transit. Carbon Neutrality ASEAN* **2025**, 173–200. [CrossRef]
72. Hariyani, D.; Hariyani, P.; Mishra, S.; Sharma, M.K. A literature review on transformative impacts of blockchain technology on manufacturing management and industrial engineering practices. *Green Technol. Sustain.* **2025**, *3*, 100169. [CrossRef]
73. Ellahi, R.M.; Wood, L.C.; Bekhit, A.E.D.A. Blockchain-Driven Food Supply Chains: A Systematic Review for Unexplored Opportunities. *Appl. Sci.* **2024**, *14*, 8944. [CrossRef]
74. Kulshrestha, N.; Agrawal, S.; Shree, D. Spare parts management in industry 4.0 era: A literature review. *J. Qual. Maint. Eng.* **2024**, *30*, 248–283. [CrossRef]
75. Madrid-Guijarro, A.; Maldonado-Guzmán, G.; Rodríguez-González, R. Unlocking resilience: The impact of Industry 4.0 technologies on manufacturing firms' response to the COVID-19 pandemic. *Manag. Decis.* **2025**, *63*, 126–154. [CrossRef]
76. Ghabak, V.; Chaugule, R. Digital Twin Framework in Supply Chain Governance. In Proceedings of the 2024 IEEE International Conference on Computing, Power and Communication Technologies (IC2PCT), Greater Noida, India, 9–10 February 2024; Volume 5, pp. 82–89.
77. Fatorachian, H.; Pawar, K. Waste efficiency in cold supply chains through industry 4.0-enabled digitalisation. *Int. J. Sustain. Eng.* **2025**, *18*, 2461564. [CrossRef]
78. Gonzalez, J.; Avelar Sosa, L.; Bravo, G.; Cruz-Mejia, O.; Mejia-Muñoz, J.M. Fog computing and Industry 4.0 for newsvendor inventory model using attention mechanism and gated recurrent unit. *Logistics* **2024**, *8*, 56. [CrossRef]
79. Alzahmi, W.; Shamayleh, A. Transformative Potential of 3D Printing in Spare Parts Management: Challenges, Benefits, and Future Directions. In Proceedings of the 2024 IEEE International Conference on Technology Management, Operations and Decisions (ICTMOD), Sharjah, United Arab Emirates, 4–6 November 2024; pp. 1–7.
80. Andres, B.; Diaz-Madroñero, M.; Soares, A.L.; Poler, R. Enabling technologies to support supply chain logistics 5.0. *IEEE Access* **2024**, *12*, 43889–43906. [CrossRef]

processes

MDPI

Article

Implementation of a Sustainable Framework for Process Optimization Through the Integration of Robotic Process Automation and Big Data in the Evolution of Industry 4.0

Leonel Patrício [1,*], Leonilde Varela [1] and Zilda Silveira [2]

1 Department of Production and Systems, Algoritmi/LASI, University of Minho, 4804-533 Guimarães, Portugal; leonilde@dps.uminho.pt
2 Department of Mechanical Engineering, Sao Carlos School of Engineering, University of Sao Paulo, Sao Paulo 13566-590, Brazil
* Correspondence: leonelfilipepatricio@gmail.com

Abstract

This study explores the integration of Robotic Process Automation (RPA) and Big Data within a sustainable framework for process optimization in the context of Industry 4.0. As industries strive to enhance operational efficiency while maintaining sustainability, the need for innovative solutions has become crucial. The research applies the PICO methodology (Population, Intervention, Comparison, Outcome) to assess the impact of combining these technologies on process optimization and sustainability. Through a real-world case study, the study demonstrates that the integration of RPA and Big Data significantly reduces execution times, minimizes operational errors, and promotes sustainable business practices. The results show that the combined framework not only enhances efficiency but also contributes to lower economic, environmental, and social impacts. The findings validate the research hypotheses, proving that the proposed framework fosters a balance between technological advancement and sustainability. This study provides valuable insights into the potential of Industry 4.0 technologies to drive both operational efficiency and corporate responsibility, offering a novel approach for industries seeking to embrace digital transformation while achieving long-term sustainability. The research contributes to the growing body of knowledge on the synergy between RPA, Big Data, and sustainability in industrial contexts.

Keywords: RPA; data science; pillars of sustainability; S-PROBIO; digital transformation

1. Introduction

Technological advancements have profoundly transformed industrial processes, driving the transition toward the so-called Industry 4.0, which is characterized by the integration of advanced technologies such as Big Data and Robotic Process Automation (RPA). This emerging paradigm heralds a new era in manufacturing and services, where digitalization, automation, and intelligent data analysis play pivotal roles in process optimization. However, sustainability, a cornerstone of contemporary development, continues to face significant challenges in the context of this industrial revolution, necessitating innovative approaches to reconcile efficiency, competitiveness, and environmental responsibility [1].

In this context, process optimization emerges as a strategic objective for organizations aiming to enhance operational efficiency and minimize waste. The utilization of Big Data enables not only the collection and storage of vast volumes of information but also the extraction of valuable insights to support data-driven decision making. Conversely, RPA stands out as a disruptive tool that facilitates the automation of repetitive tasks and the continuous improvement of workflows, fostering productivity gains. When integrated, these technologies have the potential to form a powerful framework for achieving sustainability in industrial and organizational processes [2].

The motivation for this investigation stems from the pressing need to address the demands of a globalized and highly competitive market while promoting more responsible resource management. Despite the vast potential of Industry 4.0, a significant gap persists in the scientific literature regarding the synergistic combination of Big Data and RPA to achieve sustainability objectives. This study seeks to address this gap by developing a sustainable framework that unites these technologies and explores their potential to transform industrial processes [3].

The primary aim of this research is to design and implement an innovative framework that integrates Big Data and RPA, focusing on sustainability and process optimization. Following its development, the framework's effectiveness will be validated through practical application in a real-world case study. This study intends not only to demonstrate the operational benefits of the framework but also to contribute to the advancement of knowledge in the field by offering a solution that combines technological innovation with sustainability-focused responsibility [4].

This article is structured as follows. Section 2 provides a comprehensive review of the state of the art with subsections dedicated to process optimization, Industry 4.0, Big Data, RPA, and sustainability. Section 3 outlines the methodology adopted, including the formulation of hypotheses and the application of the PICO method (Population, Intervention, Comparison, Outcome). The PICO methodology was chosen for its ability to structure and analyze complex research questions in a systematic and objective manner, ensuring precise data analysis and minimizing subjective biases. This approach, widely used in fields such as healthcare and automation processes, was adapted in this study to evaluate the integration of RPA and Big Data in process optimization and sustainability. Section 4 introduces the proposed framework, detailing its features and benefits. Section 5 demonstrates the practical application of the framework in a case study. Section 6 analyzes the results obtained, and finally, Section 7 concludes the article by summarizing the main contributions and offering suggestions for future research.

2. State of the Art

2.1. Process Optimization

Process optimization is a central concept in the fields of industrial engineering, business management, and information technology. The primary objective of optimization is to enhance the efficiency and effectiveness of processes by eliminating waste and maximizing the utilization of resources. In the literature, process optimization is frequently associated with the analysis, redesign, and reengineering of business processes with the aim of improving performance, reducing costs, and increasing the competitiveness of an organization. Its application is particularly relevant in industrial environments, where the complexity and interdependence of various systems necessitate rigorous management and a continuous search for improvement [5].

Process optimization can be approached through various techniques and methodologies. Traditionally, approaches such as continuous improvement models, the PDCA cycle (Plan–Do–Check–Act), and Six Sigma methodology have been widely used to identify and eliminate inefficiencies in processes. These approaches involve the analysis of workflows, performance measurement, and the implementation of solutions aimed at reducing variations and maximizing production with minimal resources. These methodologies have proven effective in enhancing quality and productivity, yet they often focus on a limited set of variables and the optimization of specific processes [5].

In the context of Industry 4.0, process optimization has been significantly transformed by the introduction of disruptive technologies such as automation, Big Data analysis, and artificial intelligence. The use of Big Data and the Internet of Things (IoT) has enabled the real-time collection and analysis of data from machines, sensors, and production systems, providing a holistic view of process performance. This approach allows for the identification of patterns, anomalies, and opportunities for improvement, leading to more precise and agile process optimization [6].

Concurrently, the automation of processes, particularly through Robotic Process Automation (RPA), has enabled the elimination of repetitive and manual tasks, resulting in greater operational efficiency. RPA integrates effectively into process optimization not only by automating simple tasks but also by performing complex tasks based on predefined rules and machine learning algorithms. By reducing the human workload and eliminating the possibility of human error, RPA enables organizations to focus on higher-value activities while enhancing consistency and productivity [7].

However, process optimization is not limited to cost reduction and productivity increases. Today, sustainability has emerged as a crucial variable in the optimization of industrial processes. The need to develop more environmentally friendly and responsible processes has prompted many organizations to incorporate optimization practices that not only focus on economic efficiency but also aim to reduce environmental impacts, such as energy consumption and waste generation. The integration of sustainability into process optimization strategies has given rise to new business models that seek to balance economic needs with environmental and social demands [8].

Thus, contemporary process optimization requires an integrated approach that considers not only operational efficiency but also sustainability, technological innovation, and constant adaptation to new market realities. The convergence of Big Data, RPA, and the pursuit of more sustainable processes forms the foundation of a new optimization paradigm, where the intelligent use of data and automation not only drives business performance but also promotes responsible and balanced resource management [8].

In this regard, the integration of these technologies and practices offers significant potential for creating more efficient, sustainable, and resilient processes, which are aligned with the challenges posed by Industry 4.0. Through smarter resource management, process optimization in the digital age not only contributes to continuous improvement but also strategically positions organizations to meet the demands of an ever-evolving global market.

2.2. Industry 4.0

Industry 4.0, also known as the Fourth Industrial Revolution, emerges as a fundamental milestone in the context of the digital transformation of industrial and production sectors. The term refers to the integration of advanced technologies into industrial processes with the aim of creating smart factories that are more efficient, flexible, autonomous, and interconnected. This revolution has been driven by significant technological innovations, such as the Internet of Things (IoT), artificial intelligence (AI), advanced robotics, 3D printing, cyber–physical systems, and Big Data analytics, all working in tandem to create a digitally interconnected and highly automated ecosystem [9].

The concept of Industry 4.0 is based on the transformation of traditional production processes through digitalization and the use of cutting-edge technologies that enable real-time communication between devices, machines, and systems. Through the application of cyber–

physical systems, physical devices can be monitored and controlled remotely, generating data that are processed and analyzed to optimize production performance. These systems, along with the use of intelligent sensors and actuators, enable factories to be highly adaptable, automatically adjusting to changes in production conditions and market demands [9].

One of the main features of Industry 4.0 is interconnectivity, which allows communication between all components of the production chain from machines and equipment to management and control systems. This connectivity is provided by the Internet of Things (IoT), which enables the exchange of information between physical objects and virtual systems, providing greater visibility and traceability of processes. For example, the IoT allows machines to communicate with each other to adjust production parameters, identify failures in real time, and prevent unplanned stoppages, resulting in greater operational efficiency and reduced costs [10].

Big Data analytics is another crucial pillar of Industry 4.0, as it allows the collection and processing of large volumes of data from various sources, such as sensors, machines, and management systems. By using advanced data analysis techniques and artificial intelligence, valuable insights can be gained to help organizations make more informed and accurate decisions, promoting process optimization, quality improvement, and product customization. Real-time data analysis also enables the implementation of predictive maintenance, where failures are anticipated based on patterns identified in the data, avoiding downtime and high costs associated with emergency repairs [10].

Intelligent automation is another key element of Industry 4.0 with a particular focus on Robotic Process Automation (RPA) and collaborative robots. RPA, in particular, allows for the automation of repetitive, rule-based tasks, freeing employees to focus on higher-value functions. Furthermore, the integration of collaborative robots into the production line, which work alongside human operators, provides greater flexibility and efficiency in operations, especially in production environments with high variability and complexity. The use of these robots, combined with artificial intelligence, enables machines to learn and adapt to different scenarios and working conditions, resulting in more agile and adaptable production [11].

Another relevant aspect of Industry 4.0 is mass customization, which enables the production of highly personalized goods at reduced costs. Thanks to technologies such as 3D printing and additive manufacturing, it is now possible to create on-demand products with a level of customization never before possible on a large scale. This ability to rapidly adapt to market needs and consumer desires is one of the factors that make Industry 4.0 particularly competitive in dynamic and constantly changing markets [11].

However, the transition to Industry 4.0 also presents significant challenges. The implementation of advanced technologies requires substantial investments in infrastructure and training as well as an adaptation of organizations to new work dynamics and the demands of the digital environment. Furthermore, issues related to cybersecurity and data protection are becoming increasingly relevant, as interconnectivity and real-time information exchange increase the risks of attacks and the compromise of system integrity [12].

In summary, Industry 4.0 represents a revolution in the way industrial processes are conceived, managed, and executed. The convergence of digital and physical technologies has the potential to radically transform business models, creating new levels of efficiency, flexibility, and innovation. However, the successful adoption of Industry 4.0 requires a strategic and integrated approach that considers not only the operational benefits but also the technological, economic, and social challenges associated with this transformation.

2.3. Big Data

The concept of Big Data refers to the processing and analysis of massive volumes of data which, due to their magnitude, variety, and velocity, exceed the capabilities of traditional data management and processing systems. At its core, Big Data involves the capture, storage, analysis, and visualization of data generated from various sources such as business transactions, sensors, social media, mobile devices, and others. This vast quantity of data presents a unique opportunity for organizations to derive valuable insights, which can be utilized to make more informed decisions, optimize processes, and create new value-driven solutions. However, the real challenge lies not only in managing the sheer volume of data but in the ability to extract meaningful information from it, which is a process that involves advanced data analysis techniques and artificial intelligence [13].

The nature of Big Data is often characterized by the three Vs: volume, variety, and velocity. Volume refers to the massive amount of data generated, which can reach such high levels that traditional database management systems become incapable of handling them efficiently. Variety concerns the different types of data, which may be structured, semi-structured, or unstructured, ranging from numerical data in tables to textual information on social media or videos. Velocity refers to the speed at which data are generated and must be processed with many Big Data systems operating in real time or near-real time, requiring solutions that allow for immediate analysis of data as it is produced [13].

In addition to these three fundamental aspects, a fourth V has emerged: veracity, which relates to the reliability and quality of the data. The presence of inaccurate or inconsistent data can undermine the effectiveness of analysis, making veracity a critical factor in the implementation of Big Data solutions. To ensure that analyses yield meaningful results, organizations must invest in data cleaning and validation processes, ensuring that the data used are both representative and accurate [14].

Technologies associated with Big Data have evolved significantly in recent years, offering new tools and capabilities to manage the complexity and diversity of data. Platforms such as Hadoop and Spark enable the distributed processing of large data volumes, using computer clusters to divide and process tasks simultaneously. These frameworks allow for the storage and processing of data within a scalable and cost-effective infrastructure, making Big Data management accessible to businesses of various sizes. Moreover, the use of NoSQL databases has gained prominence, as these systems are designed to handle large volumes of unstructured or semi-structured data, providing greater flexibility in managing and querying data [14].

However, Big Data analysis is not limited to the processing and storage of large quantities of data. The true value of Big Data lies in its ability to perform advanced analytics, which can range from descriptive analysis (identifying patterns and trends in historical data) to predictive analysis (using statistical models and machine learning algorithms to forecast future behaviors). Prescriptive analytics can go further, suggesting specific actions based on the results obtained, offering organizations a proactive approach to managing their resources and operations [15].

The application of Big Data has been extensively explored across various industrial sectors, such as manufacturing, healthcare, finance, marketing, and logistics. In the industrial context, the use of Big Data has enabled the implementation of predictive maintenance, where data analysis from sensors and equipment helps predict failures before they occur, minimizing costs and downtime. In supply chain management, the analysis of large data volumes can optimize delivery routes, predict demand, and improve operational efficiency. Furthermore, Big Data have been instrumental in personalizing customer experiences, analyzing purchasing behavior, and enhancing real-time marketing strategies [15].

In conjunction with Industry 4.0, Big Data have become a key driver of digital transformation within organizations, fostering innovation and competitiveness. The combination of Big Data with other technologies, such as artificial intelligence (AI) and robotics, enables the creation of intelligent systems that can learn and adapt, automating decisions and optimizing processes in real time. In particular, Big Data analytics platforms, integrated with RPA and the Internet of Things (IoT), have the potential to profoundly transform industrial operations, creating a continuous cycle of monitoring, analysis, and improvement.

However, the use of Big Data is not without challenges. Issues related to data privacy and security are increasingly pertinent, particularly with the growing volume of sensitive data being generated. The implementation of data protection policies, in accordance with regulations such as the General Data Protection Regulation (GDPR), is essential to ensure that organizations use information ethically and legally. Furthermore, interoperability between different systems and data sources remains a challenge, as data may be distributed across various platforms and in incompatible formats [16].

In summary, Big Data represent an unprecedented opportunity for organizations, enabling them to extract valuable insights and make more informed and efficient decisions. However, its successful implementation requires adequate infrastructure, the use of appropriate technologies, and a strategic approach that considers not only technical capabilities but also the ethical and legal aspects associated with the management and analysis of large volumes of data. The potential of Big Data to transform the way organizations operate is vast, and its integration with other emerging technologies, such as artificial intelligence and RPA, promises to drive innovation and competitiveness in the years to come.

2.4. *Robotic Process Automation (RPA)*

Robotic Process Automation (RPA) refers to the use of software to automate repetitive and rule-based tasks that typically require human intervention. RPA is primarily applied in business processes that involve data manipulation and the execution of sequential operations, such as form filling, querying information systems, transaction processing, and other routine activities. The main objective of RPA is to enhance efficiency as well as reduce errors and operational costs while freeing up employees to engage in higher-value tasks. This technology has gained prominence due to its ability to automate tasks in business environments swiftly and with minimal impact on existing infrastructure [17].

RPA employs "robots" or software "bots" to simulate human interaction with system interfaces, performing actions such as clicking, typing, or copying and pasting information between applications. These bots can be programmed to carry out tasks autonomously based on predefined rules and processing algorithms. The major advantage of RPA over other forms of automation lies in its ability to operate directly on user interfaces without the need for deep integration with underlying systems. This allows organizations to implement automation solutions at relatively low costs and without the need to restructure or replace existing systems [18].

RPA can be classified into different levels, depending on the complexity and intelligence involved. At the most basic level, automation relies on scripts and predefined rules to carry out well-defined and repetitive tasks. This type of automation is known as "non-intelligent" or "rule-based" RPA. At a more advanced level, RPA solutions can be combined with technologies such as artificial intelligence (AI) and machine learning (ML), creating what is referred to as intelligent RPA. Intelligent RPA allows for the automation of more complex tasks, such as decision making based on unstructured data or the interpretation of text, images, and even voice. By integrating RPA with AI, organizations can achieve a higher level of cognitive automation, where bots are capable of learning and adapting to new situations, making processes more flexible and dynamic [19].

The application of RPA across various sectors has demonstrated substantial benefits. In the financial sector, for instance, RPA has been used to automate banking reconciliation processes, manage accounts payable and receivable, and monitor suspicious transactions. In healthcare, RPA has been employed to streamline appointment scheduling, invoice processing, and patient data management. In human resources, RPA can automate administrative tasks such as processing job applications, onboarding new employees, and calculating payroll. The application of RPA in these and other sectors has led to significant reductions in operational costs, fewer human errors, and improvements in the speed and accuracy of processes [19].

One of the key advantages of RPA is its ability to scale rapidly. As the demand for automation increases, organizations can easily add more bots to handle the increased workload without the need for significant expansion of human resources. This scalability is particularly useful in scenarios involving seasonal work peaks or high volumes of transactions, such as during fiscal periods or major sales campaigns. Thus, the implementation of RPA can contribute to greater operational flexibility, enabling businesses to respond quickly to changes in market conditions and customer needs [20].

However, the adoption of RPA also faces challenges. The primary barrier to implementing RPA in many organizations is resistance to change and the lack of a corporate culture oriented toward innovation. Additionally, integrating RPA with legacy systems can be complex, as it is often necessary to ensure that bots can effectively operate on older systems not designed to work with automation. Another significant issue pertains to the security and governance of RPA. Since bots operate based on predefined rules and have access to sensitive systems, it is crucial to implement stringent controls to ensure that automated operations are conducted securely and in compliance with organizational policies and legal regulations, such as those related to data protection [21].

In terms of its impact on human work, RPA may lead to the reconfiguration of roles within organizations. While the automation of processes may result in the elimination of repetitive, low-value tasks, it can also create new job opportunities in areas such as bot management and development, data analysis, and the monitoring of automated processes. The shift from operational functions to more strategic roles can, therefore, contribute to the development of employee skills and increase overall productivity [21].

In the context of Industry 4.0, RPA plays a pivotal role in creating smarter, more connected factories and workplaces, where processes are not only automated but also continuously monitored and optimized based on real-time data. When combined with other emerging technologies such as Big Data and the IoT, RPA can help create more efficient and resilient production systems that are capable of autonomously adapting to new conditions and challenges. This integration between RPA and other Industry 4.0 technologies opens up new possibilities for the digital transformation of industrial operations, elevating levels of efficiency and flexibility [21].

In conclusion, RPA emerges as one of the most promising technologies in today's business landscape, enabling a significant transformation in the way organizations manage and execute their processes. While its implementation presents challenges, the benefits associated with automating repetitive tasks and improving productivity make RPA an essential tool in the journey toward intelligent industry and operational efficiency. The continued evolution of RPA, particularly with the integration of AI and machine learning, promises to further expand its potential, creating increasingly sophisticated solutions that are adaptable to market needs.

2.5. Sustainability

Sustainability has increasingly been recognized as a fundamental concept for the long-term development of societies, organizations, and global economies. It refers to the ability to meet the needs of the present generation without compromising the ability of future generations to meet their own needs. Practically speaking, sustainability involves a balance between three main dimensions: social, economic, and environmental, which are often referred to as the three pillars of sustainability. The integration of these pillars has gained prominence in recent years, particularly in the context of Industry 4.0, where companies seek to innovate and optimize their processes while minimizing negative impacts on people, the economy, and the planet [22].

The social pillar of sustainability relates to an organization's commitment to promoting social well-being, social justice, and equity. This pillar highlights the importance of working conditions, social inclusion, diversity, human rights, and the impact of business activities on communities. Corporate social responsibility (CSR) is one of the concepts that fall under this pillar, aiming to ensure that organizations not only seek to maximize profits but also consider the social implications of their actions by fostering a safe and fair working environment, investing in professional training and development, and respecting workers' rights. In the context of Industry 4.0, the social pillar of sustainability also requires reflection on the impact of automation and digitalization on employment and working conditions with digital inclusion and professional retraining being essential areas to ensure that technological transformations are accompanied by positive evolution for workers [23].

Economic sustainability is concerned with the ability to generate value in a lasting and profitable manner, ensuring the long-term economic viability of businesses. It goes beyond the simple maximization of short-term profits, incorporating business practices that take into account the long-term impacts of economic decisions, ensuring that economic growth is balanced and does not harm future generations. To be truly sustainable, an organization must integrate the principles of efficiency and resilience into its business models, promoting the responsible management of resources and investing in innovations that enhance productivity without causing harm to society or the environment. In the context of digital transformation and Industry 4.0, the use of technologies such as Big Data and RPA can be viewed as a way to promote economic efficiency, optimizing processes and reducing costs, while ensuring that business practices align with the needs of both the global and local economies [23].

The environmental pillar of sustainability focuses on minimizing the negative impacts of human activities on the environment. This pillar involves adopting practices that promote the conservation of natural resources, reducing pollution, and managing ecosystems sustainably with the goal of ensuring the planet's preservation for future generations. Environmental sustainability includes actions such as reducing greenhouse gas emissions, using energy resources efficiently, and managing waste sustainably. In the industrial context, the circular economy has emerged as a key concept, aiming to create cycles of production and consumption in which materials and products are reused, recycled, and extended in their useful life, thus reducing the need for new natural resources. Technologies associated with Industry 4.0, such as real-time sensing and data monitoring, play a crucial role in managing and minimizing environmental impact, allowing companies to implement more environmentally friendly and efficient solutions, such as sustainable energy usage and the optimization of material consumption [23].

The integration of the three pillars of sustainability within Industry 4.0 represents both one of the greatest challenges and a significant opportunity for innovation in the industrial sector. The use of advanced digital technologies, such as artificial intelligence (AI), Big

Data, the Internet of Things (IoT), and RPA, enables the creation of smarter, more efficient, and sustainable industrial systems that optimize production, reduce waste, and minimize environmental impact. For example, predictive maintenance, which uses sensors and data analysis to anticipate machine failures before they occur, not only improves operational efficiency but also contributes to reducing resource waste and the production of waste [24,25].

However, while Industry 4.0 technologies provide powerful tools for achieving sustainability, their implementation requires a concerted effort to ensure that companies focus not only on economic efficiency or technological innovation but also on the social and environmental impact of their operations. The responsible use of emerging technologies requires careful consideration of the effects these may have on local communities, social inclusion, and environmental protection [25].

Digital transformation and the increase in automation, often associated with Industry 4.0, may pose significant challenges to sustainability, particularly with respect to social disparities and environmental impact. Growing automation can lead to job losses, especially in sectors that rely on manual labor, and may exacerbate social inequalities unless public policies and companies invest in strategies for professional retraining and digital inclusion. At the same time, the shift to a more sustainable production model involves investments in clean technologies and production processes that minimize environmental impact, which can represent an additional cost for businesses, especially in the early stages of implementation [26].

Nevertheless, these challenges can be overcome through a strong commitment to corporate social responsibility (CSR) and long-term sustainability, where companies adopt a strategic approach that integrates technological innovation and digitalization with practices that promote social equity, environmental responsibility, and economic viability. Investment in the research and development of sustainable solutions, as well as fostering public–private partnerships, can help overcome initial barriers and transform challenges into opportunities for sustainable growth [26].

In conclusion, sustainability, through the integration of its three pillars—social, economic, and environmental—into business strategy, emerges as a key factor for the long-term success of organizations in the era of Industry 4.0. The development of innovative solutions that combine advanced technology with sustainable practices offers a pathway to a more balanced and resilient future, benefiting both businesses and society as well as the planet.

3. Methodology

3.1. Method

The methodology adopted for this study is based on the use of the PICO approach (Population, Intervention, Comparison, Outcome), which has proven to be an essential tool in scientific research due to its ability to organize and structure complex questions in a clear and systematic manner [27]. When applied rigorously, the PICO methodology allows for a detailed and objective analysis of the key elements of a study, avoiding the subjective biases often present in other approaches, such as narrative reviews. The application of the PICO methodology extends beyond fields such as healthcare and medicine, having demonstrated its relevance in areas like software engineering, automation processes, and agile development, as evidenced by recent studies. It facilitates the formulation of precise questions and the selection of relevant studies, ensuring that data are analyzed in a consistent and comparable way [27].

In this work, the PICO methodology will be applied to structure the analysis of the integration of Robotic Process Automation (RPA) and Big Data technologies within a sustainable framework with the aim of optimizing processes in the context of Industry 4.0. The adaptation of this methodology to the context of industrial process automation and the use of Big Data offers a systematic approach to understanding how the combination of these technologies impacts operational efficiency and promotes sustainable practices throughout the lifecycle of industrial operations [28].

The "population" in this study refers to industrial organizations that are in the process of adopting RPA and Big Data technologies, integrating sustainable practices within the context of Industry 4.0. The precise definition of this population is crucial to ensure that the results are relevant and applicable to organizations facing similar challenges related to process optimization, industrial automation, and the management of large-scale data. The analysis will focus on companies implementing advanced digital solutions to improve operational efficiency and ensure sustainability in their operations.

The "intervention" in this study involves the implementation of an innovative framework that combines RPA and Big Data technologies with sustainable practices, which were aimed at optimizing industrial processes while promoting both operational efficiency and corporate sustainability. This intervention is based on increasing evidence that the integration of these technologies can not only enhance process automation efficiency but also significantly contribute to reducing the economic, environmental, and social impacts of industrial operations, promoting a long-term sustainable approach.

The "comparison" will be made between traditional industrial processes, where activities are carried out manually, and processes optimized through the application of the framework proposed in this study, which integrates RPA and Big Data technologies. This analysis will allow for the assessment of differences in operational efficiencies before and after the implementation of the framework, considering gains in terms of reduced execution times, reduced operational errors, and greater sustainability in operations. The comparison aims to identify improvements in efficiency and the advantages observed after the framework's implementation.

The expected "outcome" of this study is to demonstrate that the integration of RPA and Big Data within a framework can significantly increase operational efficiency in industrial processes, promoting process optimization while contributing to more sustainable business practices. The application of the PICO methodology will allow for the formulation of solid hypotheses and a well-founded analysis, based on concrete data, that validates the effectiveness of the approach within the context of Industry 4.0.

The central research question guiding this study is as follows:

Central Research Question (CRQ)

CRQ: How can Robotic Process Automation (RPA) and Big Data technologies be integrated into a sustainable framework for process optimization in the context of Industry 4.0, maximizing operational efficiency while promoting sustainable practices?

This question seeks to understand the interaction between RPA and Big Data technologies, identifying how they can be combined within a framework that not only promotes operational efficiency but also contributes to sustainability in the industrial context, considering the challenges of Industry 4.0.

Based on the central research question, the following research hypotheses have been formulated:

Hypotheses (H)

H1: *The integration of Big Data with RPA within a framework can significantly increase efficiency in the optimization of industrial processes, reducing execution times and operational errors.*

H2: *A framework that combines RPA, Big Data, and sustainable practices in the context of Industry 4.0 contributes to the corporate sustainability of industrial operations, promoting a reduction in economic and environmental impact as well as an increase in social responsibility.*

The PICO methodology will be rigorously applied to the selection of relevant studies and articles for this work. Data analysis will be structured to ensure that the formulated hypotheses are tested and validated through reliable scientific sources with a focus on emerging technologies in the industrial sector. Source selection will be made based on a detailed bibliographic review, taking into account peer-reviewed articles, publications from 2010 to 2025, and full-text accessibility.

By applying PICO, a clear and objective path will be outlined for the analysis of the integration of RPA and Big Data technologies, focusing on how they can be applied to promote greater operational efficiency while minimizing economic, environmental, and social impacts. This process will not only validate the research hypotheses but also provide an in-depth understanding of how Industry 4.0 can be more sustainable without compromising its competitiveness and innovation.

To answer the main research question of this study, the researchers used the online scientific library made available by the Foundation for Science and Technology, focusing on four specific groups (Group 1, Group 2, Group 3 and Group 4), as detailed in Table 1.

Table 1. Groups searched through "B-on".

Group 1	Group 2	Group 3	Group 4
"RPA" OR "Robotic Process Automation" OR "Intelligent Process Automation" OR "Digital Process Automation" OR "Business Workflow Automation"	"Big Data" OR "Data Analytics" OR "Hadoop" OR "Data Science" OR "Data Visualization"	"Industry 4.0" OR "Industrial Automation" OR "Smart Manufacturing" OR "Cyber-Physical Systems" OR "Connected Factory" OR "Industrial Digitalization" OR "Advanced Manufacturing"	"Sustainability" OR "Environmental Sustainability" OR "Economic Sustainability" OR "Social Sustainability" OR "Corporate Social Responsibility" OR "Circular Economy" OR "Sustainable Development"

The research was carried out through the "B-on" platform, using the OR operator to link the Title, Keywords (KWs) or Abstract (AB) within the three established groups. Filters were then applied to the database of publications collected during the search process, and the results, regarding the total number of publications, are detailed in Table 2.

Table 2. Publications obtained through B-on, after the application of some filters.

	Set 1	Set 2	Set 3
Initial result:	173	13	7
1—Restrict to Peer-Reviewed	61	8	3
2—From 2010 to 2025	61	8	3
3—Language: English	28	7	3
4—Restrict to Full Text	28	6	3

After applying the filters, the titles, keywords and abstracts of each article were reviewed again to select those most relevant to the research topic. Initially, 193 articles were recovered. After applying the filters, 37 remained, of which only 16 were directly related to the focus of the study.

Figure 1 presents a flowchart illustrating the literature search process and screening procedure followed in this study.

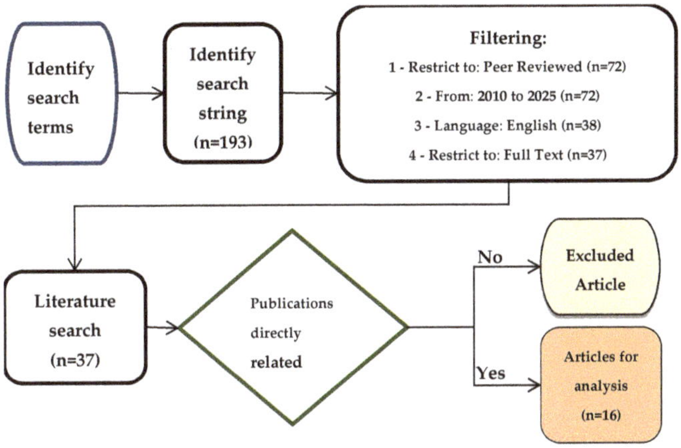

Figure 1. Flow diagram of literature search and respective screening [8].

3.2. Analysis and Synthesis of Articles

This section provides a detailed overview and analysis of the most relevant articles to the topic under investigation. Table 3, presented below, displays the 16 selected articles along with the models analyzed in each one. The table was designed to categorize the contributions of each study and was carefully developed after in-depth research in academic databases.

Table 3. Identified articles and the respective themes of the articles found.

Themes of the Articles / Articles	RPA	Big Data	Sustainability	Industry 4.0	Type
[29]	x			x	Case Study
[30]	x			x	Case Study
[31]	x		x		Model
[32]	x			x	Literature Review
[33]	x		x		Case Study
[34]	x			x	Case Study
[35]	x	x			Case Study
[36]	x			x	Literature Review
[37]	x				Model
[38]	x			x	Case Study
[39]			x	x	Case Study
[40]			x	x	Review Literature
[41]	x	x			Case Study
[42]	x	x			Case Study
[43]		x		x	Case Study
[44]		x	x	x	Case Study
% Themes p/articles	75%	31%	31%	63%	

3.3. Summary of Results

In this section, the results of the articles reviewed in the previous section are synthes-sized, highlighting the main contributions, common themes, and potential gaps in the lit-erasure. The analysis of the articles presented in Table 3 provides a clear view of the contributions to key areas such as Robotic Process Automation (RPA), Big Data, Sustainability, and Industry 4.0. The table outlines how these themes are addressed across the 16 selected articles and provides an overview of the methodologies employed. Below are the key insights from this analysis:

- RPA (Robotic Process Automation): The majority of articles (75%) focus on Robotic Process Automation, highlighting its prominent role in the research. This indicates a high relevance of RPA in the studies examined.
- Big Data: 31% of the articles discuss Big Data, suggesting that while important, it is less prevalent compared to RPA. Big Data are often associated with the analysis of large datasets, which is a topic critical in numerous industries.
- Sustainability: Similar to Big Data, sustainability is addressed in 31% of the articles. This suggests that while it is a growing area of interest, it is not the central theme in most of the studies.
- Industry 4.0: 63% of the articles include Industry 4.0, demonstrating its significance and substantial presence in the research. Industry 4.0 refers to the integration of advanced technologies in production processes, such as automation, IoT (Internet of Things), and artificial intelligence.
- Case Study: The majority of the articles (approximately 75%) employ case studies, indicating a focus on practical, real-world investigations. This suggests an applied empirical approach to the analysis of the topics under consideration.
- Model: A smaller proportion of articles (31%) use analytical models, reflecting an emphasis on theoretical frameworks and simulations to understand the dynamics of the themes.
- Literature Review: A few publications (18%) adopt literature reviews, which are useful for summarizing and evaluating the existing body of knowledge on the topics explored.
- Focus on RPA and Industry 4.0: Robotic Process Automation (RPA) and Industry 4.0 dominate the themes with the majority of articles dedicated to these aspects. This reflects the growing interest and relevance of these topics in the current research and technological development landscape.
- Less Emphasis on Big Data and Sustainability: While Big Data and sustainability are important, they are less frequently addressed in the selected articles, suggesting that these areas are explored in a more limited scope compared to RPA and Industry 4.0.
- Predominance of Case Studies: The most common methodology in the publications is the case study, reinforcing the practical and applied nature of the research. This suggests that researchers are focused on demonstrating real-world results and solutions to contemporary challenges.
- Theoretical Models and Literature Reviews: Although less common, analytical models and literature reviews remain significant, contributing to the conceptual and theoretical understanding of the themes.

In summary, the analysis of this table reveals that current research is primarily centered around automation and industry-related themes with a focus on applied studies, while Big Data and sustainability are explored to a more limited extent.

4. Framework (S-PROBIO)

4.1. Proposal for a Framework (S-PROBIO)

The S-PROBIO (Sustainability–Process Optimization–Big Data Integration with Automation) is an innovative framework created to integrate Robotic Process Automation (RPA) and Big Data technologies within the context of Industry 4.0 with the aim of optimizing processes and promoting sustainable practices across the three pillars of sustainability: environmental, social, and economic.

The proposal for this framework arises from the need to adapt industrial operations to digital transformation, seeking operational efficiency while incorporating social, environmental, and economic responsibility. The combination of RPA and Big Data offers the opportunity to automate processes, enhance data analysis, and simultaneously implement sustainable practices effectively. Below, the structure and criteria for validating the functionality of each phase of the S-PROBIO framework are presented (Figure 2).

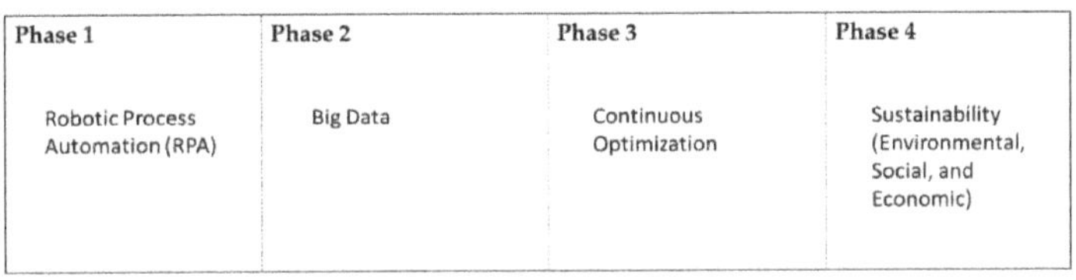

Figure 2. Framework (S-PROBIO).

The S-PROBIO framework consists of four main components that interrelate to achieve the objectives of process optimization and sustainability. The following outlines the components and phases of the framework, including the validation criteria for each phase (Table 4).

Table 4. Framework (S-PROBIO) phases.

Framework (S-PROBIO) Phases
1. Robotic Process Automation (RPA)
2. Big Data Integration
3. Continuous Optimization
4. Sustainability (Environmental, Social, and Economic)

1. Robotic Process Automation (RPA)

RPA is the first fundamental phase of the framework in which repetitive and manual processes are automated. Through the integration of digital robots, operational tasks that require time and human effort are carried out swiftly and efficiently, enabling significant gains in both efficiency and accuracy (Table 5).

Table 5. Validation criteria (RPA).

Validation Criteria (RPA)
Reduction in operational errors
Reduction in execution time
Elimination of repetitive tasks
Increase in accuracy and consistency
Employee feedback

Validation Criteria (RPA):

- Reduction in operational errors: Automation should reduce or eliminate human errors in processes.
- Reduction in execution time: The execution times of automated processes should be significantly shorter compared to manual processes.
- Elimination of repetitive tasks: Automation should replace tasks that do not add value to the operation and are repetitive.
- Increase in accuracy and consistency: Automation should ensure that process outcomes are consistent and accurate.
- Employee feedback: Employees should be consulted to verify if automation contributes to improved working conditions, allowing them to focus on more strategic tasks.

2 Big Data Integration

The integration of Big Data within the S-PROBIO framework is crucial for the analysis and optimization of automation processes. Big Data enable the collection and analysis of large volumes of data generated during automated processes, providing insights into process behavior and identifying areas for optimization (Table 6).

Table 6. Validation criteria (Big Data).

Validation Criteria (Big Data)
Quality of data analysis
Accuracy in failure prediction
Data-driven decision-making
Efficiency in data collection and storage
Application of results

We used the following validation criteria for (Big Data):

- Quality of data analysis: The data analysis should provide relevant and accurate insights for process optimization.
- Accuracy in failure prediction: Big Data should be capable of predicting failures and inconsistencies in processes before they occur.
- Data-driven decision making: Operational decisions should be based on the analyses conducted from the data.
- Efficiency in data collection and storage: Big Data integration should be efficient in terms of collecting, processing, and storing large volumes of data.
- Application of results: Big Data integration should be able to generate insights that lead to concrete actions for continuous process optimization.

3 Continuous Optimization

Continuous optimization is a vital component to ensure that automated processes are constantly evaluated and improved. This component involves the ongoing monitoring of processes and data analysis to identify opportunities for enhancement (Table 7).

Validation Criteria (Continuous Optimization):

- Efficiency in detecting inefficiencies: The system should be able to identify areas with low efficiency and suggest improvements.
- Continuous improvement of processes: The implementation of suggested improvements should result in a progressive enhancement of processes.
- Real-time performance measurement: The system should allow for real-time analysis of how adjustments and improvements affect process performance.

- Cost and time reduction: The continuous optimization process should be able to reduce operational costs and execution times.

Table 7. Validation criteria (continuous optimization).

Validation Criteria (Continuous Optimization)
Efficiency in detecting inefficiencies
Continuous improvement of processes
Real-time performance measurement
Cost and time reduction

4 Sustainability (Environmental, Social, and Economic)

Sustainability is the central pillar of the S-PROBIO framework, ensuring that automated processes not only lead to operational benefits but also promote responsible practices in environmental, social, and economic terms (Table 8).

Table 8. Validation criteria sustainability.

Validation Criteria Sustainability	
Environmental Sustainability	Reduction in energy consumption
	Reduction in paper usage
	Waste reduction
Social Sustainability	Improvement in employee wellbeing
	Increase in productivity
Economic Sustainability	Reduction in operational costs
	Return on Investment (ROI)
	Increase in competitiveness

Validation Criteria Sustainability:

- Environmental Sustainability:
    - Reduction in energy consumption: Automation should reduce energy usage either by optimizing performance or decreasing equipment operation times.
    - Reduction in paper usage: The transition from manual to digital processes should result in a significant decrease in paper and other material consumption.
    - Waste reduction: Automation and Big Data usage should contribute to waste and resource reduction, improving resource management.
- Social Sustainability:
    - Improvement in employee well-being: Automation should allow employees to focus on higher-value tasks, improving job satisfaction and well-being.
    - Increase in productivity: The reduction in repetitive tasks should lead to an increase in productivity and employee satisfaction.
- Economic Sustainability:
    - Reduction in operational costs: Automation should result in cost savings, particularly in terms of reducing errors and shortening execution times.
    - Return on Investment (ROI): The framework should allow for the analysis and assessment of ROI with the expectation that implementation costs will be quickly recovered through generated improvements.

- Increase in competitiveness: The integration of RPA and Big Data should result in more efficient processes, contributing to a long-term competitive advantage.

The implementation of the S-PROBIO framework involves a series of sequential phases each with specific criteria for validating its functionality. The following outlines the implementation steps along with the criteria for evaluating their success (Table 9):

Table 9. Criteria for evaluating their success.

Criteria for Evaluating Their Success
1. Planning and Initial Assessment
2. Automation and Big Data Implementation
3. Monitoring and Adjustments
4. Sustainability Measurement

1. Planning and Initial Assessment: This phase involves a detailed survey of manual and repetitive processes that can be automated. An evaluation will be conducted on the optimization potential through the integration of RPA and Big Data, considering environmental, social, and economic criteria.
    - Validation Criteria: Identification of processes to automate, analysis of potential automation impact, and definition of sustainability objectives.
2. Automation and Big Data Implementation: RPA will be implemented in identified processes along with the integration of Big Data solutions for real-time data collection and analysis.
    - Validation Criteria: Validation of automation impact on execution time, error reduction, and process quality improvement.
3. Monitoring and Adjustments: During and after implementation, the continuous monitoring of processes will be performed, using Big Data to adjust and improve the automation systems.
    - Validation Criteria: Constant feedback from teams and stakeholders, performance and efficiency analysis, identification of areas for continuous improvement.
4. Sustainability Measurement: After implementation, the environmental, social, and economic impacts of automation will be measured using defined KPIs.
    - Validation Criteria: Reduction in operational costs, execution times, and waste; improvement in employee well-being and positive ROI.

The S-PROBIO framework effectively integrates RPA and Big Data, providing process optimization and the promotion of sustainable practices. By following the phases and validation criteria presented, organizations can achieve higher levels of efficiency, sustainability, and profitability, contributing to the evolution of operations within the context of Industry 4.0.

4.2. Characteristics and Benefits of the Framework (S-PROBIO)

The S-PROBIO framework emerges as an innovative solution for integrating Robotic Process Automation (RPA) and Big Data technologies within the context of Industry 4.0 with the aim of optimizing industrial processes and promoting sustainable practices across the environmental, social, and economic pillars. The following presents the key characteristics and benefits derived from the implementation of this framework.

The S-PROBIO framework is characterized by its holistic and multidisciplinary approach, integrating advanced technologies to achieve effective and sustainable process optimization. Its main features include the following:

1. Integration of Emerging Technologies: The framework combines intelligent automation provided by RPA with the analysis and prediction capabilities of Big Data, enabling the optimization not only of individual processes but of the entire production chain. This integration provides enhanced visibility and more efficient data management, which are both crucial for real-time strategic decision making.
2. Emphasis on Sustainability: S-PROBIO is designed to facilitate sustainable practices, incorporating all three pillars of sustainability—environmental, social, and economic—into each phase of the automation and data analysis process. The integrated technologies enable the reduction in environmental impact (e.g., through lower energy consumption and reduced paper usage), improve employees' quality of life, and enhance the economic efficiency of the organization.
3. Process Optimization: The framework focuses on the continuous optimization of processes, enabling operations to be evaluated and adjusted over time. The integration of Big Data provides detailed analysis of each process phase, enabling rapid and precise adjustments that increase operational efficiency.
4. Focus on Operational Efficiency: Through the implementation of RPA, the framework facilitates the automation of repetitive and low-value tasks, allowing human resources to be redirected toward more strategic activities. Additionally, the integration of Big Data contributes to the reduction in errors, execution times, and operational costs.
5. Adaptation to Industry 4.0: S-PROBIO is designed to align with the principles of Industry 4.0, integrating automation and data analysis technologies in such a way as to optimize smart industry practices. This allows companies to be prepared for future challenges, maximizing the efficiency and flexibility of their operations.

The adoption of the S-PROBIO framework provides a range of benefits that directly impact operational efficiency, sustainability, and profitability for organizations. Among the main benefits, the following stand out:

1. Improvement in Operational Efficiency: Process automation through RPA leads to a significant reduction in task execution times, boosting productivity and minimizing the risk of human errors. This benefit is enhanced by the use of Big Data, which offers a detailed view of processes and facilitates the identification of continuous improvement opportunities.
2. Environmental Sustainability: The application of the framework contributes to reducing the organization's environmental impact. Automation enables a decrease in paper usage, a reduction in energy consumption, and a minimization of waste, as data analysis can optimize resource usage and promote more efficient practices. These aspects not only contribute to the company's environmental responsibility but may also result in cost reductions associated with the use of natural resources.
3. Improvement in Working Conditions: The use of RPA eliminates repetitive and low-value tasks, leading to an enhanced quality of life for employees. By allowing workers to focus on more strategic and creative activities, the framework contributes to employee motivation and well-being while optimizing the utilization of their skills.
4. Economic Benefits: The implementation of the S-PROBIO framework leads to a reduction in operational costs, particularly in areas such as resource management, failure reduction, and execution time. Process optimization and the predictive analysis capabilities of Big Data also contribute to improved economic efficiency, thus enhancing

the organization's profitability. The return on investment (ROI) analysis highlights the economic benefits of adopting the framework, demonstrating how cost reductions and efficiency improvements support the company's financial sustainability.

5. Increased Competitiveness: The implementation of the framework enables companies to gain a competitive advantage by being faster and more accurate in executing their operations. The use of Big Data and RPA allows for more informed and agile decision making, optimizing operational flexibility and the capacity to adapt to market changes.
6. Continuous Monitoring and Adjustment Capability: The use of Big Data within the framework enables the real-time monitoring of processes, providing valuable data for continuous adjustments. Through data analysis, companies can quickly identify inefficiencies, failures, or areas for improvement, allowing proactive adjustments that result in the ongoing evolution of processes.
7. Simplified Adoption of Advanced Technologies: S-PROBIO facilitates the implementation and adoption of emerging technologies such as robotic automation and Big Data analytics in industrial environments. Through a structured approach adapted to the context of Industry 4.0, organizations can benefit from technological innovation in an incremental and effective manner.

S-PROBIO represents an innovative framework that integrates RPA and Big Data technologies with the aim of optimizing industrial processes while promoting sustainable practices. Its features, such as technological integration, a focus on sustainability, and operational efficiency, offer a holistic approach to addressing the challenges of Industry 4.0. Furthermore, the benefits resulting from its implementation, ranging from cost reduction to improvements in working conditions and sustainability, make this framework a strategic solution for organizations seeking continuous and sustainable progress in their performance.

5. Case Study

5.1. Case Study Presentation

The case study presented in this article refers to the implementation of the S-PROBIO framework within an industrial sector organization with a particular focus on the administrative department. This department plays a critical role in managing administrative processes such as document management, invoice approvals, and financial report processing. The case study presented in this article focuses on the administrative department of an organization in the industrial sector. This department plays a fundamental role in managing internal processes related to business administration, being responsible for essential activities such as document management, the validation and approval of invoices, financial processing and the generation of accounting reports. These functions are crucial to ensuring the company's operational integrity, ensuring that financial processes are carried out accurately, in accordance with legal requirements and aligned with organizational goals. Before applying the S-PROBIO framework, the department faced significant challenges, such as a high manual workload, high rate of human errors, low process efficiency and difficulty in tracking financial data. These limitations impacted not only internal performance but also the agility in strategic decision making. The department was chosen to implement S-PROBIO due to its strategic importance and the clear need for digital transformation, becoming a representative example of the benefits that the integration of RPA and Big Data technologies can offer in complex administrative processes. Prior to the integration of Robotic Process Automation (RPA) and Big Data technologies, these processes were carried out manually, resulting in significant time consumption, a high error rate, and a lack of transparency in operations. The following outlines the manual process, the need for RPA and Big Data integration, and the justification for the development of a structured framework, such as S-PROBIO, to optimize and sustain this process.

The process under review involves the management of invoices and financial documents in the administrative department. The manual workflow entailed several stages, such as receiving and verifying invoices, validating tax information, entering data into financial systems, and tracking payments. Each of these stages required human intervention, which resulted in several limitations:

1. Invoice Receipt and Processing: Invoices were received physically or via email and were manually checked by an operator, who compared the invoice details with purchase or service records.
2. Validation and Approval: After verification, the operator cross-checked tax information and values, subsequently uploading the data into the ERP (Enterprise Resource Planning) system. The invoice approval process involved multiple validation stages, which led to considerable delays and relied on the availability of various parties involved.
3. Data Entry and Archiving: Data entry into the systems was performed manually with operators transcribing invoice information into the accounting system. Additionally, there was the physical and digital archiving of invoices, which is a task that, despite its importance, was prone to errors such as document duplication or indexing mistakes.
4. Tracking and Reporting: After data entry and processing, payment tracking was completed via manual reports, which made it difficult to analyze accounts payable and track the progress of financial transactions in real time.

This manual process led to several limitations, including long execution times, frequent human errors, a lack of agility in decision making, and difficulty in data traceability and analysis. Furthermore, the heavy use of paper and the need for both physical and digital archiving of invoices had a significant environmental impact and resulted in high operational costs.

Given the context of the described manual process, the organization identified a clear need to transform the workflow to make processes faster, more accurate, and more efficient. The integration of RPA and Big Data was seen as the ideal solution to address the challenges of the process, including the following:

1. RPA (Robotic Process Automation): The use of RPA allowed for the automation of repetitive and low-value tasks such as extracting data from invoices, entering these data into financial systems, and sending them for validation and approval. Automating these stages not only eliminated human error but also sped up the execution time, enabling faster and more efficient invoice processing.
2. Big Data: The use of Big Data introduced a new layer of intelligence to the process. The analysis of large volumes of financial data enabled the organization to conduct predictive analysis on payments and cash flow, improving decision making and risk management. Additionally, Big Data facilitated the identification of invoice patterns, highlighting potential anomalies or inconsistencies, which contributed to better tax compliance.
3. Sustainability: The use of RPA and Big Data also offered the organization an opportunity to adopt more sustainable practices, such as reducing paper usage and decreasing energy consumption associated with document storage and processing. Digital automation allowed the organization to eliminate manual tasks, improving energy efficiency and reducing the environmental impact.

The need to develop a structured framework, such as S-PROBIO, to integrate RPA and Big Data technologies into the administrative process arose due to the complexity of the scenario and the multiple benefits expected from its implementation. The integration of these two technologies required a systematic approach that not only optimized the process but also ensured its sustainability across environmental, social, and economic dimensions.

1. Environmental Sustainability: The reduction in paper usage and energy consumption were directly supported by automation. The S-PROBIO framework enabled the elimination of physical and manual processes, contributing to the organization's environmental sustainability.
2. Social Sustainability: Automation through RPA freed employees from repetitive and low-value tasks, allowing them to focus on more impactful strategic activities, such as financial data analysis or process improvement. Furthermore, the increased accuracy and speed of processes enhanced employees' quality of life by reducing workload and boosting job satisfaction.
3. Economic Sustainability: RPA and Big Data led to significant reductions in operational costs, improved cash flow, and increased return on investment (ROI). Process automation and predictive analysis provided financial efficiency, optimizing processing costs and increasing the organization's profitability.

The combination of RPA and Big Data was pivotal in transforming an inefficient manual process into an agile, effective, and sustainable one, while the S-PROBIO framework ensured that the implementation of these technologies was aligned with the organization's objectives of optimization and sustainability. The structure of S-PROBIO provided a robust approach to managing the integration of the technologies, addressing the complexities involved and ensuring that all benefits were realized across the relevant sustainability dimensions.

5.2. Application of the Framework to the Case Study

The application of the S-PROBIO framework to the case study developed for the automation of the administrative process of managing invoices and financial documents within the company was a critical step in validating the effectiveness of the integration between Robotic Process Automation (RPA) and Big Data in the context of Industry 4.0. The implementation of the framework aimed to optimize the process, improving operational efficiency and promoting sustainable practices in line with the three pillars of sustainability: environmental, social, and economic.

The implementation was divided into several phases, in accordance with the S-PROBIO framework, which integrates automation processes and data analysis. Each phase was structured to address a specific aspect of the administrative process and ensure that the technologies were applied effectively and sustainably.

1. Phase 1: Mapping the Current Process and Identifying Improvement Points

The first stage of the implementation involved mapping the manual administrative process of invoice management, as it was carried out before the integration of the technologies. Through a detailed analysis of the process stages, critical points in terms of execution time, human error, and environmental impact were identified. From this analysis, it became evident that the process was inefficient and unsustainable, as it relied on multiple manual interventions, consumed large amounts of resources (paper, energy), and had a high risk of data validation errors.

Criteria for validation:

- Total execution time of the process before automation.

- Number of operational errors and failures in invoice processing.
- Consumption of physical resources (paper) and energy in the manual process.

2 Phase 2: Implementation of Robotic Process Automation (RPA)

Based on the process mapping, the first intervention of the framework was developed, which involved the implementation of RPA to automate repetitive, low-value tasks. RPA technology was applied to the stages of data extraction from invoices, entering information into ERP systems, validating tax data, and forwarding invoices for approval. This automation not only reduced execution times but also eliminated human errors that frequently occurred during manual data transcription.

RPA was implemented gradually, initially in a limited set of processes, to validate the effectiveness of the solution. The automation allowed invoices to be processed more quickly, error-free, and without the need for human intervention in repetitive stages.

The following criteria were used for validation:

- Reduction in execution time for automated processes.
- Elimination of human errors in automated tasks.
- Increased capacity for processing invoices without additional staff.

3 Phase 3: Integration of Big Data for Predictive Analysis and Process Management

The second technology integrated into the process was Big Data, which were used to aggregate, analyze, and generate valuable insights from the large volumes of financial data generated by the invoice management process. Big Data were employed to conduct predictive analysis on cash flow, forecast potential payment delays, and identify patterns in invoices that could indicate anomalies or fraud. Moreover, Big Data enabled the creation of real-time dashboards, facilitating the monitoring of invoice status and payment progress.

The implementation of Big Data brought significant improvements in decision making, as the organization gained a clearer view of the financial impact of invoices and the behavior of accounts payable. The integration of financial data with other business data sources also contributed to better risk management and optimization of cash flow.

The following criteria were used for validation:

- Accuracy of predictive analysis related to cash flow and payments.
- Identification of anomalies and fraud through Big Data analysis.
- Effectiveness of dashboards in visualizing and controlling invoices and payments.

4 Phase 4: Evaluation of Results and Sustainability

The final phase of the S-PROBIO framework application involved evaluating the results obtained from the integration of the technologies. The evaluation was conducted from the perspective of the three pillars of sustainability: environmental, social, and economic.

1. Environmental Sustainability: Automation and data analysis significantly contributed to reducing paper and energy usage. The digitalization of processes, combined with the elimination of the need for physical storage of invoices, resulted in a substantial reduction in environmental impact. A comparison of paper and energy consumption before and after automation revealed a considerable reduction in these resources.
   Criteria for validation:
   - Quantification of reductions in paper and energy consumption.
   - Analysis of the environmental footprint of the process before and after automation.
2. Social Sustainability: The implementation of RPA allowed administrative department employees to focus on more strategically valuable tasks, such as financial data analysis and the development of management reports. The automation of repetitive

tasks improved the work environment, as employees engaged in more challenging, impactful activities, which contributed to higher job satisfaction and reduced stress associated with monotonous tasks.

The following criteria were used for validation:

- Improvement in employee satisfaction and well-being.
- Reduction in time spent on repetitive tasks.
- Increased motivation and productivity at work.

3. Economic Sustainability: The combination of RPA and Big Data led to a significant reduction in operational costs, such as a decrease in errors and a faster processing time for invoices. The economic impact was also measured through the return on investment (ROI), which demonstrated that in the short term, the organization had recouped its initial costs associated with automation and data analysis. Additionally, the use of these technologies contributed to improved profitability, optimization of cash flow, and greater efficiency in financial processes.
   The following criteria were used for validation:
   - Calculation of ROI following the implementation of the technologies.
   - Reduction in operational costs.
   - Increased profitability and improved cash flow.

The application of the S-PROBIO framework to the administrative invoice management process proved highly effective in optimizing the process, improving operational efficiency, and promoting sustainable practices. The combination of RPA and Big Data not only allowed for the automation of repetitive stages but also enhanced financial data analysis, resulting in a positive impact on the three dimensions of sustainability. The implementation of the framework provided the organization with greater agility, cost reductions, a lower environmental impact, and better utilization of human resources, demonstrating the effectiveness of this approach in the context of Industry 4.0.

6. Results Analysis and Discussion

6.1. Results Analysis

The implementation of the S-PROBIO framework, which integrated Robotic Process Automation (RPA) and Big Data technologies into the administrative process of managing invoices and financial documents, resulted in significant improvements across various metrics. These improvements were evaluated based on the time required for each task, the reduction in errors, resource consumption, and the overall financial impact.

The Table 10 summarizes the key metrics before and after the implementation of RPA and Big Data, providing a clear comparison of the improvements achieved:

- Execution Time Reduction: The integration of RPA enabled the processing time for each invoice to be significantly reduced from 240 min to just 30 min. This dramatic reduction in time enabled the company to handle a larger volume of invoices with the same number of employees; see Figure 3.
- Elimination of Human Errors: By automating tasks such as data entry and validation, RPA eliminated human errors, which were previously a common issue. This contributed to greater accuracy in invoice processing, reducing errors to zero.
- Resource Consumption: The adoption of digital processes resulted in significant reductions in paper and energy consumption. Paper usage was reduced by 80%, and energy consumption decreased by the same percentage, contributing to environmental sustainability; see Figure 4.

- Invoice Processing Capacity: The automated system greatly enhanced the organization's capacity to process invoices. The monthly capacity increased from 100 invoices to 1000 invoices, which is a tenfold increase in processing power; see Figure 5.
- Cost Reduction: The reduction in human errors, paper usage, and time spent on manual tasks led to a significant decrease in operational costs, which fell by 55%. This cost reduction, combined with improved processing efficiency, resulted in a positive ROI within just six months.
- Employee Satisfaction: Automation allowed employees to shift their focus from repetitive tasks to more value-added activities such as data analysis and process improvement. As a result, employee satisfaction improved by 50% with staff reporting less stress and higher motivation.
- Cash Flow Prediction: The integration of Big Data allowed the organization to enhance its ability to predict cash flow, improving accuracy from 65% to 90%. This was crucial for optimizing financial management and ensuring timely payments; see Figure 6.

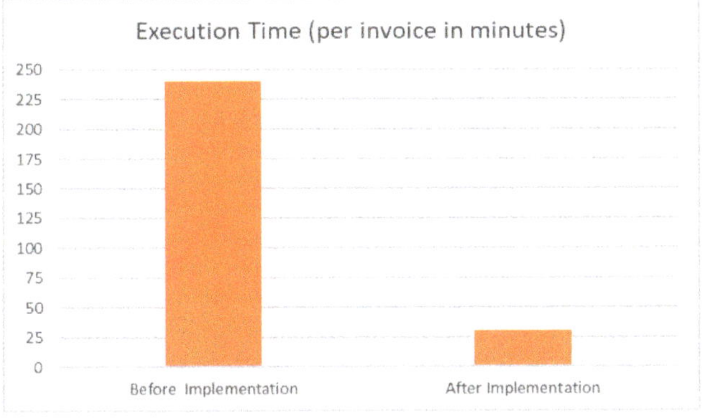

Figure 3. Execution time (per invoice in minutes).

Table 10. Comparison of metrics before and after implementing the Framework.

Metric	Before Implementation	After Implementation	Improvement
Execution Time (per invoice)	240 min	30 min	87.5% reduction
Human Errors (per 100 invoices)	15 errors	0 errors	100% reduction
Paper Consumption (per month)	1000 sheets	200 sheets	80% reduction
Energy Consumption (per month)	500 kWh	100 kWh	80% reduction
Invoice Processing Capacity	100 invoices/month	1000 invoices/month	900% increase
Operational Costs (per month)	$10,000	$4500	55% reduction
ROI (Return on Investment)	N/A	120% (6-month period)	Positive ROI
Employee Satisfaction (Survey)	60% satisfaction rate	90% satisfaction rate	50% improvement
Time Spent on Repetitive Tasks (per month)	120 h	30 h	75% reduction
Cash Flow Prediction Accuracy	65% accuracy	90% accuracy	25% improvement

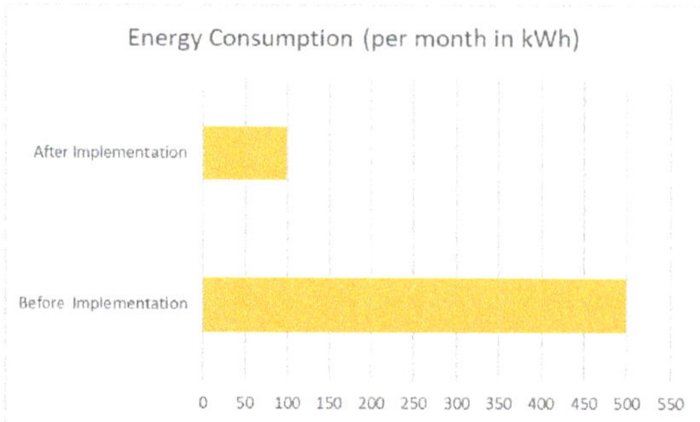

Figure 4. Energy consumption (per month in kWh).

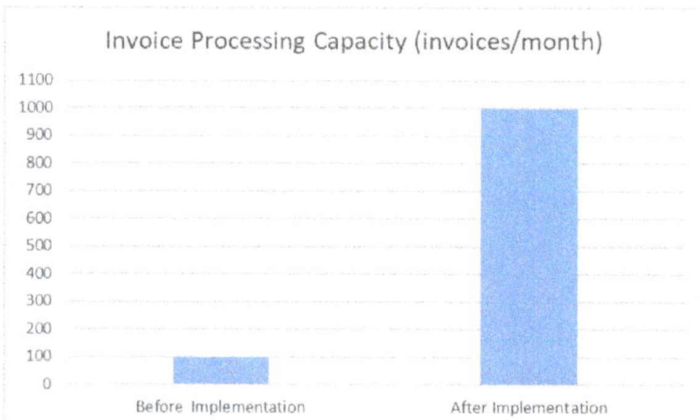

Figure 5. Invoice processing capacity (invoices/month).

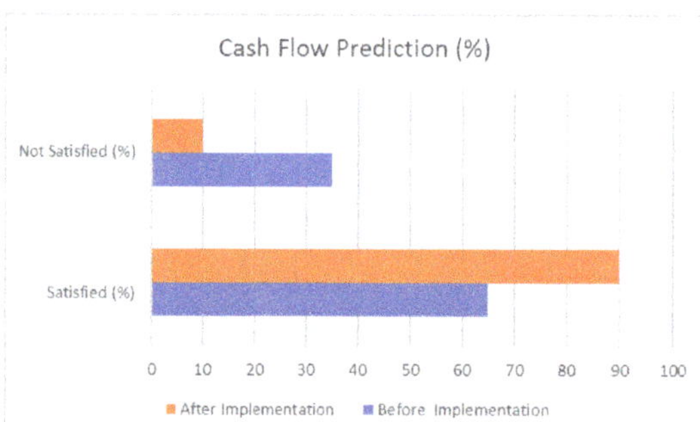

Figure 6. Cash flow prediction (%).

6.2. Discussion

The results from the application of the S-PROBIO framework demonstrate a clear and successful transformation of the administrative invoice management process. Several key observations can be made from these outcomes:

- Operational Efficiency: The most notable benefit of the implementation of RPA and Big Data is the significant improvement in operational efficiency. The reduction in execution time, errors, and operational costs highlights how the integration of automation and data analytics can revolutionize traditional, manual business processes. By eliminating the bottlenecks caused by human intervention and speeding up repetitive tasks, the organization achieved a level of operational efficiency that would have been impossible with manual processing alone.

- Environmental Impact: The reduction in paper and energy consumption underscores the important role of digital technologies in promoting environmental sustainability. With less reliance on physical documents and storage, the organization made substantial strides in reducing its environmental footprint, aligning with broader sustainability goals. These reductions are particularly important in industries that still rely heavily on paper for their operations, demonstrating that digital transformation can contribute to long-term environmental benefits.
- Social Implications: The shift from manual to automated processes had a profound impact on employee satisfaction and well-being. By eliminating repetitive and low-value tasks, employees were able to focus on higher-level activities that required more cognitive input, such as data analysis and financial reporting. This change not only led to improved job satisfaction but also had the potential to increase employee retention and motivation. In addition, by streamlining the process, employees experienced less stress, which contributed to a healthier work environment.
- Economic Sustainability: The reduction in operational costs, combined with the positive ROI, illustrates how technology can drive financial sustainability. The cost savings from fewer human errors, reduced paper consumption, and improved process speed contributed directly to a healthier bottom line. Furthermore, the ability to process more invoices with fewer resources allowed the organization to scale without incurring additional costs. The increased predictive accuracy in cash flow management, enabled by Big Data, also helped the company make more informed financial decisions, ensuring better liquidity and profitability.
- Scalability and Agility: The results also show that the technologies implemented are highly scalable and capable of supporting future growth. The organization's ability to process invoices at a tenfold increase while maintaining accuracy and efficiency demonstrates that the system can easily adapt to higher volumes of work without requiring proportional increases in staff. This scalability is crucial in a rapidly evolving business landscape where organizations must remain agile to stay competitive.
- Predictive Analytics: Big Data's integration into the invoice processing workflow proved instrumental in enhancing decision-making processes. By providing real-time visions into cash flow, payment patterns, and potential fraud risks, Big Data not only optimized operational efficiency but also provided a strategic advantage in managing financial risks. The predictive analysis capabilities further allowed the organization to stay ahead of cash flow issues, reducing the likelihood of late payments and improving vendor relationships.

In conclusion, the case study demonstrates the significant benefits of integrating RPA and Big Data into the administrative processes of invoice management. By automating repetitive tasks and leveraging data-driven insights, the organization achieved greater efficiency, sustainability, and financial success. These outcomes highlight the potential of Industry 4.0 technologies to transform business processes and contribute to the long-term sustainability and profitability of organizations.

7. Conclusions

This study focused on the integration of Robotic Process Automation (RPA) and Big Data technologies within a sustainable framework for process optimization in the context of Industry 4.0. The analysis and implementation of the framework aimed not only to increase operational efficiency but also to promote sustainable practices across economic, social, and environmental dimensions. Through the implementation of this framework, the interaction between these emerging technologies was explored, demonstrating the benefits of their combination in industrial processes.

The primary conclusion drawn from this research is that the integration of RPA and Big Data within a sustainable framework can have a significant impact on industrial process optimization, enhancing operational efficiency and promoting sustainability. The automation of repetitive tasks using RPA and the predictive data analysis capabilities of Big Data were found to considerably reduce execution times, minimize operational errors, and improve decision making. Furthermore, the integration of these technologies contributed to the reduction in environmental impacts, such as lower paper usage and energy consumption, and enhanced social responsibility, as employees were able to focus on more strategic and value-added tasks.

The following benefits were identified:

- Operational Efficiency: The application of RPA enabled the automation of repetitive tasks, speeding up processes and reducing human errors. Big Data provided a more accurate analysis of operations, helping to identify patterns and predict behaviors.
- Sustainability: The use of RPA and Big Data contributed to reducing environmental impact by decreasing the need for physical document storage and lowering energy consumption. From a social perspective, automation liberated employees from monotonous tasks, improving their job satisfaction and productivity. Economically, the framework led to lower operational costs and improved cash flow.

The central research question of this study, "How can Robotic Process Automation (RPA) and Big Data technologies be integrated into a sustainable framework for process optimization in the context of Industry 4.0, maximizing operational efficiency while promoting sustainable practices?" was answered affirmatively. The combination of these technologies not only improved industrial process efficiency but also fostered sustainability within the organization. The developed framework proved effective in integrating the technologies, achieving a balance between process optimization and corporate responsibility. After analyzing the data presented in Table 10 and Figures 3–6, it is possible to clearly and objectively observe the significant improvements achieved after implementing the S-PROBIO framework. These results confirm the effectiveness of the integration of RPA and Big Data technologies, supporting the conclusion that the adoption of the framework brought tangible benefits both in terms of operational efficiency and in terms of sustainability. The detailed analysis of the numbers and graphs fully justifies the implementation of S-PROBIO, highlighting its positive impact on the organization's operations.

The hypotheses formulated at the outset of the study were validated through the practical implementation of the framework. Hypothesis H1, which proposed that the integration of Big Data with RPA could significantly increase efficiency in industrial process optimization, was substantiated with reductions in execution times and operational errors observed. Hypothesis H2, which suggested that a framework combining RPA, Big Data, and sustainable practices would contribute to corporate sustainability, was also confirmed, as there was a notable reduction in economic and environmental impacts and an improvement in social responsibility.

The PICO methodology was successfully applied in this study, providing a structured and objective approach to data analysis. The systematic approach formed a solid foundation for the formulation of hypotheses, the selection of relevant studies, and the application of the framework. The implementation of the framework, which combined RPA, Big Data, and sustainable practices, proved effective in optimizing processes and fostering sustainability, as evidenced by the results obtained during the evaluation phase.

While the results of the framework implementation are promising, this study does have some limitations. The analysis was carried out within a single case study, which may limit the generalizability of the findings to other organizations or industries. Additionally, the adoption of RPA and Big Data may be challenging for companies with financial constraints or insufficient technological infrastructure. It is also worth considering that although automation had a positive impact on human labor in this study, it could potentially generate resistance in companies that are unprepared for such a transition.

Future research could expand the application of the framework to different industrial sectors to evaluate its effectiveness in more diverse contexts. Conducting a longitudinal study would provide a deeper understanding of the long-term benefits of integrating RPA and Big Data particularly in terms of sustainability. Furthermore, it would be valuable to explore the social impact of automation in greater detail, assessing its implications for employment and professional training within the context of Industry 4.0.

This work also opens avenues for the development of new frameworks and solutions that combine emerging technologies with sustainable practices. The continuous evolution of RPA and Big Data technologies requires the ongoing adaptation of the proposed frameworks to ensure they not only optimize processes but also promote a more responsible and environmentally conscious industry.

In conclusion, this study contributes to the understanding of how the integration of RPA and Big Data can be leveraged to achieve both sustainability and process optimization in Industry 4.0, offering a practical and innovative approach that can serve as a reference for future implementations across different industrial contexts.

Author Contributions: In this paper, a proposal for a framework for integrating robotic process automation and big data to optimize processes in the evolution of Industry 4.0, with a focus on sustainability, was proposed and presented by L.P., L.V. and Z.S. The main investigation, encompassing the development and implementation of the framework, was conducted by L.P.; the problem discussion, evaluation of the topic's significance, literature review, and manuscript preparation were collaboratively undertaken by L.P., L.V. and Z.S. The overall supervision of this work was provided by L.V. and Z.S., while project administration and funding acquisition were coordinated by L.V. All authors have read and agreed to the published version of the manuscript.

Funding: The project was merged by FCT—Foundation for Science and Technology through the 1354 scope of the R&D units project: UIDB/00319/2020.

Data Availability Statement: The original contributions presented in the study are included in the article, further inquiries can be directed to the corresponding author.

Conflicts of Interest: The authors declare no conflicts of interest.

References

1. Oláh, J.; Aburumman, N.; Popp, J.; Khan, M.; Haddad, H.; Kitukutha, N. Impact of Industry 4.0 on Environmental Sustainability. *Sustainability* **2020**, *12*, 4674. [CrossRef]
2. Singh, S.; El-Kassar, A. Role of Big Data Analytics in Developing Sustainable Capabilities. *J. Clean. Prod.* **2019**, *213*, 1264–1273. [CrossRef]
3. Gholami, H.; Lee, J.K.Y.; Ali, A. Big Data Analytics for Sustainable Products: A State-of-the-Art Review and Analysis. *Sustainability* **2023**, *15*, 12758. [CrossRef]
4. Venigandla, K.; Vemuri, N.; Aneke, E. Empowering Smart Cities with AI and RPA: Strategies for Intelligent Urban Management and Sustainable Development. *Int. J. Sci. Res. Manag. (IJSRM)* **2024**, *12*, 1117–1125. [CrossRef]
5. Chukasova-Ilyushkina, E. Approaches to Analysis and Optimization of Business Processes of an Industrial Enterprise. *Vestn. Polotsk State Univ. Part D Econ. Leg. Sci.* **2024**, 50–53. [CrossRef]
6. Shysh, A.; Kovtun, I.; Sheveria, Y. Business Information Management and Innovative Approaches to Business Process Optimization. *Bus. Inform.* **2023**. [CrossRef]

7. Yoqubjonov, J.; Gibadullin, R.; Nuriev, M. Advanced Robotic Process Automation for Enterprise Efficiency. *E3S Web Conf.* **2023**, *431*, 07011. [CrossRef]
8. Patrício, L.; Varela, L.; Silveira, Z. Integration of Artificial Intelligence and Robotic Process Automation: Literature Review and Proposal for a Sustainable Model. *Appl. Sci.* **2024**, *14*, 9648. [CrossRef]
9. Bonilla, S.; Silva, H.; Da Silva, M.; Gonçalves, R.; Sacomano, J. Industry 4.0 and Sustainability Implications: A Scenario-Based Analysis of the Impacts and Challenges. *Sustainability* **2018**, *10*, 3740. [CrossRef]
10. Autor, D. Industry 4.0 and Its Impact on Indian Economy. *EPRA Int. J. Res. Dev. (IJRD)* **2022**, *7*, 51–54. [CrossRef]
11. Jamwal, A.; Agrawal, R.; Sharma, M.; Giallanza, A. Industry 4.0 Technologies for Manufacturing Sustainability: A Systematic Review and Future Research Directions. *Appl. Sci.* **2021**, *11*, 5725. [CrossRef]
12. Amjad, M.; Rafique, M.; Khan, M. Leveraging Optimized and Cleaner Production through Industry 4.0. Sustain. *Prod. Consum.* **2021**, *26*, 859–871. [CrossRef]
13. Shu, H. Big Data Analytics: Six Techniques. *Geo-Spatial Inf. Sci.* **2016**, *19*, 119–128. [CrossRef]
14. Wu, X.; Zhu, X.; Wu, G.; Ding, W. Data Mining with Big Data. *IEEE Trans. Knowl. Data Eng.* **2016**, *26*, 97–107. [CrossRef]
15. Prajapati, A.; Gupta, A.; Mishra, S.; Raj, D.; Singh, M.; Goyal, M. An Exploration on Big Data Analytical Techniques: A Review. In Proceedings of the 2024 11th International Conference on Computing for Sustainable Global Development (INDIACom), New Delhi, India, 28 February–1 March 2024; pp. 123–128. [CrossRef]
16. Mahmud, M.; Huang, J.; Salloum, S.; Emara, T.; Sadatdiynov, K. A Survey of Data Partitioning and Sampling Methods to Support Big Data Analysis. *Big Data Min. Anal.* **2020**, *3*, 85–101. [CrossRef]
17. Patrício, L.; Varela, L.; Silveira, Z. Proposal for a Sustainable Model for Integrating Robotic Process Automation and Machine Learning in Failure Prediction and Operational Efficiency in Predictive Maintenance. *Appl. Sci.* **2025**, *15*, 854. [CrossRef]
18. Patrício, L.; Costa, L.; Varela, L.; Ávila, P. Sustainable Implementation of Robotic Process Automation Based on a Multi-Objective Mathematical Model. *Sustainability* **2023**, *15*, 15045. [CrossRef]
19. Patrício, L.F.S.; de Sousa Costa, C.R.; Fernandes, L.P.; Varela, M.L.R. Structure for the Implementation and Control of Robotic Process Automation Projects. In *Advanced Network Technologies and Intelligent Computing. ANTIC 2022. Communications in Computer and Information Science*; Woungang, I., Dhurandher, S.K., Pattanaik, K.K., Verma, A., Verma, P., Eds.; Springer: Cham, Switzerland, 2023; Volume 1798. [CrossRef]
20. Patrício, L.; Ávila, P.; Varela, L.; Cruz-Cunha, M.M.; Ferreira, L.P.; Bastos, J.; Castro, H.; Silva, J. Literature Review of Decision Models for the Sustainable Implementation of Robotic Process Automation. *Procedia Comput. Sci.* **2023**, *219*, 870–878. [CrossRef]
21. Sousa Costa, C.R.; Patrício, L.F.S.; Varela, M.L.R.; Ferreira, P.V. The Contribution of Robotic Process Automation (RPA) in Improving Energy Efficiency: Case Study. In *Proceedings of the Eighth International Congress on Information and Communication Technology. ICICT 2023; Lecture Notes in Networks and Systems*; Yang, X.S., Sherratt, R.S., Dey, N., Joshi, A., Eds.; Springer: Singapore, 2024; Volume 695. [CrossRef]
22. Varela, L.; Araujo, A.; Ávila, P.; Castro, H.; Putnik, G. Evaluation of the Relation between Lean Manufacturing, Industry 4.0, and Sustainability. *Sustainability* **2019**, *11*, 1439. [CrossRef]
23. Naderi, M.; Ares, E.; Peláez, G.; Prieto, D.; Araújo, M. Sustainable Operations Management for Industry 4.0 and Its Social Return. *IFAC-PapersOnLine* **2019**, *52*, 457–462. [CrossRef]
24. Barata, J.; Silva, F.; Almeida, M. Ceramic Industry 4.0. Res. Anthol. Cross-Ind. Chall. Ind. 4.0. 2019. Available online: https://www.igi-global.com/gateway/chapter/210488 (accessed on 10 February 2025).
25. Hojnik, J.; Biloslavo, R.; Cicero, L.; Cagnina, M. Sustainability Indicators for the Yachting Industry: Empirical Conceptualization. *J. Clean. Prod.* **2020**, *249*, 119368. [CrossRef]
26. Ramísio, P.; Pinto, L.; Gouveia, N.; Costa, H.; Arezes, D. Sustainability Strategy in Higher Education Institutions: Lessons Learned from a Nine-Year Case Study. *J. Clean. Prod.* **2019**, *222*, 300–309. [CrossRef]
27. Cumpston, M.S.; McKenzie, J.E.; Thomas, J.; Brennan, S.E. The Use of 'PICO for Synthesis' and Methods for Synthesis Without Meta-Analysis: Protocol for a Survey of Current Practice in Systematic Reviews of Health Interventions. *F1000Research* **2021**, *9*, 678. [CrossRef] [PubMed]
28. Nishikawa-Pacher, A. Research Questions with PICO: A Universal Mnemonic. *Publications* **2022**, *10*, 21. [CrossRef]
29. Badmus, A. Implementing Technology Manufacturing with Robotics Process Automation (RPA). *J. Knowl. Learn. Sci. Technol.* **2023**, *2*, 103. [CrossRef]
30. Karn, S.; Kotecha, R. RPA-based Implementation of IoT. *SSRN Electron. J.* **2021**. [CrossRef]
31. Daase, C.; Pandey, A.; Staegemann, D.; Turowski, K. Sustainability in Robotic Process Automation: Proposing a Universal Implementation Model. In *Automation and Robotics—Volume 1: ICINCO, Proceedings of the 20th International Conference on Informatics in Control, Rome, Italy, 13–15 November 2023*; SciTePress: Setúbal, Portugal, 2023; pp. 770–779. Available online: https://www.scitepress.org/Link.aspx?doi=10.5220/0012260200003543 (accessed on 10 February 2025).

32. Ribeiro, J.; Lima, R.; Eckhardt, T.; Paiva, S. Robotic Process Automation and Artificial Intelligence in Industry 4.0—A Literature Review. *Procedia Comput. Sci.* **2020**, *181*, 51–58. [CrossRef]
33. Alshehri, A.; Aljarbou, M.; Elaraki, Y.; Alsehaimi, A. Evaluating Robotic Process Automation (RPA) Technology for Human-Centric and Future-Oriented Automated Tracking and Adherence to Sustainable Regulations in Construction. *Nanotechnology Perceptions* **2024**, *20*, 430–449. [CrossRef]
34. Joseph, O. Sustainable Banking through Robotic Process Automation: What Role does ESG and Cognitive AI Play? *J. Digitovation Inf. Syst.* **2023**, *3*, 116–140. [CrossRef]
35. Malini, G. Automation of Big Data Analytics Using Robotic Process Automation. *Int. J. Sci. Res. Comput. Sci. Eng. Inf. Technol.* **2021**, *7*, 602–605. [CrossRef]
36. Pramod, D. Robotic Process Automation for Industry: Adoption Status, Benefits, Challenges, and Research Agenda. *Benchmarking Int. J.* **2021**, *29*, 1562–1586. [CrossRef]
37. Quille, R.; Almeida, F.; Borycz, J.; Corrêa, P.; Filgueiras, L.; Machicao, J.; Almeida, G.; Midorikawa, E.; Demuner, V.; Bedoya, J.; et al. Performance Analysis Method for Robotic Process Automation. *Sustainability* **2023**, *15*, 3702. [CrossRef]
38. Ciufudean, C. Industry 4.0's Product Development Process Aided by Robotic Process Automation. *Robot. Autom. Eng. J.* **2018**, *3*, 555614. [CrossRef]
39. Mahariya, S.; Kumar, A.; Singh, R.; Gehlot, A.; Akram, S.; Twala, B.; Iqbal, M.; Priyadarshi, N. Smart Campus 4.0: Digitalization of University Campus with Assimilation of Industry 4.0 for Innovation and Sustainability. *J. Adv. Res. Appl. Sci. Eng. Technol.* **2023**, *32*, 120–138. [CrossRef]
40. Ejsmont, K.; Gładysz, B.; Kluczek, A. Impact of Industry 4.0 on Sustainability—Bibliometric Literature Review. *Sustainability* **2020**, *12*, 5650. [CrossRef]
41. Wei, Q. Task Allocation of Financial Robots Based on RPA and Big Data Algorithms. In Proceedings of the 2024 Third International Conference on Distributed Computing and Electrical Circuits and Electronics (ICDCECE), Ballari, India, 26–27 April 2024; pp. 1–5. [CrossRef]
42. Sobczak, A. Robotic Process Automation as a Digital Transformation Tool for Increasing Organizational Resilience in Polish Enterprises. *Sustainability* **2022**, *14*, 1333. [CrossRef]
43. Zheng, S. Research on the Transformation of Digital Intelligence of Accounting Profession Based on Big Data Technology. *Appl. Math. Nonlinear Sci.* **2024**, *9*. [CrossRef]
44. Raut, R.; Mangla, S.; Narwane, V.; Gardas, B.; Priyadarshinee, P.; Narkhede, B. Linking Big Data Analytics and Operational Sustainability Practices for Sustainable Business Management. *J. Clean. Prod.* **2019**, *224*, 10–24. [CrossRef]

processes

MDPI

Article

Prototype of a Multimodal Platform Including EEG and HRV Measurements Intended for Neuroergonomics Applications

Awad M. Aljuaid

Department of Industrial Engineering, College of Engineering, Taif University, P.O. Box 11099, Taif 21944, Saudi Arabia; amjuaid@tu.edu.sa

Abstract

Drowsiness and stress greatly influence worker health and productivity and workplace safety. Conflict between workplace expectations and employee control results in stress, which causes mental and physical reactions that affect performance and raise the risk of accidents at work. A common precursor to inadvertent drowsiness increases workplace risks and costs due to lost productivity and accidents. Developments in the interdisciplinary subject of neuroergonomics enable the creation of novel systems to track and minimize these issues. This work introduces prototype testing to demonstrate the system's ability to detect stress and drowsiness. Along with other indicators such as body temperature, heart rate (HR), and SpO_2 levels, the system incorporates electroencephalography (EEG) and heart rate variability (HRV). By analyzing these biosignals, the system detects stress and drowsiness in real time, providing alerts to both users and the supervisor's BI dashboard. The design is flexible, offering two wearable forms: a headband and an armband. Prototype testing demonstrates the system's ability to detect stress and drowsiness effectively, paving the way for safer and more productive workplace environments.

Keywords: neuroergonomics; BI; safety; stress; attention; drowsiness; EEG; PPG; HRV; SpO_2

1. Introduction

Stress can be defined as "the harmful physical and emotional responses that can result from conflicts between job demands on the employee and the amount of control an employee has over meeting these demands" [1]. Stress in the workplace has been revealed to have negative effects on employee health and performance and institutional profits [2], and it increases the number of job-related accidents [3–5]. Stress in the workplace has consequences for productivity, as found in several studies [6–9]. In addition to production and construction lines, many hospitals have investigated the impact of stress on health care staff [10–12]. Undeniable problems were highlighted during the COVID-19 outbreak, which include problems related to the mental health of health care staff; stress was one of the main mental issues that affected performance and increased the potential for medical mistakes [13–17]. Additionally, drowsiness, which refers to the early stages of sleeping or unintentional sleep at work, can have a negative impact on safety, productivity, and the education environment [18,19]. Statistics show that job-related accidents caused by sleeping at work cost up to USD 50 billion per year [20]. Moreover, serious workplace accidents and injuries are more likely to occur when people are drowsy due to sleep issues, including insomnia and sleep loss [21,22].

New non-invasive systems that can detect human biosignals have been made possible by developments in advanced engineering. These technologies enable organizations to better understand human behavior and improve working conditions and design by gathering and analyzing biological data. Furthermore, they allow scientists to study countless interpretations of human intentions as a form of controlling commands that are translated into machine actions.

In order to improve performance and lower risks, neuroergonomics combines ergonomics and neurology to obtain a deeper understanding of the human brain at work [23–26]. Neuroscience is a scientific field that studies the nervous system and human brain's function and structure. Because it employs biosignals to investigate how people use technology in their daily lives and at work, neuroergonomics has grown into a significant area of study. The knowledge it provides aids in the creation of improved work environments that take into account both the physical and psychological components of human capabilities [27–30].

This work introduces a portable biosignal-based prototype of a system intended for monitoring stress and drowsiness that builds on previous research [31]. Along with other indicators such as body temperature, HR, and SpO_2 levels, the system incorporates EEG and HRV. By analyzing these biosignals, the system detects stress and drowsiness in real time, providing alerts to both users' and supervisors' BI dashboards. The design is flexible, offering two wearable forms: a headband and an armband.

2. Overview

2.1. Stress- and Drowsiness-Related Biosignals

Stress and its related emotional and physical responses have been studied comprehensively over the years. There are central (brain and spinal cord) and peripheral parts in the human nervous system. Both somatic and autonomic components make up the peripheral system; the autonomic system regulates emotional states and physiological processes through its parasympathetic (relaxation) and sympathetic (stress reaction) branches [32,33]. Research has shown that stress and emotional signals can be measured from the human nervous system using both direct and indirect methods [34]. Skin response measures, heart activity, and brain imaging can all be used to track stress. According to scientific studies, human error rates are directly related to sleepiness, which is regulated by neurobiological processes [35].

2.1.1. Brain EEG

EEG measures voltage oscillations that fall into five frequency bands: delta (1–4 Hz), theta (4–8 Hz), alpha (8–12 Hz), beta (13–25 Hz), and gamma (>25 Hz). It records electrical activity from brain neurons along the scalp [36,37].

Hee Seo [33] found that there was a significant difference in the right prefrontal cortex in the beta (β)-band activity between two groups: stressed and unstressed. Brain electrical power activity increases with stress. This result is supported in [38–40]; recently, [41–43] showed that, in stress tasks, the beta-band increased while the alpha (α)-band decreased. In a recent study by Minguillon et al. [44], the scholars found that the ratio of the average power gamma divided by the average power theta and alpha, $\frac{\overline{\gamma}}{\overline{\theta} \text{ and } \overline{\alpha}}$, increases with stress, allowing stress to be detected. In another study using the convolutional neural network, Joong proposed a device that detected and monitored stress over time [45]. As reported by Putman, the θ/β ratio moderately influences the effects of stress on attention [46].

The central nervous system has been utilized to study sleep stages; EEG is the most powerful tool for exploring the brain's biosignals and patterns during sleep and before it in sleepiness/drowsiness.

Previous studies indicate that an increase or decrease in the brain's waves/rhythms can predict drowsiness. Borghini, Shabani, and Ohsuga found that the EEG power in alpha, delta, and theta increased during mental fatigue and drowsiness in the frontal lobe [47,48]. The results presented by Nikhil and Agustina demonstrate that the power spectrum density (PSD) of alpha and theta rhythms is positively correlated with drowsiness [49,50]. Kaida et al. found that the alpha and theta PSD significantly increases with an increase in the Karolinska sleepiness scale (KSS) [51]. Recently, Mikito et al. used a single electrode at the prefrontal cortex to detect drowsiness and found the same results: that alpha and theta rhythms were increased [52]. Nguyen also found an increase in delta rhythms [53].

2.1.2. Heart Rate Variability (HRV)

Through the vagus nerve, the autonomic nervous system regulates heart function; stress causes parasympathetic suppression and sympathetic activation [54]. This changes under a physiological reaction called "fight-or-flight", which leads to several changes in HR and HRV [55]. An increase in HRV is associated with positive emotions, while a decrease in HRV is related to negative emotions. HRV can be analyzed through time and frequency domains [34,56] using root mean square of the successive differences (RMSSD) calculations on the R-R intervals in ECG measurements, according to Equation (1) [57]:

$$\mathrm{RMSSD} = \sqrt{\frac{\sum_{i=1}^{n-1}(\mathrm{RR}_{i+1} - \mathrm{RR}_i)^2}{n-1}} \quad (1)$$

HRV frequency domain analysis uses the LF/HF ratio (0.04–0.15 Hz/0.15–0.4 Hz). Research shows that stress decreases HRV as measured with the RMSSD (time domain), but increases the LF/HF ratio (frequency domain) [55,57–67].

Drowsiness can also be determined from HRV; Rodriguez et al. conducted an experiment on participants while they were driving in a real-world setting to identify signs of drowsiness and discovered a notable difference [68]. Shan Hu et al., M. Awais, and Furman reported a decrease in HRV LF/HF as subjects transitioned from awake to drowsy [69–71]. MURATA also reported that HRV (RRV3) increased under drowsy states [72]. Another study found a decrease in the time domain measures such as RMSSD at the beginning of drowsiness using KSS [73].

2.1.3. Other Physiological Indicators

Stress and drowsiness can also be determined using other physiological indicators such as the oxygen saturation level (SpO_2), HR, and body temperature. Relaxation and decreased sympathetic nervous system activity are frequently linked to drowsiness; the HR decreases as a result and contrasts with that under stress responses [74]. A drop in SpO_2 level can be associated with both stress and drowsiness due to hyperventilation [75]. In order to prepare for sleep, the hypothalamus, which controls body temperature, lowers the core temperature when drowsiness sets in [76–78].

3. Methodology: Conceptual and Algorithm Design

The portable system's design, components, and configuration are shown in block diagrams and pictorial representations in Figures 1 and 2. The working principle and algorithm flowchart are shown in Figure 3. The algorithm begins by assessing the worker's current state through multiple measurements, starting with the HRV feature LF/HF. If this ratio falls outside the normal range of $1.5 \leq LF/HF \leq 3$ [79,80], the algorithm proceeds to evaluate other stress and drowsiness features including RR intervals and the RMSSD. It then checks these values against established thresholds to confirm whether the worker is experiencing stress or drowsiness.

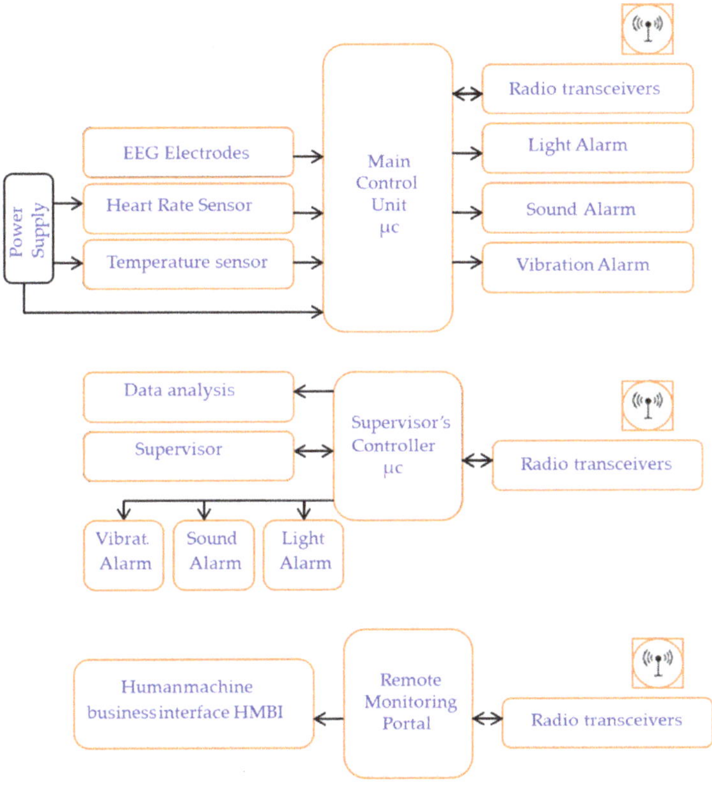

Figure 1. The system block diagram representation.

The primary components include two control units, sensors, ECG and EEG electrodes, sound and light indicators (an LED and buzzer), a vibratory motor, a wireless transceiver, and a battery power supply. The system consists of three main parts: the user part with a bio-data acquisition and signal processing system, the supervisor's part that performs data streaming and analysis, and the alert part and management system. The system is designed to acquire physiological biosignal measures to predict human mental and physical states in a working environment from two human organs: the head and the upper arm. Then, it sends the data to the control unit, which, based on the control algorithm, evaluates the information by analyzing the acquired data. If a critical state is registered, it notifies and alerts the user, as well as continuously and wirelessly transmitting the acquired data and resulting analysis to supervisors. The system continuously records the human biosignals, i.e., the EEG, HRV, and body's most basic functions and signs: temperature, HR, and SpO_2 level. A total of 27 healthy participants aged 32 ± 12 (22 males and 5 females) participated in the system's testing and evaluation.

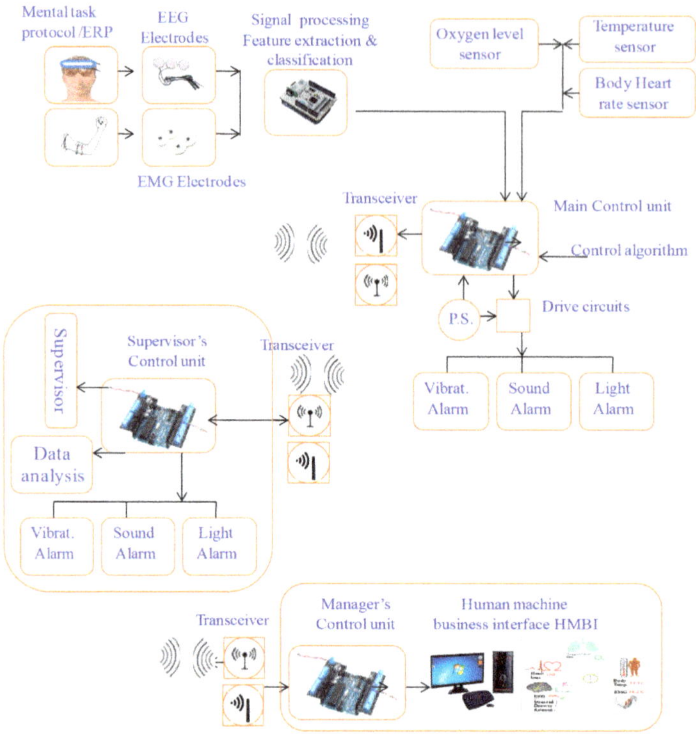

Figure 2. Overall system design: pictorial and functional diagram representation.

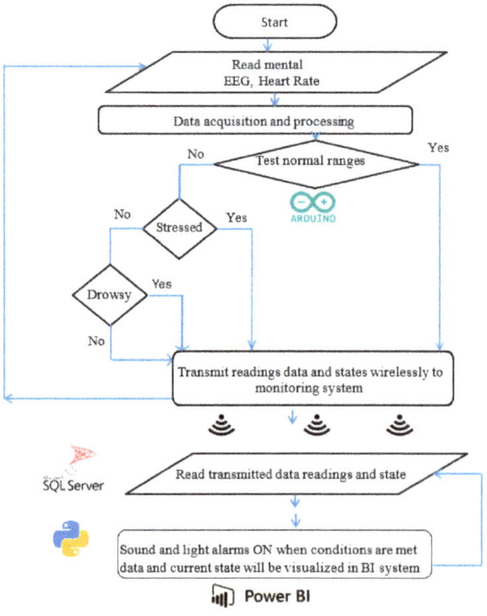

Figure 3. Working principle and algorithm flowchart representation.

4. System Design and Hardware Integration

The design features two setup forms: one to be worn on the upper arm and another to be worn on the head. This allows employees to move freely throughout their office and avoid any annoyance, impairments, or issues that may emerge from wearing the portable system. The gadget has integrated electronic components and is designed similarly to a sports headband. The sensors, including dry EEG electrodes above each eye to detect brain impulses, and temperature sensors in the forehead and temple regions, are in contact with spots on the skull. With sensors placed on the triceps region of the upper arm, the armband form of the gadget uses electronic components to measure a variety of biosignals. Depending on the purpose, the design can be modified to create several bracelet styles. Biosignal sensors (EEG, temperature, HR, and oxygen saturation), processor chips, indicators, a wireless communication component, a data visualization component, interfaces, and a battery power source are some of the physical elements that make up the system.

4.1. EEG and Signal Processing Modules

Dry comb snap silver–silver chloride-coated EEG electrodes that are compatible with the system's design can be used. Another option is the principal brainwave sensor of the OpenBCI 4-channel Ganglion Biosensing Board (OpenBCI 4-channel Ganglion, Brooklyn, NY, USA), as illustrated in Figure 4. This module measures eye blinks, attention, and meditation using basic dry electrodes.

Figure 4. OpenBCI Ganglion with EEG electrodes.

The scalp impedance and other noise interference cause the EEG electrodes to pick up tiny electrical impulses from the human head, ranging from 1 μV to 100 mV. The signal processing circuit filters the raw EEG signals, as shown in Figure 5, using both low-pass and high-pass filters, which eliminate high-frequency and low-frequency noise, respectively. A notch filter selectively blocks a narrow frequency band while allowing other frequencies to pass, after which a post-amplifier boosts the signal by 1000 times using a non-inverting amplifier, as calculated in Equation (2). The processed analog EEG signal is then converted to digital format using an analog-to-digital converter (ADC).

$$K = 1 + R_f/R_{in} \quad (2)$$

Body temperature sensors: A suitable choice is the waterproof DS18B20 temperature sensor, with alternative options including the DS18B20 and LM35 sensors, and a circuit diagram is shown in Figure 6.

Heart rate sensor: The HR represents the frequency of cardiac contractions, measuring the number of beats occurring within a one-minute period. An optical HR sensor is a suitable choice for reading heartbeats, with various design forms shown in Figure 7. The sensor uses photoplethysmography (PPG), which uses light to monitor blood flow and provide HR information. A transmitter (IRTx) and receiver (IRRx) are part of the sensor unit. The optical HR sensor's components, circuit, block diagram, and operating principle are also shown. The fraction of oxygen-saturated hemoglobin compared to total hemoglobin in the blood, or simply the amount of oxygen in the blood, is the blood oxygen saturation level (SpO_2). The usual range for arterial SpO_2 is 95–100%; values above 95% are considered normal. Hypoxemia, or an abnormally low blood oxygen concentration, means SpO_2 levels below 90%. The most effective choice for measuring blood oxygen saturation levels is the MAX30102 Maxim integrated high-sensitivity pulse oximeter and heart rate sensor, which is also displayed in Figure 7.

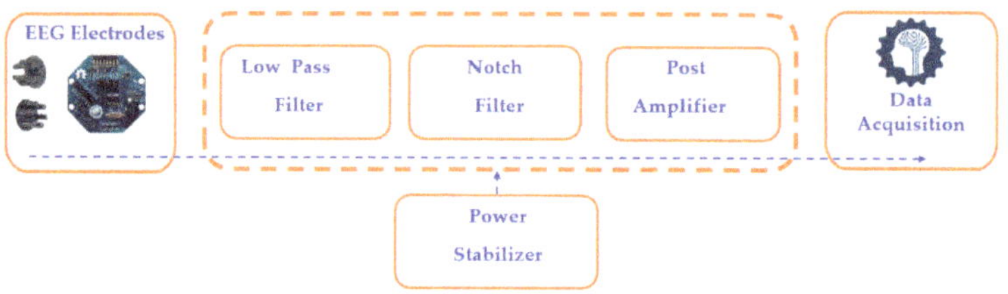

Figure 5. Block diagram of a single-channel EEG system.

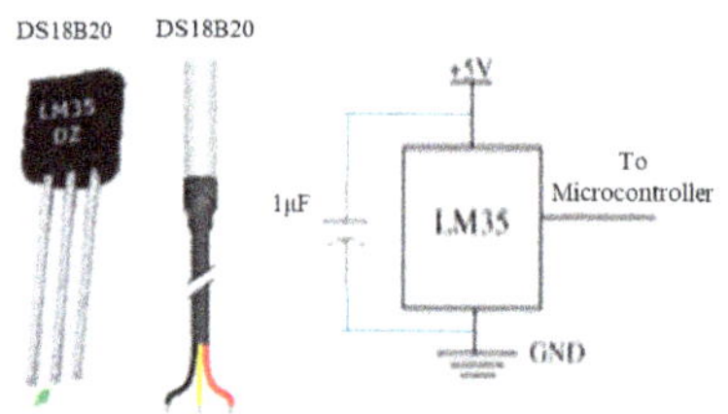

Figure 6. Temperature sensors and circuit diagram.

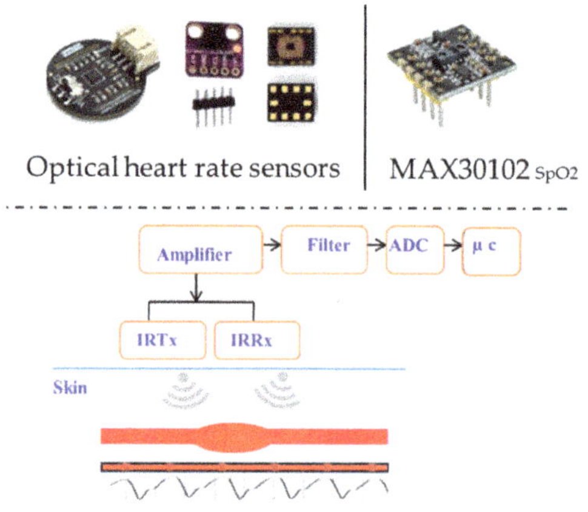

Figure 7. HR and SpO_2 sensors.

Sound and light alert indicators: A passive low-level buzzer alarm with a high-intensity LED is an example of appropriate hardware. Figure 8 shows the hardware integration and circuit diagram. The vibration motor's circuit diagram and hardware integration are shown in Figure 9, presenting an adequate choice for non-audible alerts.

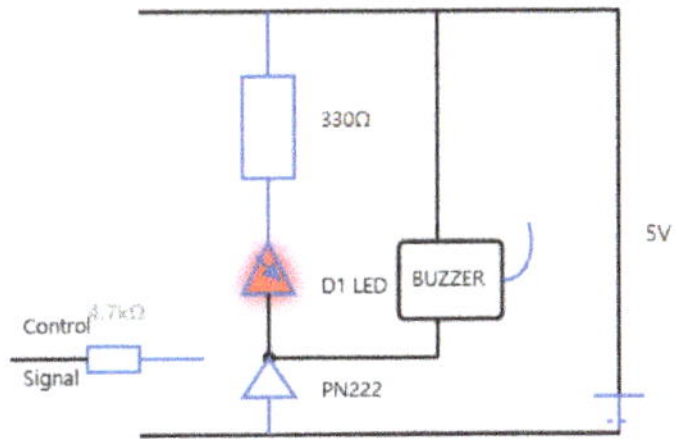

Figure 8. Sound and light hardware integration and circuit diagram.

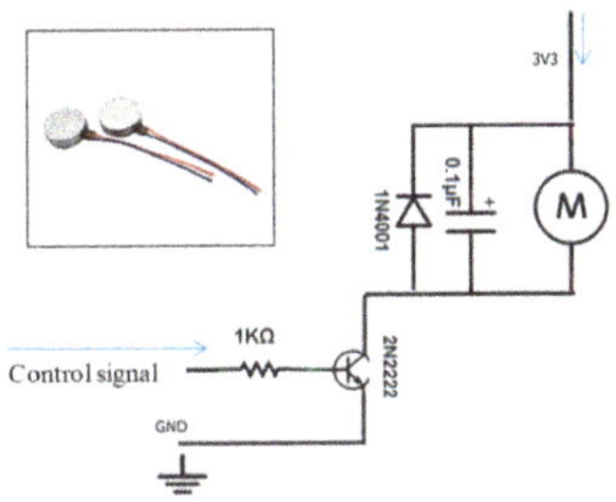

Figure 9. Vibration motor and its circuit diagram.

As shown in Figure 10, the 433 MHz wireless serial transceiver module is a good choice for wireless communication because it can transmit up to 1000 m.

Figure 10. A 433 MHz wireless serial module.

4.2. The Control Unit and Power Supply

Due to the design requirements, including the processing capabilities, a small size, the number of utilized input analog sensors, the required digital input pins, and transceiver module compatibility, the most suitable control unit is the ATmega328p on the Arduino Nano board or the ATmega2560 on the Arduino Mega board; both boards are shown in Figure 11.

ATmega328p Mega 2560

Figure 11. The control unit Arduino (ATmega328p/ATmega2560).

A suitable power supply choice is the Lithium Polymer Battery Li-Po li-ion, 3.7 V, 180 mAH, Rechargeable Battery, and another potential power supply option is the Button Cell—Lithium Ion 3V Multipurpose Battery. To achieve the required power level, designers can use a combination of parallel series connections.

5. Prototype Validation, Overall System and Subsystem Physical Prototyping, and Experimental Setup

Figure 12 depicts the overall system's hardware design integration in pictorial form. The EEG electrode was wired to the signal processing interface module, which, in turn, interfaced to the control unit. Temperature and HR sensors were interfaced to the control unit. The alerting circuit with a transistor, light, and sound alarm was interfaced as an actuator.

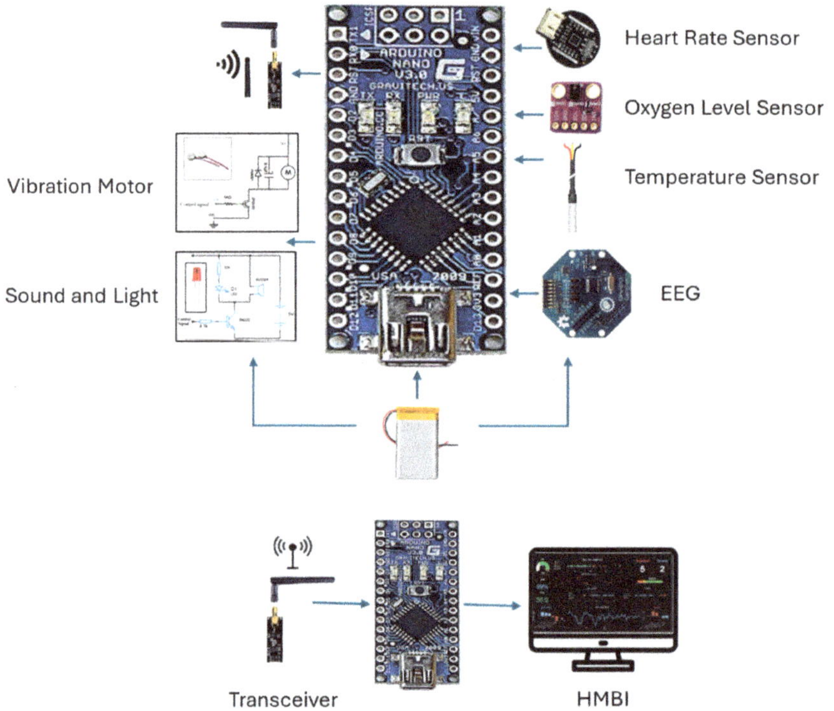

Figure 12. The pictorial form represents the overall portable system hardware design integration.

The wireless communication unit was built utilizing an NRF24L01 transceiver module; both the WiFi and GSM modules for long-distance transmission can be utilized. The data are wirelessly transmitted to the SQL server and displayed on supervisors' PCs; they are received and displayed using a special human–machine business interface (HMBI) design.

Finally, two forms of the overall system prototype were built by integrating all the subsystems into one overall unit with two different designs.

5.1. EEG Signal Subsystem Wiring, Reading, and Processing

The EEG signals were acquired from the user's head and read by the control unit. The EEG data were acquired and preprocessed using the OpenBCI GUI software (OpenBCI GUI v6.0.0-beta.1), and the power spectra of all the bands were acquired, as shown in Figure 13.

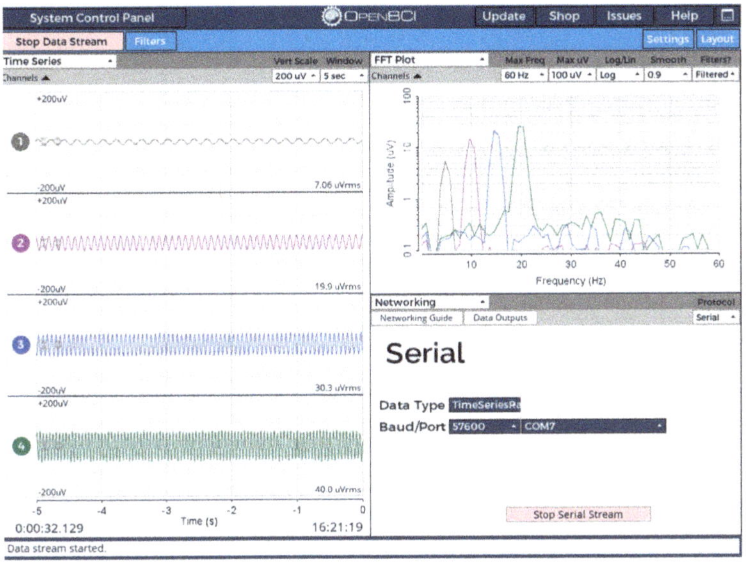

Figure 13. Acquired EEG data from OpenBCI_Ganglion.

5.2. The Sensor Subsystem Wiring and Integration

To read the body's most basic functions and signs, the three applied sensors (oxygen level, temperature, and heart rate) are integrated in one input module. The sensors' integration and testing setup is shown in Figure 14. The reading from each of these sensors is one numerical value that is increased depending on the human mental and physical state in an industrial setting, for example. The average normal temperature is around 37 °C. For a normal human, the oxygen level (SpO_2) is typically above 95%, and below 90% is considered low. Body temperature increases when humans are focused with high attention, and the oxygen level and HR similarly increase; the body requires more oxygen when focused or stressed, and there is also an increase in HR in this situation. The acquired values are displayed on the supervisor's HMBI design as numerical values.

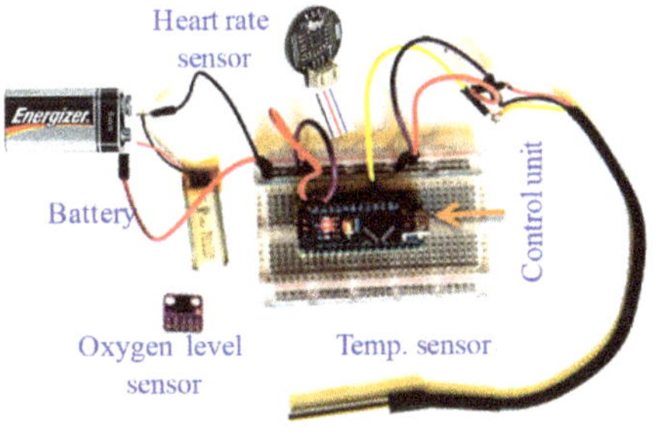

Figure 14. Integrated module consisting of the three applied sensors for testing purposes.

5.3. The Actuator Subsystem Wiring and Integration

5.3.1. The User's Alarm and Notification Subsystem

In the design, the alarm and notification subsystem, consisting of vibration, sound, and light alarms, is applied to notify both the user (e.g., worker/operator) and the supervisor.

The vibrating alarm was developed utilizing the selected vibrating motor and built as shown in Figure 15, in accordance with the circuit diagram.

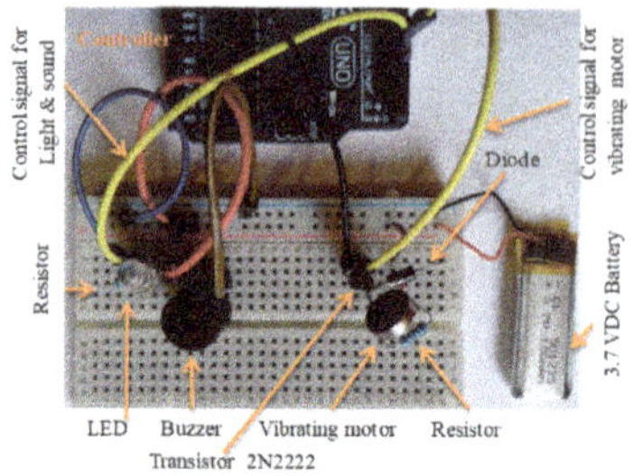

Figure 15. The two circuits developed for testing, including light, sound, and vibrating alarms.

The light and sound alarms were physically built with a high-intensity white LED and buzzer, as shown in Figure 16, according to the circuit diagram.

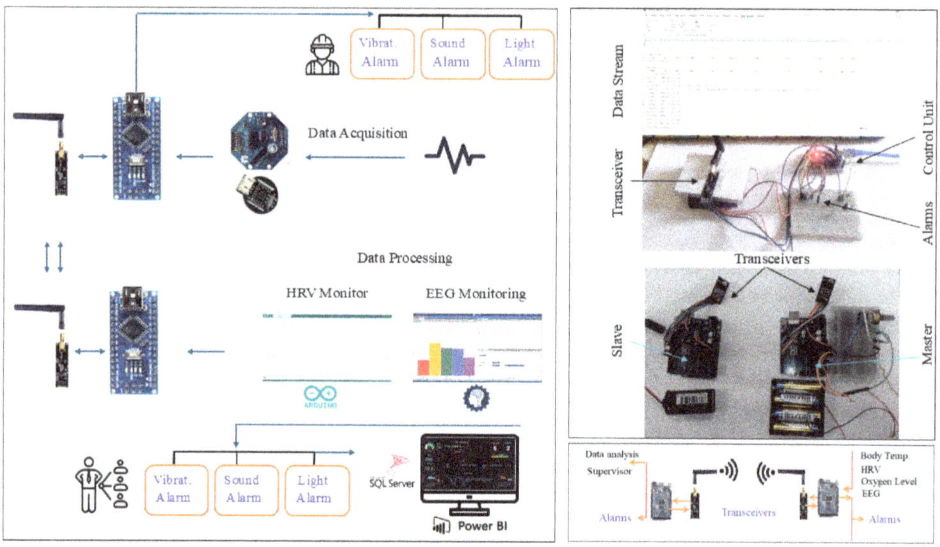

Figure 16. The supervisor's alerting and notification subsystem.

5.3.2. The Wireless Communication Subsystem

To wirelessly pair up/connect the master and slave control units, a pair of transceiver modules, nRF24L01, is utilized. The wireless communication subsystem was physically developed. The system's design was developed to alert and notify both the user and the supervisor. The alerting and notification subsystem was designed as shown in Figure 16, consisting of a wireless communications unit, a PC displaying the data, and three types of alarms (light, sound, and vibratory motor) to be activated when an indicator's value drops below or increases beyond specific thresholds.

5.4. Building the Overall System Prototype

Both design setups (head and upper arm) were built by integrating all the physically built circuits, modules, components, and subsystems into one overall system prototype. The developed prototypes, with the hardware integration and placement of all the components and subsystems, including the EEG module, sensors, actuators, alarms, wireless communication module, and power supply subsystems, are shown in Figure 17.

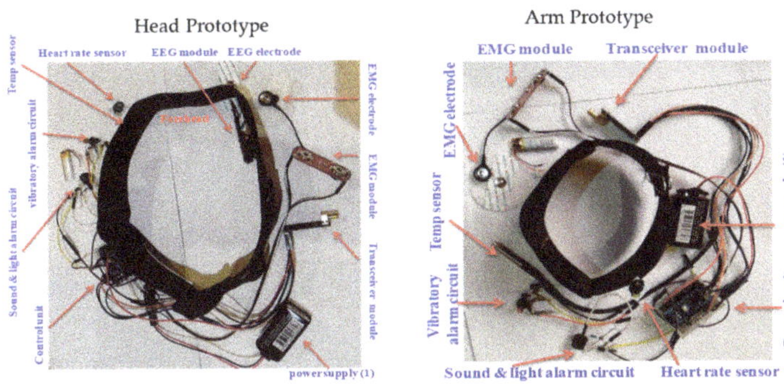

Figure 17. Hardware integration and placement of components in developed prototype.

6. Results and Discussion

6.1. Results

The Arduino IDE (Version 2.3.4) was used to program and upload code to the microcontroller in C++. The features of HRV (HR, R-R, RMSSD, LF, HF, and LF/HF) were extracted using the designed code in real time. A detailed description of the code is provided in the Supplementary Materials. In addition, the EEG PSD data were extracted in all bands—delta, theta, alpha, beta, and gamma—using OpenBCI GUI software and then streamed to the network.

A total of 27 healthy participants volunteered to test the system and the HRV features extracted, as listed in Table 1.

Figure 18 summarizes the average HRV features across all participants, comparing males and females. The female sample shows a lower RMSSD (approximately 23.5 ms) compared to the males (37.2 ms) and a higher LF/HF ratio (1.5 compared to 1.3); this sample of females may have experienced more stress than the males.

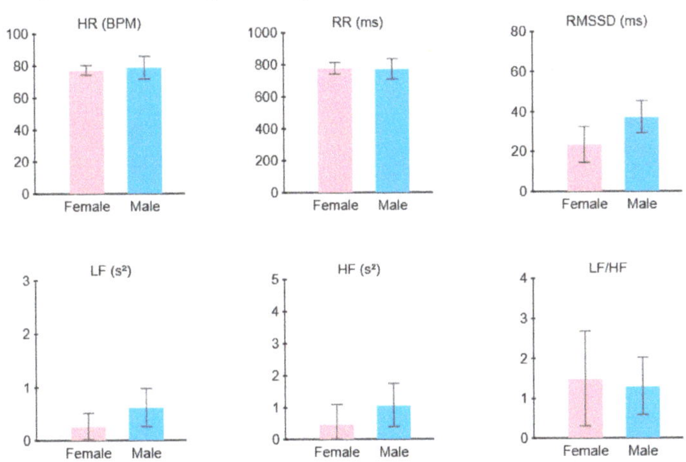

Figure 18. Average HRV features across all participants: comparison between males and females.

Table 1. HRV features from a sample of 27 participants.

	Participant Number	HR (BPM)	RR (ms)	RMSSD (ms)	LF (s^2)	HF (s^2)	LF/HF
Female	F1	91 ± 4	657.81 ± 29.04	20.55 ± 12.08	0.15 ± 0.14	0.14 ± 0.19	1.53 ± 1.10
	F2	74 ± 3	801.60 ± 31.27	32.45 ± 14.84	0.34 ± 0.46	0.80 ± 1.29	0.87 ± 0.69
	F3	76 ± 5	789.54 ± 47.22	31.62 ± 6.36	0.49 ± 0.25	0.87 ± 0.84	1.08 ± 1.02
	F4	72 ± 2	825.46 ± 50.06	11.28 ± 5.27	0.12 ± 0.19	0.39 ± 0.77	2.45 ± 2.27
	F5	73 ± 2	812.06 ± 22.03	21.37 ± 6.25	0.23 ± 0.21	0.13 ± 0.06	1.54 ± 0.86
Male	M1	95 ± 11	634.24 ± 79.26	40.95 ± 11.75	0.81 ± 0.03	0.76 ± 0.10	1.07 ± 0.10
	M2	89 ± 6	671.08 ± 46.39	29.97 ± 7.17	0.48 ± 0.40	0.45 ± 0.62	2.00 ± 2.06
	M3	85 ± 12	713.33 ± 96.59	54.68 ± 8.82	1.65 ± 0.50	4.22 ± 0.52	0.40 ± 0.06
	M4	74 ± 8	818.26 ± 83.85	56.20 ± 11.54	1.19 ± 1.31	2.88 ± 1.67	0.70 ± 0.52
	M5	77 ± 1	776.95 ± 12.60	12.21 ± 0.43	0.04 ± 0.00	0.01 ± 0.01	6.87 ± 0.71
	M6	76 ± 11	794.86 ± 109.40	60.98 ± 11.72	2.21 ± 0.20	3.09 ± 1.41	0.61 ± 0.47
	M7	72 ± 9	845.21 ± 96.32	52.28 ± 6.72	1.36 ± 1.01	2.21 ± 1.78	0.97 ± 0.57
	M8	78 ± 6	771.91 ± 65.22	48.90 ± 6.30	0.73 ± 0.47	1.24 ± 0.12	0.57 ± 0.33
	M9	66 ± 5	907.13 ± 67.69	51.97 ± 1.98	0.69 ± 0.52	0.82 ± 0.12	0.86 ± 0.55
	M10	94 ± 5	634.01 ± 37.45	15.24 ± 8.49	0.43 ± 0.16	0.43 ± 0.95	1.60 ± 1.20
	M11	76 ± 4	790.55 ± 38.42	32.37 ± 4.34	0.24 ± 0.17	0.16 ± 0.03	1.54 ± 0.86
	M12	77 ± 8	780.67 ± 70.27	37.73 ± 5.91	0.05 ± 0.01	0.18 ± 0.11	0.28 ± 0.04
	M13	68 ± 5	878.43 ± 57.68	27.85 ± 4.82	0.41 ± 0.30	0.49 ± 0.22	1.02 ± 0.70
	M14	75 ± 16	819.52 ± 119.91	34.69 ± 25.08	0.14 ± 0.03	0.41 ± 0.03	0.34 ± 0.05
	M15	78 ± 9	766.77 ± 57.90	19.78 ± 8.59	0.12 ± 0.08	0.06 ± 0.03	2.23 ± 1.29
	M16	74 ± 2	803.29 ± 21.00	9.73 ± 4.38	0.16 ± 0.15	0.04 ± 0.02	2.66 ± 2.75
	M17	79 ± 5	759.18 ± 46.30	27.79 ± 9.41	0.35 ± 0.19	0.43 ± 0.27	1.02 ± 0.66
	M18	70 ± 3	846.83 ± 42.25	28.16 ± 9.40	0.23 ± 0.16	1.12 ± 2.59	1.00 ± 0.75
	M19	91 ± 10	659.79 ± 81.36	39.05 ± 10.42	0.80 ± 0.56	0.74 ± 0.64	1.42 ± 1.32
	M20	93 ± 6	648.30 ± 44.44	26.47 ± 9.58	0.31 ± 0.17	0.63 ± 0.38	0.72 ± 0.59
	M21	70 ± 6	863.06 ± 64.85	73.93 ± 7.46	0.47 ± 0.50	1.51 ± 1.38	0.30 ± 0.12
	M22	75 ± 7	804.52 ± 69.67	36.94 ± 3.44	0.74 ± 0.97	1.60 ± 2.05	0.41 ± 0.12

The criteria used to classify the worker state (stressed, drowsy, or normal) were built based on previous studies and norms, as shown in Table 2.

Table 2. Worker state classification criteria based on prior research.

State	HRV		EEG
Stressed	RMSSD ≤ 27 [79]	1.7 ≤ LF/HF ≤ 3 [79]	30% increase in beta and gamma [33,38–40]
Drowsy	RMSSD ≥ 50 [80]	1.5 ≤ LF/HF * ≤ 3 [80]	30% increase in theta and alpha [41–43]

* The LF/HF condition in drowsiness will be activated if the HR is less than 65 (BPM).

6.2. System Visualization

Power BI software (Power BI 2.140.1577) was used to develop the supervisor dashboard for HRV, including time-domain measures such as the RMSSD and frequency-domain measures such as the LF, HF, and LF/HF. Additionally, the dashboard displays the EEG power spectral density (PSD) across all bands, as well as the EEG signal in the time domain. Other parameters, such as SpO_2 levels, were integrated and also shown on the dashboard, as illustrated in Figure 19.

6.3. Discussion

Wearable headbands and armbands make up the current system design. The headband weighs 109.6 g, and the armband weighs 79 g; this is effective, but could benefit from some ergonomic improvements to guarantee comfort throughout extended use, as confirmed by feedback from the user. For wearable technology, factors such as weight, material flexibility, and adaptability are crucial because prolonged use may cause discomfort. Accurate data collection depends on the quality of biosignal sensors, yet issues with signal dependability and noise interference remain. Stress states and HRV measurements (RMSSD and LF/HF ratio) do not typically exhibit consistent correlations, although stress generally reduces the RMSSD and raises the LF/HF. Accurately interpreting physiological data, which can differ depending on age, gender, and medical conditions, is essential to the system's functionality. The system's capacity to remember individual baselines, anticipate the start of stress or drowsiness, and examine complicated multimodal data for subtle patterns may be improved by incorporating machine learning (ML) methods.

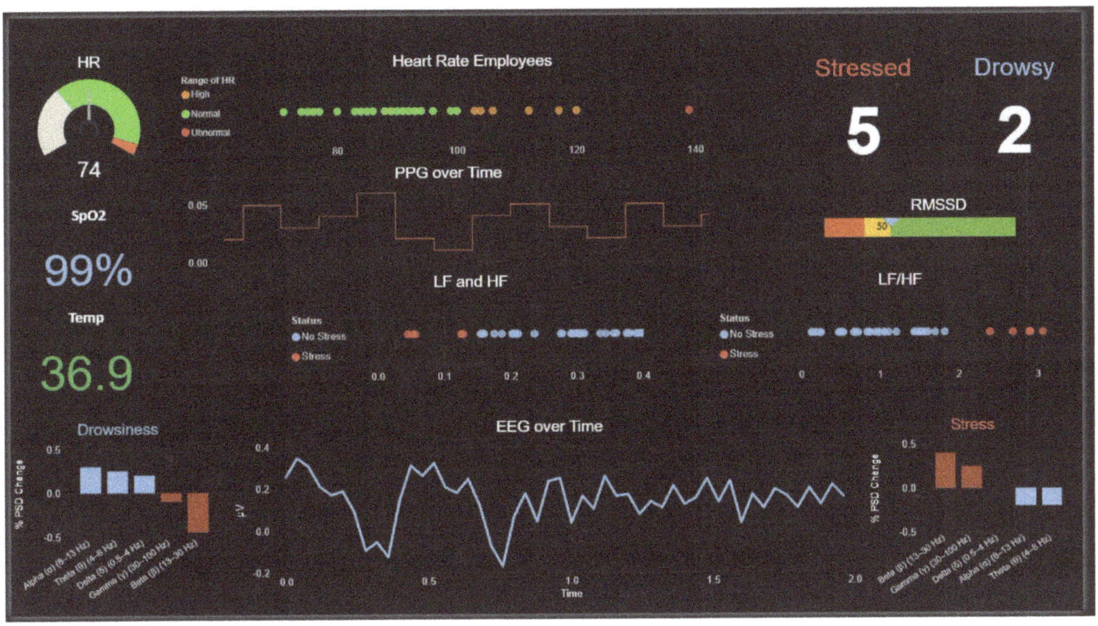

Figure 19. The supervisor's HMBI (BI dashboard).

7. Conclusions

Stress and drowsiness at work are serious issues that have a negative impact on worker safety, productivity, and well-being. This study developed a complete prototype system for monitoring stress and drowsiness. Advanced signal processing, wireless connection, and warning mechanisms are all incorporated into the system to alert supervisors and users to critical physical and mental situations. Through changes in temperature, SpO_2 levels, HRV measures, and EEG brainwave patterns, the design's adaptability, which includes headband and armband combinations, ensures comfort and usefulness in a range of work environments. This technology offers a non-invasive way to lower workplace accidents, increase productivity, and boost employee well-being by incorporating neuroergonomics concepts. To further evaluate the system's effectiveness, future research should concentrate on improving the design, including machine learning for predictive analytics, and using the system in various industrial and health care settings. Additionally, advanced sensors with higher accuracy and better noise-cancelation capabilities can be integrated. Furthermore, machine learning models that can learn and adapt to individual baseline physiological metrics can be developed and incorporated. Adaptive thresholding techniques that personalize state detection parameters can be utilized. In addition, efforts should be made to redesign wearable components (headbands and armbands) to make them more ergonomic, lightweight, and comfortable for extended use. Flexible materials and adjustable fittings should be used. This system can be adapted to critical work environments, such as operating rooms in health care facilities, where human errors can have severe consequences.

Supplementary Materials: The following supporting information can be downloaded at https://www.mdpi.com/article/10.3390/pr13041074/s1.

Funding: This research was funded by Taif University, Saudi Arabia, Project No. TU-DSPP-2024-240.

Institutional Review Board Statement: The study was conducted in accordance with the guidelines of the Ethics Committee of Taif University for studies involving humans, from which it received approval (HAO-02-T-105; 12 July 2021).

Informed Consent Statement: Informed consent was obtained from all the subjects involved in the study.

Data Availability Statement: The data are contained within the article.

Acknowledgments: The authors extend their appreciation to Taif University, Saudi Arabia, for supporting this work through project number TU-DSPP-2024-240.

Conflicts of Interest: The author declares no conflicts of interest.

References

1. Bickford, M. *Stress in the Workplace: A General Overview of the Causes, the Effects, and the Solution*; Canadian Mental Health Association Newfoundland and Labrador Division: St. John's, NL, Canada, 2005; Volume 8, pp. 1–3.
2. Wu, X.; Li, Y.; Yao, Y.; Luo, X.; He, X.; Yin, W. Development of construction workers job stress scale to study and the relationship between job stress and safety behavior: An empirical study in Beijing. *Int. J. Environ. Res. Public Health* **2018**, *15*, 2409. [CrossRef] [PubMed]
3. Leung, M.-Y.; Liang, Q.; Olomolaiye, P. Impact of job stressors and stress on the safety behavior and accidents of construction workers. *J. Manag. Eng.* **2016**, *32*, 04015019.
4. Barkhordari, A.; Malmir, B.; Malakoutikhah, M. An analysis of individual and social factors affecting occupational accidents. *Saf. Health Work* **2019**, *10*, 205–212.
5. Day, A.J.; Brasher, K.; Bridger, R.S. Accident proneness revisited: The role of psychological stress and cognitive failure. *Accid. Anal. Prev.* **2012**, *49*, 532–535.
6. Cowan, R.; Sanditov, B.; Weehuizen, R. Productivity effects of innovation, stress and social relations. *J. Econ. Behav. Organ.* **2011**, *79*, 165–182.
7. Ekienabor, E. Impact of job stress on employees' productivity and commitment. *Int. J. Res. Bus. Manag. Account.* **2016**, *2*, 124–133.
8. Mark, G.; Smith, A.P. Effects of occupational stress, job characteristics, coping, and attributional style on the mental health and job satisfaction of university employees. *Anxiety Stress Coping* **2012**, *25*, 63–78.
9. Fernandez, R. Help your team manage stress, anxiety, and burnout. *Harv. Bus. Rev.* **2016**, *53*.
10. Park, Y.-M.; Kim, S.Y. Impacts of job stress and cognitive failure on patient safety incidents among hospital nurses. *Saf. Health Work* **2013**, *4*, 210–215.
11. Lee, W.Y. The interacting effects of cognitive failure, consciousness and job stress on safety behavior and accidents. *Korean J. Ind. Organ. Psychol.* **2006**, *19*, 475–497.
12. Mastoras, G.; Poulin, C.; Norman, L.; Weitzman, B.; Pozgay, A.; Frank, J.R.; Posner, G. Stress Testing the Resuscitation Room: Latent Threats to Patient Safety Identified During Interprofessional In Situ Simulation in a Canadian Academic Emergency Department. *AEM Educ. Train.* **2020**, *4*, 254–261.
13. Chen, Q.; Liang, M.; Li, Y.; Guo, J.; Fei, D.; Wang, L.; He, L.; Sheng, C.; Cai, Y.; Li, X. Mental health care for medical staff in China during the COVID-19 outbreak. *Lancet Psychiatry* **2020**, *7*, e15–e16. [PubMed]
14. Adams, J.G.; Walls, R.M. Supporting the health care workforce during the COVID-19 global epidemic. *JAMA* **2020**, *323*, 1439–1440.

15. Xiao, H.; Zhang, Y.; Kong, D.; Li, S.; Yang, N. The effects of social support on sleep quality of medical staff treating patients with coronavirus disease 2019 (COVID-19) in January and February 2020 in China. *Med. Sci. Monit. Int. Med. J. Exp. Clin. Res.* **2020**, *26*, e923549-1.
16. Park, S.-C.; Park, Y.C. Mental health care measures in response to the 2019 novel coronavirus outbreak in Korea. *Psychiatry Investig.* **2020**, *17*, 85. [PubMed]
17. Lai, J.; Ma, S.; Wang, Y.; Cai, Z.; Hu, J.; Wei, N.; Wu, J.; Du, H.; Chen, T.; Li, R. Factors associated with mental health outcomes among health care workers exposed to coronavirus disease 2019. *JAMA Netw. Open* **2020**, *3*, e203976.
18. Tsou, M.-T.; Chang, B.C.-C. Association of depression and excessive daytime sleepiness among sleep-deprived college freshmen in northern Taiwan. *Int. J. Environ. Res. Public Health* **2019**, *16*, 3148. [CrossRef] [PubMed]
19. Ramey, S.; MacQuarrie, A.; Cochrane, A.; McCann, I.; Johnston, C.W.; Batt, A.M. Drowsy and dangerous? Fatigue in paramedics: An overview. *Irish Journal of Paramedicine.* **2019**, *4*, 1–9. [CrossRef]
20. Åkerstedt, T.; Knutsson, A.; Westerholm, P.; Theorell, T.; Alfredsson, L.; Kecklund, G. Work organisation and unintentional sleep: Results from the WOLF study. *Occup. Environ. Med.* **2002**, *59*, 595–600.
21. Åkerstedt, T.; Fredlund, P.; Gillberg, M.; Jansson, B. A prospective study of fatal occupational accidents–relationship to sleeping difficulties and occupational factors. *J. Sleep Res.* **2002**, *11*, 69–71.
22. Melamed, S.; Oksenberg, A. Excessive daytime sleepiness and risk of occupational injuries in non-shift daytime workers. *Sleep* **2002**, *25*, 315–322. [CrossRef] [PubMed]
23. Parasuraman, R.; Rizzo, M. *Neuroergonomics: The Brain at Work*; Oxford University Press: Oxford, UK, 2008; Volume 3.
24. Karwowski, W. The discipline of ergonomics and human factors. In *Handbook of Human Factors and Ergonomics*; Wiley: Hoboken, NJ, USA, 2006; Volume 3.
25. Karwowski, W. *International Encyclopedia of Ergonomics and Human Factors*; CRC Press: Boca Raton, FL, USA, 2006; Volume 3.
26. Karwowski, W. *Handbook of Standards and Guidelines in Ergonomics and Human Factors*; CRC Press: Boca Raton, FL, USA, 2005.
27. Mehta, R.K.; Parasuraman, R. Neuroergonomics: A review of applications to physical and cognitive work. *Front. Hum. Neurosci.* **2013**, *7*, 889.
28. Rahman, M.; Karwowski, W.; Fafrowicz, M.; Hancock, P.A. Neuroergonomics applications of electroencephalography in physical activities: A systematic review. *Front. Hum. Neurosci.* **2019**, *13*, 182. [CrossRef] [PubMed]
29. Momin, B.; Kalas, M.S. Study and implementation of advanced neuroergonomic techniques. *Adv. Comput.* **2012**, *3*, 9. [CrossRef]
30. Karwowski, W.; Siemionow, W.; Gielo-Perczak, K. Physical neuroergonomics: The human brain in control of physical work activities. *Theor. Issues Ergon. Sci.* **2003**, *4*, 175–199.
31. Aljuaid, A.M. Theoretical design of EEG-based Neuroergonomics integrated portable system, applying direct Psychophysiological indicators. In Proceedings of the 2019 Industrial & Systems Engineering Conference (ISEC), Jeddah, Saudi Arabia, 19–20 January 2019.
32. Mai, J.K.; Paxinos, G. *The Human Nervous System*; Academic Press: Cambridge, MA, USA, 2011.
33. Seo, S.-H.; Lee, J.-T. Stress and EEG. *Converg. Hybrid Inf. Technol.* **2010**, *1*, 413–424.
34. Kreibig, S.D. Autonomic nervous system activity in emotion: A review. *Biol. Psychol.* **2010**, *84*, 394–421.
35. Dinges, D.F. An overview of sleepiness and accidents. *J. Sleep Res.* **1995**, *4*, 4–14.
36. Niedermeyer, E.; da Silva, F.H.L. *Electroencephalography: Basic Principles, Clinical Applications, and Related Fields*; Wolters Kluwer Health: Philadelphia, PA, USA, 2005.
37. Khalifa, W.; Salem, A.; Roushdy, M.; Revett, K. A survey of EEG based user authentication schemes. In Proceedings of the 2012 8th International Conference on Informatics and Systems (INFOS), Giza, Egypt, 14–16 May 2012.
38. Al-Shargie, F.; Kiguchi, M.; Badruddin, N.; Dass, S.C.; Hani, A.F.M.; Tang, T.B. Mental stress assessment using simultaneous measurement of EEG and fNIRS. *Biomed. Opt. Express* **2016**, *7*, 3882–3898.
39. Saeed, S.M.U.; Anwar, S.M.; Majid, M.; Bhatti, A.M. Psychological stress measurement using low cost single channel EEG headset. In Proceedings of the 2015 IEEE International Symposium on Signal Processing and Information Technology (ISSPIT), Abu Dhabi, United Arab Emirates, 7–10 December 2015.
40. Saeed, S.M.U.; Anwar, S.M.; Majid, M. Quantification of human stress using commercially available single channel eeg headset. *IEICE Trans. Inf. Syst.* **2017**, *100*, 2241–2244.
41. Zanetti, M.; Faes, L.; Nollo, G.; De Cecco, M.; Pernice, R.; Maule, L.; Pertile, M.; Fornaser, A. Information dynamics of the brain, cardiovascular and respiratory network during different levels of mental stress. *Entropy* **2019**, *21*, 275. [CrossRef]
42. Hafeez, M.A.; Shakil, S.; Jangsher, S. Stress effects on exam performance using EEG. In Proceedings of the 2018 14th International Conference on Emerging Technologies (ICET), Islamabad, Pakistan, 21–22 November 2018.

43. Sulaiman, N.; Ying, B.S.; Mustafa, M.; Jadin, M.S. Offline LabView-based EEG Signals Analysis for Human Stress Monitoring. In Proceedings of the 2018 9th IEEE Control and System Graduate Research Colloquium (ICSGRC), Shah Alam, Malaysia, 3–4 August 2018.
44. Minguillon, J.; Perez, E.; Lopez-Gordo, M.A.; Pelayo, F.; Sanchez-Carrion, M.J. Portable system for real-time detection of stress level. *Sensors* **2018**, *18*, 2504. [CrossRef]
45. Lee, J.H.; Gamper, H.; Tashev, I.; Dong, S.; Ma, S.; Remaley, J.; Holbery, J.D.; Yoon, S.H. Stress Monitoring using Multimodal Bio-sensing Headset. In Proceedings of the Extended Abstracts of the 2020 CHI Conference on Human Factors in Computing Systems Extended Abstracts, Honolulu HI USA, 25–30 April 2020. [CrossRef]
46. Putman, P.; Verkuil, B.; Arias-Garcia, E.; Pantazi, I.; van Schie, C. EEG theta/beta ratio as a potential biomarker for attentional control and resilience against deleterious effects of stress on attention. *Cogn. Affect. Behav. Neurosci.* **2014**, *14*, 782–791. [PubMed]
47. Borghini, G.; Astolfi, L.; Vecchiato, G.; Mattia, D.; Babiloni, F. Measuring neurophysiological signals in aircraft pilots and car drivers for the assessment of mental workload, fatigue and drowsiness. *Neurosci. Biobehav. Rev.* **2014**, *44*, 58–75. [PubMed]
48. Ohsuga, M.; Kamakura, Y.; Inoue, Y.; Noguchi, Y.; Shimada, K.; Mishiro, M. Estimation of driver's arousal state using multi-dimensional physiological indices. In Proceedings of the International Conference on Engineering Psychology and Cognitive Ergonomics, Orlando, FL, USA, 9–14 July 2011; Springer: Berlin/Heidelberg, Germany, 2011.
49. Pal, N.R.; Chuang, C.-Y.; Ko, L.-W.; Chao, C.-F.; Jung, T.-P.; Liang, S.-F.; Lin, C.-T. EEG-based subject-and session-independent drowsiness detection: An unsupervised approach. *EURASIP J. Adv. Signal Process.* **2008**, *2008*, 519480.
50. Correa, A.G.; Orosco, L.; Laciar, E. Automatic detection of drowsiness in EEG records based on multimodal analysis. *Med. Eng. Phys.* **2014**, *36*, 244–249.
51. Kaida, K.; Takahashi, M.; Åkerstedt, T.; Nakata, A.; Otsuka, Y.; Haratani, T.; Fukasawa, K. Validation of the Karolinska sleepiness scale against performance and EEG variables. *Clin. Neurophysiol.* **2006**, *117*, 1574–1581.
52. Ogino, M.; Mitsukura, Y. Portable drowsiness detection through use of a prefrontal single-channel electroencephalogram. *Sensors* **2018**, *18*, 4477. [CrossRef]
53. Nguyen, T.; Ahn, S.; Jang, H.; Jun, S.C.; Kim, J.G. Utilization of a combined EEG/NIRS system to predict driver drowsiness. *Sci. Rep.* **2017**, *7*, 43933.
54. Akselrod, S.; Gordon, D.; Ubel, F.A.; Shannon, D.C.; Berger, A.; Cohen, R.J. Power spectrum analysis of heart rate fluctuation: A quantitative probe of beat-to-beat cardiovascular control. *Science* **1981**, *213*, 220–222.
55. Taelman, J.; Vandeput, S.; Spaepen, A.; Van Huffel, S. Influence of mental stress on heart rate and heart rate variability. In Proceedings of the 4th European Conference of the International Federation for Medical and Biological Engineering, Antwerp, Belgium, 23–27 November 2008; Springer: Berlin/Heidelberg, Germany, 2009.
56. Pumprla, J.; Howorka, K.; Groves, D.; Chester, M.; Nolan, J. Functional assessment of heart rate variability: Physiological basis and practical applications. *Int. J. Cardiol.* **2002**, *84*, 1–14. [CrossRef] [PubMed]
57. Malik, M.; Bigger, J.T.; Camm, A.J.; Kleiger, R.E.; Malliani, A.; Moss, A.J.; Schwartz, P.J. Heart rate variability: Standards of measurement, physiological interpretation, and clinical use. *Eur. Heart J.* **1996**, *17*, 354–381. [CrossRef]
58. Kim, D.; Seo, Y.; Salahuddin, L. Decreased long term variations of heart rate variability in subjects with higher self reporting stress scores. In Proceedings of the 2008 Second International Conference on Pervasive Computing Technologies for Healthcare, Tampere, Finland, 30 January–1 February 2008.
59. Sun, F.-T.; Kuo, C.; Cheng, H.-T.; Buthpitiya, S.; Collins, P.; Griss, M. Activity-aware mental stress detection using physiological sensors. In *Mobile Computing, Applications, and Services, Proceedings of the Second International ICST Conference, MobiCASE 2010, Santa Clara, CA, USA, 25–28 October 2010*; Springer: Berlin/Heidelberg, Germany, 2010.
60. Kim, H.-G.; Cheon, E.-J.; Bai, D.-S.; Lee, Y.H.; Koo, B.-H. Stress and heart rate variability: A meta-analysis and review of the literature. *Psychiatry Investig.* **2018**, *15*, 235. [CrossRef] [PubMed]
61. Li, Z.; Snieder, H.; Su, S.; Ding, X.; Thayer, J.F.; Treiber, F.A.; Wang, X. A longitudinal study in youth of heart rate variability at rest and in response to stress. *Int. J. Psychophysiol.* **2009**, *73*, 212–217. [CrossRef]
62. Taelman, J.; Vandeput, S.; Vlemincx, E.; Spaepen, A.; Van Huffel, S. Instantaneous changes in heart rate regulation due to mental load in simulated office work. *Eur. J. Appl. Physiol.* **2011**, *111*, 1497–1505. [CrossRef]
63. Schubert, C.; Lambertz, M.; Nelesen, R.; Bardwell, W.; Choi, J.-B.; Dimsdale, J. Effects of stress on heart rate complexity—A comparison between short-term and chronic stress. *Biol. Psychol.* **2009**, *80*, 325–332. [CrossRef] [PubMed]
64. Hjortskov, N.; Rissén, D.; Blangsted, A.K.; Fallentin, N.; Lundberg, U.; Søgaard, K. The effect of mental stress on heart rate variability and e during computer work. *Eur. J. Appl. Physiol.* **2004**, *92*, 84–89. [CrossRef]
65. Papousek, I.; Nauschnegg, K.; Paechter, M.; Lackner, H.K.; Goswami, N.; Schulter, G. Trait and state positive affect and cardiovascular recovery from experimental academic stress. *Biol. Psychol.* **2010**, *83*, 108–115. [CrossRef]
66. Traina, M.; Cataldo, A.; Galullo, F.; Russo, G. Heart rate variability in healthy subjects. *Minerva Psichiatr* **2011**, *227*, 31.

67. Castaldo, R.; Melillo, P.; Bracale, U.; Caserta, M.; Triassi, M.; Pecchia, L. Acute mental stress assessment via short term HRV analysis in healthy adults: A systematic review with meta-analysis. *Biomed. Signal Process. Control* **2015**, *18*, 370–377. [CrossRef]
68. Rodriguez-Ibañez, N.; García-Gonzalez, M.A.; de la Cruz, M.A.F.; Fernández-Chimeno, M.; Ramos-Castro, J. Changes in heart rate variability indexes due to drowsiness in professional drivers measured in a real environment. In Proceedings of the 2012 Computing in Cardiology, Krakow, Poland, 9–12 September 2012.
69. Hu, S.; Bowlds, R.L.; Gu, Y.; Yu, X. Pulse wave sensor for non-intrusive driver's drowsiness detection. In Proceedings of the 2009 Annual International Conference of the IEEE Engineering in Medicine and Biology Society, Minneapolis, MN, USA, 3–6 September 2009.
70. Awais, M.; Badruddin, N.; Drieberg, M. A non-invasive approach to detect drowsiness in a monotonous driving environment. In Proceedings of the TENCON 2014-2014 IEEE Region 10 Conference, Bangkok, Thailand, 22–25 October 2014.
71. Furman, G.D.; Baharav, A.; Cahan, C.; Akselrod, S. Early detection of falling asleep at the wheel: A heart rate variability approach. In Proceedings of the 2008 Computers in Cardiology, Bologna, Italy, 14–17 September 2008.
72. Murata, A.; Hiramatsu, Y. Evaluation of drowsiness by HRV measures-basic study for drowsy driver detection. In Proceedings of the Fourth International Workshop on Computational Intelligence & Applications, Shanghai, China, 15–18 September 2008; IEEE SMC Hiroshima Chapter.
73. Mahachandra, M.; Sutalaksana, I.Z.; Suryadi, K. Sensitivity of heart rate variability as indicator of driver sleepiness. In Proceedings of the 2012 Southeast Asian Network of Ergonomics Societies Conference (SEANES), Langkawi, Malaysia, 9–12 July 2012.
74. Zhuo, K.; Gao, C.; Wang, X.; Zhang, C.; Wang, Z. Stress and sleep: A survey based on wearable sleep trackers among medical and nursing staff in Wuhan during the COVID-19 pandemic. *Gen. Psychiatry* **2020**, *33*, e100260.
75. Shabani, H.; Mikaili, M.; Noori, S.M.R. Assessment of recurrence quantification analysis (RQA) of EEG for development of a novel drowsiness detection system. *Biomed. Eng. Lett.* **2016**, *6*, 196–204.
76. Kistler, A.; Mariauzouls, C.; von Berlepsch, K. Fingertip temperature as an indicator for sympathetic responses. *Int. J. Psychophysiol.* **1998**, *29*, 35–41.
77. Ohsuga, M.; Shimono, F.; Genno, H. Assessment of phasic work stress using autonomic indices. *Int. J. Psychophysiol.* **2001**, *40*, 211–220. [PubMed]
78. Bando, S.; Oiwa, K.; Nozawa, A. Evaluation of dynamics of forehead skin temperature under induced drowsiness. *IEEJ Trans. Electr. Electron. Eng.* **2017**, *12*, S104–S109.
79. Zubair, M.; Yoon, C. Multilevel mental stress detection using ultra-short pulse rate variability series. *Biomed. Signal Process. Control* **2020**, *57*, 101736.
80. Bušek, P.; Vaňková, J.; Opavský, J.; Salinger, J.; Nevšímalová, S. Spectral analysis of heart rate variability in sleep. *Physiol. Res.* **2005**, *54*, 369–376.

processes

MDPI

Article

MAL-XSEL: Enhancing Industrial Web Malware Detection with an Explainable Stacking Ensemble Model

Ezz El-Din Hemdan [1,2], Samah Alshathri [3,*], Haitham Elwahsh [4,5], Osama A. Ghoneim [6] and Amged Sayed [7,8,*]

1 Structure and Materials Research Lab, Prince Sultan University, Riyadh 11586, Saudi Arabia; ezzeldinhemdan@el-eng.menofia.edu.eg
2 Department of Computer Science and Engineering, Faculty of Electronic Engineering, Menoufia University, Menouf 32952, Egypt
3 Department of Information Technology, College of Computer and Information Sciences, Princess Nourah bint Abdulrahman University, P.O. Box 84428, Riyadh 11671, Saudi Arabia
4 Faculty of Information Technology, Applied Science Private University, Amman 11931, Jordan; h_elwahsh@asu.edu.jo
5 Department of Computer Science, Faculty of Computers and Information, Kafrelsheikh University, Kafrelsheikh 33511, Egypt
6 Department of Computer Science, Faculty of Computers and Information, Tanta University, Tanta 31527, Egypt; osamaghoneim@ics.tanta.edu.eg
7 Department of Electrical Energy Engineering, College of Engineering & Technology, Arab Academy for Science Technology & Maritime Transport, Smart Village Campus, Giza 12577, Egypt
8 Industrial Electronics and Control Engineering Department, Faculty of Electronic Engineering, Menoufia University, Menouf 32952, Egypt
* Correspondence: sealshathry@pnu.edu.sa (S.A.); amged.sayed@aast.edu (A.S.)

Abstract

The escalating global incidence of malware presents critical cybersecurity threats to manufacturing, automation, and industrial process control systems. Given the fast-developing web applications and IoT devices in use by industry operations, securing a transparent and effective malware detection mechanism has become imperative to operational resilience and data integrity. Classical methods of malware detection are conventionally opaque "black boxes" with limited transparency, thus eroding trust and hindering deployment in security-sensitive contexts. In this respect, this research proposes MAL-XSEL—a malware detection framework using an explainable stacking ensemble learning approach for performing high-accuracy classification and interpretable decision-making. MAL-XSEL explicates the model predictions through Shapley additive explanations (SHAP) and local interpretable model-agnostic explanations (LIME), which enable security analysts to validate how the detection logic works and prioritize the features contributing to the most critical threats. Evaluated on two benchmark datasets, MAL-XSEL outperformed conventional machine learning models, achieving top accuracies of 99.62% (ClaMP dataset) and 99.16% (MalwareDataSet). Notably, it surpassed state-of-the-art algorithms such as LightGBM (99.52%), random forest (99.33%), and decision trees (98.89%) across both datasets while maintaining computational efficiency. A unique interaction of ensemble learning and XAI is employed for detection, not only with improved accuracy but also with interpretable insight into the behavior of malware, thereby allowing trust to be substantiated in an automated system. By closing the divide between performance and interpretability, MAL-XSEL enables cybersecurity practitioners to deploy transparent and auditable defenses against an ever-growing resource of threats. This work demonstrates how there can be no compromise on explainability in security-critical applications and, as such, establishes a roadmap for future research on industrial malware analysis tools.

other conventional machine learning models. Moreover, [22] proposed an efficient malware detection framework using traditional machine learning classifiers, including random forest, KNNs, and extra trees, applied to the UNSW-NB15 dataset. By employing feature encoding and an entropy feature selection technique, their approach achieved notable classification performance with random forest.

Despite these advancements, a clear research gap remains. It is observed that a substantial amount of research has been conducted on malware detection using machine learning. However, there are only a few contributions to the literature that explore the use of genetic algorithms in conjunction with current machine learning algorithms for effectively identifying malware in web applications over insecure networks (Table 1).

Table 1. Overview of significant research on learning-based malware detection.

Work	Focus of Research	Method	Description
[16]	Portable executable header features	Feature selection	Excluded non-contributory features to improve model performance
[17]	Detecting unknown malware	Frequency of opcode occurrence	Developed a framework using machine learning techniques
[18]	Cloud-based malware detection	Dataset with GAN-based augmentation	Dynamic and static feature-rich malware classifier using deep learning
[19]	Multimodal malware detection and robustness	CNN-based fusion of grayscale and adversarial defense	Proposed a fusion model leveraging multiple visual malware features
[20]	Enduring malware detection methods	Concept drift	Established that discussing concept drift enhances detection methods
[21]	Android malware classification	Optimizer ensemble learning with GAs	Used GAs to optimize RF parameters for maximum classifier accuracy
[22]	Network-based malware detection	Multiple ML algorithms with TFIDF feature selection	Used KNNs, RF, ET, DT, LR, and nnMLP

3. Proposed Methodology

In this study, we develop a new intelligent malware detection framework specifically designed for secure industrial web applications. Unlike most traditional methods that apply static features or even fall back to black-box models, this approach (MAL-XSEL) combines explainable artificial intelligence and a stacking-based ensemble learning strategy to produce higher accuracy and interpretability. The novelty of this framework consists of a combination of various machine learning models, feature-specific interpretability techniques, such as SHAP and LIME, and a modular workflow that considers the unique structure of portable executables (PEs), as illustrated in Figure 3. A PE file is the standard format for Windows desktop applications. This study targets malware detection by analyzing the PE header, which contains essential metadata and operational information needed by the system to execute the file. By focusing on this structure, the proposed approach aims to detect previously unseen malware that often escapes conventional antivirus systems.

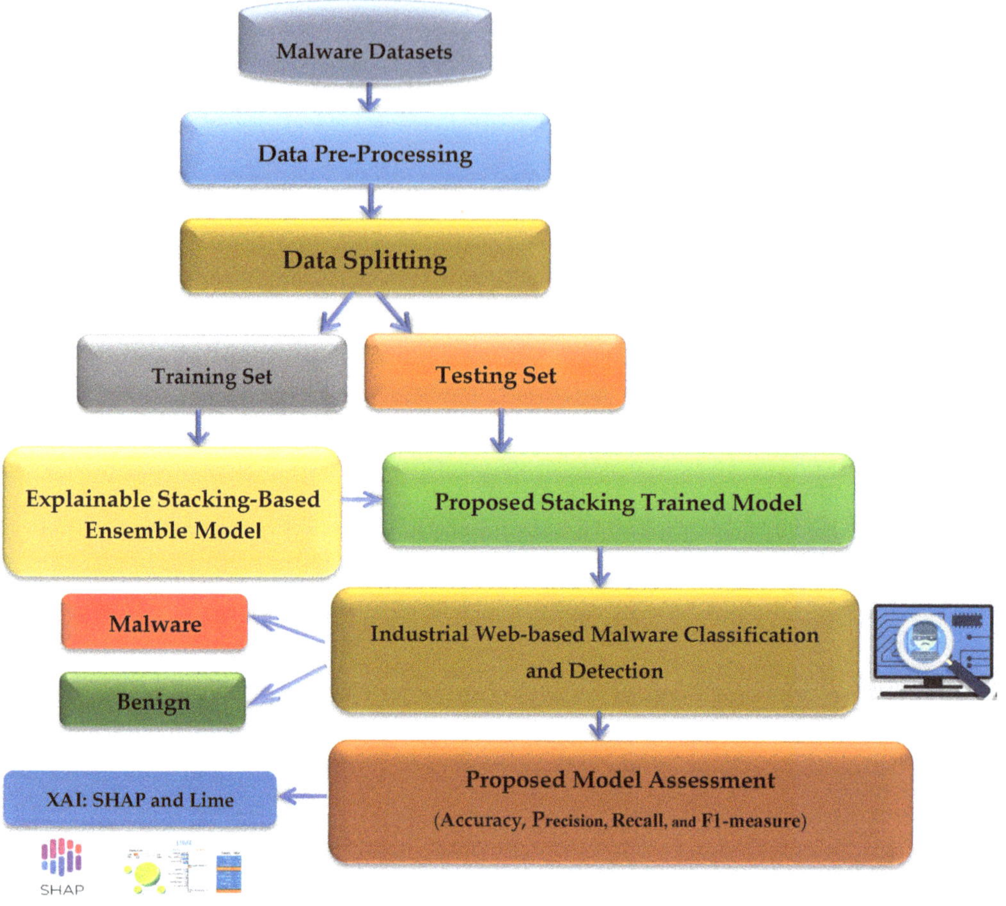

Figure 3. Explainable stacking-based ensemble model for detecting malware.

Furthermore, the proposed MAL-XSEL framework encompasses several key steps essential for the malware recognition process, which are outlined as follows:

1. **Data Collection and Preparation:** Web-based malware datasets are collected from various sources, ensuring a diverse and comprehensive dataset for training and evaluation. The collected data undergo preprocessing, including cleaning, handling missing values, and standardization, to ensure compatibility with the framework's processing requirements.
2. **Data Formatting:** The two selected malware datasets are structured and formatted to align with the requirements of the explainable stacking-based ensemble model. Proper formatting ensures data consistency and enhances feature extraction for improved classification accuracy.
3. **Dataset Partitioning:** The datasets are split into training and testing sets using an 80/20 ratio, where 80% of the data are used to train the model and 20% are reserved for evaluating its performance. This split ensures an effective balance between training the model and assessing its generalization ability.

4. **Model Training (Explainable Stacking-Based Ensemble Learning):** The core of the proposed approach involves training an ensemble model using a stacking-based strategy. This model integrates multiple machine learning classifiers, such as LightGBM, random forest (RF), CatBoost, decision tree (DT), AdaBoost, k-nearest neighbors (KNN), and linear discriminant analysis (LDA), to leverage their strengths and enhance malware detection accuracy. The training process optimally combines these classifiers to improve predictive performance.
5. **Testing and Classification:** Once trained, the model classifies web-based malware in real time using the 20% test data. The performance of individual base classifiers and the stacking ensemble is analyzed to determine their effectiveness in identifying malicious web-based threats.
6. **Performance Evaluation:** A comprehensive evaluation of the proposed system is conducted using classification-oriented metrics, including accuracy, precision, recall, F1-score, and AUC-ROC. The results are compared against conventional machine learning methods to validate the superiority of the stacking-based ensemble approach.
7. **Explainable AI (XAI) for Model Interpretability:** To enhance transparency and provide insights into the model's decision-making, SHAP is employed to offer both global and local interpretability, explaining each feature's contribution to predictions. Additionally, LIME assesses the impact of individual input features, even identifying minor factors influencing malware detection. This interpretability empowers cybersecurity professionals to better understand and trust the model's classification decisions.

This comprehensive pipeline enables accurate, explainable, and generalizable malware detection, setting MAL-XSEL apart from prior work focused on isolated models or non-interpretable detection strategies.

4. Experimental Study

This section provides an overview of the experimental setup, including details about the two malware datasets utilized. Additionally, it presents the analysis and discussion of the results obtained from the proposed framework, which focuses on classifying malware in executable files within web applications over unsecured systems and assessing the efficacy of the Python (3.11.12)-coded framework used in this study for malware detection.

4.1. Malware Datasets

The projected approach is instigated using numerous PE header datasets containing both malware and benign files sourced from multiple repositories. These datasets are as follows:

1. **Dataset 1 (Classification of Malware (CLaMP)):** This dataset was obtained from the Mendeley data repository. This dataset comprises 5210 file samples, with 2722 classified as malware and 2488 as benign, featuring a total of 69 attributes [23].
2. **Dataset 2 (MalwareDataSet):** This dataset was sourced from GitHub, containing 137,444 samples, of which 40,918 are malware and 96,526 are benign. This dataset includes nine extracted features [24].

4.2. Assessment Criteria

The evaluation of the suggested approach is based on precision, recall, and F-measures, which serve as performance metrics for the classifiers. These metrics are calculated for both positive ("P") and negative ("N") classifications across the test dataset. Table 2 illustrates a confusion matrix (CM), a fundamental tool for evaluating the classification model's performance. It involves four essential parameters: true positive (TP), true negative

(TN), false positive (FP), and false negative (FN). Understanding these parameters aids in comprehensively assessing the classification model's efficacy in detecting malware within web applications in an effective way.

Table 2. CM for malware detection.

	Predicted Benign	Predicted Malware
Actual Benign	*TP*	*FN*
Actual Malware	*FP*	*TN*

The following formulas may be used to compute assessment measures such as accuracy, precision, recall, and F1-score once the confusion matrix's parameters have been established [25,26]:

$$Accuracy = \frac{TP + TN}{TP + FP + FN + TN} \quad (1)$$

$$\text{Precision} = \frac{TP}{TP + FP} \quad (2)$$

$$\text{Recall} = \frac{TP}{TP + FN} \quad (3)$$

$$\text{F1} - \text{score} = 2 * \frac{Precision * Recall}{Precision + Recall} \quad (4)$$

4.3. Results Analysis

This section discusses the results of the proposed approach for detecting malware. Table 3 presents the performance evaluation of different machine learning models used in the MAL-XSEL system on CLaMP (Dataset 1), based on four key metrics: accuracy, precision, recall, and F1-score. The results indicate that the stacking ensemble model achieves the highest performance, with an accuracy of 0.9962, confirming the effectiveness of combining multiple base learners for improved malware detection. Among the individual models, LightGBM (0.9952 accuracy), CatBoost (0.9942 accuracy), and random forest (RF) (0.9933 accuracy) perform exceptionally well, showcasing the strength of tree-based ensemble techniques in malware classification. Decision tree (DT) (0.9789 accuracy) and AdaBoost (0.9683 accuracy) demonstrate moderate performance but are outperformed by more advanced boosting models such as LightGBM and CatBoost. On the other hand, k-nearest neighbors (KNN) (0.928 accuracy) and linear discriminant analysis (LDA) (0.7639 accuracy) achieve the lowest accuracy, suggesting that traditional models may struggle with complex malware detection patterns. Overall, the results confirm that the stacking-based approach significantly enhances classification accuracy, making MAL-XSEL a highly effective and interpretable solution for web-based malware detection.

Table 4 presents the performance evaluation of the MAL-XSEL system on Malware-DataSet (Dataset 2), utilizing accuracy, precision, recall, and F1-score as key performance metrics. The findings reveal that the stacking ensemble model achieves the highest accuracy (0.9916), reinforcing its effectiveness in improving malware detection. Among individual models, decision tree (DT) (0.9889 accuracy), LightGBM (0.9881 accuracy), and CatBoost (0.9866 accuracy) exhibit strong classification performance, highlighting their robustness in malware detection. Meanwhile, random forest (RF) (0.9758 accuracy) and AdaBoost (0.9694 accuracy) perform moderately well but fall short of the accuracy levels achieved by more advanced boosting techniques. K-nearest neighbors (KNN) (0.9844 accuracy) provides competitive results, though it remains slightly less effective than tree-based models. Linear discriminant analysis (LDA) (0.8548 accuracy) records the lowest accuracy, suggesting its limitations in handling complex malware classification tasks. Overall, these results confirm that the stacking-based ensemble model delivers the most accurate and balanced performance.

Table 3. Results of the proposed system on Dataset 1.

Model	Accuracy	Precision	Recall	F1-Score
LDA	0.7639	0.7706	0.766	0.7633
KNN	0.928	0.928	0.9279	0.928
DT	0.9789	0.9793	0.9786	0.9789
RF	0.9933	0.9935	0.9931	0.9933
AdaBoost	0.9683	0.9685	0.9682	0.9683
LightGBM	0.9952	0.9953	0.9951	0.9952
CatBoost	0.9942	0.9943	0.9942	0.9942
Stacking	**0.9962**	**0.9962**	**0.9961**	**0.9962**

Table 4. Results of the proposed system on Dataset 2.

Model	Accuracy	Precision	Recall	F1-Score
LDA	0.8547	0.8253	0.8388	0.8314
KNN	0.9843	0.9824	0.9804	0.9814
DT	0.9889	0.9870	0.9867	0.9868
RF	0.9758	0.9698	0.9730	0.9714
AdaBoost	0.9694	0.9686	0.9584	0.9633
LightGBM	0.9881	0.9857	0.9861	0.9859
CatBoost	0.9866	0.9844	0.9838	0.9841
Stacking	**0.9915**	**0.9897**	**0.9902**	**0.9899**

Table 5 presents the class-wise performance of the proposed MAL-XSEL system on Dataset 1, evaluating each model's precision, recall, and F1-score for Class 0 (benign files) and Class 1 (malware files). The stacking ensemble model achieves the highest performance across both classes, with a precision of 0.998 (Class 0) and 0.9944 (Class 1), recall of 0.9941 (Class 0) and 0.9981 (Class 1), and F1-scores of 0.996 (Class 0) and 0.9963 (Class 1). These results highlight the stacking model's superior ability to correctly classify both benign and malware files with minimal misclassification. Among individual classifiers, LightGBM, CatBoost, and random forest (RF) perform exceptionally well, with precision, recall, and F1-scores above 0.99 for both classes, indicating strong generalization and robustness. Decision tree (DT) and AdaBoost follow closely behind, achieving slightly lower but still competitive scores, particularly in recall and F1-score. On the other hand, k-nearest neighbors (KNN) and linear discriminant analysis (LDA) exhibit lower performance. LDA has the weakest performance, particularly in recall for Class 1 (0.6922), indicating that it struggles to correctly detect malware samples. KNN performs better but is still outperformed by tree-based models, reinforcing the effectiveness of ensemble learning for malware classification.

Generally, these results confirm that the stacking-based approach significantly enhances malware detection accuracy for both classes, reducing the risk of false positives (incorrectly classifying benign files as malware) and false negatives (failing to detect actual malware). This further solidifies MAL-XSEL as a highly reliable and interpretable solution for web malware detection.

Table 5. Results of the proposed system on Dataset 1 for Class 0 and Class 1.

Model	Precision		Recall		F1-Score	
	Class 0	Class 1	Class 0	Class 1	Class 0	Class 1
LDA	0.7203	0.8208	0.8399	0.6922	0.7755	0.751
KNN	0.9267	0.9292	0.9249	0.931	0.9258	0.9301
DT	0.9879	0.9707	0.9684	0.9888	0.978	0.9797
RF	0.998	0.9889	0.9881	0.9981	0.993	0.9935
AdaBoost	0.9721	0.9649	0.9625	0.9739	0.9672	0.9694
LightGBM	0.998	0.9926	0.9921	0.9981	0.995	0.9953
CatBoost	0.996	0.9926	0.9921	0.9963	0.9941	0.9944
Stacking	**0.998**	**0.9944**	**0.9941**	**0.9981**	**0.996**	**0.9963**

Table 6 provides a class-wise evaluation of the MAL-XSEL system on Dataset 2, analyzing the precision, recall, and F1-score of each model for Class 0 (benign files) and Class 1 (malware files). The stacking ensemble model demonstrates the highest overall performance, achieving a precision of 0.9943 (Class 0) and 0.9852 (Class 1), recall of 0.99358 (Class 0) and 0.9868 (Class 1), and F1-scores of 0.9939 (Class 0) and 0.9860 (Class 1). These results highlight the stacking model's superior capability in accurately distinguishing between benign and malware samples, reinforcing its reliability as a robust classification approach. Among individual classifiers, LightGBM, CatBoost, and random forest (RF) deliver outstanding results, with precision, recall, and F1-scores exceeding 0.98 for both classes, demonstrating strong generalization across different malware variations. Decision tree (DT) and AdaBoost also perform well, though AdaBoost exhibits a slightly lower recall for Class 1 (0.9306), suggesting a tendency to misclassify some malware samples. In contrast, k-nearest neighbors (KNN) and linear discriminant analysis (LDA) show comparatively weaker performance. LDA, in particular, records the lowest scores, particularly for Class 1 (precision: 0.7407, recall: 0.79874, F1-score: 0.7686), indicating its challenges in accurately classifying malware. While KNN outperforms LDA, it remains less effective than ensemble-based models, reaffirming the advantage of stacking and boosting techniques for malware detection. Overall, these findings confirm that the stacking-based ensemble model significantly enhances malware detection for both benign and malicious files, reducing misclassification risks. The results further establish MAL-XSEL as a powerful and interpretable cybersecurity solution, offering enhanced security for web applications against evolving malware threats.

Table 6. Results of the proposed system on Dataset 2 for Class 0 and Class 1.

Model	Precision		Recall		F1-Score	
	Class 0	Class 1	Class 0	Class 1	Class 0	Class 1
LDA	0.9098	0.7407	0.87902	0.79874	0.8941	0.7686
KNN	0.9852	0.9545	0.98008	0.96603	0.9826	0.9602
DT	0.9872	0.9775	0.99035	0.97049	0.9888	0.9740
RF	0.9918	0.9821	0.99228	0.9812	0.9920	0.9816
AdaBoost	0.9704	0.9668	0.98618	0.9306	0.9782	0.9483
LightGBM	0.9917	0.9797	0.99124	0.9809	0.9915	0.9803
CatBoost	0.9899	0.9788	0.99087	0.9767	0.9904	0.9778
Stacking	**0.9943**	**0.9852**	**0.99358**	**0.9868**	**0.9939**	**0.9860**

Likewise, Figures 4 and 5 show the confusion matrices for Dataset 1 and Dataset 2, respectively. Figure 6 presents a comparison of the results for Dataset 1 and Dataset 2 using the proposed approach and other machine learning models. Stacking outperforms all other models on both datasets, achieving the highest accuracy (0.9915 on Dataset 1 and 0.9962 on Dataset 2), along with superior precision, recall, and F1-scores. This confirms that the stacking model is the most effective approach compared to LDA, KNN, DT, RF, AdaBoost, LightGBM, and CatBoost.

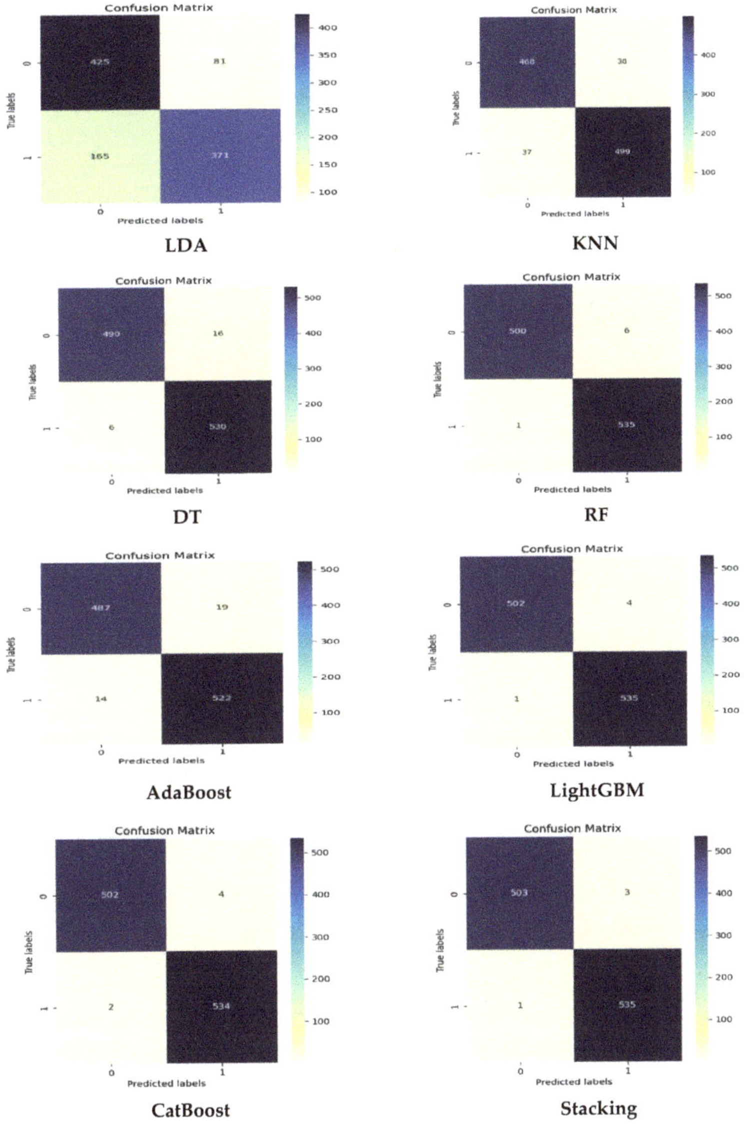

Figure 4. Confusion matrix for Dataset 1.

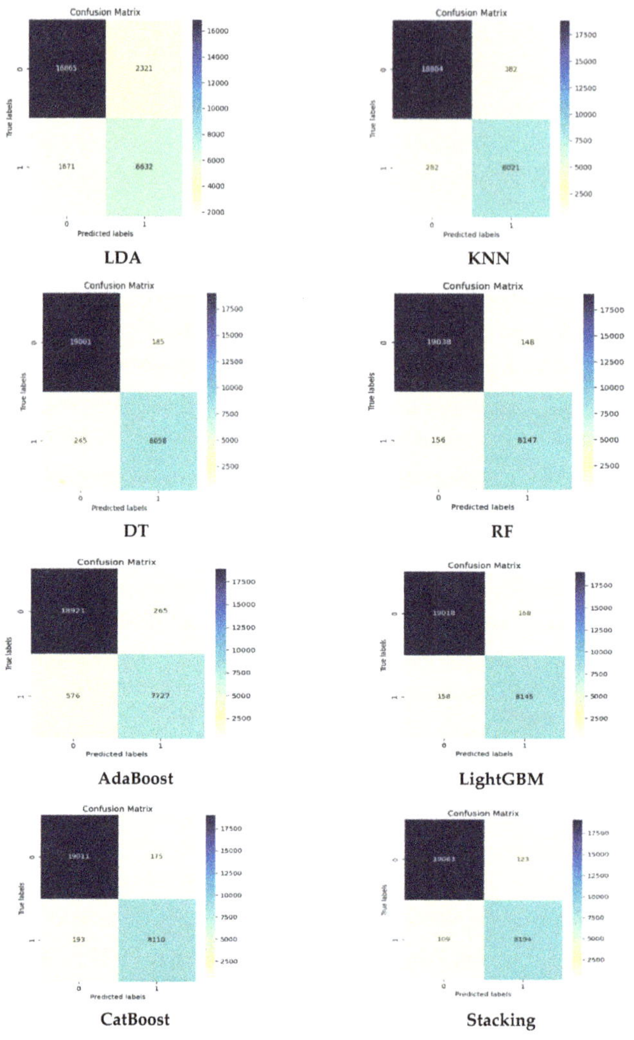

Figure 5. Confusion matrix for Dataset 2.

Interpretability enhances confidence in AI models by clarifying predictions and exposing model weaknesses. In this work, it is wanted to achieve a transparent analysis of the classification decisions by assessing individual features by means of XAI techniques—SHAP and LIME—as shown in Figures 7–9. These visual explanations help cybersecurity analysts better understand why a sample was classified as malware versus benign, allowing for better decision-making and thereby facilitating policy adjustments. For example, Figure 8 features SHAP summary plots that highlight specific features, for instance, Feature 3 or Feature 5, that greatly affect the classification decision in both datasets. In the same way, Figure 9 ranks feature importance based on average impact so that it guides the analysts on which behaviors or signals to watch most closely. Our framework enhances real-time applications by providing such insights (e.g., adaptive firewall rules, automatic alert thresholds, or even policy writing for industrial intrusion prevention systems (IPSs)). In using such interpretability tools, we are effectively converting the raw predictions into actionable intelligence, which will enhance the usability of the system, making it more practical for real-world cybersecurity implementations.

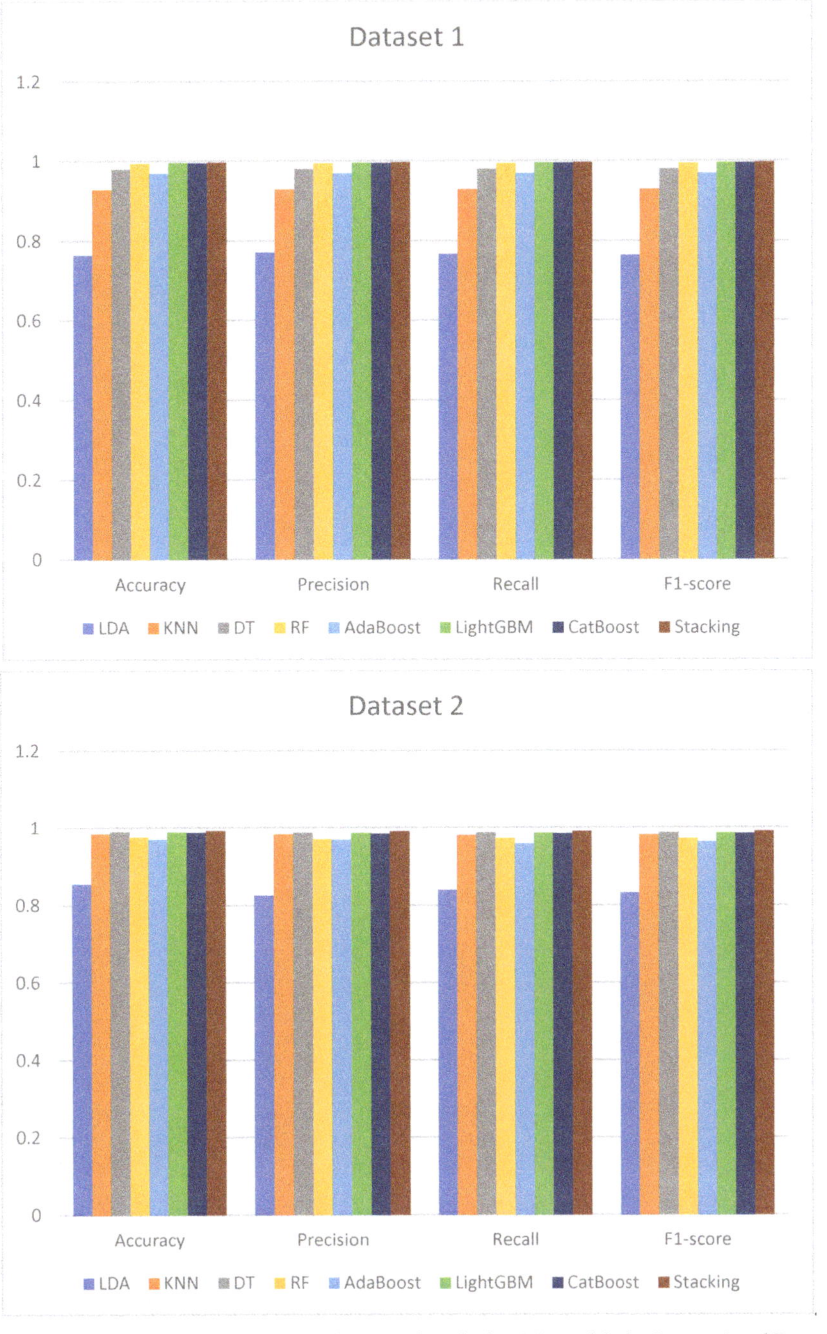

Figure 6. A comparison of the proposed approach and other ML models for Dataset 1 and Dataset 2 for malware detection.

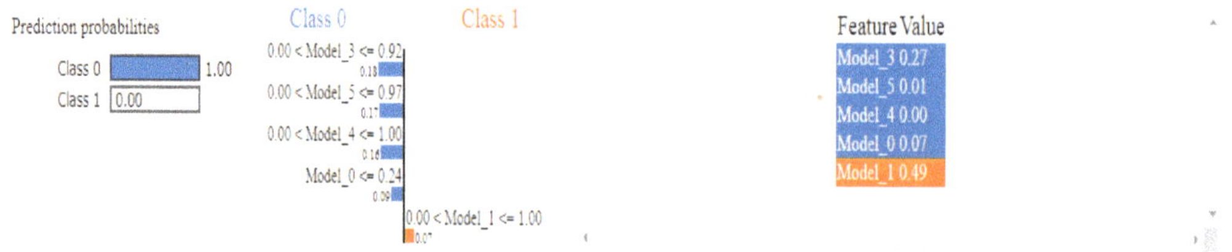

(a) Dataset 1

(b) Dataset 2

Figure 7. Lime for Dataset 1 and Dataset 2 with the proposed stacking approach.

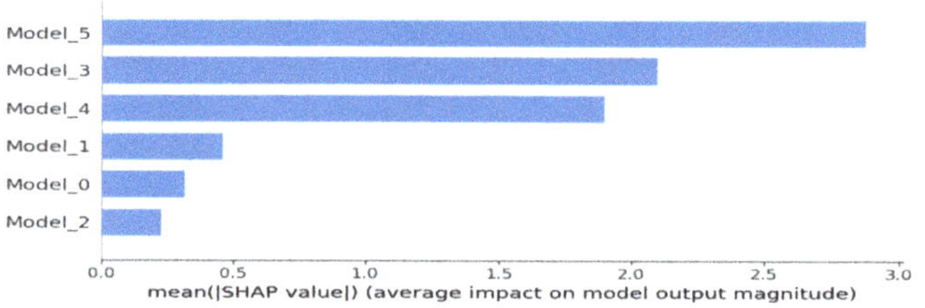

(a) Dataset 1

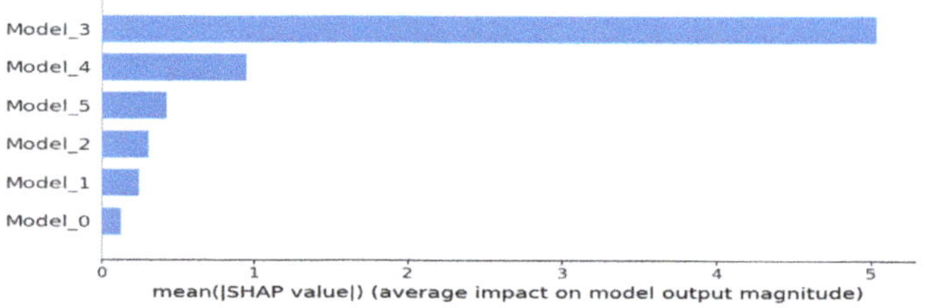

(b) Dataset 2

Figure 8. Shape for model impact for Dataset 1 and Dataset 2 with the proposed stacking approach.

Table 7 and Figure 10 present a comparative analysis between the proposed MAL-XSEL framework and an existing malware detection model [27] based on their performance on two datasets: ClaMP (Dataset 1) and MalwareDataSet (Dataset 2). The comparison evaluates accuracy as the primary performance metric. The existing approach utilizes a 1D-CNN-LSTM model, a deep learning-based hybrid architecture, achieving an accuracy of 98.75% on the ClaMP dataset and 97.07% on the MalwareDataSet dataset. While these results indicate strong performance, deep learning models often operate as "black boxes", lacking interpretability, which can limit their adoption in security-critical applications.

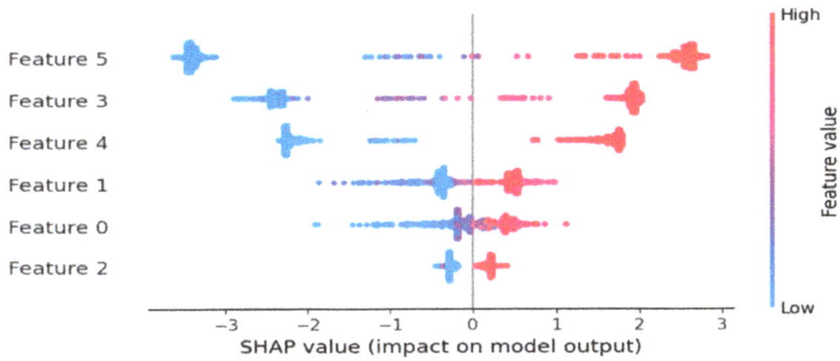

(a) Dataset 1

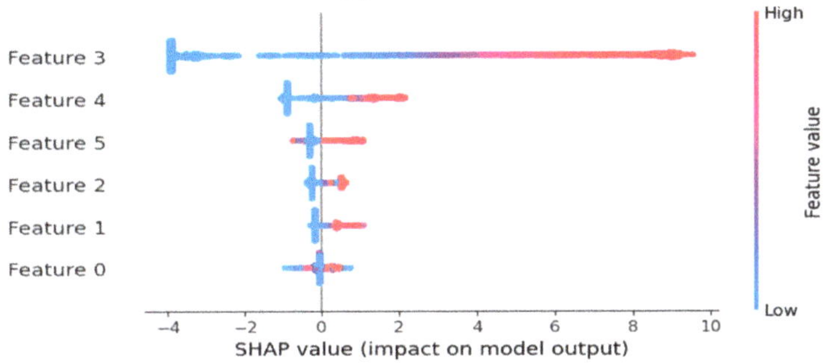

(b) Dataset 2

Figure 9. Shape for feature importance for Dataset 1 and Dataset 2 with the proposed stacking approach.

Table 7. Comparison between the proposed framework and existing work.

Work	Dataset	Model	Outcomes
[27]	ClaMP	1D-CNN-LSTM Model	Accuracy is 98.75%
	MalwareDataSet		Accuracy is 97.07%
Proposed Approach	ClaMP (Dataset 1)	Stacking Ensemble Approach	Accuracy is 99.62%
	MalwareDataSet (Dataset 2)		Accuracy is 99.15%

In contrast, the proposed approach employs a hybrid stacking-based ensemble of machine learning models, significantly outperforming the existing work. It achieves an accuracy of 99.62% on the ClaMP dataset and 99.15% on the MalwareDataSet dataset, demonstrating a notable improvement in malware classification. The stacking ensemble approach leverages multiple base learners, optimizing classification performance by combining the strengths of different machine learning models. In general, this comparison highlights the effectiveness of the proposed MAL-XSEL framework, which not only improves accuracy but also incorporates explainability through XAI techniques. By enhancing transparency in decision-making, MAL-XSEL provides a more robust and interpretable solution for malware detection in industrial web applications, ensuring greater security for critical infrastructure, industrial control systems, and enterprise networks.

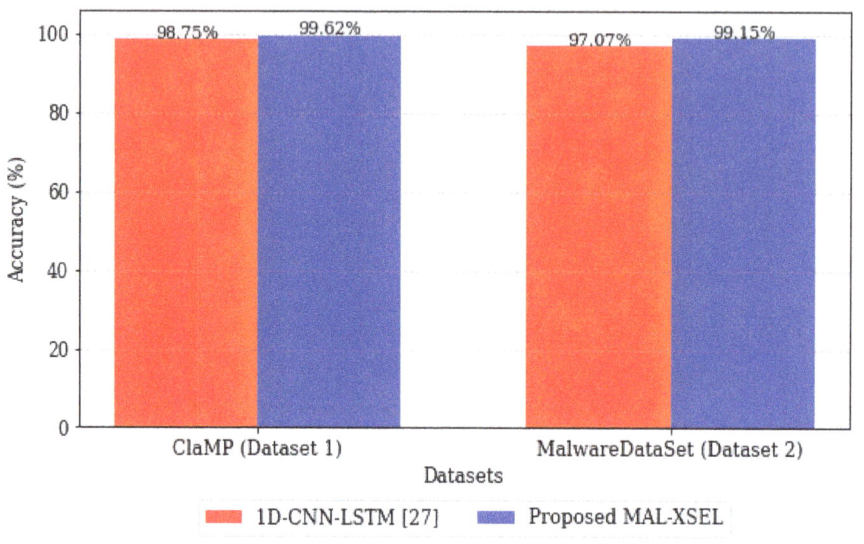

Figure 10. Accuracy comparison of the proposed technique with related work.

4.4. Limitations and Challenges

Despite the promising results achieved by the proposed **MAL-XSEL** framework, several limitations and practical challenges remain that warrant further investigation for real-world deployment, particularly in industrial control systems (ICSs) and IIoT environments. These limitations can be summarized as follows:

- **Computational Complexity:** The stacking ensemble model, while achieving high accuracy, requires significant computational resources during training and inference. In industrial environments, where real-time threat detection is crucial for operational continuity, the integration of multiple base learners and an additional meta-learner increases processing time, making immediate malware detection in resource-constrained industrial control systems (ICSs) and IIoT networks challenging.
- **Feature Dependence and Data Variability:** The model's performance is influenced by the quality and diversity of the datasets used for training. In industrial sectors, malware targeting ICSs and enterprise networks often exhibit distinct characteristics compared to traditional IT environments. Differences in feature representation across various malware datasets may affect generalizability, requiring extensive feature engineering and adaptation for new or unseen malware types.
- **Interpretability vs. Performance Trade-off:** While SHAP and LIME enhance interpretability, their computational cost can be high, especially when explaining complex ensemble models. Generating local and global explanations for large-scale datasets can slow down decision-making in industrial security operation centers. This trade-off must be carefully balanced to maintain operational efficiency.
- **Dataset Generalization:** The datasets used in this study vary in size and composition, which may limit the generalizability of the model across all real-world malware threats. Additionally, zero-day malware or novel attack vectors may not be well-represented, which poses challenges for detection. Addressing this requires ongoing model adaptation and the inclusion of more diverse and dynamic data sources.
- **Adversarial Evasion and Concept Drift:** Malware variants continuously evolve, and adversarial attacks may attempt to bypass detection models by obfuscating features.

Industrial environments, such as power grids, smart factories, and IIoT systems, are particularly vulnerable to these evolving threats. Additionally, concept drift in malware characteristics over time necessitates frequent retraining and adaptation to maintain model effectiveness.

- **Scalability and Deployment Challenges:** Deploying the MAL-XSEL system in large-scale, real-world ICS or IIoT environments introduces several practical challenges that must be addressed to ensure operational feasibility. To maintain adaptability in dynamic threat landscapes, the system should support incremental learning or modular retraining to effectively handle newly emerging malware variants. Additionally, compliance with cybersecurity standards and interoperability with various industrial components are essential to ensure the system's credibility, security, and scalability. Low-latency, real-time inference is also critical for effective threat detection in industrial systems, where delays can result in significant operational risks. To address the computational burden, optimization techniques such as model pruning, quantization, and knowledge distillation can be applied to reduce the model's footprint. In this context, edge computing presents a promising solution by enabling local decision-making with real-time responsiveness, thereby reducing network latency and ensuring timely malware detection.

5. Conclusions and Future Work

In this study, we proposed MAL-XSEL, an explainable stacking-based ensemble learning model for web malware classification. By integrating explainable artificial intelligence (XAI) techniques, the proposed framework enhances both detection accuracy and interpretability, addressing the limitations of traditional "black-box" malware detection models. In industrial environments, where cybersecurity is critical for protecting industrial control systems (ICSs), IIoT networks, and enterprise web applications, ensuring transparency in malware detection is essential. Through the application of SHAP and LIME, security professionals can gain deeper insights into the decision-making process, improving trust and adoption in security-critical environments. The experimental evaluation on two benchmark datasets demonstrated the superior performance of MAL-XSEL compared to conventional machine learning models. The stacking ensemble approach consistently achieved higher accuracy, with a peak accuracy of 0.9962 in Dataset 1 and 0.9916 in Dataset 2, confirming its effectiveness in malware detection. These results validate the potential of ensemble learning and XAI techniques in strengthening cybersecurity defenses for web applications. By combining high detection accuracy, transparency, and interpretability, MAL-XSEL represents a practical and robust solution for fighting the evolving malware threats in industrial systems.

Future research can focus on reducing complexity for real-time deployment in resource-constrained environments or improving the adaptability of models to malware behavior changes due to concept drift. However, developing lightweight models that are interpretable yet maintain a high detection accuracy will also be critical for industrial-scale implementation. In the future, to increase the efficiency of systems against zero-day threats, one can integrate live traffic streaming datasets or live data streams with real network simulations. Applying model simplification approaches can be studied with the intention of minimizing the resource requirements of the model while ensuring performance and transparency. In the same context, implementing post hoc XAI techniques during the critical analysis rather than during real-time applications would also strike a balance between interpretability and latency. These paths are going to help reduce the gap between detection accuracy and operational deployment, thereby enhancing the prospect of explainable AI for applications within the realms of cybersecurity in practice.

Author Contributions: Conceptualization, E.E.-D.H., S.A., O.A.G. and A.S.; Formal analysis, E.E.-D.H. and A.S.; Funding acquisition, S.A.; Investigation, H.E., O.A.G. and A.S.; Methodology, H.E.; Software, E.E.-D.H. and O.A.G.; Validation, H.E.; Writing—original draft, E.E.-D.H. and A.S.; Writing—review and editing, S.A., H.E. and O.A.G. All authors have read and agreed to the published version of the manuscript.

Funding: This research was funded by the Princess Nourah bint Abdulrahman University Researchers Supporting Project number (PNURSP2025R197), Princess Nourah bint Abdulrahman University, Riyadh, Saudi Arabia.

Data Availability Statement: The original contributions presented in this study are included in the article; further inquiries can be directed to the corresponding author.

Acknowledgments: Princess Nourah bint Abdulrahman University Researchers Supporting Project number (PNURSP2025R197), Princess Nourah bint Abdulrahman University, Riyadh, Saudi Arabia.

Conflicts of Interest: The authors declare no conflicts of interest.

Abbreviation

AI	Artificial Intelligence
AUC-ROC	Area Under the Curve—Receiver Operating Characteristic
PE	Portable Executables
CM	Confusion Matrix
CNN	Convolutional Neural Network
DT	Decision Tree
FN	False Negative
FP	False Positive
ICS	Industrial Control System
IIoT	Industrial Internet of Things
KNN	K-Nearest Neighbors
LDA	Linear Discriminant Analysis
LIME	Local Interpretable Model-Agnostic Explanations
RF	Random Forest
SHAP	SHapley Additive exPlanations
XAI	Explainable Artificial Intelligence

References

1. Egele, M.; Scholte, T.; Kirda, E.; Kruegel, C. A survey on automated dynamic malware-analysis techniques and tools. *ACM Comput. Surv. (CSUR)* **2008**, *44*, 1–42. [CrossRef]
2. Falowo, I.; Ozer, M.; Li, C.; Abdo, J.B. Evolving Malware and DDoS Attacks: Decadal Longitudinal Study. *IEEE Access* **2024**, *12*, 39221–39237. [CrossRef]
3. Knapp, E.D. *Industrial Network Security: Securing Critical Infrastructure Networks for Smart Grid, SCADA, and Other Industrial Control Systems*; Elsevier: Amsterdam, The Netherlands, 2024.
4. Al-Hawawreh, M.; Alazab, M.; Ferrag, M.A.; Hossain, M.S. Securing the Industrial Internet of Things against ransomware attacks: A comprehensive analysis of the emerging threat landscape and detection mechanisms. *J. Netw. Comput. Appl.* **2024**, *223*, 103809. [CrossRef]
5. Zhang, X.; Wu, K.; Chen, Z.; Zhang, C. MalCaps: A Capsule Network Based Model for the Malware Classification. *Processes* **2021**, *9*, 929. [CrossRef]
6. Zhao, S.; Tuan, L.A.; Fu, J.; Wen, J.; Luo, W. Exploring Clean Label Backdoor Attacks and Defense in Language Models. *IEEE/ACM Trans. Audio Speech Lang. Process.* **2024**, *32*, 3014–3024. [CrossRef]
7. Takemoto, K. All in How You Ask for It: Simple Black-Box Method for Jailbreak Attacks. *Appl. Sci.* **2024**, *14*, 3558. [CrossRef]

8. Cui, X.; Aparcedo, A.; Jang, Y.K.; Lim, S.N. On the Robustness of Large Multimodal Models against Image Adversarial Attacks. In Proceedings of the IEEE/CVF Conference on Computer Vision and Pattern Recognition, Seattle, WA, USA, 17–18 June 2024; pp. 24625–24634.
9. Jiang, W.; Han, H.; He, M.; Gu, W. ML-based pre-deployment SDN performance prediction with neural network boosting regression. *Expert. Syst. Appl.* **2024**, *241*, 122774. [CrossRef]
10. Jiang, W. Graph-based deep learning for communication networks: A survey. *Comput. Commun.* **2022**, *185*, 40–54. [CrossRef]
11. Lu, H.; Uddin, S. Explainable stacking-based model for predicting hospital readmission for diabetic patients. *Information* **2022**, *13*, 436. [CrossRef]
12. Kovanen, T.; Nuojua, V.; Lehto, M. Cyber threat landscape in energy sector. In Proceedings of the ICCWS 2018 13th International Conference on Cyber Warfare and Security, Washington, DC, USA, 8–9 March 2018; Academic Conferences and Publishing Limited: Cambridge, MA, USA, 2018; p. 353.
13. Huang, Y.; Liu, J.; Xiang, X.; Wen, P.; Wen, S.; Chen, Y.; Chen, L.; Zhang, Y. Malware Identification Method in Industrial Control Systems Based on Opcode2vec and CVAE-GAN. *Sensors* **2024**, *24*, 5518. [CrossRef]
14. Zheng, Y.; Na, Z.; Ji, W.; Lu, Y. An Adaptive Fuzzy SIR Model for Real-Time Malware Spread Prediction in Industrial Internet of Things Networks. *IEEE Internet Things J.* **2025**. [CrossRef]
15. Naeem, H.; Ullah, F.; Naeem, M.R.; Khalid, S.; Vasan, D.; Jabbar, S.; Saeed, S. Malware detection in industrial internet of things based on hybrid image visualization and deep learning model. *Ad. Hoc Netw.* **2020**, *105*, 102154. [CrossRef]
16. Hussain, A.; Asif, M.; Ahmad, M.B.; Mahmood, T.; Raza, M.A. Malware detection using machine learning algorithms for windows platform. In Proceedings of the International Conference on Information Technology and Applications: ICITA 2021, Dubai, United Arab Emirates, 13–14 November 2021; Springer Nature: Singapore, 2022; pp. 619–632.
17. Sharma, S.; Rama Krishna, C.; Sahay, S.K. Detection of advanced malware by machine learning techniques. In *Soft Computing: Theories and Applications: Proceedings of the SoCTA 2017*; Springer: Singapore, 2019; pp. 333–342.
18. Nguyen, P.S.; Huy, T.N.; Tuan, T.A.; Trung, P.D.; Long, H.V. Hybrid Feature Extraction and Integrated Deep Learning for Cloud-Based Malware Detection. *Comput. Secur.* **2025**, *150*, 104233. [CrossRef]
19. Johny, J.A.; Asmitha, K.A.; Vinod, P.; Radhamani, G.; Rehiman, K.A.R.; Conti, M. Deep learning fusion for effective malware detection: Leveraging visual features. *Clust. Comput.* **2025**, *28*, 135. [CrossRef]
20. Guerra-Manzanares, A.; Bahsi, H.; Luckner, M. Leveraging the first line of defense: A study on the evolution and usage of android security permissions for enhanced android malware detection. *J. Comput. Virol. Hacking Tech.* **2023**, *19*, 65–96. [CrossRef]
21. Taha, A.; Barukab, O. Android malware classification using optimized ensemble learning based on genetic algorithms. *Sustainability* **2022**, *14*, 14406. [CrossRef]
22. Azeem, M.; Khan, D.; Iftikhar, S.; Bawazeer, S.; Alzahrani, M. Analyzing and Comparing the Effectiveness of Malware Detection: A Study of Machine Learning Approaches. *Heliyon* **2024**, *10*, e24008. [CrossRef] [PubMed]
23. Kumar, A. ClaMP (Classification of Malware with PE Headers). Mendeley Data 2020, V1. Available online: https://data.mendeley.com/datasets/xvyv59vwvz/1 (accessed on 18 January 2025).
24. Yıldırım, E. MalwareDataSet. Available online: https://github.com/emr4h/Malware-Detection-Using-Machine-Learning (accessed on 18 January 2025).
25. Alshathri, S.; Sayed, A.; Hemdan, E.E.-D. An Intelligent Attack Detection Framework for the Internet of Autonomous Vehicles with Imbalanced Car Hacking Data. World Electr. *Veh. J.* **2024**, *15*, 356. [CrossRef]
26. Sayed, A.; Alshathri, S.; Hemdan, E.E.-D. Conditional Generative Adversarial Networks with Optimized Machine Learning for Fault Detection of Triplex Pump in Industrial Digital Twin. *Processes* **2024**, *12*, 2357. [CrossRef]
27. Alqahtani, A.; Azzony, S.; Alsharafi, L.; Alaseri, M. Web-Based Malware Detection System Using Convolutional Neural Network. *Digital* **2023**, *3*, 273–285. [CrossRef]

processes

MDPI

Article

A Type-2 Fuzzy Logic Expert System for AI Selection in Solar Photovoltaic Applications Based on Data and Literature-Driven Decision Framework

Citlaly Pérez-Briceño [1], Pedro Ponce [1,*], Qipei Mei [2] and Aminah Robinson Fayek [2]

[1] Institute of Advanced Materials for Sustainable Manufacturing, Tecnologico de Monterrey, Monterrey 64849, NL, Mexico; a01336766@tec.mx
[2] Department of Civil and Environmental Engineering, University of Alberta, 7-203 Donadeo Innovation Centre for Engineering, 9211 116 St NW, Edmonton, AB T6G 1H9, Canada; qipei.mei@ualberta.ca (Q.M.); aminah.robinson@ualberta.ca (A.R.F.)
* Correspondence: pedro.ponce@tec.mx

Abstract

Artificial intelligence (AI) has emerged as a transformative tool for optimizing photovoltaic (PV) systems, enhancing energy efficiency, predictive maintenance, and fault detection. This study presents a systematic literature review and bibliometric analysis to identify the most commonly used AI techniques and their applications in PV systems. The review provides details on the advantages, limitations, and optimal use cases of various review techniques, such as Artificial Neural Networks, Fuzzy Logic, Convolutional Neural Networks, Long-Short Term Memory, Support Vector Machines, Decision Trees, Random Forest, k-Nearest Neighbors, and Particle Swarm Optimization. The findings highlight that maximum power point tracking (MPPT) optimization is the most widely researched AI application, followed by solar power forecasting, parameter estimation, fault detection and classification, and solar radiation forecasting. The bibliometric analysis reveals a growing trend in AI-PV research from 2018 to 2024, with China, the United States, and European countries leading in contributions. Furthermore, a type-2 fuzzy logic system is developed in MATLAB R2023b for automating AI technique selection based on the problem type, offering a practical tool for researchers, industry professionals, and policymakers. The study also discusses the practical implications of adopting AI in PV systems and provides future directions for research. This work serves as a comprehensive reference for advancing AI-driven solar PV technologies, contributing to a more efficient, reliable, and sustainable energy future.

Keywords: artificial intelligence; solar photovoltaic energy; fuzzy logic system; literature review

1. Introduction

Solar photovoltaic (PV) systems have become a cornerstone of the global shift towards renewable energy, providing a sustainable solution to meet growing energy demands while addressing environmental concerns. Their ability to capture abundant solar radiation and convert it into electricity positions them as a leading renewable energy technology [1]. However, several key challenges continue to hinder their widespread adoption and optimal performance. These include efficiency losses caused by environmental variability, shading, and temperature fluctuation; complexities in maintenance; and difficulties integrating with existing grid systems [2].

Artificial intelligence (AI) has emerged as a game-changing technology for addressing the challenges faced by solar PV systems. By applying AI-driven algorithms for predictive maintenance, efficiency optimization, and system monitoring, researchers and industry professionals are enhancing the operational reliability and economic viability of PV systems [3]. Interest in integrating AI with PV technologies has grown significantly in recent years, highlighting their immense potential. For example, advancements in machine learning and big data analytics have greatly improved the accuracy of energy yield predictions and system fault diagnosis [4].

By facilitating advanced solutions such as predictive maintenance, energy forecasting, and real-time fault detection, AI can enhance the efficiency and reliability of PV systems while reducing operational costs and downtime. This integration is especially vital in the context of the global shift toward renewable energy, as the demand for sustainable energy solutions continues to rise. However, despite rapid advancements, the current body of literature often lacks a comprehensive method of synthesis that captures the full range of AI techniques and their applications throughout the PV system lifecycle. Additionally, challenges such as scalability, generalizability, and integration with grid and energy storage systems remain underexplored, highlighting the need for in-depth review to guide future research efforts.

This study aims to address these gaps by systematically reviewing the current body of research on AI applications in solar PV systems. Utilizing the PRISMA (Preferred Reporting Items for Systematic Review and Meta-Analyses) methodology, 163 papers published between 2020 and 2024 were selected from an initial dataset of 774 studies for detailed analysis. The review examines the AI techniques employed, their applications across various stages of the PV system lifecycle, and the challenges and opportunities shaping this field. Furthermore, it introduces an AI selection system.

This study makes several contributions to the field of AI applications in PV systems, providing a comprehensive analysis of AI-driven advancements for improving PV performance, efficiency, and reliability. The key contributions of this work are as follows:

- Presenting a systematic literature review of the most commonly used AI techniques, including Artificial Neural Networks (ANNs), fuzzy logic (FL), Convolutional Neural Networks (CNNs), Long Short-Term Memory (LSTM) networks, Support Vector Machines (SVMs), Decision Trees (DT), Random Forest (RF), k-Nearest Neighbors (kNN), and Particle Swarm Optimization (PSO). It highlights the advantages and limitations of each technique in PV applications;
- Categorizing the most relevant AI applications in PV systems, identifying Maximum Power Point Tracking (MPPT) optimization, solar power forecasting, parameter estimation, fault detection, and solar radiation forecasting as the most extensively researched areas. The review quantifies the impact of AI on each of these applications;
- This work includes a bibliometric analysis, identifying publication trends, leading countries, and research hotspots;
- Identifying current challenges in AI-driven PV optimization and outlining key future research directions;
- Structuring a decision-making framework for selecting AI techniques based on problem complexity, data availability, and computational requirements;
- Introducing a MATLAB-based type-2 fuzzy system to assist the user in selecting the most suitable AI technique for PV applications by accounting for uncertainty in input variables and expert evaluations;

- Finally, this study serves as a reference for researchers, engineers, and policymakers, as it can contribute to the standardization of AI methodologies in PV systems, helping the stakeholders involved align their works with emerging trends and technological advancements.

Unlike prior reviews that focus either on technical performance or specific AI models, this study integrates a PRISMA-based literature review with a bibliometric analysis and a Type-2 Fuzzy Logic Expert System to guide AI model selection in solar PV applications. This hybrid methodology offers both a systematic synthesis of current research and a decision-support framework, addressing both academic and practical gaps in the field.

The remainder of this paper is organized as follows. Section 2 outlines the methodology for the literature review, including the use of the PRISMA framework and bibliometric tools. Section 3 presents the findings from the bibliometric analysis and literature review, focusing on key AI techniques and their applications. Section 4 offers recommendations for future research directions. Section 5 shows the recommended decision-making framework for AI selection in PV systems, including the developed MATLAB-based type-2 fuzzy system, while Section 6 presents a discussion of the main insights of this work. Finally, the paper concludes with a summary of the findings and their implications for the field.

2. Methodology: PRISMA

To ensure transparency, methodological rigor, and replicability, this study used the PRISMA framework. This approach offers a structured, step-by-step process for conducting a systematic review, ensuring a thorough and impartial analysis of the existing literature. As established in prior research [5], the methodology is divided into two primary phases:

1. Identification and screening. This initial phase involves defining research questions, creating a review protocol, performing extensive searches across various databases, and assessing the relevance of retrieved studies by reviewing their titles, abstracts, and full texts;
2. Data extraction and synthesis. In this stage, relevant data are systematically gathered using standardized formats, organized effectively, and analyzed through narrative descriptions, statistical methods, or meta-analytical techniques.

By adhering to this methodology, this review aims to address the following research questions:

1. What are the primary AI techniques applied to solar PV systems, and how have they evolved over the years?
2. What are the most common application areas for AI in the lifecycle of solar PV systems, and what challenges do they address?
3. What gaps exist in the current body of research, and what future directions are proposed for integrating AI into solar PV systems?

The systematic literature review workflow is outlined in Figure 1, and it consists of the following stages:

1. **Identification.** A total of 774 papers were retrieved from the Web of Science database. The search strategy targeted titles, abstracts, and keywords using terms such as "artificial intelligence", "AI", "solar photovoltaic", "photovoltaic system", and "photovoltaic panel". The inclusion criteria for this stage were journal articles published between 2020 and 2024 in English or Spanish;
2. **Screening.** After applying filters for document type, publication year, language, and category, 307 papers were retained. At this stage, duplicates and unrelated studies were excluded, resulting in 193 papers. Titles and abstracts were reviewed to eliminate

studies irrelevant to the scope of the chosen title. Table 1 provides a detailed summary of the inclusion and exclusion criteria applied during this phase.

3. **Eligibility.** The full texts of the remaining 213 papers were reviewed. Studies that lacked an explicit application of AI or did not focus on solar PV systems were excluded, leaving 163 eligible papers. Exclusions included 50 studies, with 26 full texts unavailable and 6 review papers not meeting inclusion criteria;
4. **Inclusion.** The final dataset comprised 163 papers deemed relevant for inclusion in the review. These papers were selected based on their adherence to the defined criteria and their focus on the application of AI in solar PV systems.

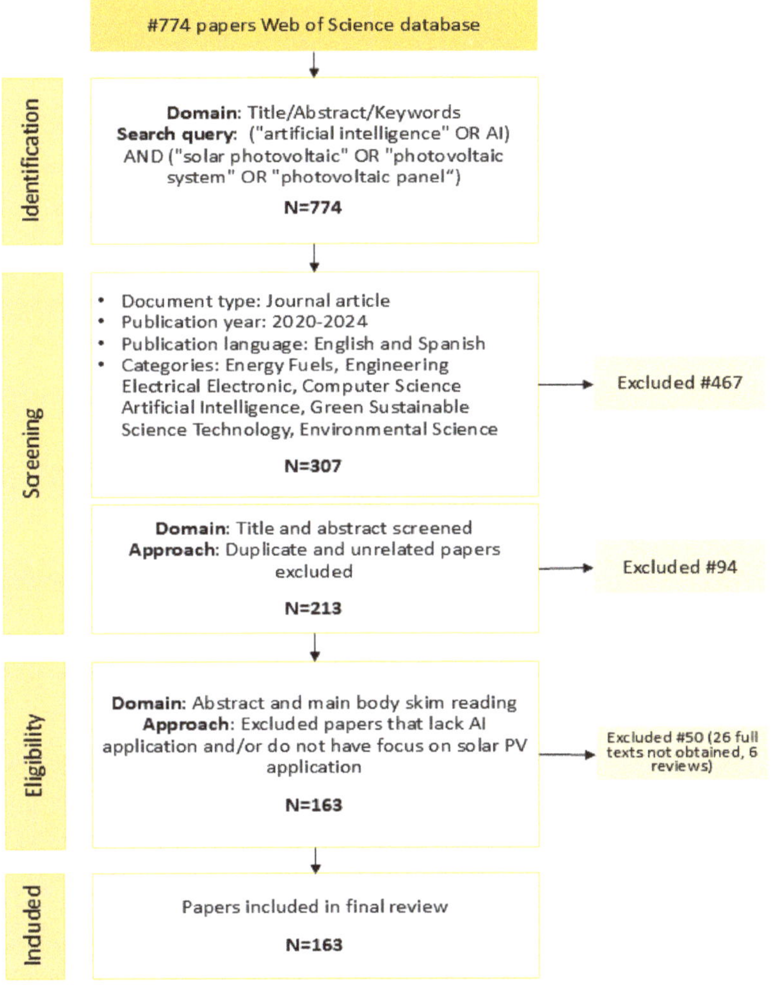

Figure 1. Methodology of the systematic literature search used in this study.

Data from the selected studies were carefully extracted and compiled into a structured database (Table A1—Appendix A). This included details such as author(s), title, publication year, journal name, AI techniques employed, and their specific applications in solar PV systems. The analysis emphasized identifying key trends, research gaps, and potential opportunities within the field. Furthermore, a bibliometric analysis was performed to better analyze trends in publication frequency, country contributions, and keyword co-occurrence. This methodological approach facilitated an in-depth exploration of the intersection between AI and solar photovoltaic systems, offering valuable insights and highlighting potential directions for future research.

Table 1. Inclusion and exclusion criteria used in the PRISMA review.

Category	Inclusion Criteria	Exclusion Criteria	Justification
Language	English, Spanish	Languages other than English or Spanish (e.g., Chinese, German, French)	Spanish was included due to its relevance in Latin American contexts and the authors' proficiency. Other languages were excluded to ensure consistency in full-text review and due to translation limitations.
Document type	Peer-reviewed journal articles	Editorials, letters, notes, abstracts only, non-peer-reviewed sources	To ensure scientific rigor and reliability of reported results.
Timeframe	Publications from 2020–2024	Publications before 2020	The timeframe reflects the period of significant growth in AI and PV research.
Subject area	AI techniques applied to solar PV systems	Studies on AI unrelated to solar PV or those focusing on other renewable technologies only (e.g., wind, hydro)	Focused on the intersection of AI and solar PV systems to maintain relevance.
Availability	Full-text available online via institutional access or open access	Inaccessible full texts	Full-text access was necessary for accurate content analysis.
Content focus	Application of AI methods (e.g., ML, FL, DL, optimization) to solar PV forecasting, MPPT, fault detection, grid integration, among other applications	Reviews or studies that do not apply or evaluate AI techniques for solar PV systems	The review aimed to synthesize applied AI techniques in the solar PV context specifically.

3. Results

This section presents the results from the bibliometric analysis and the literature review findings.

3.1. Bibliometric Analysis

To provide a comprehensive overview of the research landscape on the application of AI in solar PV systems, a bibliometric analysis was conducted using the 774 papers initially retrieved from the Web of Science database prior to applying inclusion and exclusion criteria. This analysis was performed using the Bibliometrix 4.0.1 [6] tool and VOSviewer 1.6.18 software to examine annual trends, geographical distribution, keyword co-occurrences, and thematic evolution within the field.

The annual distribution of publications on AI applications in solar PV systems, illustrated in Figure 2, shows a significant increase in research activity over the past two decades, with an especially sharp rise in recent years. The number of publications remained relatively low until 2010, after which steady growth was observed. A more pronounced increase is evident from 2018 onward, reflecting the accelerating interest in AI applications for PV systems. The peak in recent years suggests that AI-driven innovations in solar energy are gaining momentum, aligning with global efforts to enhance the efficiency, reliability, and integration of renewable energy technologies. It is important to highlight that the decrease in the scientific production data for the year 2025 is because that year is not included in this study, and the results appearing on the search are "early access".

The geographical distribution of scientific contributions, shown in Figure 3, indicates that research in this field is primarily driven by a few key countries. The United States leads with the highest number of publications, followed by significant contributions from China, India, and several European nations. These countries have invested heavily in AI-driven renewable energy solutions, often supported by governmental policies promoting sustainable energy research. Regions such as South America, the Middle East, and parts of Africa show comparatively lower research activity, pointing to opportunities for greater academic and practical involvement in these areas.

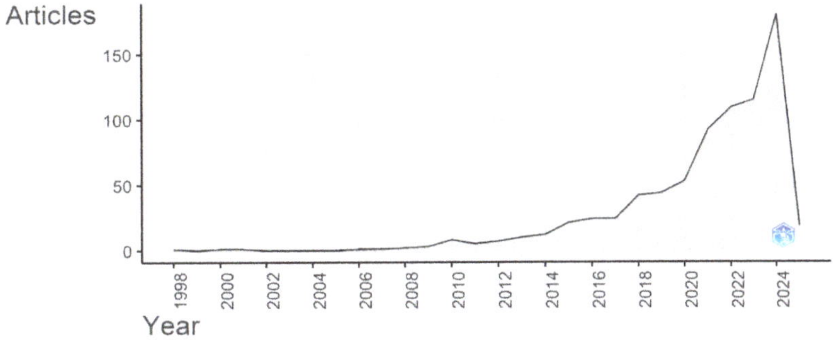

Figure 2. Annual scientific production generated using Bibliometrix 4.0.1.

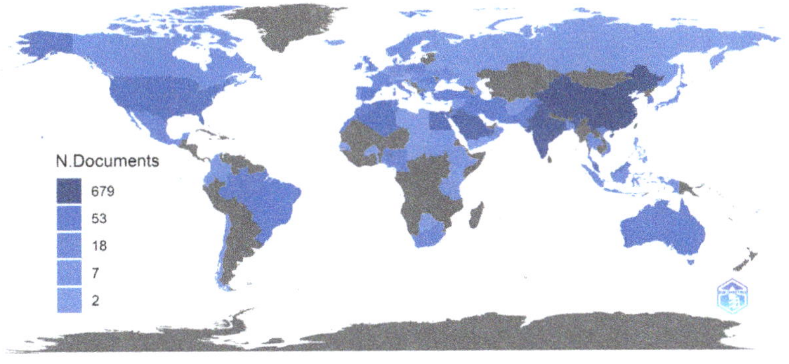

Figure 3. Country scientific production generated using Bibliometrix 4.0.1.

The keyword co-occurrence analysis (Figure 4) highlights the interconnectedness of key themes within this research area. Prominent terms such as "artificial intelligence", "solar energy", "photovoltaic systems", and "optimization" reflect the central focus of the literature. Clusters of related keywords, including "machine learning", "fault detection", and "energy forecasting", showcase the diverse applications of AI in the field. The network further identifies emerging trends such as "predictive maintenance", "energy management", and "real-time monitoring", signaling the evolving directions of research and innovation.

The keyword density visualization, shown in Figure 5, provides additional insight into the intensity and primary focus areas within the research field. The color gradient indicates the frequency and relevance of keyword appearances across the analyzed literature. Warmer colors (yellow and green) represent higher density and stronger co-occurrence between keywords, whereas cooler colors (blue and purple) indicate lower frequency and weaker associations. Dominant keywords like "artificial intelligence", "optimization", and "photovoltaic systems" stand out, surrounded by related topics such as "power forecasting", "fault diagnosis", and "efficiency". The less-dense areas of the map indicate emerging topics, such as AI applications in hybrid energy systems, real-time monitoring, and the integration of AI with smart grids. These patterns reinforce the notion that while certain AI applications in PV systems are well-established, there remains substantial room for exploration in areas such as intelligent grid coordination and advanced energy management systems.

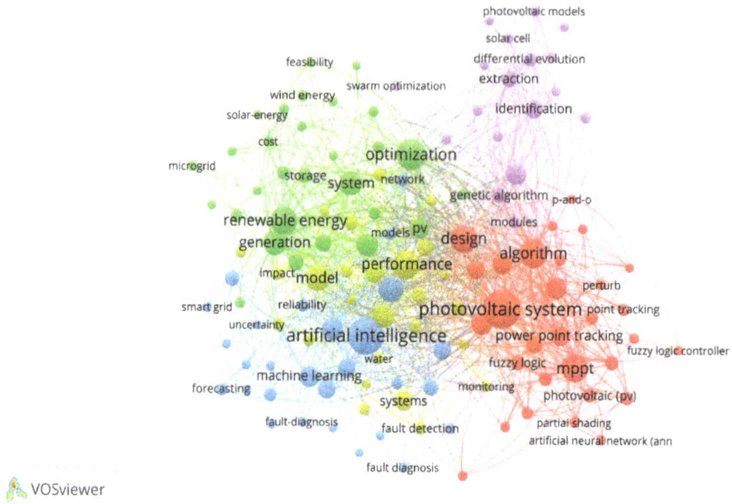

Figure 4. Keyword co-occurrence network generated using VOSviewer 1.6.18.

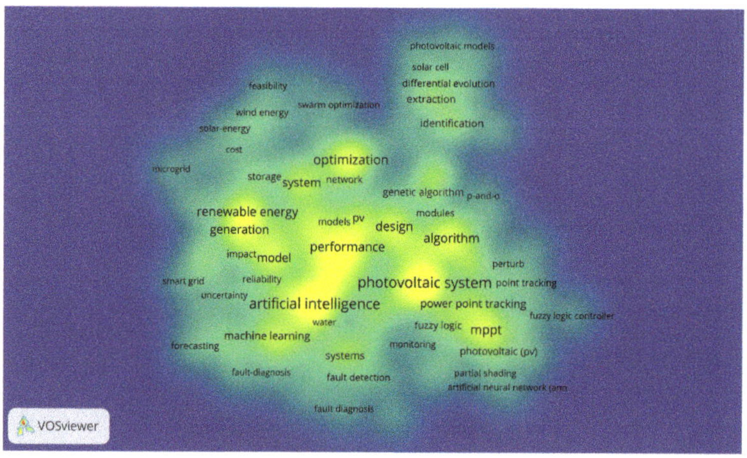

Figure 5. Keyword density network generated using VOSviewer 1.6.18. Warmer colors (yellow/green) indicate higher keyword co-occurrence density, while cooler colors (blue/purple) indicate lower density.

The thematic evolution network (Figure 6) illustrates the progression of research themes over time. This Sankey diagram visualizes how key research themes and keywords have evolved. The left side represents the dominant themes and keywords used in the literature between 1998 and 2019, while the right side shows the themes that emerged or remained relevant from 2020 to 2025. The connecting lines (or flows) indicate the semantic or conceptual continuity between themes across the two periods, based on co-word and co-occurrence analysis.

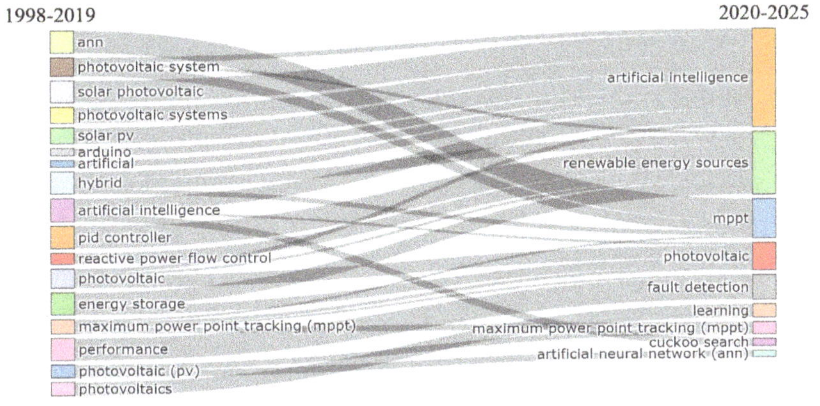

Figure 6. Thematic evolution network generated using Bibliometrix 4.0.1.

The line thickness represents the strength of the connection, that is, how strongly a topic in the earlier period is linked to a theme in the later period, based on shared keywords and context in the reviewed publications.

For example, themes like "photovoltaic system", "solar PV", and "photovoltaic (PV)" evolved into more specific or refined concepts such as "photovoltaic", "MPPT", and "fault detection". Similarly, broad early topics like "artificial intelligence" and "hybrid" transitioned into more applied and granular research themes like "renewable energy sources", "learning", and "artificial neural network (ANN)" in the later period.

The bibliometric analysis underscores the rapid expansion and global reach of research on AI applications in solar PV systems. It highlights leading contributors, key research themes, and their development over time. These insights offer a valuable foundation for understanding the current state of the field and pinpointing areas for future exploration.

3.2. Literature Review Findings

This subsection shows the insights from the 163 papers chosen (Table A1—Appendix A) for in-depth analysis, emphasizing the application of AI techniques in solar PV systems. The review highlights the variety of AI methodologies, their application across different stages of the PV system's lifecycle, and the emerging trends, challenges, and gaps within the existing research.

3.2.1. AI Techniques Applied in Solar PV Systems

The reviewed studies highlight a variety of AI techniques designed to tackle specific challenges in solar PV systems. The reviewed techniques can be broadly classified into categories such as supervised learning (e.g., ANN, SVM, DT, RF, kNN), deep learning (e.g., CNN, LSTM), optimization methods (e.g., PSO), and fuzzy logic-based reasoning systems (e.g., FL). The most common include the following.

- **Artificial Neural Networks (ANNs)**

ANNs are popular for modeling nonlinear relationships and learning from historical data. The topology of ANNs, including the number of hidden layers, neurons per layer, and activation functions, plays a crucial role in its performance. Improper topology selection can lead to underfitting or overfitting, reducing the model's ability to generalize to new data [7,8]. To address this, optimal topologies are often identified empirically or through

optimization algorithms, such as particle swarm optimization (PSO) and genetic algorithms (GA), particularly in applications like PV system forecasting and fault detection [9].
Several studies demonstrated that ANN-based MPPT controllers significantly enhance power extraction efficiency, with methods such as wavelet-assisted ANN MPPT achieving a tracking efficiency of 99.9%, outperforming traditional Perturb and Observe methods (P&O) [10]. Similarly, ANN-driven adaptive MPPT for grid-connected PV systems resulted in a 12% efficiency improvement, ensuring stable voltage regulation despite fluctuating solar conditions [11]. ANN-based fault detection and diagnosis models also proved highly effective, with hybrid ANN-Fireworks Algorithm (FWA) techniques achieving 99.98% accuracy and significantly reducing diagnostic time [12]. In PV inverter fault tolerance, ANN-based real-time control reduced total harmonic distortion (THD) to 8.24%, thereby enhancing system stability and resilience [13].
ANN-based forecasting models also demonstrated high accuracy in predicting solar energy generation and urban energy demand, with Decision Support Systems achieving 99% predictive accuracy [14]. Hybrid ANN approaches, such as ANN-FLC and variable step P&O, further improved MPPT performance under partial shading conditions, tracking power with >99.65% efficiency while reducing steady-state oscillations [15]. These findings underscore ANN's ability to optimize PV efficiency, enhance grid reliability, and support intelligent energy management in smart cities.

- **Fuzzy Logic (FL)**

FL systems are used for decision-making under uncertainty by applying rule-based reasoning. Studies have demonstrated that FL-based MPPT algorithms outperform conventional techniques like P&O by providing faster adaptation to varying irradiance and temperature conditions. A comparative study showed that an FL-enhanced P&O method reduced oscillations in steady-state conditions while achieving a higher power yield [16]. Additionally, a hybrid FL-MPPT system integrating genetic algorithms (GA) and PSO improved tracking accuracy and reduced power loss, leading to a tracking efficiency above 99.5% [17].
A study comparing FL and ANNs used in real-time PV fault detection demonstrated that FL achieved 99.2% accuracy in identifying six common fault types, including partial shading, bypass diode failures, and soiling [18]. Another study integrating FL-based voltage source control with a multilevel inverter showed a significant reduction in THD, with voltage THD as low as 1.13% and current THD below 1.52%, ensuring smoother grid integration [19].
The emergence of hybrid AI controllers, such as FL–ANN and FL–Genetic Algorithm combinations, presents a promising avenue for further research. These approaches leverage FL's rule-based adaptability and ANN's pattern recognition capabilities, offering higher precision and robustness in dynamic PV environments [20].

- **Convolutional Neural Networks (CNNs)**

CNNs are powerful in processing spatial data, and have been widely used for fault detection through image classification.
A key application of CNNs is in fault detection and the classification of PV panel defects. One study employed a Visual Geometry Group (VGG-16)-based CNN model for detecting physical and electrical anomalies in PV systems using radiometric infrared thermography. The system achieved 91.46% accuracy in classifying common defects [21]. Another investigation implemented CNN-based aerial imaging to detect hotspots and cracks in PV panels, demonstrating 93.3% classification accuracy [22]. A hybrid CNN–Multilayer Perceptron (MLP) outperformed traditional statistical models by effectively predicting global solar radiation with high spatial accuracy [23]. Furthermore, CNN-driven geospatial mapping techniques have been used to detect and quantify rooftop

installations, achieving an average precision rate of 93% [24]. These findings indicate that CNN-based approaches significantly enhance PV system reliability, efficiency, and integration into energy grids.

- **Long Short-Term Memory (LSTM)**

LSTM networks are a type of recurrent neural network designed to capture temporal dependencies in time-series data.
LSTM-based solar power forecasting models have proven superior to traditional statistical methods, particularly in handling nonlinear meteorological data. A hybrid CNN-LSTM model was developed for PV power estimation, achieving R^2 values above 0.98 across varying weather conditions, and thus demonstrating high reliability in forecasting [25]. The R^2 value, also known as the coefficient of determination, shows how well the model's predictions match the actual data—values closer to 1.0 mean better accuracy. Similarly, an LSTM-based smart grid recommender system optimized energy harvesting strategies, reducing the gap between predicted and actual PV energy generation, and improving grid demand response management [26].
An LSTM model was compared against ANNs and regression-based approaches for tracking global maximum power points. The results indicate that LSTM reduced power tracking errors by 31% compared to ANN models, proving its efficiency in the real-time optimization of PV systems under partial shading conditions [27]. Furthermore, hybrid wavelet–LSTM models have been successfully applied to solar tracking systems, achieving higher accuracy in power estimation for dual-axis PV trackers, enabling real-time position adjustments based on forecasted solar intensity [28].
These findings highlight the robustness of LSTM-based models in addressing solar energy intermittency, optimizing power generation, and enhancing MPPT efficiency.

- **Support Vector Machine (SVM)**

SVMs are supervised learning models effective for use in classification tasks with small datasets. A study utilizing a Least Squares SVM (LSSVM) model combined with Variational Mode Decomposition (VMD) and Whale Optimization Algorithm (WOA) achieved a 17.17% reduction in Mean Absolute Percentage Error compared to conventional SVM approaches [29]. Another study implemented an SVM-based prediction framework for residential PV power estimation in Saudi Arabia, achieving higher accuracy than conventional regression models when applied to real-time weather and irradiance datasets [30]. A cloud-computing-integrated SVM fault detection system was validated using MATLAB/Simulink simulations and achieved 97.4% classification accuracy in differentiating various PV system faults [31]. Another study proposed an ensemble learning model combining SVM with k-Nearest Neighbors (kNN) and Decision Trees, achieving high precision in detecting snail trail faults, microcracks, and panel delamination issues [32]. An SVM-based islanding detection technique integrated with Gaussian Radial Basis Function kernels achieved 99.2% detection accuracy while minimizing false alarms to 0.2%, significantly outperforming traditional passive and active islanding detection methods [33]. These findings confirm that SVM-based approaches are highly effective in PV fault diagnostics, short-term energy forecasting, and grid stability assessments.

- **Decision Trees (DT) and Random Forest (RF)**

DTs offer interpretable rule-based structures for decision-making, while RF is an ensemble learning technique that constructs a collection of decision trees and averages their outputs to enhance accuracy and generalization.

A major application of RF and DT models is in solar PV power forecasting, where they outperform traditional statistical models in predicting short-term and long-term energy generation. A study integrating Random Forest with Data Envelopment Analysis for PV site selection demonstrated that RF models effectively predict solar panel efficiency across various locations, helping policymakers and energy planners optimize solar farm placement [34]. Another investigation in PV power forecasting under soiling conditions showed that RF models incorporating a Cleanness Index significantly improved accuracy, reducing the Mean Absolute Error from 1.24% to 0.22%, highlighting the effectiveness of feature selection in RF models [35].

A DT-based diagnostic framework for PV systems successfully classified inverter failures, bypass diode malfunctions, and partial shading faults with an accuracy exceeding 95% [36]. Additionally, an ensemble learning model combining DT, SVM, and kNN achieved 99.8% accuracy in detecting line-to-line and open-circuit faults [32].

A recent study on building-integrated PV (BIPV) systems combined RF and LSTM models, reducing the Root Mean Square Error (RMSE) from 4.75 to 2.97 for horizontal surfaces and significantly improving forecasting accuracy [37]. These findings highlight the strength of tree-based machine learning models in enhancing PV system performance, improving fault detection accuracy, and optimizing energy forecasting.

- **k-Nearest Neighbor (kNN)**

The kNN algorithm classifies instances based on similarity to neighboring data points.

A key application of kNN is in solar energy forecasting, where it has been employed to estimate solar radiation and PV power generation. One study applied a kNN-based forecasting model to predict short-term PV power output, achieving 10% to 25% improvement compared to reference persistence methods [38].

kNN models have also been utilized in the fault detection and diagnosis of PV systems. A hybrid kNN–SVM–DT ensemble learning approach was successfully implemented to classify PV panel faults, including snail trail degradation and microcracks, achieving 97.4% fault classification accuracy [32]. Additionally, a study on cloud computing-integrated PV monitoring employed kNN for real-time fault detection, reducing false alarms and misclassifications [31]. In PV site selection and power estimation, kNN has been applied in combination with geospatial analysis and clustering techniques to determine optimal locations for solar panel installation, improving solar harvesting potential while considering environmental conditions [39].

These findings highlight kNN's versatility and accuracy in forecasting, fault detection, and system optimization for PV applications.

- **Particle Swarm Optimization**

PSO is a bio-inspired optimization algorithm that mimics the behavior of bird flocks or fish schools to identify optimal solutions.

One study introduced a fuzzy adaptive PSO-based MPPT approach that dynamically adjusts PSO parameters, leading to a 14% faster convergence under shading conditions and 30% faster under uniform irradiation compared to conventional PSO [40]. Similarly, another study implemented a hybrid PSO-based MPPT for electric vehicles, optimizing power transfer from triple-junction solar cells to a DC–DC converter. The proposed approach outperformed P&O and other heuristic algorithms in terms of response time and efficiency [41].

In solar PV modeling and parameter estimation, PSO has been utilized to accurately identify PV cell parameters for both single-diode and double-diode models. An enhanced PSO method achieved higher parameter estimation accuracy with lower computational complexity than traditional numerical methods [42]. Additionally, a hybrid PSO–Grey Wolf Optimization

(PSO-GWO) algorithm demonstrated superior accuracy in extracting PV panel parameters, minimizing RMSE values compared to standalone optimization techniques [43].
A PSO-based two-axis solar tracker optimized panel positioning without requiring a mathematical sun movement model, achieving an increase in energy capture efficiency while reducing computational load [44]. These findings highlight the effectiveness of PSO-based approaches in enhancing MPPT performance, improving parameter estimation accuracy, and optimizing solar tracking systems.

A variety of software tools and platforms are commonly used to implement AI algorithms in solar PV applications, as can be observed in Table 2.

Table 2. Common software platforms used for implementing AI techniques in solar PV applications.

AI Technique	Common Software Tools
ANN, SVM, DT, RF	Python (Scikit-learn), MATLAB (ML Toolbox) [45,46]
CNN, LSTM	Python (TensorFlow, Keras, PyTorch), MATLAB (Deep Learning Toolbox) [21,47]
PSO, GA	MATLAB (Global Optimization Toolbox), Python (PyGAD, DEAP) [48]
FL	MATLAB (Fuzzy Logic Toolbox), Python (scikit-fuzzy) [16]
Reinforcement learning	Python (OpenAI Gym, Stable Baselines), MATLAB (Reinforcement Learning Toolbox) [49]

3.2.2. Applications Across the PV System Lifecycle

AI techniques have been utilized effectively across different stages of the solar PV system lifecycle. The five most common applications include the following.

- **Maximum Power Point Tracking (MPPT)**

MPPT is a technique used in PV systems to continuously adjust operating conditions and maximize power output under varying environmental conditions such as irradiance and temperature.
Among the reviewed studies, MPPT emerged as the most common application of AI in PV systems, with 39 studies dedicated to this topic. MPPT plays a critical role in optimizing energy yield from PV systems, particularly under fluctuating environmental conditions. The surveyed literature highlights a broad spectrum of AI-enhanced MPPT techniques, including machine learning, FL, ANNs, and metaheuristic optimization approaches. Metaheuristic optimization refers to flexible, nature-inspired algorithms used to solve complex optimization problems where traditional methods are inefficient or fail [50].
The studies revealed that hybrid AI-based MPPT strategies provide superior performance compared to traditional P&O or incremental conductance methods. For example, a neuro-fuzzy MPPT control strategy demonstrated an 8.2% increase in average power generation and 60% reduction in tracking time compared to conventional techniques [51]. Additionally, an AI-enhanced MPPT algorithm leveraging PSO and FL achieved a tracking efficiency of 99%, outperforming traditional approaches in dynamic irradiance scenarios [52]. Another study introduced a novel metaheuristic-based MPPT algorithm that significantly improved convergence speed and power stability under partial shading conditions [53].
Furthermore, studies integrating deep learning models for MPPT highlighted the potential for predictive control strategies, which anticipate changes in solar irradiance and adjust the duty cycle of the power converter in real time [27].

- **Power forecasting**

Solar power forecasting estimates the actual electricity output of a PV system, incorporating variables such as system configuration, inverter efficiency, and temperature.

Among the reviewed papers, solar power forecasting emerged as the second most common AI application in PV systems, with 36 studies dedicated to this topic. The accurate forecasting of solar PV power plays a critical role in grid stability, energy management, and economic planning, particularly given the intermittency and nonlinearity of solar energy generation. The reviewed literature showcases a diverse range of AI-driven methodologies aimed at improving forecasting accuracy, resilience to missing data, and adaptability to weather fluctuations.

The studies highlight that deep learning-based models, particularly LSTM networks and hybrid CNN-LSTM architectures, offer superior performance over traditional statistical methods. A notable study developed an attention-based CNN-LSTM model, achieving significant reductions in forecast error compared to ARIMA and SVR models, improving short-term forecasting accuracy [54]. Another paper introduced a hybrid transformer-based probabilistic forecasting model, which demonstrated superior uncertainty quantification capabilities over conventional ANN and gradient boosting approaches [55].

A missing-data tolerant LSTM model demonstrated robust performance in handling incomplete datasets while maintaining high forecasting accuracy across different weather conditions [56]. Similarly, a randomized learning ensemble method combining Extreme Learning Machines (ELM), Stochastic Configuration Networks (SCN), and Randomized Vector Functional Links (RVFL) improved probabilistic forecasting precision, reducing mean absolute percentage error (MAPE) by up to 35% [57].

Overall, the literature confirms that hybrid AI-based forecasting models outperform traditional methods in handling solar energy variability, uncertainty, and real-time adaptability.

- **Parameter estimation**

Parameter estimation involves identifying the optimal values of a PV system's internal model parameters (e.g., diode factors, resistances) to accurately simulate or control system behavior.

Parameter estimation emerged as the third most common application of AI in PV systems, with 27 studies dedicated to this topic. Accurate parameter estimation is essential for optimizing PV models, improving energy yield, and enhancing system efficiency under varying environmental conditions. The reviewed literature highlights a range of AI-based optimization approaches, including metaheuristic algorithms, deep learning models, and hybrid AI frameworks, aimed at accurately extracting key parameters of PV models.

The studies reveal that metaheuristic optimization algorithms significantly enhance parameter estimation accuracy. A hybrid Chimp-Sine Cosine Algorithm (HCSCA) demonstrated superior performance in estimating single-diode and double-diode model parameters, achieving an error reduction below 10^{-10} across multiple execution runs [58]. Similarly, an enhanced Slime Mould Algorithm (SMA) incorporating random learning and Nelder–Mead simplex methods exhibited higher convergence speed and robustness compared to traditional optimization approaches [59].

Deep learning models have also been integrated into parameter estimation frameworks, where a Multilayer Perceptron (MLP) model optimized using Marine Predators Optimization (MPO) successfully predicted PV system parameters with high accuracy, demonstrating its potential for real-time system tuning [60]. Another study employed a Gradient-Based Optimizer (GBO), which outperformed traditional swarm intelligence algorithms in extracting PV parameters, particularly under noisy and uncertain data conditions [61].

The findings indicate that hybrid AI-based parameter estimation models provide superior accuracy, faster convergence, and greater robustness in PV system modeling.

- **Fault detection/diagnosis/classification**

Fault detection refers to the identification and diagnosis of abnormal conditions or failures (e.g., shading, inverter faults) in PV systems to ensure reliability, safety, and performance. Fault detection, diagnosis, and classification in PV systems emerged as the fourth most common application of AI, with 14 studies dedicated to this topic. Accurate and automated fault detection is crucial for maintaining PV system reliability, optimizing performance, and preventing energy losses due to faulty modules, shading, and inverter failures. The reviewed literature highlights a range of AI-driven fault detection techniques, including deep learning models, hybrid AI frameworks, and statistical learning approaches, to improve the accuracy and efficiency of PV fault classification.

A hybrid CNN–Generative Adversarial Network model significantly enhanced PV fault detection by generating high-quality synthetic fault data, improving classification accuracy by 23% on small datasets [62]. Additionally, a multi-scale CNN-based deep learning model was developed to classify 11 different types of PV defects, achieving a fault classification accuracy of 97.32% [63].

Machine learning models such as SVM, DT, and RF were also extensively used in real-time monitoring and predictive maintenance. A DT-based classification model for grid-connected PV systems achieved 99.5% fault detection accuracy, effectively distinguishing between grid anomalies, inverter malfunctions, and module failures [64].

Furthermore, hybrid models have demonstrated robust real-time performance in fault detection and predictive maintenance. An adaptive neuro-fuzzy inference system (ANFIS) integrated with SVM achieved a classification accuracy of 95% in detecting partial shading, open-circuit, and bypass diode faults [65].

Overall, the literature confirms that AI-driven fault detection and classification techniques significantly enhance the reliability, accuracy, and efficiency of PV system monitoring.

- **Solar radiation forecasting/prediction**

Solar radiation forecasting focuses on predicting the intensity of solar irradiance reaching the Earth's surface, which is primarily used for input resource planning.

Among the reviewed papers, solar radiation forecasting has emerged as the fifth most common application of AI in PV systems, with seven studies dedicated to this topic. Accurate solar radiation prediction is crucial for optimizing PV system performance, energy management, and grid stability, as solar irradiance variations significantly impact PV power generation. The reviewed literature highlights a range of machine learning (ML) and deep learning (DL) models, integrating satellite imagery, meteorological data, and hybrid AI frameworks for improved forecasting accuracy.

One study proposed a hybrid CNN-MLP model, which successfully integrated global climate model data with observational meteorological inputs. The proposed model outperformed traditional statistical and standalone AI models, achieving lower RMSE and higher forecasting accuracy across multiple time scales [23]. Another study employed satellite imagery with a Convolutional Long Short-Term Memory (ConvLSTM) model, demonstrating a 3% improvement in RMSE compared to traditional time-series models [66].

A study integrating wavelet decomposition with feedforward neural networks (FFNNs) for intra-hour solar radiation forecasting achieved a forecast deviation of less than 4% in 90.6% of test cases, significantly outperforming persistence models [67]. Additionally, a hybrid ensemble learning approach combining extreme gradient boosting, light gradient boosting, and categorical boosting methods demonstrated high adaptability for real-time solar radiation forecasting in smart grid applications, further reducing forecasting errors [68].

The findings highlight that hybrid AI approaches, integrating deep learning, ensemble learning, and metaheuristic optimization techniques, significantly enhance solar radiation forecasting accuracy.

3.2.3. Trends over Time

The analysis of the reviewed literature highlights a clear progression in research focus over time, as follows:

- **2020–2022**—Early studies concentrated on exploratory applications of AI, primarily focusing on energy forecasting and fault detection;
- **2023–2024**—More recent research shows a shift towards integrated solutions, including hybrid AI models for MPPT, AI-enabled IoT system for real-time monitoring, and methods for managing environmental factors such as dust and shading.

This progressive evolution was identified using two main criteria. The first one is the thematic evolution network, which shows how keywords and research themes have shifted between the periods of 1998–2019 and 2020–2025, based on co-word analysis and keyword clustering. The second criterion is the keyword frequency and co-occurrence analysis, since the shift in research focus was further supported by the frequency of keywords in the reviewed publications.

4. Future Directions

The findings from this literature review point to several emerging trends poised to shape the future of AI applications in solar PV systems. As research progresses, these trends indicate a movement toward more integrated, resilient, and sustainable solutions aimed at tackling both technical and environmental challenges within renewable energy systems.

4.1. Hybrid AI Models for Enhanced Performance

The growing use of hybrid AI models is expected to persist as researchers aim to address the limitations of single-method approaches. By integrating techniques like fuzzy logic, neural networks, and optimization algorithms, hybrid models provide the flexibility and adaptability required to manage the inherent complexities of PV systems. These models are anticipated to become increasingly sophisticated, harnessing advancements in ensemble learning and metaheuristic optimization to enhance accuracy and efficiency in critical tasks such as MPPT and fault detection.

4.2. Integration of AI with IoT and Edge Computing

The integration of AI with IoT technologies and edge computing is poised to transform real-time monitoring and decision-making in PV systems. IoT-enabled sensors and edge devices can deliver continuous data streams, allowing AI models to process and respond to these data instantly. This approach will significantly improve predictive maintenance, fault detection, and energy management while minimizing the latency and computational demand of cloud-based systems. These advancements are expected to lead to autonomous PV systems capable of seamlessly adapting to dynamic environmental and operational conditions.

4.3. AI-Driven Energy Storage and Grid Integration

With the growing demand for renewable energy, integrating AI into energy storage systems and grid management will become increasingly vital. Future research is expected to prioritize AI models that optimize battery charging and discharging cycles, enhance energy storage efficiency, and ensure stable grid operations. AI's ability to forecast energy demand

and generation patterns will play a crucial role in managing the variability of solar power and maintaining a balanced energy supply. This development is particularly significant for regions transitioning to smart grids and renewable energy-based infrastructures.

Despite its potential, the integration of AI with energy storage and grid systems remains relatively underexplored due to several challenges, including the lack of standardized real-time data, limited interoperability between PV, storage, and grid components, and the complexity of managing hybrid systems under uncertainty. Practical barriers such as regulatory constraints and cybersecurity concerns also limit deployment. Future research should focus on developing open-access datasets, simulation tools like digital twins, and modular AI frameworks adaptable to diverse energy infrastructures.

4.4. *Increased Focus on Environmental and Operation Robustness*

Future research also will focus on developing AI models capable of operating reliably under a wide range of climatic operational conditions. Improving the generalizability and robustness of these systems will ensure consistent performance despite variations in solar irradiance, temperature, and dust accumulation. This emphasis will be crucial for successfully deploying PV systems in geographically and environmentally diverse regions, including areas with harsh or unpredictable conditions.

4.5. *Generative AI and Advanced Neural Architectures*

Generative AI and advanced neural architectures, such as transformers, are poised to revolutionize the future of PV systems. These models hold the potential to improve the design and optimization of PV components, accurately predict long-term systems performance, and simulate intricate interactions within energy systems. Their applications may also include optimizing solar panel manufacturing processes, promoting both economic efficiency and environmental sustainability.

4.6. *Standardization and Open-Source Data Sharing*

The absence of standardized datasets remains a significant challenge in the field. Future efforts are likely to prioritize the development of open-source data repositories and the establishment of benchmarking standards for AI models in solar PV systems. These initiatives will foster collaboration, enhance model training, and improve the reproducibility of research outcomes. Standardization will also allow for fair comparisons between different AI approaches, driving faster innovation and more efficient deployment.

4.7. *Localized and Context-Specific Solutions*

With an increasing emphasis on equitable energy access, future research is expected to emphasize localized AI solutions designed to meet the unique needs of developing regions. These solutions can tackle challenges such as affordability, infrastructure constraints, and energy poverty. By creating context-sensitive AI models, researchers can help ensure that the advantages of AI-powered PV systems reach underserved communities, contributing to global sustainability objectives.

The future of AI in solar PV systems is set for transformative advancements, fueled by hybrid modeling, IoT integration, energy storage optimization, and the adoption of cutting-edge AI techniques. These developments promise to improve the efficiency and reliability of PV systems while addressing broader challenges in renewable energy adoption and sustainability. By prioritizing robust, scalable, and context-sensitive solutions, the research community can help drive the transition to a cleaner and more sustainable energy future.

5. Selecting the Right AI Technique Depending on the Problem

AI offers a diverse range of techniques for optimizing PV systems, each with distinct strengths and limitations. Selecting the right AI technique depends on the specific problem to be addressed, such as power optimization, energy forecasting, fault detection, or predictive maintenance. This section serves as a practical guide for identifying PV challenges and matching them with the most suitable AI solutions.

Different AI models exhibit varying performance characteristics depending on the complexity, data availability, and real-time processing requirements of PV applications. Table 3 summarizes AI techniques, as well as their advantages, disadvantages, and ideal use cases.

Table 3. Advantages and disadvantages of AI techniques for PV challenges.

PV Challenge	Recommended AI Technique(s)	Advantages	Disadvantages
MPPT optimization	ANNs, FL, PSO	High adaptability to changing conditions, improved tracking efficiency, reduced oscillations [69–71].	High computational complexity for ANN, FL requires well-defined rules, PSO may get stuck in local optima [69–71].
Solar power forecasting	LSTM networks, CNN-LSTM hybrid, SVMs, DT	Captures temporal dependencies, effective for time-series data, with high accuracy in long-term prediction [72–74].	LSTM requires large datasets, SVMs struggle with high-dimensional input, DT may overfit [72–74].
Fault detection and diagnosis	CNNs, SVMs, R, kNN	High classification accuracy, suitable for image-based defect detection, real-time monitoring [38,72,75,76].	CNNs require labeled datasets and high processing power, SVMs slow with large data; kNN sensitive to noise [38,72,75,76].
Parameter estimation for PV models	PSO, ANNs, DT, GA	Fast optimization, efficient for non-linear problems, useful in PV model tuning [14,44,77,78].	PSO prone to local optima, ANN requires extensive training, GA computationally expensive [14,44,77,78].
Solar radiation prediction	LSTM, CNNs, Ensemble Learning RF, gradient boosting, ANFIS	Handles multiple meteorological variables, improves PV system planning [35,37,65,79,80].	Computationally intensive for deep learning models, require high-quality datasets [35,37,65,79,80].
Real-time grid stability and island detection	DT, RF, RL	Fast execution, effective for real-time grid monitoring, interpretable models [81–83].	DT prone to overfitting, RL requires extensive training and simulation [81–83].

Selecting the right AI model also depends on the nature of the data and computational constraints. The following guidelines can help determine the most suitable approach.

- **For small datasets or limited computational power** (rule-based AI and traditional ML models):
    - FL is ideal when expert knowledge can define system behavior (e.g., MPPT control);
    - SVMs and kNN work well when labeled datasets are small but need accurate classification (e.g., fault detection);
    - DT and RF provide fast, interpretable results for fault classification and grid monitoring.
- **For large, complex datasets with time dependencies** (DL models):
    - LSTM networks excel in forecasting applications where historical patterns impact future outcomes (e.g., power generation prediction);
    - CNNs are ideal for image-based defect detection (e.g., microcracks, panel degradation);
    - Hybrid CNN-LSTM models combine the strengths of spatial and temporal learning, improving solar energy prediction.

- **For real-time, dynamic optimization** (swarm intelligence and RL):
  - PSO, as part of swarm intelligence, mimics the collective behavior of decentralized systems such as bird flocks to efficiently explore and optimize solutions. It is widely used for MPPT control and parameter estimation, optimizing PV performance in changing conditions;
  - RL operates through trial-and-error learning, where an agent interacts with its environment to learn the best actions over time based on feedback. Thus, it is an emerging technique for adaptive grid management and real-time decision-making.

Furthermore, recent advancements highlight the importance of hybrid AI techniques, which combine multiple models to enhance performance and mitigate individual weaknesses. Some common hybrid approaches include the following:

- **FL-ANN for MPPT optimization**—Combines the adaptability of ANN with the interpretability of FL for real-time control;
- **CNN-SVM for fault detection**—Uses CNN for feature extraction and SVM for efficient classification;
- **LSTM-PSO for energy forecasting**—Leverages PSO for hyperparameter tuning in solar power prediction models;
- **RF + RL for learning for grid stability**—RF helps classify grid anomalies, while RL adapts to dynamic energy fluctuations.

To assist in selecting the appropriate AI techniques, the decision-making framework observed in Figure 7 can be used.

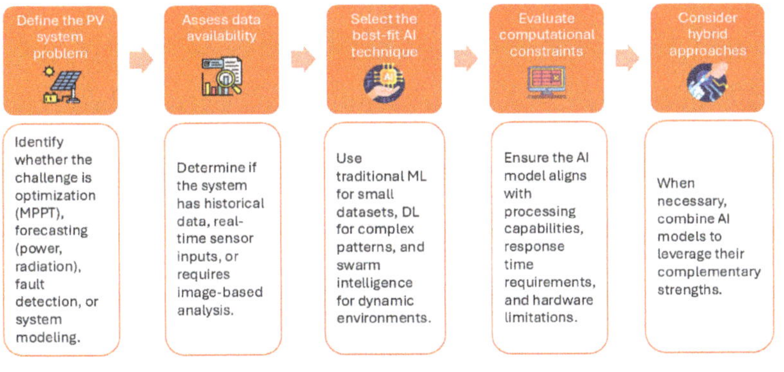

Figure 7. Decision-making framework for selecting AI technique.

5.1. Fuzzy Expert System for AI Selection in Solar PV Energy Applications

To enhance decision-making in selecting the most suitable AI technique for PV system optimization, a fuzzy expert system was developed in MATLAB using the Fuzzy Logic Designer tool based on the AI selection framework outlined in the literature review. This system automates the AI selection process based on problem type and system characteristics, enabling users—regardless of their AI expertise—to obtain precise recommendations for fault detection, energy forecasting, MPPT optimization, and other solar energy applications. These problem types correspond to the major areas of AI research in solar PV systems, ensuring that the fuzzy logic system is designed to address real-world challenges.

This system provides multiple benefits that make it a powerful and flexible decision-support tool for the solar PV energy sector. The system eliminates the need for manual AI selection by automatically recommending the most suitable technique based on input parameters, which saves time and ensures the most effective AI selection. The system also allows users to modify membership functions and rule sets in MATLAB's Fuzzy Logic Designer, enabling customization as new AI techniques emerge or as system requirements evolve. Furthermore, unlike static decision models, this fuzzy system evaluates multiple AI techniques to provide the most effective approach for each problem.

The fuzzy system operates by taking a problem definition as input and recommending an AI technique as output, as can be observed in Figure 8.

For the inputs, the user specifies the type of PV system problem to be solved. For the outputs, based on the problem type, the fuzzy system provides a recommendation for the most effective AI technique, as shown in Table 4.

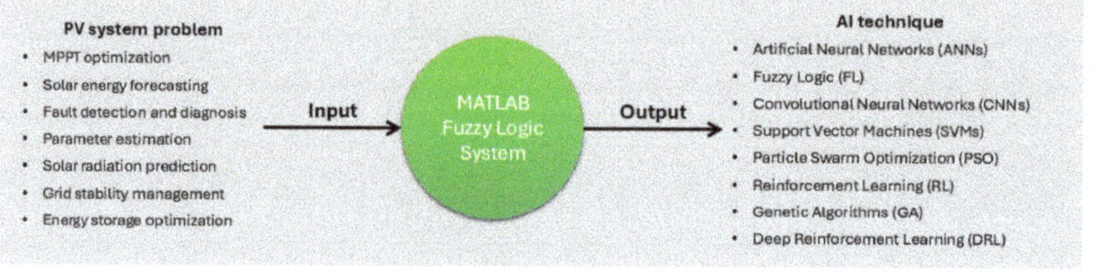

Figure 8. Fuzzy logic system—input and output diagram.

Table 4. PV systems' most common problems and AI techniques recommended to solve them.

PV System Problem	Recommended AI Technique(s)
• **MPPT optimization:** Selecting AI techniques that improve energy harvesting efficiency. • **Solar energy forecasting:** Identifying suitable models for predicting solar power generation trends. • **Fault detection and diagnosis**: Selecting AI approaches that enhance PV system monitoring and predictive maintenance. • **Parameter estimation:** Recommending techniques that tune system models for better performance. • **Solar radiation prediction:** Identifying AI-driven weather prediction and irradiance modeling solution. • **Grid stability management:** Choosing AI techniques for balancing PV output with grid demands. • **Energy storage optimization:** Selecting methods that improve battery management and energy dispatching.	• **ANNs:** Suitable for pattern recognition, forecasting, and MPPT optimization. • **FL:** Ideal for handling uncertainty in fault detection and real-time energy adjustments. • **CNNs:** Best for image-based fault detection (e.g., thermal imaging, microcrack analysis). • **SVMs:** Recommended for classification-based fault detection and efficiency diagnosis. • **PSO:** Effective in finding optimal MPPT settings and tuning model parameters. • **RL:** Applied to self-learning grid management and dynamic energy optimization. • **GA:** used in system optimization tasks where multiple variables need fine-tuning. • **DRL:** Applied in autonomous grid management and self-optimizing PV systems.

The Fuzzy Logic Designer in MATLAB enables users to modify the system based on evolving AI advancements. By adjusting the membership functions, rule sets, and decision parameters, the system can accommodate new AI techniques and adapt to different PV applications. This ensures that the model remains relevant and scalable for future developments in solar energy and AI research. Figure 9 illustrates the fuzzy logic system developed in MATLAB. To run the system, the user provides an input (e.g., "MPPT Optimization") and obtains an AI recommendation (e.g., "PSO or ANN").

The fuzzy system applies rule-based logic derived from findings in the literature to match input conditions with AI technique recommendations. These rules integrate knowledge from multiple studies evaluating AI models in terms of accuracy, computational efficiency, and adaptability. For example, if the problem complexity is high and data availability is low, FL or DT is recommended due to the lower computational demands of both. If the system requires MPPT optimization, the literature suggests that PSO and ANNs are the most effective, as they have demonstrated superior performance in tracking the maximum power point. Similarly, for fault detection applications requiring image-based analysis, CNNs are prioritized due to their high classification accuracy.

A total of 25 if–then rules were developed, linking specific PV challenges to recommended AI approaches. All rules were assigned equal weight in the current system configuration, reflecting equal importance across decision criteria. Table 5 shows the rules integrated into the system.

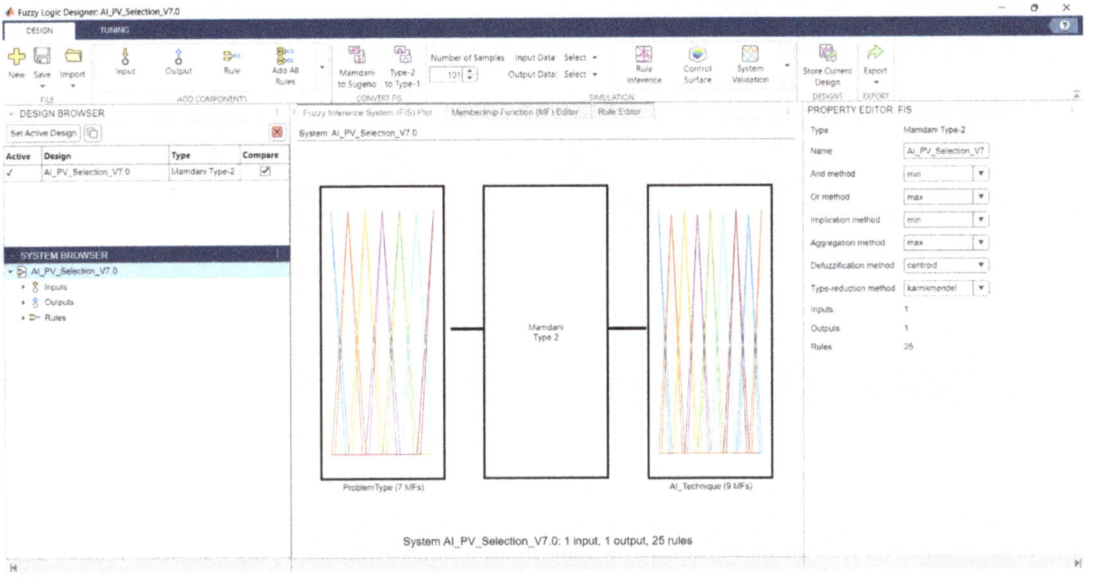

Figure 9. Fuzzy logic system for solar PV energy problems.

Table 5. If–then rules for PV energy problems and AI techniques.

Rule No.	IF Problem Type	THEN AI Technique
1	MPPT Optimization	ANN
2	MPPT Optimization	Fuzzy Logic
3	MPPT Optimization	PSO
4	MPPT Optimization	Hybrid AI
5	Solar Power Forecasting	Fuzzy Logic
6	Solar Power Forecasting	CNN
7	Solar Power Forecasting	Hybrid AI
8	Solar Power Forecasting	Genetic Algorithm (GA)
9	Fault Detection	CNN
10	Fault Detection	SVM
11	Fault Detection	Hybrid AI
12	Fault Detection	Genetic Algorithm (GA)
13	Parameter Estimation	SVM
14	Parameter Estimation	PSO
15	Parameter Estimation	Hybrid AI
16	Parameter Estimation	Genetic Algorithm (GA)
17	Solar Radiation Prediction	PSO
18	Solar Radiation Prediction	Reinforcement Learning
19	Solar Radiation Prediction	Hybrid AI
20	Solar Radiation Prediction	Deep Reinforcement Learning
21	Grid Stability Management	Reinforcement Learning
22	Grid Stability Management	Hybrid AI
23	Grid Stability Management	Deep Reinforcement Learning
24	Energy Storage Optimization	Energy Storage Optimization
25	Energy Storage Optimization	Deep Reinforcement Learning

The membership functions were designed using Gaussian shapes to model uncertainty smoothly across input and output variables. Placement was determined based on the categorization of PV problems and AI techniques identified in the review. Each function was centered on a specific problem type or technique, with overlapping ranges to allow flexible reasoning during fuzzy inference. Fine-tuning was performed manually using MATLAB's Fuzzy Logic Designer.

To further illustrate the structure of the developed fuzzy expert system, Figure 10 presents the conceptual architecture, outlining the main components involved in processing input data to generate AI technique recommendations.

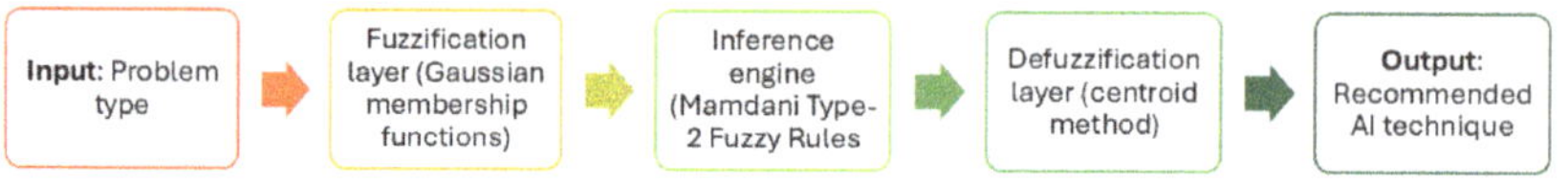

Figure 10. Conceptual architecture of the Type-2 Fuzzy Logic Expert System developed for AI technique selection in solar PV applications.

The fuzzy system does not provide a single crisp output, but instead ranks AI techniques based on membership values, allowing for flexible decision-making. If a user specifies MPPT optimization with high computational constraints, the system might rank PSO as the most suitable, followed by ANN and FL, with decreasing membership values. This ranking process is based on literature-backed evaluations of AI techniques rather than arbitrary selection, ensuring that the system's recommendations align with empirical research.

5.1.1. Type-2 Fuzzy Logic System

The literature review highlights that solar PV optimization involves high uncertainty due to environmental variability, incomplete data, and fluctuating operational conditions. Traditional type-1 fuzzy logic assigns fixed membership values, which may not fully capture the uncertainty present in AI selection for PV applications [84]. To address this limitation, the developed fuzzy expert system utilizes type-2 fuzzy logic, which incorporates Upper and Lower Membership Functions (UMF and LMF) to create a Footprint of Uncertainty (FOU) [85]. This approach allows the system to handle data variability more effectively, ensuring robust AI recommendations even under uncertain conditions [86]; thus, the additional degree of flexibility makes it highly suitable for complex and dynamically changing environments, such as solar PV systems. Studies in the literature emphasize the importance of adaptable AI selection frameworks, making type-2 fuzzy logic a suitable choice for this decision-support tool [87,88].

The system is designed to reflect real-world PV system problems identified in research, ensuring that its AI selection methodology is based on scientific evidence rather than subjective choice. The rule-based structure of the fuzzy system is supported by empirical studies on AI performance in different PV applications, making it a reliable tool for researchers, industry professionals, and policymakers. By incorporating type-2 fuzzy logic, the system further enhances decision-making by addressing the inherent uncertainty in AI model selection. The integration of knowledge from 163 reviewed studies ensures that the fuzzy logic system is scientifically grounded, practical for industry applications, and adaptable to future advancements in AI for solar PV optimization.

One of the main reasons why type-2 fuzzy logic is a better option is its ability to manage the high uncertainty in AI selection for PV applications. The selection of an AI model depends on multiple uncertain factors, all of which may fluctuate in real-world scenarios. In a type-1 fuzzy system, membership values are fixed, which means the system may not accurately capture variations in AI suitability due to unpredictable changes in solar irradiance, temperature, or cloud cover. A type-2 fuzzy system, on the other hand, introduces an extra layer of flexibility by considering a range of possible membership values, ensuring the AI selection remains robust even when input conditions are highly uncertain [84].

Another key advantage of type-2 fuzzy logic is its enhanced ability to handle noisy and imprecise data. In solar PV systems, sensor readings may have errors, missing data, or inconsistencies due to environmental disturbances. Type-1 fuzzy logic struggles in these cases since it assigns a single membership value, which may not adequately reflect the uncertainty present in the data. Type-2 fuzzy logic compensates for this by allowing multiple possible membership values, providing a more reliable decision-making process. This ensures that even if sensor data fluctuate or contain noise, the AI selection remains stable and less affected by anomalies [89].

Furthermore, the type-2 fuzzy logic system improves the generalization ability in decision-making frameworks. Since it inherently accounts for uncertainty, it provides more stable AI recommendations across different solar PV installations, locations, and datasets [90]. In contrast, type-1 fuzzy logic may be overfitted to specific conditions, requiring manual tuning when applied to different environments [91]. Type-2 fuzzy logic enables a more universal AI selection model, reducing the need for constant recalibration when applied to new datasets or operational conditions [86].

In the context of AI selection for PV applications, a type-2 fuzzy system enhances decision reliability by ensuring that AI recommendations are not overly sensitive to variations in input data. If different AI techniques have overlapping suitability levels due to fluctuating conditions, a type-1 system may struggle to differentiate them precisely. A type-2 fuzzy system resolves this issue by accommodating a range of potential values, allowing for more nuanced AI ranking and selection.

Overall, type-2 fuzzy logic is a superior option for AI selection in solar PV applications because it handles uncertainty more effectively, reduces sensitivity to noise, enhances adaptability to changing conditions, and improves generalization across different environments. Its ability to model imprecise and dynamic membership values ensures that AI selection remains robust, flexible, and optimized for real-world PV system challenges. By implementing type-2 fuzzy logic in AI selection frameworks, researchers and industry professionals can achieve more reliable, scalable, and adaptive AI-driven solar PV optimization.

5.1.2. Fuzzy Values

The fuzzy values in the fuzzy expert system for AI selection in solar PV applications handle uncertainty and imprecise information in decision-making. Unlike crisp values, which provide binary or exact outcomes, fuzzy values represent degrees of membership within a defined range, typically between 0 and 1. This allows the system to determine the suitability of different AI techniques rather than providing a single fixed answer.

Recent studies have extended fuzzy logic with multi-criteria decision-making methods, such as fuzzy AHP for process optimization in laser cutting applications [92], and health system monitoring [93]. These works highlight the versatility of fuzzy frameworks in structured decision problems, and support the relevance of our expert system approach in energy domains.

In the first step, fuzzification is applied to convert input parameters, such as problem complexity, data availability, and computational cost, into fuzzy linguistic terms such as "low", "medium", or "high". For instance, if the computational cost is at 75%, the system may classify it as "high", with a membership value of 0.8, and "medium" with a membership value of 0.2. This classification ensures that the AI selection process considers multiple possible techniques instead of defaulting to a single approach.

Once the inputs have been fuzzified, the system evaluates predefined fuzzy rules to determine the best AI technique for a given problem. A typical rule might state that if problem complexity is high and data availability is low, then FL or DT should be prioritized over computationally intensive models such as CNNs or LSTM networks. The membership functions play a key role in this process by ensuring that different AI techniques are ranked according to their suitability rather than being strictly classified as either acceptable or unacceptable.

Following the rule evaluation, the fuzzy system aggregates the results in the defuzzification stage. This process translates fuzzy outputs into a ranked list of AI techniques, where each technique receives a numerical score indicating its relative appropriateness. If, for example, the system evaluates AI techniques for a fault detection problem and assigns membership values of 0.90 to FL, 0.80 to DT, 0.75 to PSO, 0.65 to ANN, and 0.50 to CNN, then FL and DT would be recommended first. At the same time, ANN and CNN would be deprioritized due to their lower suitability scores.

The fuzzy system's ability to process and rank AI techniques based on membership values rather than absolute classification provides several advantages. It accommodates uncertainty in AI selection by considering multiple AI techniques rather than forcing a strict binary choice. It also provides adaptability by ensuring that AI recommendations dynamically update as system conditions evolve. Additionally, it enhances decision-making transparency by providing a structured ranking of AI techniques based on well-defined criteria.

In practice, when selecting an AI technique for fault detection in a PV system, the fuzzy system processes input conditions such as problem complexity, computational cost, and data availability. It assigns membership values to these parameters, influencing the ranking of AI techniques. Suppose problem complexity is classified as high with a membership value of 0.75, computational cost as medium with a membership value of 0.60, and data availability as low with a membership value of 0.40. In that case, the system evaluates these inputs against predefined fuzzy rules. As a result, the fuzzy system may determine that FL is the best choice with a membership value of 0.85, followed by SVM at 0.75, ANN at 0.60, and CNN at 0.40. This outcome ensures that FL and SVM are prioritized for fault detection in this specific scenario, while CNN is deprioritized due to its higher computational cost and lower suitability under the given conditions.

6. Discussion

The bibliometric analysis revealed key trends in AI applications for PV systems, with a steady increase in research publications over the past decade, reflecting the growing academic and industrial interest in AI-driven PV optimization. The highest number of publications appeared between 2018 and 2023, indicating a surge in AI research for renewable energy integration. Geographic trends in the analysis showed that China, the United States, and European countries contributed the largest share of publications, likely due to their investments in AI research and solar energy projects.

The systematic literature review highlights the increasing role of AI in enhancing PV systems by improving efficiency, reliability, and operational stability. The analysis identified ANNs, FL, CNNs, LSTM networks, SVMs, DT, RF, kNN, and PSO as the most frequently employed AI techniques in PV applications. Among these, ANNs were the most commonly used, particularly in MPPT, fault detection, and power forecasting, achieving power tracking efficiencies of up to 99.9% and fault diagnosis accuracies exceeding 99%. The use of hybrid AI models (e.g., ANN-FL, CNN-LSTM, PSO-FL, DT-RF) has demonstrated superior performance, consistently outperforming standalone techniques by leveraging multiple optimization and learning strategies.

Metaheuristic approaches such as PSO and GA have also been widely applied, particularly for parameter estimation, MPPT optimization, and solar tracking, with studies showing significant reductions in convergence time and improved tracking accuracy under dynamic environmental conditions. Hybrid models that integrate metaheuristic algorithms with machine learning approaches have further enhanced PV system adaptability, ensuring higher resilience against uncertainties in solar power generation. The literature confirms that combining AI with optimization techniques significantly improves energy efficiency and fault-tolerant capabilities, making these methodologies highly relevant for real-world PV system applications.

The review has also identified the most common AI applications in PV systems, with MPPT emerging as the most extensively researched area, followed by solar power forecasting, parameter estimation, fault detection, and solar radiation forecasting.

The fuzzy expert system developed for AI selection in solar PV applications is based on the systematic literature review, which identifies key AI techniques used for solar PV optimization, categorizes their applications, and evaluates their effectiveness in different scenarios. The literature review provides a foundation by analyzing 163 studies that highlight the strengths and limitations of various AI methods, such as ANNs for MPPT optimization and power forecasting, FL for handling uncertain conditions and fault detection, CNNs for image-based defect identification, LSTM networks for time-series-based solar forecasting, and SVMs for classification-based efficiency diagnosis. Other methods, including DT, RF, kNN, PSO, and RL, have also been identified as useful for different aspects of PV system optimization.

Applying the fuzzy expert system in AI selection for solar PV applications ensures that rigid classifications do not restrict decision-making, but that this latter is instead guided by a nuanced and adaptable ranking system. This approach significantly improves the efficiency of AI selection by considering multiple influencing factors simultaneously and adjusting recommendations based on real-time conditions. The fuzzy system enhances the reliability, accuracy, and scalability of AI-driven decision-making in solar energy applications through membership values.

Future work will involve the practical validation of the fuzzy expert system through case studies in industrial solar PV projects. This will include applying the system to real-world scenarios, collecting feedback from domain experts, and comparing its output with existing expert decision-making practices. Such validation will be essential to assessing the system's applicability, trustworthiness, and value in operational environments.

6.1. Challenges and Gaps

Despite the significant progress made in applying AI to solar PV systems, several challenges and gaps remain. A key issue identified in the reviewed literature is the lack of standardized, high-quality datasets. This limitation hampers the training, validation, and benchmarking of AI models, making it difficult to compare results across studies or implement solutions on a larger scale. Furthermore, the scalability and real-time performance of AI systems present notable challenges, particularly for large-scale PV installations. While advanced models offer high accuracy, their substantial computational demands often impede their use in real-time applications.

Another important gap lies in the limited exploration of AI-driven integration with energy storage and grid management systems. While significant progress has been made in standalone PV systems, research on coordinating solar energy with storage solutions or ensuring grid stability remains scarce. Additionally, many AI models face challenges in generalizing across diverse operational conditions and climatic environments. Variations in solar irradiance, temperature, and other contextual factors can greatly impact the performance of these algorithms, highlighting the need for robust and adaptable models. Addressing these challenges is crucial for fully harnessing AI's potential to accelerate the shift toward sustainable energy systems.

Another significant insight is the prominence of fault detection and predictive maintenance as leading application areas. This focus underscores the industry's priority on improving system reliability and lowering operational costs—key factors for the broader adoption of solar PV technology. In recent years, there has been a notable shift toward real-time AI applications, fueled by the integration of advanced AI techniques with IoT and edge computing. This trend emphasizes the increasing importance of rapid decision-making and continuous system monitoring in the operation of modern PV systems.

Tailored AI applications in these areas could address unique challenges, including energy access, affordability, and infrastructure limitations, paving the way for more inclusive advancements in renewable energy.

In selecting the appropriate AI technique for PV system optimization, it is crucial to consider the specific problem, data availability, and computational constraints, as observed in Figure 7. A problem-driven approach to AI selection ensures that PV systems benefit from the most effective computational intelligence techniques, leading to higher efficiency, reliability, and scalability in real-world applications.

A crucial consideration in adopting AI techniques for solar PV system applications is the trustworthiness and reliability of these methods. AI models, particularly those developed through machine learning and generative approaches, inherently depend on the quality and representativeness of the data used for training. The validation of AI models typically involves rigorous methodologies such as cross-validation, comparison with real operational data, and sensitivity analyses to ensure robustness and accuracy. In this work, the authors have selected techniques that have been extensively validated in the literature, and prioritize those with proven accuracy and reliability in empirical applications. Additionally, they acknowledge that Generative AI models can sometimes produce inaccurate or misleading information if not properly trained or guided. To address this, all AI-generated content in this study was manually verified against original literature sources and validated using expert judgement to ensure correctness and relevance. Deep learning models, in particular, often require large, high-quality datasets and are prone to overfitting if not carefully regularized, which can limit their generalizability in real-world PV scenarios. Moreover, the proposed type-2 fuzzy logic expert system inherently accounts for uncertainty, enhancing decision reliability under variable operational and environmental conditions. However, users should remain cautious, continually validating and updating AI models with new data, particularly when they are deployed in dynamic real-world scenarios.

6.2. *Practical Implications for Researchers, Industry, and Policymakers*

The integration of AI in PV systems has significant practical implications for various stakeholders, including researchers, industry professionals, and policymakers. As AI-driven techniques continue to advance, understanding their real-world applicability is essential for bridging the gap between academic research, industrial deployment, and regulatory frameworks.

For researchers, the findings of this study highlight the importance of selecting the right AI technique depending on the specific PV application. Future research should focus on developing lightweight AI models that can operate in real time on edge computing devices, reducing reliance on cloud processing. Additionally, the lack of standardized datasets and benchmarking frameworks for AI-based PV optimization remains a challenge. Open-access datasets, standardized testing environments, and collaborative AI modeling efforts can enhance reproducibility and accelerate innovation in the field.

For industry, AI presents significant opportunities for enhancing efficiency, reducing maintenance costs, and improving grid integration. AI-based fault detection and predictive maintenance systems can enable proactive monitoring, minimizing energy losses and reducing downtime. Companies investing in AI-driven MPPT optimization can achieve higher energy yields, improving the financial viability of PV installations. Moreover, AI-powered forecasting models can help PV operators better integrate renewable energy into the grid, reducing fluctuations and improving demand–response strategies. However, to fully leverage AI, the industry must invest in scalable AI solutions that can be seamlessly integrated with existing PV monitoring systems and energy management platforms.

Policymakers play a crucial role in facilitating AI adoption in the solar energy sector by implementing supportive regulations and incentives. Data-sharing policies should be established to promote collaborative AI model development, allowing researchers and industry stakeholders to access real-world PV system data. Governments can also encourage investment in AI research for renewable energy through grants, subsidies, and regulatory frameworks that incentive AI-based energy optimization. Finally, ethical and environmental considerations, such as AI's energy consumption and data privacy concerns, must be addressed to ensure sustainable and responsible AI deployment in the PV sector.

7. Conclusions

This review underscores the transformative impact of AI in advancing solar PV systems, highlighting its ability to tackle critical challenges such as efficiency optimization, fault detection, energy forecasting, and system reliability. Through a systematic analysis of 163 studies published between 2020 and 2024, this study explores a diverse range of AI techniques—including machine learning, neural networks, fuzzy logic, and hybrid approaches—applied across different stages of the PV system lifecycle. The findings highlight key trends, such as the increasing use of hybrid AI models, the integration of AI with IoT for real-time applications, and a growing focus on addressing practical, system-level challenges.

This work offers valuable contributions by presenting a comprehensive synthesis of global research trends, emerging themes, and existing gaps in the field. Key challenges identified include the absence of standardized datasets, scalability issues for large-scale deployments, and the limited integration of AI with energy storage and grid systems. While these challenges pose obstacles, they also create opportunities for further research and innovation. By tackling these issues, researches and practitioners can harness AI's full potential to enhance the efficiency and sustainability of PV systems. To maximize the benefits of AI in PV systems, stronger collaboration between researchers, industry leaders, and policymakers is needed.

A major contribution of this work is the development of a type-2 fuzzy logic system in MATLAB, designed to automate AI selection for PV applications. This system enables researchers, engineers, and industry professionals to identify the most appropriate AI technique for a given PV problem, reducing manual effort and improving decision-making efficiency. This approach significantly improves the efficiency of AI selection by considering multiple influencing factors simultaneously and adjusting recommendations based on real-time conditions.

Looking forward, future research should prioritize the development of scalable, real-time AI models for large-scale PV systems, the integration of AI with energy storage and grid management technologies, and the creation of standardized datasets to enable benchmarking and collaboration. In parallel, increasing attention should be paid to explainable

AI (XAI) techniques, which can improve the transparency and interpretability of AI-driven decisions, particularly in critical infrastructure such as solar energy systems. Furthermore, designing localized AI solutions tailored to the specific needs of developing regions can be instrumental in enhancing energy access and infrastructure. Addressing these challenges will allow AI to further drive advancements in solar PV systems and play a pivotal role in the global transition to a sustainable energy future.

Author Contributions: Conceptualization, C.P.-B. and P.P.; methodology, C.P.-B. and P.P.; software, C.P.-B. and P.P.; validation, P.P.; formal analysis, C.P.-B.; investigation, C.P.-B. and P.P.; resources, P.P.; data curation, C.P.-B.; writing—original draft preparation, C.P.-B. and P.P.; writing—review and editing, P.P., Q.M. and A.R.F.; visualization, C.P.-B.; supervision, P.P., A.R.F. and Q.M.; project administration, P.P.; funding acquisition, P.P. All authors have read and agreed to the published version of the manuscript.

Funding: This research project is supported by Institute of Advanced Materials for Sustainable Manufacturing, Tecnologico de Monterrey.

Data Availability Statement: Data will be made available on request.

Acknowledgments: The authors would like to acknowledge the technical support Tecnologico de Monterrey, and the University of Alberta in the production of this work.

Conflicts of Interest: The authors declare no conflicts of interest.

Abbreviations

The following abbreviations are used in this manuscript:

AI	Artificial Intelligence
PV	Photovoltaic
PRISMA	Preferred Reporting Items for Systematic Review and Meta-Analyses
ANNs	Artificial Neural Networks
FL	Fuzzy Logic
CNNs	Convolutional Neural Networks
LSTM	Long Short-Term Memory
SVMs	Support Vector Machines
DT	Decision Trees
RF	Random Forest
kNN	k-Nearest Neighbors
PSO	Particle Swarm Optimization
MPPT	Maximum Power Point Tracking
PCA	Principal Component Analysis
PDet	Progressive Deformable Transformer
Extra Trees	Extremely Randomized Trees
LR	Linear Regression
ABOD	Angle-Based Outlier Detection
O&M	Operation and Maintenance
NN	Neural Networks
GAI	Generative AI
AOA	Arithmetic Optimized Algorithm
ELM	Extreme Learning Machine
CSA	Crow Search Algorithm
SENMSSA	Super-Evolutionary Nelder–Mead Salp Swarm Algorithm
GRU	Gate Recurrent Unit
CVO	Coronavirus Optimization

ANFHC	Adaptive Neuro–Fuzzy Hybrid Controller
FFNN	Feedforward Neural Network
MsSCNN	Multiscale Siamese Convolutional Neural Network
P&O	Perturb and Observe
P2P	Peer-to-peer
DAGRU	Dual-attention Gated Recurrent Units
LCA	Life Cycle Assessment
FES	Fuzzy Expert System
DRL	Deep Reinforcement Learning
PSM	Proportional Selection Method
RSM	Random Selection Method (RSM)
DSF	Dense and Successive Features
HFPE	Hierarchical Feature Precision and Extraction
CCEAF	Contextual Characteristics Extraction and Attribute Fusion
MFO	Moth Flame Optimization
MLP	Multilayer Perceptron
TSFS	Two-stage Feature Selection
ICMODA	Improved Chaotic Multi-objective Dragonfly Algorithm
SMFTS	Seasonal Multivariable Fuzzy Time Series
ORELM	Outlier-robust Extreme Learning Machine
1D-CNN	One-dimensional Convolutional Neural Network
MPC	Model Predictive Control
MFSO	Modified Fluid Search Optimization
ANFIS	Adaptive Neural-fuzzy Inference System
FLC	Fuzzy logic controller
GA	Genetic Algorithm
ABC	Artificial Bee Colony
TLBO	Teaching Learning-based Optimization
DTBO	Driving Training-based Optimization
HYTREM	Hybrid Tree-based Ensemble Learning Model
ELADE	Elite Learning Adaptive Differential Evolution
NFN	Neuro Fuzzy Network
NF	Nonlinear Function
SMC	Sliding Mode Control
FPA	Flower Pollination
CSA	Cuckoo Search Algorithm
WGAN	Wasserstein Generative Adversarial Network
RERNN	Recalling-enhanced Recurrent Neural Network
SGO	Shell Game Optimization
MRAC	Model Reference Adaptive Control
FCM	Fuzzy C-means
WOA	Whale Optimization Algorithm
LSSVM	Least Squares Support Vector Machine
ISSA	Improved Sparrow Search Algorithm
DNN	Deep Neural Network
ECANet	Efficient Channel Attention Module
TCN	Temporal Convolutional Network
VGG	Visual Geometry Group
RESNET	Residual Neural Network
ICSA	Chameleon Swarm Algorithm
RPNN	Recurrent Perceptron Neural Network
MLP	Multi-layer Perceptron
BSA	Backtracking Search Optimization Algorithm

GWO	Grey Wolf Optimizer
HHO	Harris Hawks Optimization
LGBM	Light Gradient Boosting
UPQC	Unified Power Quality Conditioner
EHO	Elephant Herding Optimization
ABWO	Adaptive Black Widow Optimization Algorithm
SDA	Similar Day Analysis
ECPA	Enhanced Colony Predation Algorithm
LNMHGS	Laplacian Nelder–Mead Hunger Games Search
TFT	Temporal Fusion Transformer
HBA	Honey Badger Algorithm
DLH	Dimensional Learning Hunting
DE	Differential Evolution
SAR	Search and Rescue Optimization Algorithm
HGS	Hunger Games Search
ASMA	Adaptive Slime Mould Algorithm
ET	Extra Tress
GB	Gradient Boosting
HCSCA	Hybrid Chimp–Sine Cosine Algorithm
FWA	Evolutionary Fireworks Algorithm
LsSVR	Least Square Support Vector Regression
TLABC	Teaching–Learning-based Artificial Bee Colony
IWA	Invasive Weed Algorithm
ISMA	Improved Slime Mould Algorithm
DOA	Dragonfly Optimization Algorithm
LASSO	Least Absolute Shrinkage and Selection Operator
PR	Polynomial Regression
DL	Deep Learning
EMSFLA	Ensemble Multi-strategy-driven Shuffled Frog Leading Algorithm
ALSM	Attention-based Long-term and Short-term Temporal NN Prediction Model
MRTPP	Multiple Relevant and Target Variables Prediction Pattern
SFLBS	Shuffled Frog-leaping Algorithm
ELM	Extreme Learning Machine
RLHE	Randomized Learning-based Hybrid Ensemble
CCO	Crisscross Optimizer
EO	Equilibrium Optimizer
SSA	Salp Swarm Optimization
MFO	Orthogonal Moth Flame Optimization
THD	Total Harmonic Distortion
VMD	Variational Mode Decomposition
BIPV	Building-integrated PV
RMSE	Root Mean Square Error
ELM	Extreme Learning Machines
SCN	Stochastic Configuration Networks
RVFL	Randomized Vector Functional Links
MAPE	Mean Absolute Percentage Error
SMA	Slime Mould Algortihm
MPO	Marine Predators Optimization
GBO	Gradient-based Optimizer
UMF	Upper Membership Functions
LMF	Lower Membership Functions
FOU	Footprint of Uncertainty

Appendix A

Table A1. Papers included in the final review.

Reference	Title	Year	AI technique	Application
[94]	An innovative hybrid model combining informer and K-Means clustering methods for invisible multisite solar power estimation	2024	MissForest, K-means, Principal Component Analysis (PCA)	Location selection
[95]	Levenberg-Marquardt algorithm-based solar PV energy integrated internet of home energy management system	2024	Levenberg-Marquardt, Bayesian Regularization, Scaled Conjugate Gradient	Energy management
[21]	Enhanced Fault Detection in Photovoltaic Panels Using CNN-Based Classification with PyQt5 Implementation	2024	CNN and VGG16 architecture	Fault detection
[96]	PDeT: A Progressive Deformable Transformer for Photovoltaic Panel Defect Segmentation	2024	Progressive Deformable Transformer (PDeT)	Defect segmentation
[64]	Explainable artificial intelligence of tree-based algorithms for fault detection and diagnosis in grid-connected photovoltaic systems	2024	Extremely Randomized Trees (Extra Trees)	Fault detection and diagnosis
[97]	Research on a Photovoltaic Panel Dust Detection Algorithm Based on 3D Data Generation	2024	YOLOv8 model, SENetV2, AKConv, and DySample	Dust detection
[98]	Non-invasive health status diagnosis of solar PV panel using ensemble classifier	2024	Ensemble Classifier	Health monitoring
[18]	Comparative study of real-time photovoltaic fault diagnosis using artificial intelligence: Fuzzy logic and neural network approaches	2024	FL and ANN	Fault diagnosis
[99]	An adaptive method for real-time photovoltaic power forecasting utilizing mathematics and statistics: Case studies in Australia and Vietnam	2024	Linear regression (LR)	Power forecasting
[32]	An ensemble learning framework for snail trail fault detection and diagnosis in photovoltaic modules	2024	SVM, KNN, and DT	Fault detection and diagnosis
[36]	Novel data-driven health-state architecture for photovoltaic system failure diagnosis	2024	XGBoost, DT, KNN, and Angle-Based Outlier Detection (ABOD)	Fault diagnosis and predictive O&M
[100]	A feature space class balancing strategy-based fault classification method in solar photovoltaic modules	2024	CNN, Feature Space Class Balancing, PatchUp-based Feature Mixing:	Fault classification
[101]	Backpropagation artificial neural network-based maximum power point tracking controller with image encryption inspired solar photovoltaic array reconfiguration	2024	ANN	MPPT
[25]	A novel hybrid intelligent approach for solar photovoltaic power prediction considering UV index and cloud cover	2024	LST and CNN	Power forecasting
[102]	Battery-less uncertainty-based control of a stand-alone PV-electrolyzer system	2024	NN and FL	Power forecasting
[22]	Radiometric Infrared Thermography of Solar Photovoltaic Systems: An Explainable Predictive Maintenance Approach for Remote Aerial Diagnostic Monitoring	2024	CNN	Fault detection and diagnosis
[103]	Multi-step photovoltaic power forecasting using transformer and recurrent neural networks	2024	Transformer networks and LSTM	Power forecasting
[82]	Examining nonlinear effects of socioecological drivers on urban solar energy development in China using machine learning and high-dimensional data	2024	SVM—Recursive Feature Elimination, RF, DT, XGBoost	Nonlinear effects examination

Table A1. *Cont.*

Reference	Title	Year	AI technique	Application
[83]	Two-Stage Neural Network Optimization for Robust Solar Photovoltaic Forecasting	2024	NNs and RF	Power forecasting
[104]	SkyGPT: Probabilistic ultra-short-term solar forecasting using synthetic sky images from physics-constrained VideoGPT	2024	GAI	Sky forecasting
[105]	Implementation of optimized extreme learning machine-based energy storage scheme for grid connected photovoltaic system	2024	Arithmetic Optimized Algorithm (AOA) based Extreme Learning Machine (ELM)	Power forecasting
[106]	A Life-Long Learning XAI Metaheuristic-Based Type-2 Fuzzy System for Solar Radiation Modeling	2024	XAI Metaheuristic	Solar radiation forecasting
[107]	Analyzing grid connected shaded photovoltaic systems with steady state stability and crow search MPPT control	2024	Crow Search Algorithm (CSA)	MPPT
[108]	Super-evolutionary mechanism and Nelder-Mead simplex enhanced salp swarm algorithm for photovoltaic model parameter estimation	2024	Super-Evolutionary Nelder-Mead Salp Swarm Algorithm (SENMSSA) and Nelder-Mead simplex method	Parameter estimation
[109]	Novel applications of various neural network models for prediction of photovoltaic system power under outdoor condition of mountainous region	2024	NN	Power forecasting
[110]	Developing a Deep Learning and Reliable Optimization Techniques for Solar Photovoltaic Power Prediction	2024	CNN and LSTM	Power forecasting
[111]	A new dust detection method for photovoltaic panel surface based on Pytorch and its economic benefit analysis	2024	Adam algorithm	Dust detection
[112]	Short-term photovoltaic prediction based on CNN-GRU optimized by improved similar day extraction, decomposition noise reduction and SSA optimization	2024	Convolution Neural Network-Gate Recurrent Unit (CNN-GRU) and Sparrow Search Algorithm	Power forecasting
[113]	A Coronavirus Optimization (CVO) algorithm to harvest maximum power from PV systems under partial and complex partial shading conditions	2024	Coronavirus Optimization (CVO) algorithm	MPPT
[114]	Improved YOLOv8-GD deep learning model for defect detection in electroluminescence images of solar photovoltaic modules	2024	YOLOv8-GD.	Fault detection
[115]	Multi-objective based Hybrid Artificial Intelligence Controlled Parallel Inverter in Islanded and Grid Connected Operations	2024	adaptive neuro-fuzzy hybrid controller (ANFHC)	Inverters
[116]	Explainable Deep Learning Model for Grid-Connected Photovoltaic System Performance Assessment for Improving System Reliability	2024	feedforward neural network (FFNN)	Performance assessment
[117]	Photovoltaic Panel Defect Detection via Multiscale Siamese Convolutional Fusion Network With Information Bottleneck Theory	2024	multiscale Siamese convolutional neural network (MsSCNN)	Fault detection
[10]	Wavelet and Signal Analyzer Based High-Frequency Ripple Extraction in the Context of MPPT Algorithm in Solar PV Systems	2024	ANN and P&O	MPPT
[55]	Enhancing One-Day-Ahead Probabilistic Solar Power Forecast With a Hybrid Transformer-LUBE Model and Missing Data Imputation	2024	XGBoost	Power forecasting

Table A1. *Cont.*

Reference	Title	Year	AI technique	Application
[118]	Energy Community Management Based on Artificial Intelligence for the Implementation of Renewable Energy Systems in Smart Homes	2024	Multi-agent deep reinforcement learning, Markov Decision Process, ANN	Peer-to-peer (P2P) markets
[119]	Photovoltaic power forecasting: A dual-attention gated recurrent unit framework incorporating weather clustering and transfer learning strategy	2024	dual-attention gated recurrent units (DAGRU)	Power forecasting
[120]	Forecasting meteorological impacts on the environmental sustainability of a large-scale solar plant via artificial intelligence-based life cycle assessment	2024	ANN	Life Cycle Assessment (LCA)
[121]	An efficient power extraction using artificial intelligence based machine learning model for SPV array reconfiguration in solar industries	2024	Fuzzy Expert System (FES)	PV arrays configurations
[81]	Energy management of buildings with energy storage and solar photovoltaic: A diversity in experience approach for deep reinforcement learning agents	2024	Deep reinforcement learning (DRL), K-means Clustering, Proportional Selection Method (PSM), Random Selection Method (RSM)	Energy management
[122]	Quadratic interpolation and a new local search approach to improve particle swarm optimization: Solar photovoltaic parameter estimation	2024	Particle Swarm Optimization (PSO)	Parameter estimation
[123]	SEiPV-Net: An Efficient Deep Learning Framework for Autonomous Multi-Defect Segmentation in Electroluminescence Images of Solar Photovoltaic Modules	2023	encoder-decoder networks, Dense and Successive Features (DSF), Hierarchical Feature Precision and Extraction (HFPE), Contextual Characteristics Extraction and Attribute Fusion (CCEAF), attention mechanisms, loss functions	Defects detection
[124]	Accurate and generalizable photovoltaic panel segmentation using deep learning for imbalanced datasets	2023	Deep learning (GenPV)	PV panel segmentation
[125]	Moth flame optimization for the maximum power point tracking scheme of photovoltaic system under partial shading conditions	2023	Moth Flame Optimization (MFO)	MPPT
[126]	Open-Circuit Fault Diagnosis for Three-Phase Inverter in Photovoltaic Solar Pumping System Using Neural Network and Neuro-Fuzzy Techniques	2023	NNs and neuro-fuzzy networks	Inverter fault detection
[127]	Application of Artificial Intelligence Algorithms in Multilayer Perceptron and Elman Networks to Predict Photovoltaic Power Plant Generation	2023	MLP (Multilayer Perceptron) and Elman networks	Power forecasting
[128]	A solar radiation intelligent forecasting framework based on feature selection and multivariable fuzzy time series	2023	two-stage feature selection (TSFS), improved chaotic multi-objective dragonfly algorithm (ICMODA), seasonal multivariable fuzzy time series (SMFTS), and outlier-robust extreme learning machine (ORELM)	Solar radiation prediction

Table A1. *Cont.*

Reference	Title	Year	AI technique	Application
[129]	Deep learning based model predictive control of active filter inverter as interface for photovoltaic generation	2023	one-dimensional convolutional neural network (1D-CNN) based model predictive control (MPC)	Inverters
[130]	A novel global MPPT technique to enhance maximum power from PV systems under variable atmospheric conditions	2023	modified fluid search optimization (MFSO) and adaptive neural-fuzzy inference system (ANFIS)	MPPT
[15]	New Hybrid MPPT Technique Including Artificial Intelligence and Traditional Techniques for Extracting the Global Maximum Power from Partially Shaded PV Systems	2023	ANN, Variable Step P&O, and FL Controller (FLC)	MPPT
[44]	A new two-axis solar tracker based on the online optimization method: Experimental investigation and neural network modeling	2023	genetic algorithm (GA), PSO, Artificial Bee Colony (ABC), teaching learning based optimization (TLBO), and ANN	Solar tracker
[131]	Assessing the impact of soiling on photovoltaic efficiency using supervised learning techniques	2023	LR, RF, DT, Multilayer Perceptron and LSTM	Soiling
[53]	Driving training-based optimization (DTBO) for global maximum power point tracking for a photovoltaic system under partial shading condition	2023	driving training-based optimization (DTBO)	MPPT
[68]	A Hybrid Ensemble Model for Solar Irradiance Forecasting: Advancing Digital Models for Smart Island Realization	2023	XGBoost, light gradient boosting machine, categorical boosting, RF, and hybrid tree-based ensemble learning model (HYTREM)	Solar radiation forecasting
[31]	Cloud Computing and Machine Learning-based Electrical Fault Detection in the PV System	2023	SVM, Naive Bayes, KNN, DT and RF	Fault detection
[132]	Extracting accurate parameters of photovoltaic cell models via elite learning adaptive differential evolution	2023	Elite Learning Adaptive Differential Evolution (ELADE)	Parameter estimation
[51]	Optimizing Large-Scale PV Systems with Machine Learning: A Neuro-Fuzzy MPPT Control for PSCs with Uncertainties	2023	neuro fuzzy network (NFN)	MPPT
[77]	Novel extreme seeking control framework with ordered excitation and nonlinear function based PSO: Method and application	2023	nonlinear function (NF) based PSO	MPPT
[41]	A robust MPPT approach based on first-order sliding mode for triple-junction photovoltaic power system supplying electric vehicle	2023	sliding mode control (SMC), P&O, PSO, flower pollination (FPA) and the cuckoo search algorithm (CSA)	MPPT
[62]	Efficient fault diagnosis approach for solar photovoltaic array using a convolutional neural network in combination of generative adversarial network under small dataset	2023	Wasserstein generative adversarial network (WGAN) and CNN	Fault detection and diagnosis
[133]	Energy Efficiency Improvement in Photovoltaic Installation Using a Twin-Axis Solar Tracking Mechanism with LDR Sensors Compared with Neuro-Fuzzy Adaptive Inference Structure	2023	Neuro fuzzy inference method	MPPT
[134]	A RERNN-SGO Technique for Improved Quasi-Z-Source Cascaded Multilevel Inverter Topology for Interfacing Three Phase Grid-Tie Photovoltaic System	2023	recalling-enhanced recurrent neural network (RERNN) and Shell Game Optimization (SGO)	MPPT
[37]	Energy forecasting of the building-integrated photovoltaic façade using hybrid LSTM	2023	RF and LSTM	Power forecasting

Table A1. *Cont.*

Reference	Title	Year	AI technique	Application
[135]	Design and implementation of a new adaptive MPPT controller for solar PV systems	2023	model reference adaptive control (MRAC)	MPPT
[13]	Real-time hardware-in-loop based open circuit fault diagnosis and fault tolerant control approach for cascaded multilevel inverter using artificial neural network	2023	ANN	Inverter fault detection
[29]	Short-Term Power Prediction by Using Least Square Support Vector Machine With Variational Mode Decomposition in a Photovoltaic System	2023	fuzzy C-means (FCM), whale optimization algorithm (WOA), Least squares support vector machine (LSSVM), and improved sparrow search algorithm (ISSA)	Power forecasting
[136]	Intelligent maximum power point tracking for coastal photovoltaic system concerning the corrosion and aging of modules	2023	deep neural network (DNN)	MPPT
[137]	A Short-Term Power Prediction Method Based on Temporal Convolutional Network in Virtual Power Plant Photovoltaic System	2023	efficient channel attention module (ECANet), and temporal convolutional network (TCN)	Power forecasting
[138]	Performance Analysis of Classification and Detection for PV Panel Motion Blur Images Based on Deblurring and Deep Learning Techniques	2023	visual geometry group-16 (VGG-16), VGG-19, residual neural network-18 (RESNET-18), RESNET-50, RESNET-101, and CNN	Snow detection
[34]	Combining data envelopment analysis and Random Forest for selecting optimal locations of solar PV plants	2023	RF	Location selection
[139]	Variable boundary reinforcement learning for maximum power point tracking of photovoltaic grid-connected systems	2023	variable boundary reinforcement learning	MPPT
[140]	A Novel Nonisolated Quasi Z-Source Multilevel Inverter for Solar Photovoltaic Energy System Using Robust Technique: An ICSA–RPNN Technique	2023	chameleon swarm algorithm (ICSA) and recurrent perceptron neural network (RPNN)	Inverters
[23]	Hybrid Convolutional Neural Network-Multilayer Perceptron Model for Solar Radiation Prediction	2023	CNN and multi-layer perceptron (MLP)	Solar radiation forecasting
[141]	Boosted backtracking search optimization with information exchange for photovoltaic system evaluation	2023	backtracking search optimization algorithm (BSA) and TLBO	Parameter estimation
[52]	Maximum power point tracking technique based on variable step size with sliding mode controller in photovoltaic system	2023	Grey wolf optimizer (GWO)	MPPT
[142]	Experimental Modeling of a New Multi-Degree-of-Freedom Fuzzy Controller Based Maximum Power Point Tracking from a Photovoltaic System	2022	multi-degree-of-freedom FLC	MPPT
[143]	Optimized RNN-oriented power quality enhancement and THD reduction for micro grid integration of PV system with MLI: Crow Search-based Harris Hawks Optimization concept	2022	Optimized Recurrent NN, CSA and Harris Hawks Optimization (HHO)	Grid integration
[60]	Photovoltaic System Parameter Estimation Using Marine Predators Optimization Algorithm Based on Multilayer Perceptron	2022	GWO and marine predators optimization algorithms and multilayer perceptron	Parameter estimation
[144]	Geospatial assessment of rooftop solar photovoltaic potential using multi-source remote sensing data	2022	CNN and MLP	Geospatial assessment

Table A1. *Cont.*

Reference	Title	Year	AI technique	Application
[145]	A Novel Approach to Achieve MPPT for Photovoltaic System Based SCADA	2022	CSA and ABC	MPPT
[78]	Photovoltaic Energy Production Forecasting through Machine Learning Methods: A Scottish Solar Farm Case Study	2022	LR, kNN, DT, XGBoost, Light Gradient Boosting (LGBM), MLP, Elman NN and LSTM	Power forecasting
[146]	Optimal Design of an Artificial Intelligence Controller for Solar-Battery Integrated UPQC in Three Phase Distribution Networks	2022	soccer league algorithm and ANN	unified power quality conditioner (UPQC)
[49]	Deep Reinforcement Learning for the Optimal Angle Control of Tracking Bifacial Photovoltaic Systems	2022	DRL	Angle control
[56]	An Integrated Missing-Data Tolerant Model for Probabilistic PV Power Generation Forecasting	2022	recursive LSTM	Power forecasting
[147]	Generation of Maximum Power in Grid Connected PV System Based MPPT Control Using Hybrid Elephant Herding Optimization Algorithm	2022	Elephant Herding Optimization (EHO) and ANN	MPPT
[66]	Solar radiation forecasting with deep learning techniques integrating geostationary satellite images	2022	3D-CNN and ConvLSTM	Solar radiation forecasting
[148]	Information sharing search boosted whale optimizer with Nelder-Mead simplex for parameter estimation of photovoltaic models	2022	WOA	Parameter estimation
[149]	Prediction of photovoltaic power output based on similar day analysis using RBF neural network with adaptive black widow optimization algorithm and K-means clustering	2022	radial basis function NNs, adaptive black widow optimization algorithm (ABWO), similar day analysis (SDA) and K-means clustering	Power forecasting
[150]	Neural Network Controlled Solar PV Battery Powered Unified Power Quality Conditioner for Grid Connected Operation	2022	ANN	Power quality
[151]	Extremal Nelder–Mead colony predation algorithm for parameter estimation of solar photovoltaic models	2022	enhanced colony predation algorithm (ECPA)	Parameter estimation
[152]	A Novel Hybrid MPPT Technique Based on Harris Hawk Optimization (HHO) and Perturb and Observer (P&O) under Partial and Complex Partial Shading Conditions	2022	HHO	MPPT and shading
[153]	Parameter estimation of static solar photovoltaic models using Laplacian Nelder-Mead hunger games search	2022	Laplacian Nelder-Mead hunger games search (LNMHGS)	Parameter estimation
[154]	Application of Temporal Fusion Transformer for Day-Ahead PV Power Forecasting	2022	Temporal Fusion Transformer (TFT)	Power forecasting
[155]	Modified honey badger algorithm based global MPPT for triple-junction solar photovoltaic system under partial shading condition and global optimization	2022	Honey Badger Algorithm (HBA) and Dimensional Learning Hunting (DLH)	MPPT and shading
[156]	A population diversity-controlled differential evolution for parameter estimation of solar photovoltaic models	2022	differential evolution (DE)	Parameter estimation
[63]	An efficient fault classification method in solar photovoltaic modules using transfer learning and multi-scale convolutional neural network	2022	CNN	Fault detection and classification
[157]	Modified search and rescue optimization algorithm for identifying the optimal parameters of high efficiency triple-junction solar cell/module	2022	Search and Rescue optimization algorithm (SAR)	Parameter estimation

Table A1. *Cont.*

Reference	Title	Year	AI technique	Application
[158]	Quantum Nelder-Mead Hunger Games Search for optimizing photovoltaic solar cells	2022	Hunger Games Search (HGS)	Parameter estimation
[67]	Forecasting intra-hour solar photovoltaic energy by assembling wavelet based time-frequency analysis with deep learning neural networks	2022	feedforward NN	Solar radiation forecasting
[159]	Hardware-In-the-Loop Validation of Direct MPPT Based Cuckoo Search Optimization for Partially Shaded Photovoltaic System	2022	CSO	MPPT and shading
[160]	Constraint estimation in three-diode solar photovoltaic model using Gaussian and Cauchy mutation-based hunger games search optimizer and enhanced Newton–Raphson method	2022	HGS	Parameter estimation
[61]	Improved gradient-based optimizer for parameters extraction of photovoltaic models	2022	Gradient-based optimizer (GBO)	Parameter estimation
[161]	Short-Term Forecasting of Energy Production for a Photovoltaic System Using a NARX-CVM Hybrid Model	2022	nonlinear autoregressive with exogenous inputs (NARX)	Power forecasting
[162]	Power Generation Prediction of Building-Integrated Photovoltaic System with Colored Modules Using Machine Learning	2022	NN	Power forecasting
[163]	Adaptive slime mould algorithm for optimal design of photovoltaic models	2022	adaptive slime mould algorithm (ASMA)	Parameter estimation
[35]	Tree-based machine learning models for photovoltaic output power forecasting that consider photovoltaic panel soiling	2022	RF, extra trees (ET) and gradient boosting (GB)	Power forecasting and soiling
[58]	Solar photo voltaic module parameter extraction using a novel Hybrid Chimp-Sine Cosine Algorithm	2022	hybrid Chimp-Sine cosine algorithm (HCSCA)	Parameter estimation
[164]	Analyzing the performance of photovoltaic systems using support vector machine classifier	2022	SVM	Monitoring
[12]	A Novel Hybrid Method Based on Fireworks Algorithm and Artificial Neural Network for Photovoltaic System Fault Diagnosis	2022	ANN and Evolutionary Fireworks Algorithm (FWA)	Fault detection
[165]	Data driven approach to forecast the next day aggregate production of scattered small rooftop solar photovoltaic systems without meteorological parameters	2022	least square support vector regression (LsSVR)	Power forecasting
[166]	A powerful meta-heuristic search algorithm for solving global optimization and real-world solar photovoltaic parameter estimation problems	2022	teaching-learning-based artificial bee colony (TLABC)	Parameter estimation
[167]	Modeling of solar photovoltaic power using a two-stage forecasting system with operation and weather parameters	2022	ANN	Power forecasting
[39]	Artificial Intelligence Applications in Estimating Invisible Solar Power Generation	2022	KNN	Power forecasting
[79]	Hourly Forecasting of Solar Photovoltaic Power in Pakistan Using Recurrent Neural Networks	2022	LSTM and bidirectional LSTM	Power forecasting
[27]	Comparison of deep learning and regression-based MPPT algorithms in PV systems	2022	LSTM	MPPT
[168]	Large-scale photovoltaic system in green building: MPPT control based on deep neural network and dynamic time-window	2022	DNN	MPPT
[169]	Fuzzy State-Dependent Riccati Equation (FSDRE) Control of the Reverse Osmosis Desalination System With Photovoltaic Power Supply	2022	invasive weed algorithm (IWA)	MPPT

Table A1. *Cont.*

Reference	Title	Year	AI technique	Application
[26]	AI-Empowered Recommender System for Renewable Energy Harvesting in Smart Grid System	2022	LSTM	Power forecasting
[38]	Fuzzy Based MPPT and Solar Power Forecasting Using Artificial Intelligence	2022	KNN	MPPT
[170]	Dynamic Model and Intelligent Maximum Power Point Tracking Approach for Large-Scale Building-Integrated Photovoltaic System	2021	DE)	MPPT
[24]	GeoAI for detection of solar photovoltaic installations in the Netherlands	2021	CNN (TernausNet)	PV systems detection
[59]	An evolutionary Nelder–Mead slime mould algorithm with random learning for efficient design of photovoltaic models	2021	improved slime mould algorithm (ISMA)	Parameter estimation
[171]	A Dragonfly Optimization Algorithm for Extracting Maximum Power of Grid-Interfaced PV Systems	2021	dragonfly optimization algorithm (DOA)	MPPT
[30]	Small-Scale Solar Photovoltaic Power Prediction for Residential Load in Saudi Arabia Using Machine Learning	2021	least absolute shrinkage and selection operator (LASSO), RF, LR, polynomial regression (PR), XGBoost, SVM, and deep learning (DL)	Power forecasting
[172]	Deep-Learning-Based Probabilistic Estimation of Solar PV Soiling Loss	2021	backbone networks	Soiling
[173]	Random reselection particle swarm optimization for optimal design of solar photovoltaic modules	2021	PSO and cuckoo search	Parameter estimation
[174]	Development of Artificial Intelligence Techniques for Solar PV Power Forecasting for Dehradun Region of India	2021	MLP, ridge regression, DT, RF, SVM and KNN	Powe forecasting
[80]	CNN-based deep learning technique for improved H7 TLI with grid-connected photovoltaic systems	2021	CNN	Inverters
[175]	Predicting diurnal outdoor performance and degradation of organic photovoltaics via machine learning; relating degradation to outdoor stress conditions	2021	ANNs	Power forecasting and degradation
[176]	Evaluation of constraint in photovoltaic cells using ensemble multi-strategy shuffled frog leading algorithms	2021	ensemble multi strategy-driven shuffled frog leading algorithm (EMSFLA)	Parameter estimation
[40]	Design and Evaluation of Fuzzy Adaptive Particle Swarm Optimization Based Maximum Power Point Tracking on Photovoltaic System Under Partial Shading Conditions	2021	fuzzy adaptive PSO and conventional PSO	MPPT
[14]	Implementation of solar energy in smart cities using an integration of artificial neural network, photovoltaic system and classical Delphi methods	2021	ANN	Power forecasting
[43]	Optimal Parameter Estimation of Solar PV Panel Based on Hybrid Particle Swarm and Grey Wolf Optimization Algorithms	2021	PSO and GWO	Parameter estimation
[54]	Day-ahead hourly photovoltaic power forecasting using attention-based CNN-LSTM neural network embedded with multiple relevant and target variables prediction pattern	2021	attention-based long-term and short-term temporal neural network prediction model (ALSM), CNN, LSTM, and attention mechanism under the multiple relevant and target variables prediction pattern (MRTPP)	Power forecasting

Table A1. *Cont.*

Reference	Title	Year	AI technique	Application
[20]	A Novel Hybrid Maximum Power Point Tracking Controller Based on Artificial Intelligence for Solar Photovoltaic System Under Variable Environmental Conditions	2021	FL	MPPT
[65]	Proposed ANFIS Based Approach for Fault Tracking, Detection, Clearing and Rearrangement for Photovoltaic System	2021	ANFIS	Fault detection
[177]	Real-time implementation of MPPT for renewable energy systems based on Artificial intelligence	2021	fuzzy NN	MPPT
[178]	Modeling Renewable Energy Systems by a Self-Evolving Nonlinear Consequent Part Recurrent Type-2 Fuzzy System for Power Prediction	2021	fuzzy NN	Modeling
[179]	Effective Segmentation Approach for Solar Photovoltaic Panels in Uneven Illuminated Color Infrared Images	2021	k-means clustering	Segmentation
[17]	Promising MPPT Methods Combining Metaheuristic, Fuzzy-Logic and ANN Techniques for Grid-Connected Photovoltaic	2021	FL, PSO, and ANN	MPPT
[180]	Neural Network Approach for Global Solar Irradiance Prediction at Extremely Short-Time-Intervals Using Particle Swarm Optimization Algorithm	2021	PSO and NN	Solar radiation prediction
[181]	Evolutionary shuffled frog leaping with memory pool for parameter optimization	2021	shuffled frog-leaping algorithm (SFLBS)	Parameter estimation
[182]	Novel AI Based Energy Management System for Smart Grid With RES Integration	2021	GWO	Energy management
[183]	Artificial Intelligence Method for the Forecast and Separation of Total and HVAC Loads With Application to Energy Management of Smart and NZE Homes	2021	LSTM	Power forecasting
[11]	Adaptive maximum power point tracking using neural networks for a photovoltaic systems according grid	2021	ANN	MPPT
[184]	Robust configuration and intelligent MPPT control for building integrated photovoltaic system based on extreme learning machine	2021	extreme learning machine (ELM)	MPPT
[42]	Parameter Identification of Photovoltaic Cell Model Based on Enhanced Particle Swarm Optimization	2021	enhanced PSO	Parameter estimation
[16]	Analysis of a Traditional and a Fuzzy Logic Enhanced Perturb and Observe Algorithm for the MPPT of a Photovoltaic System	2021	FL	MPPT
[57]	Randomised learning-based hybrid ensemble model for probabilistic forecasting of PV power generation	2020	randomized learning-based hybrid ensemble (RLHE) model	Power forecasting
[28]	Hybrid deep learning for power generation forecasting in active solar trackers	2020	LSTM	Power forecasting
[185]	Horizontal and vertical crossover of Harris hawk optimizer with Nelder-Mead simplex for parameter estimation of photovoltaic models	2020	HHO and Crisscross Optimizer (CCO)	Parameter estimation
[186]	Solar photovoltaic parameter estimation using an improved equilibrium optimizer	2020	Equilibrium Optimizer (EO)	Parameter estimation
[187]	Orthogonally adapted Harris hawks optimization for parameter estimation of photovoltaic models	2020	HHO	Parameter estimation
[19]	Artificial intelligent controller-based power quality improvement for microgrid integration of photovoltaic system using new cascade multilevel inverter	2020	FL	MPPT

Table A1. *Cont.*

Reference	Title	Year	AI technique	Application
[33]	Data Description Technique-Based Islanding Classification for Single-Phase Grid-Connected Photovoltaic System	2020	SVM	Islanding detection
[188]	A particle swarm optimisation-trained feedforward neural network for predicting the maximum power point of a photovoltaic array	2020	ANN and PSO	MPPT
[189]	Photovoltaic power forecast using empirical models and artificial intelligence approaches for water pumping systems	2020	FFNN and ANFIS	Power forecasting
[190]	Multiple scenarios multi-objective salp swarm optimization for sizing of standalone photovoltaic system	2020	Salp Swarm Optimization (SSA)	Sizing
[191]	Orthogonal Nelder-Mead moth flame method for parameters identification of photovoltaic modules	2020	orthogonal moth flame optimization (MFO)	Parameter estimation
[192]	Forecast uncertainty-based performance degradation diagnosis of solar PV systems	2020	ensemble method based on the dropout technique	Degradation diagnosis
[193]	Optimal Sizing of Standalone Photovoltaic System Using Improved Performance Model and Optimization Algorithm	2020	DE multi-objective optimization	Sizing
[194]	Characterization of a polycrystalline photovoltaic cell using artificial neural networks	2020	ANN	Parameter estimation
[195]	Implementation of ANFIS-mptc for 20 kwp spv power generation and comparison with FLMPPT under dissimilar conditions	2020	ANFIS	MPPT

References

1. Ramos, A.; Chatzopoulou, M.A.; Guarracino, I.; Freeman, J.; Markides, C.N. Hybrid Photovoltaic-Thermal Solar Systems for Combined Heating, Cooling and Power Provision in the Urban Environment. *Energy Convers. Manag.* **2017**, *150*, 838–850. [CrossRef]
2. Li, S.; Haskew, T.A.; Li, D.; Hu, F. Integrating Photovoltaic and Power Converter Characteristics for Energy Extraction Study of Solar PV Systems. *Renew. Energy* **2011**, *36*, 3238–3245. [CrossRef]
3. Awasthi, A.; Shukla, A.K.; Sr, M.M.; Dondariya, C.; Shukla, K.N.; Porwal, D.; Richhariya, G. Review on Sun Tracking Technology in Solar PV System. *Energy Rep.* **2020**, *6*, 392–405. [CrossRef]
4. Usman, Z.; Tah, J.; Abanda, H.; Nche, C. A Critical Appraisal of PV-Systems' Performance. *Buildings* **2020**, *10*, 192. [CrossRef]
5. Elshater, A.; Abusaada, H. Developing Process for Selecting Research Techniques in Urban Planning and Urban Design with a PRISMA-Compliant Review. *Soc. Sci.* **2022**, *11*, 471. [CrossRef]
6. Aria, M.; Cuccurullo, C. *Bibliometrix*: An R-Tool for Comprehensive Science Mapping Analysis. *J. Informetr.* **2017**, *11*, 959–975. [CrossRef]
7. Yousaf, S.; Shafi, I.; Ahmad, J. Features Selection and Optimized Neural Network Architecture for Modelling Flows in Solar Collectors. In Proceedings of the 2017 International Conference on Frontiers of Information Technology (FIT), Islamabad, Pakistan, 18–20 December 2017; pp. 247–252.
8. Abid, S.; Chtourou, M.; Djemel, M. Pruning Approaches for Selection of Neural Networks Structure. In Proceedings of the 10th International Multi-Conferences on Systems, Signals & Devices 2013 (SSD13), Hammamet, Tunisia, 18–21 March 2013; pp. 1–4.
9. Ji, T.W.; Gao, J.; Zhao, G.Q.; Zhang, C.R. Research on GA-ANN Integration and Its Applications to Cold Extrusion Process Design. *Mater. Sci. Forum* **2006**, *532–533*, 897–900. [CrossRef]
10. Taher, M.A.; Behnamfar, M.; Sarwat, A.I.; Tariq, M. Wavelet and Signal Analyzer Based High- Frequency Ripple Extraction in the Context of MPPT Algorithm in Solar PV Systems. *IEEE Access* **2024**, *12*, 113726–113740. [CrossRef]
11. Sahraoui, H.; Mellah, H.; Drid, S.; Chrifi-Alaoui, L. Adaptive Maximum Power Point Tracking Using Neural Networks for a Photovoltaic Systems According Grid. *Electr. Eng. Electromech.* **2021**, *5*, 57–66. [CrossRef]
12. Sebbane, S.; Akchioui, N.E. A Novel Hybrid Method Based on Fireworks Algorithm and Artificial Neural Network for Photovoltaic System Fault Diagnosis. *Int. J. Renew. Energy Res.* **2022**, *12*, 239–247.

13. Sivapriya, A.; Kalaiarasi, N.; Vishnuram, P.; Abou Houran, M.; Bajaj, M.; Pushkarna, M.; Kamel, S. Real-Time Hardware-in-Loop Based Open Circuit Fault Diagnosis and Fault Tolerant Control Approach for Cascaded Multilevel Inverter Using Artificial Neural Network. *Front. Energy Res.* **2023**, *10*, 1083662. [CrossRef]
14. Ghadami, N.; Gheibi, M.; Kian, Z.; Faramarz, M.G.; Naghedi, R.; Eftekhari, M.; Fathollahi-Fard, A.M.; Dulebenets, M.A.; Tian, G. Implementation of Solar Energy in Smart Cities Using an Integration of Artificial Neural Network, Photovoltaic System and Classical Delphi Methods. *Sustain. Cities Soc.* **2021**, *74*, 103149. [CrossRef]
15. Masry, M.Z.-E.; Mohammed, A.; Amer, F.; Mubarak, R. New Hybrid MPPT Technique Including Artificial Intelligence and Traditional Techniques for Extracting the Global Maximum Power from Partially Shaded PV Systems. *Sustainability* **2023**, *15*, 10884. [CrossRef]
16. Remoaldo, D.; Jesus, I. Analysis of a Traditional and a Fuzzy Logic Enhanced Perturb and Observe Algorithm for the MPPT of a Photovoltaic System. *Algorithms* **2021**, *14*, 24. [CrossRef]
17. Ali, M.N.; Mahmoud, K.; Lehtonen, M.; Darwish, M.M.F. Promising MPPT Methods Combining Metaheuristic, Fuzzy-Logic and ANN Techniques for Grid-Connected Photovoltaic. *Sensors* **2021**, *21*, 1244. [CrossRef]
18. Bacha, M.; Terki, A.; Bacha, M. Comparative Study of Real-Time Photovoltaic Fault Diagnosis Using Artificial Intelligence: Fuzzy Logic and Neural Network Approaches. *Energy Sources Part A Recovery Util. Environ. Eff.* **2024**, *46*, 13536–13560. [CrossRef]
19. Mahendravarman, I.; Elankurisil, S.A.; Venkateshkumar, M.; Ragavendiran, A.; Chin, N. Artificial Intelligent Controller-Based Power Quality Improvement for Microgrid Integration of Photovoltaic System Using New Cascade Multilevel Inverter. *Soft Comput.* **2020**, *24*, 18909–18926. [CrossRef]
20. Khan, M.J. Pushparaj A Novel Hybrid Maximum Power Point Tracking Controller Based on Artificial Intelligence for Solar Photovoltaic System Under Variable Environmental Conditions. *J. Electr. Eng. Technol.* **2021**, *16*, 1879–1889. [CrossRef]
21. Ledmaoui, Y.; El Maghraoui, A.; El Aroussi, M.; Saadane, R. Enhanced Fault Detection in Photovoltaic Panels Using CNN-Based Classification with PyQt5 Implementation. *Sensors* **2024**, *24*, 7407. [CrossRef]
22. Qureshi, U.R.; Rashid, A.; Altini, N.; Bevilacqua, V.; La Scala, M. Radiometric Infrared Thermography of Solar Photovoltaic Systems: An Explainable Predictive Maintenance Approach for Remote Aerial Diagnostic Monitoring. *Smart Cities* **2024**, *7*, 1261–1288. [CrossRef]
23. Ghimire, S.; Nguyen-Huy, T.; Prasad, R.; Deo, R.C.; Casillas-Pérez, D.; Salcedo-Sanz, S.; Bhandari, B. Hybrid Convolutional Neural Network-Multilayer Perceptron Model for Solar Radiation Prediction. *Cogn. Comput.* **2023**, *15*, 645–671. [CrossRef]
24. Kausika, B.B.; Nijmeijer, D.; Reimerink, I.; Brouwer, P.; Liem, V. GeoAI for Detection of Solar Photovoltaic Installations in the Netherlands. *Energy AI* **2021**, *6*, 100111. [CrossRef]
25. Aman, R.; Rizwan, M.; Kumar, A. A Novel Hybrid Intelligent Approach for Solar Photovoltaic Power Prediction Considering UV Index and Cloud Cover. *Electr. Eng.* **2024**, *107*, 1203–1224. [CrossRef]
26. Patel, R.K.; Kumari, A.; Tanwar, S.; Hong, W.-C.; Sharma, R. AI-Empowered Recommender System for Renewable Energy Harvesting in Smart Grid System. *IEEE Access* **2022**, *10*, 24316–24326. [CrossRef]
27. Karabinaoğlu, M.; Çakir, B.; Başoğlu, M.; Kazdaloğlu, A.; Güneroğlu, A. Comparison of Deep Learning and Regression-Based MPPT Algorithms in PV Systems. *Turk. J. Electr. Eng. Comput. Sci.* **2022**, *30*, 2319–2338. [CrossRef]
28. Frizzo Stefenon, S.; Kasburg, C.; Nied, A.; Rodrigues Klaar, A.C.; Silva Ferreira, F.C.; Waldrigues Branco, N. Hybrid Deep Learning for Power Generation Forecasting in Active Solar Trackers. *IET Gener. Transm. Distrib.* **2020**, *14*, 5667–5674. [CrossRef]
29. Cai, Y.; Hao, S.; Wen, X.; Li, H.; He, X.; Chen, L.; Ren, J. Short-Term Power Prediction by Using Least Square Support Vector Machine with Variational Mode Decomposition in a Photovoltaic System. *IEEE Access* **2023**, *11*, 143486–143500. [CrossRef]
30. Mohana, M.; Saidi, A.S.; Alelyani, S.; Alshayeb, M.J.; Basha, S.; Anqi, A.E. Small-Scale Solar Photovoltaic Power Prediction for Residential Load in Saudi Arabia Using Machine Learning. *Energies* **2021**, *14*, 6759. [CrossRef]
31. Ragul, S.; Tamilselvi, S.; Rengarajan, S.; Guna Sundari, S. Cloud Computing and Machine Learning-Based Electrical Fault Detection in the PV System. *IETE J. Res.* **2023**, *69*, 8735–8752. [CrossRef]
32. Sepúlveda-Oviedo, E.H.; Travé-Massuyès, L.; Subias, A.; Pavlov, M.; Alonso, C. An Ensemble Learning Framework for Snail Trail Fault Detection and Diagnosis in Photovoltaic Modules. *Eng. Appl. Artif. Intell.* **2024**, *137*, 109068. [CrossRef]
33. Haque, A.; Alshareef, A.; Khan, A.I.; Alam, M.M.; Kurukuru, V.S.B.; Irshad, K. Data Description Technique-Based Islanding Classification for Single-Phase Grid-Connected Photovoltaic System. *Sensors* **2020**, *20*, 3320. [CrossRef] [PubMed]
34. Cattani, G. Combining Data Envelopment Analysis and Random Forest for Selecting Optimal Locations of Solar PV Plants. *Energy AI* **2023**, *11*, 100222. [CrossRef]
35. Tovilović, D.M.; Đurišić, Ž.R. Tree-Based Machine Learning Models for Photovoltaic Output Power Forecasting That Consider Photovoltaic Panel Soiling. *Int. J. Sustain. Energy* **2022**, *41*, 1279–1302. [CrossRef]
36. Montes-Romero, J.; Heinzle, N.; Livera, A.; Theocharides, S.; Makrides, G.; Sutterlueti, J.; Ransome, S.; Georghiou, G.E. Novel Data-Driven Health-State Architecture for Photovoltaic System Failure Diagnosis. *Sol. Energy* **2024**, *279*, 112820. [CrossRef]

37. Sarkar, S.; Karthick, A.; Kumar Chinnaiyan, V.; Patil, P.P. Energy Forecasting of the Building-Integrated Photovoltaic Façade Using Hybrid LSTM. *Environ. Sci. Pollut. Res.* **2023**, *30*, 45977–45985. [CrossRef]
38. Geethamahalakshmi, G.; Kalaiarasi, N.; Nageswari, D. Fuzzy Based MPPT and Solar Power Forecasting Using Artificial Intelligence. *Intell. Autom. Soft Comput.* **2021**, *32*, 1667–1685. [CrossRef]
39. Wu, Y.-K.; Lai, Y.-H.; Huang, C.-L.; Phuong, N.T.B.; Tan, W.-S. Artificial Intelligence Applications in Estimating Invisible Solar Power Generation. *Energies* **2022**, *15*, 1312. [CrossRef]
40. Guo, L.; Abdul, N.M.M. Design and Evaluation of Fuzzy Adaptive Particle Swarm Optimization Based Maximum Power Point Tracking on Photovoltaic System Under Partial Shading Conditions. *Front. Energy Res.* **2021**, *9*, 712175. [CrossRef]
41. Hamed, S.B.; Abid, A.; Hamed, M.B.; Sbita, L.; Bajaj, M.; Ghoneim, S.S.M.; Zawbaa, H.M.; Kamel, S. A Robust MPPT Approach Based on First-Order Sliding Mode for Triple-Junction Photovoltaic Power System Supplying Electric Vehicle. *Energy Rep.* **2023**, *9*, 4275–4297. [CrossRef]
42. Wang, R. Parameter Identification of Photovoltaic Cell Model Based on Enhanced Particle Swarm Optimization. *Sustainability* **2021**, *13*, 840. [CrossRef]
43. Rezk, H.; Arfaoui, J.; Gomaa, M.R. Optimal Parameter Estimation of Solar PV Panel Based on Hybrid Particle Swarm and Grey Wolf Optimization Algorithms. *Int. J. Interact. Multimed. Artif. Intell.* **2021**, *6*, 145. [CrossRef]
44. Masoumi, A.P.; Bagherian, V.; Tavakolpour-Saleh, A.R.; Masoomi, E. A New Two-Axis Solar Tracker Based on the Online Optimization Method: Experimental Investigation and Neural Network Modeling. *Energy AI* **2023**, *14*, 100284. [CrossRef]
45. Abdel-Nasser, M.; Mahmoud, K. Accurate Photovoltaic Power Forecasting Models Using Deep LSTM-RNN. *Neural Comput. Applic.* **2019**, *31*, 2727–2740. [CrossRef]
46. Marino, D.L.; Amarasinghe, K.; Manic, M. Building Energy Load Forecasting Using Deep Neural Networks. In Proceedings of the IECON 2016—42nd Annual Conference of the IEEE Industrial Electronics Society, Florence, Italy, 24–27 October 2016; pp. 7046–7051.
47. Rathore, A.; Gupta, P.; Sharma, R.; Singh, R. Day Ahead Solar Forecast Using Long Short Term Memory Network Augmented with Fast Fourier Transform-Assisted Decomposition Technique. *Renew. Energy* **2025**, *247*, 123021. [CrossRef]
48. Eseye, A.T.; Zhang, J.; Zheng, D. Short-Term Photovoltaic Solar Power Forecasting Using a Hybrid Wavelet-PSO-SVM Model Based on SCADA and Meteorological Information. *Renew. Energy* **2018**, *118*, 357–367. [CrossRef]
49. Tsuchida, S.; Nonaka, H.; Yamada, N. Deep Reinforcement Learning for the Optimal Angle Control of Tracking Bifacial Photovoltaic Systems. *Energies* **2022**, *15*, 8083. [CrossRef]
50. Talbi, E.-G. *Metaheuristics: From Design to Implementation*; Wiley: Hoboken, NJ, USA, 2009; Volume 74, ISBN 978-0-470-27858-1.
51. Asif; Ahmad, W.; Qureshi, M.B.; Khan, M.M.; Fayyaz, M.A.B.; Nawaz, R. Optimizing Large-Scale PV Systems with Machine Learning: A Neuro-Fuzzy MPPT Control for PSCs with Uncertainties. *Electronics* **2023**, *12*, 1720. [CrossRef]
52. Hai, T.; Zain, J.M.; Nakamura, H. Maximum Power Point Tracking Technique Based on Variable Step Size with Sliding Mode Controller in Photovoltaic System. *Soft Comput.* **2023**, *27*, 3829–3845. [CrossRef]
53. Rehman, H.; Sajid, I.; Sarwar, A.; Tariq, M.; Bakhsh, F.I.; Ahmad, S.; Mahmoud, H.A.; Aziz, A. Driving Training-Based Optimization (DTBO) for Global Maximum Power Point Tracking for a Photovoltaic System under Partial Shading Condition. *IET Renew. Power Gener.* **2023**, *17*, 2542–2562. [CrossRef]
54. Qu, J.; Qian, Z.; Pei, Y. Day-Ahead Hourly Photovoltaic Power Forecasting Using Attention-Based CNN-LSTM Neural Network Embedded with Multiple Relevant and Target Variables Prediction Pattern. *Energy* **2021**, *232*, 120996. [CrossRef]
55. Phan, Q.-T.; Wu, Y.-K.; Phan, Q.-D. Enhancing One-Day-Ahead Probabilistic Solar Power Forecast with a Hybrid Transformer-LUBE Model and Missing Data Imputation. *IEEE Trans. Ind. Appl.* **2024**, *60*, 1396–1408. [CrossRef]
56. Li, Q.; Xu, Y.; Chew, B.S.H.; Ding, H.; Zhao, G. An Integrated Missing-Data Tolerant Model for Probabilistic PV Power Generation Forecasting. *IEEE Trans. Power Syst.* **2022**, *37*, 4447–4459. [CrossRef]
57. Liu, W.; Xu, Y. Randomised Learning-Based Hybrid Ensemble Model for Probabilistic Forecasting of PV Power Generation. *IET Gener. Transm. Distrib.* **2020**, *14*, 5909–5917. [CrossRef]
58. Vandrasi, R.K.; Sravana Kumar, B.; Devarapalli, R. Solar Photo Voltaic Module Parameter Extraction Using a Novel Hybrid Chimp-Sine Cosine Algorithm. *Energy Sources Part A Recovery Util. Environ. Eff.* **2022**, 1–20. [CrossRef]
59. Weng, X.; Heidari, A.A.; Liang, G.; Chen, H.; Ma, X. An Evolutionary Nelder–Mead Slime Mould Algorithm with Random Learning for Efficient Design of Photovoltaic Models. *Energy Rep.* **2021**, *7*, 8784–8804. [CrossRef]
60. Colak, M.; Balci, S. Photovoltaic System Parameter Estimation Using Marine Predators Optimization Algorithm Based on Multilayer Perceptron. *Electr. Power Compon. Syst.* **2022**, *50*, 1087–1099. [CrossRef]
61. Jiang, Y.; Luo, Q.; Zhou, Y. Improved Gradient-Based Optimizer for Parameters Extraction of Photovoltaic Models. *IET Renew. Power Gener.* **2022**, *16*, 1602–1622. [CrossRef]

62. Lu, X.; Lin, Y.; Lin, P.; He, X.; Fang, G.; Cheng, S.; Chen, Z.; Wu, L. Efficient Fault Diagnosis Approach for Solar Photovoltaic Array Using a Convolutional Neural Network in Combination of Generative Adversarial Network under Small Dataset. *Sol. Energy* **2023**, *253*, 360–374. [CrossRef]
63. Korkmaz, D.; Acikgoz, H. An Efficient Fault Classification Method in Solar Photovoltaic Modules Using Transfer Learning and Multi-Scale Convolutional Neural Network. *Eng. Appl. Artif. Intell.* **2022**, *113*, 104959. [CrossRef]
64. Noura, H.N.; Allal, Z.; Salman, O.; Chahine, K. Explainable Artificial Intelligence of Tree-Based Algorithms for Fault Detection and Diagnosis in Grid-Connected Photovoltaic Systems. *Eng. Appl. Artif. Intell.* **2025**, *139*, 109503. [CrossRef]
65. Bendary, A.F.; Abdelaziz, A.Y.; Ismail, M.M.; Mahmoud, K.; Lehtonen, M.; Darwish, M.M.F. Proposed ANFIS Based Approach for Fault Tracking, Detection, Clearing and Rearrangement for Photovoltaic System. *Sensors* **2021**, *21*, 2269. [CrossRef] [PubMed]
66. Gallo, R.; Castangia, M.; Macii, A.; Macii, E.; Patti, E.; Aliberti, A. Solar Radiation Forecasting with Deep Learning Techniques Integrating Geostationary Satellite Images. *Eng. Appl. Artif. Intell.* **2022**, *116*, 105493. [CrossRef]
67. Rodríguez, F.; Azcárate, I.; Vadillo, J.; Galarza, A. Forecasting Intra-Hour Solar Photovoltaic Energy by Assembling Wavelet Based Time-Frequency Analysis with Deep Learning Neural Networks. *Int. J. Electr. Power Energy Syst.* **2022**, *137*, 107777. [CrossRef]
68. So, D.; Oh, J.; Leem, S.; Ha, H.; Moon, J. A Hybrid Ensemble Model for Solar Irradiance Forecasting: Advancing Digital Models for Smart Island Realization. *Electronics* **2023**, *12*, 2607. [CrossRef]
69. Hota, L.; Dash, P. Comparative Analysis of Stock Price Prediction by ANN and RF Model. *Comput. Intell. Mach. Learn.* **2021**, *2*, 1–9. [CrossRef]
70. Sy, D.; Nguyen, S. *ANFIS: Establishing and Applying to Managing Online Damage*; IntechOpen: London, UK, 2019; ISBN 978-1-78984-231-9.
71. Sanchez, B.; Duran, D.; Martinez, K.; Viera, W. Advances of Artificial Intelligence in Aeronautics. *Athenea Eng. Sci. J.* **2023**, *4*, 34–42. [CrossRef]
72. Kayabasi, A.; Akdagli, A. A Comparative Study on ANN, ANFIS and SVM Methods for Computing Resonant Frequency of A-Shaped Compact Microstrip Antennas. In Proceedings of the ICCITE 2015:17th International Conference on Communication and Information Technology and Engineering, PyeongChang, Republic of Korea, 1–3 July 2015.
73. Mohammed, R.; Abdulrazzaq, M.; Mohammed, A.; Mardikyan, K.; Çevık, M. Enhancing Smart Grid Efficiency: A Modified ANN-LSTM Approach for Energy Storage and Distribution Optimization. In Proceedings of the 2023 7th International Symposium on Multidisciplinary Studies and Innovative Technologies (ISMSIT), Ankara, Turkey, 26–28 October 2023; p. 5.
74. Wang, Y.; Wang, B.; Di, J.; Sun, J.; Wang, X. Research on Diabetes Prediction Analysis Based on Improved ANN Algorithm. *J. Phys. Conf. Ser.* **2021**, *2010*, 012118. [CrossRef]
75. Bao, C.; Li, Q.; Shen, Z.; Tai, C.; Wu, L.; Xiang, X. Approximation Analysis of Convolutional Neural Networks. *East Asian J. Appl. Math.* **2023**, *13*, 524–549. [CrossRef]
76. Essam, Y.; Ahmed, A.N.; Ramli, R.; Chau, K.-W.; Idris Ibrahim, M.S.; Sherif, M.; Sefelnasr, A.; El-Shafie, A. Investigating Photovoltaic Solar Power Output Forecasting Using Machine Learning Algorithms. *Eng. Appl. Comput. Fluid. Mech.* **2022**, *16*, 2002–2034. [CrossRef]
77. Liu, G.; Zhu, L.; Li, H.; Li, J.; Lv, Q. Novel Extreme Seeking Control Framework with Ordered Excitation and Nonlinear Function Based PSO: Method and Application. *Sol. Energy* **2023**, *255*, 126–137. [CrossRef]
78. Cabezón, L.; Ruiz, L.G.B.; Criado-Ramón, D.; Gago, E.J.; Pegalajar, M.C. Photovoltaic Energy Production Forecasting through Machine Learning Methods: A Scottish Solar Farm Case Study. *Energies* **2022**, *15*, 8732. [CrossRef]
79. Khan, S.; Shaikh, F.; Siddiqui, M.M.; Hussain, T.; Kumar, L.; Nahar, A. Hourly Forecasting of Solar Photovoltaic Power in Pakistan Using Recurrent Neural Networks. *Int. J. Photoenergy* **2022**, *2022*, 7015818. [CrossRef]
80. Ramasamy, S.; Perumal, M. CNN-Based Deep Learning Technique for Improved H7 TLI with Grid-Connected Photovoltaic Systems. *Int. J. Energy Res.* **2021**, *45*, 19851–19868. [CrossRef]
81. Hussain, A.; Musilek, P. Energy Management of Buildings with Energy Storage and Solar Photovoltaic: A Diversity in Experience Approach for Deep Reinforcement Learning Agents. *Energy AI* **2024**, *15*, 100313. [CrossRef]
82. Zhao, Y.; Ge, W.; Sun, Y.; Qiao, G.; Zhu, D.; Ai, H. Examining Nonlinear Effects of Socioecological Drivers on Urban Solar Energy Development in China Using Machine Learning and High-Dimensional Data. *J. Environ. Manag.* **2024**, *360*, 121092. [CrossRef] [PubMed]
83. Oh, J.; So, D.; Jo, J.; Kang, N.; Hwang, E.; Moon, J. Two-Stage Neural Network Optimization for Robust Solar Photovoltaic Forecasting. *Electronics* **2024**, *13*, 1659. [CrossRef]
84. John, R.; Coupland, S. Type-2 Fuzzy Logic and the Modelling of Uncertainty in Applications. In *Human-Centric Information Processing Through Granular Modelling*; Bargiela, A., Pedrycz, W., Eds.; Springer: Berlin/Heidelberg, Germany, 2009; pp. 185–201, ISBN 978-3-540-92916-1.
85. Castillo, O.; Melin, P. Interval Type-2 Fuzzy Logic Applications. In *Human-Centric Information Processing Through Granular Modelling*; Bargiela, A., Pedrycz, W., Eds.; Springer: Berlin/Heidelberg, Germany, 2009; pp. 203–231, ISBN 978-3-540-92916-1.

86. Chimatapu, R.; Hagras, H.; Kern, M.; Owusu, G. Enhanced Deep Type-2 Fuzzy Logic System For Global Interpretability. In Proceedings of the 2021 IEEE International Conference on Fuzzy Systems (FUZZ-IEEE), Luxembourg, 11–14 July 2021; pp. 1–8.
87. Ontiveros-Robles, E.; Melin, P.; Castillo, O. Comparative Analysis of Noise Robustness of Type 2 Fuzzy Logic Controllers. *Kybernetika* **2018**, *54*, 175–201. [CrossRef]
88. Janarthanan, R.; Maheshwari, R.U.; Shukla, P.K.; Shukla, P.K.; Mirjalili, S.; Kumar, M. Intelligent Detection of the PV Faults Based on Artificial Neural Network and Type 2 Fuzzy Systems. *Energies* **2021**, *14*, 6584. [CrossRef]
89. Aminifar, S.; Marzuki, A. Uncertainty in Interval Type-2 Fuzzy Systems. *Math. Probl. Eng.* **2013**, *2013*, 452780. [CrossRef]
90. Comas, D.S.; Pastore, J.I.; Bouchet, A.; Ballarin, V.L.; Meschino, G.J. Type-2 Fuzzy Logic in Decision Support Systems. In *Soft Computing for Business Intelligence*; Espin, R., Pérez, R.B., Cobo, A., Marx, J., Valdés, A.R., Eds.; Springer: Berlin/Heidelberg, Germany, 2014; pp. 267–280, ISBN 978-3-642-53737-0.
91. Kondratenko, N.; Snihur, O. A Method of Building Type-2 Fuzzy Logic Systems in Multidimensional Objects Identification Problems. *East. Eur. J. Enterp. Technol.* **2017**, *3*, 38–45. [CrossRef]
92. Basar, G.; Der, O. Multi-Objective Optimization of Process Parameters for Laser Cutting Polyethylene Using Fuzzy AHP-Based MCDM Methods. *Proc. Inst. Mech. Eng. Part E* **2025**. [CrossRef]
93. Pant, S.; Garg, P.; Kumar, A.; Ram, M.; Kumar, A.; Sharma, H.K.; Klochkov, Y. AHP-Based Multi-Criteria Decision-Making Approach for Monitoring Health Management Practices in Smart Healthcare System. *Int. J. Syst. Assur. Eng. Manag.* **2024**, *15*, 1444–1455. [CrossRef]
94. Phan, Q.-T.; Wu, Y.-K.; Phan, Q.-D. An Innovative Hybrid Model Combining Informer and K-Means Clustering Methods for Invisible Multisite Solar Power Estimation. *IET Renew. Power Gener.* **2024**, *18*, 4318–4333. [CrossRef]
95. Rokonuzzaman, M.; Rahman, S.; Hannan, M.A.; Mishu, M.K.; Tan, W.-S.; Rahman, K.S.; Pasupuleti, J.; Amin, N. Levenberg-Marquardt Algorithm-Based Solar PV Energy Integrated Internet of Home Energy Management System. *Appl. Energy* **2025**, *378*, 124407. [CrossRef]
96. Zhou, P.; Fang, H.; Wu, G. PDeT: A Progressive Deformable Transformer for Photovoltaic Panel Defect Segmentation. *Sensors* **2024**, *24*, 6908. [CrossRef]
97. Xie, C.; Li, Q.; Yang, Y.; Zhang, L.; Liu, X. Research on a Photovoltaic Panel Dust Detection Algorithm Based on 3D Data Generation. *Energies* **2024**, *17*, 5222. [CrossRef]
98. Krishna Veni, K.S.; Senthil Kumar, N.; Gnanavadivel, J. Non-Invasive Health Status Diagnosis of Solar PV Panel Using Ensemble Classifier. *Electr. Eng.* **2024**, 1–13. [CrossRef]
99. Nguyen-Duc, T.; Vu-Xuan-Son, H.; Do-Dinh, H.; Nguyen-Vu-Nhat, N.; Fujita, G.; Tran-Thanh, S. An Adaptive Method for Real-Time Photovoltaic Power Forecasting Utilizing Mathematics and Statistics: Case Studies in Australia and Vietnam. *IET Renew. Power Gener.* **2024**, *18*, 2589–2604. [CrossRef]
100. Wu, S.; Kong, Y.; Xu, R.; Guo, Y.; Chen, Z.; Zheng, X. A Feature Space Class Balancing Strategy-Based Fault Classification Method in Solar Photovoltaic Modules. *Eng. Appl. Artif. Intell.* **2024**, *136*, 108991. [CrossRef]
101. Kumaraswamy, M.; Naik, K.A. Backpropagation Artificial Neural Network-Based Maximum Power Point Tracking Controller with Image Encryption Inspired Solar Photovoltaic Array Reconfiguration. *Eng. Appl. Artif. Intell.* **2024**, *136*, 108979. [CrossRef]
102. Martinez Lopez, V.A.; Isabella, O.; Zeman, M.; Ziar, H. Battery-Less Uncertainty-Based Control of a Stand-Alone PV-Electrolyzer System. *J. Power Sources* **2024**, *614*, 234934. [CrossRef]
103. Kim, J.; Obregon, J.; Park, H.; Jung, J.-Y. Multi-Step Photovoltaic Power Forecasting Using Transformer and Recurrent Neural Networks. *Renew. Sustain. Energy Rev.* **2024**, *200*, 114479. [CrossRef]
104. Nie, Y.; Zelikman, E.; Scott, A.; Paletta, Q.; Brandt, A. SkyGPT: Probabilistic Ultra-Short-Term Solar Forecasting Using Synthetic Sky Images from Physics-Constrained VideoGPT. *Adv. Appl. Energy* **2024**, *14*, 100172. [CrossRef]
105. Ganesh, R.; Saha, T.K.; Sai Kumar, M.L.S. Implementation of Optimized Extreme Learning Machine-Based Energy Storage Scheme for Grid Connected Photovoltaic System. *J. Energy Storage* **2024**, *88*, 111611. [CrossRef]
106. Almaraashi, M.; Abdulrahim, M.; Hagras, H. A Life-Long Learning XAI Metaheuristic-Based Type-2 Fuzzy System for Solar Radiation Modeling. *IEEE Trans. Fuzzy Syst.* **2024**, *32*, 2102–2115. [CrossRef]
107. Ben Smida, M.; Azar, A.T.; Sakly, A.; Hameed, I.A. Analyzing Grid Connected Shaded Photovoltaic Systems with Steady State Stability and Crow Search MPPT Control. *Front. Energy Res.* **2024**, *12*, 1381376. [CrossRef]
108. Wu, H.; Chen, Y.; Cai, Z.; Heidari, A.A.; Chen, H.; Zhang, Y. Super-Evolutionary Mechanism and Nelder-Mead Simplex Enhanced Salp Swarm Algorithm for Photovoltaic Model Parameter Estimation. *IET Renew. Power Gener.* **2024**, *18*, 2209–2237. [CrossRef]
109. Yadav, A.K.; Khargotra, R.; Lee, D.; Kumar, R.; Singh, T. Novel Applications of Various Neural Network Models for Prediction of Photovoltaic System Power under Outdoor Condition of Mountainous Region. *Sustain. Energy Grids Netw.* **2024**, *38*, 101318. [CrossRef]

110. Shivashankaraiah, K.; Shashibhushan, G.; Prema Kirubakaran, A.; Anjanappa, C.; Sharma, P.K.; Mayiladuthurai Vaidyanathan, I.; Taqui, S.N.; Obaid, S.A.; Alfarraj, S. Developing a Deep Learning and Reliable Optimization Techniques for Solar Photovoltaic Power Prediction. *Electr. Power Compon. Syst.* **2024**, 1–15. [CrossRef]
111. Shao, Y.; Zhang, C.; Xing, L.; Sun, H.; Zhao, Q.; Zhang, L. A New Dust Detection Method for Photovoltaic Panel Surface Based on Pytorch and Its Economic Benefit Analysis. *Energy AI* **2024**, *16*, 100349. [CrossRef]
112. Li, R.; Wang, M.; Li, X.; Qu, J.; Dong, Y. Short-Term Photovoltaic Prediction Based on CNN-GRU Optimized by Improved Similar Day Extraction, Decomposition Noise Reduction and SSA Optimization. *IET Renew. Power Gener.* **2024**, *18*, 908–928. [CrossRef]
113. Sarwar, S.; Javed, M.Y.; Asghar, A.B.; Iqbal, W.; Ejsmont, K.; Jaffery, M.H. A Coronavirus Optimization (CVO) Algorithm to Harvest Maximum Power from PV Systems under Partial and Complex Partial Shading Conditions. *Energy Rep.* **2024**, *11*, 1693–1710. [CrossRef]
114. Cao, Y.; Pang, D.; Zhao, Q.; Yan, Y.; Jiang, Y.; Tian, C.; Wang, F.; Li, J. Improved YOLOv8-GD Deep Learning Model for Defect Detection in Electroluminescence Images of Solar Photovoltaic Modules. *Eng. Appl. Artif. Intell.* **2024**, *131*, 107866. [CrossRef]
115. Gaddameedhi, S.; Srinivas, P.; Sagar, E.V. Multi-Objective Based Hybrid Artificial Intelligence Controlled Parallel Inverter in Islanded and Grid Connected Operations. Available online: http://www.eurekaselect.com (accessed on 13 December 2024).
116. Hassan, I.; Alhamrouni, I.; Younes, Z.; Azhan, N.H.; Mekhilef, S.; Seyedmahmoudian, M.; Stojcevski, A. Explainable Deep Learning Model for Grid-Connected Photovoltaic System Performance Assessment for Improving System Reliability. *IEEE Access* **2024**, *12*, 120729–120746. [CrossRef]
117. Ma, W.; Chen, B.; Wang, B.; Chen, B. Photovoltaic Panel Defect Detection via Multiscale Siamese Convolutional Fusion Network with Information Bottleneck Theory. *IEEE Trans. Instrum. Meas.* **2024**, *73*, 1–15. [CrossRef]
118. Khayyat, M.M.; Sami, B. Energy Community Management Based on Artificial Intelligence for the Implementation of Renewable Energy Systems in Smart Homes. *Electronics* **2024**, *13*, 380. [CrossRef]
119. Tang, Y.; Yang, K.; Zhang, S.; Zhang, Z. Photovoltaic Power Forecasting: A Dual-Attention Gated Recurrent Unit Framework Incorporating Weather Clustering and Transfer Learning Strategy. *Eng. Appl. Artif. Intell.* **2024**, *130*, 107691. [CrossRef]
120. Wan, M.J.; Phuang, Z.X.; Hoy, Z.X.; Dahlan, N.Y.; Azmi, A.M.; Woon, K.S. Forecasting Meteorological Impacts on the Environmental Sustainability of a Large-Scale Solar Plant via Artificial Intelligence-Based Life Cycle Assessment. *Sci. Total Environ.* **2024**, *912*, 168779. [CrossRef]
121. Sharma, M.; Pareek, S.; Singh, K. An Efficient Power Extraction Using Artificial Intelligence Based Machine Learning Model for SPV Array Reconfiguration in Solar Industries. *Eng. Appl. Artif. Intell.* **2024**, *129*, 107516. [CrossRef]
122. Qaraad, M.; Amjad, S.; Hussein, N.K.; Farag, M.A.; Mirjalili, S.; Elhosseini, M.A. Quadratic Interpolation and a New Local Search Approach to Improve Particle Swarm Optimization: Solar Photovoltaic Parameter Estimation. *Expert Syst. Appl.* **2024**, *236*, 121417. [CrossRef]
123. Eesaar, H.; Joe, S.; Rehman, M.U.; Jang, Y.; Chong, K.T. SEiPV-Net: An Efficient Deep Learning Framework for Autonomous Multi-Defect Segmentation in Electroluminescence Images of Solar Photovoltaic Modules. *Energies* **2023**, *16*, 7726. [CrossRef]
124. Guo, Z.; Zhuang, Z.; Tan, H.; Liu, Z.; Li, P.; Lin, Z.; Shang, W.-L.; Zhang, H.; Yan, J. Accurate and Generalizable Photovoltaic Panel Segmentation Using Deep Learning for Imbalanced Datasets. *Renew. Energy* **2023**, *219*, 119471. [CrossRef]
125. Chang, C.C.W.; Ding, T.J.; Han, W.; Chai, C.C.; Yam, C.M.; Yian, H.C.; Xin, L.H. Moth Flame Optimization for the Maximum Power Point Tracking Scheme of Photovoltaic System under Partial Shading Conditions. *Energy Rep.* **2023**, *9*, 374–379. [CrossRef]
126. Bengharbi, A.; Laribi, S.; Allaoui, T.; Amina, M. Open-Circuit Fault Diagnosis for Three-Phase Inverter in Photovoltaic Solar Pumping System Using Neural Network and Neuro-Fuzzy Techniques. *Electrica* **2023**, *23*, 505–515. [CrossRef]
127. Drałus, G.; Mazur, D.; Kusznier, J.; Drałus, J. Application of Artificial Intelligence Algorithms in Multilayer Perceptron and Elman Networks to Predict Photovoltaic Power Plant Generation. *Energies* **2023**, *16*, 6697. [CrossRef]
128. Gao, Y.; Li, P.; Yang, H.; Wang, J. A Solar Radiation Intelligent Forecasting Framework Based on Feature Selection and Multivariable Fuzzy Time Series. *Eng. Appl. Artif. Intell.* **2023**, *126*, 106986. [CrossRef]
129. Rasoulian, A.; Saghafi, H.; Abbasian, M.; Delshad, M. Deep Learning Based Model Predictive Control of Active Filter Inverter as Interface for Photovoltaic Generation. *IET Renew. Power Gener.* **2023**, *17*, 3151–3162. [CrossRef]
130. Hai, T.; Zain, J.M.; Muranaka, K. A Novel Global MPPT Technique to Enhance Maximum Power from PV Systems under Variable Atmospheric Conditions. *Soft Comput.* **2023**, 1–14. [CrossRef]
131. Silva, L.A.C.; Ruiz, L.G.B.; Criado-Ramón, D.; Bessa, J.G.; Micheli, L.; Jiménez, M.d.C.P. Assessing the Impact of Soiling on Photovoltaic Efficiency Using Supervised Learning Techniques. *Expert. Syst. Appl.* **2023**, *231*, 120816. [CrossRef]
132. Gu, Z.; Xiong, G.; Fu, X.; Mohamed, A.W.; Al-Betar, M.A.; Chen, H.; Chen, J. Extracting Accurate Parameters of Photovoltaic Cell Models via Elite Learning Adaptive Differential Evolution. *Energy Convers. Manag.* **2023**, *285*, 116994. [CrossRef]

133. Hadroug, N.; Iratni, A.; Hafaifa, A.; Boudjemline, A.; Alshammari, O.S.; Jerbi, H.; Colak, I.; Chen, X. Energy Efficiency Improvement in Photovoltaic Installation Using a Twin-Axis Solar Tracking Mechanism with LDR Sensors Compared with Neuro-Fuzzy Adaptive Inference Structure. *J. Electr. Eng. Technol.* **2023**, *18*, 2943–2967. [CrossRef]
134. Bhavani, M.; Manoharan, P.S. A RERNN-SGO Technique for Improved Quasi-Z-Source Cascaded Multilevel Inverter Topology for Interfacing Three Phase Grid-Tie Photovoltaic System. *Int. J. Artif. Intell. Tools* **2023**, *32*, 2350009. [CrossRef]
135. Manna, S.; Singh, D.K.; Akella, A.K.; Kotb, H.; AboRas, K.M.; Zawbaa, H.M.; Kamel, S. Design and Implementation of a New Adaptive MPPT Controller for Solar PV Systems. *Energy Rep.* **2023**, *9*, 1818–1829. [CrossRef]
136. An, Q.; Gao, C.; Deng, Q. Intelligent Maximum Power Point Tracking for Coastal Photovoltaic System Concerning the Corrosion and Aging of Modules. *J. Intell. Fuzzy Syst.* **2023**, *45*, 4051–4070. [CrossRef]
137. Zhou, X.; Pang, C.; Zeng, X.; Jiang, L.; Chen, Y. A Short-Term Power Prediction Method Based on Temporal Convolutional Network in Virtual Power Plant Photovoltaic System. *IEEE Trans. Instrum. Meas.* **2023**, *72*, 1–10. [CrossRef]
138. Al-Dulaimi, A.A.; Guneser, M.T.; Hameed, A.A.; Márquez, F.P.G.; Fitriyani, N.L.; Syafrudin, M. Performance Analysis of Classification and Detection for PV Panel Motion Blur Images Based on Deblurring and Deep Learning Techniques. *Sustainability* **2023**, *15*, 1150. [CrossRef]
139. Gao, F.; Hu, R.; Yin, L. Variable Boundary Reinforcement Learning for Maximum Power Point Tracking of Photovoltaic Grid-Connected Systems. *Energy* **2023**, *264*, 126278. [CrossRef]
140. Santhi, R.; Srinivasan, A. A Novel Nonisolated Quasi Z-Source Multilevel Inverter for Solar Photovoltaic Energy System Using Robust Technique: An ICSA–RPNN Technique. *J. Circuit Syst. Comp.* **2023**, *32*, 2350104. [CrossRef]
141. Weng, X.; Liu, Y.; Heidari, A.A.; Cai, Z.; Lin, H.; Chen, H.; Liang, G.; Alsufyani, A.; Bourouis, S. Boosted Backtracking Search Optimization with Information Exchange for Photovoltaic System Evaluation. *Energy Sci. Eng.* **2023**, *11*, 267–298. [CrossRef]
142. El-Khatib, M.F.; Sabry, M.-N.; Abu El-Sebah, M.I.; Maged, S.A. Experimental Modeling of a New Multi-Degree-of-Freedom Fuzzy Controller Based Maximum Power Point Tracking from a Photovoltaic System. *Appl. Syst. Innov.* **2022**, *5*, 114. [CrossRef]
143. Praveena, A.; Sathishkumar, K. Optimized RNN-Oriented Power Quality Enhancement and THD Reduction for Micro Grid Integration of PV System with MLI: Crow Search-Based Harris Hawks Optimization Concept. *Front. Energy Res.* **2022**, *10*, 1038533. [CrossRef]
144. Jiang, H.; Yao, L.; Lu, N.; Qin, J.; Liu, T.; Liu, Y.; Zhou, C. Geospatial Assessment of Rooftop Solar Photovoltaic Potential Using Multi-Source Remote Sensing Data. *Energy AI* **2022**, *10*, 100185. [CrossRef]
145. Alhasnawi, B.N.; Jasim, B.H.; Alhasnawi, A.N.; Sedhom, B.E.; Jasim, A.M.; Khalili, A.; Bureš, V.; Burgio, A.; Siano, P. A Novel Approach to Achieve MPPT for Photovoltaic System Based SCADA. *Energies* **2022**, *15*, 8480. [CrossRef]
146. Srilakshmi, K.; Sujatha, C.N.; Balachandran, P.K.; Mihet-Popa, L.; Kumar, N.U. Optimal Design of an Artificial Intelligence Controller for Solar-Battery Integrated UPQC in Three Phase Distribution Networks. *Sustainability* **2022**, *14*, 13992. [CrossRef]
147. Vijayakumar, S.; Govindaraju, C. Generation of Maximum Power in Grid Connected PV System Based MPPT Control Using Hybrid Elephant Herding Optimization Algorithm. *Electr. Power Compon. Syst.* **2022**, *50*, 301–317. [CrossRef]
148. Peng, L.; He, C.; Heidari, A.A.; Zhang, Q.; Chen, H.; Liang, G.; Aljehane, N.O.; Mansour, R.F. Information Sharing Search Boosted Whale Optimizer with Nelder-Mead Simplex for Parameter Estimation of Photovoltaic Models. *Energy Convers. Manag.* **2022**, *270*, 116246. [CrossRef]
149. Liu, H.; Zhou, Y.; Luo, Q.; Huang, H.; Wei, X. Prediction of Photovoltaic Power Output Based on Similar Day Analysis Using RBF Neural Network with Adaptive Black Widow Optimization Algorithm and K-Means Clustering. *Front. Energy Res.* **2022**, *10*, 990018. [CrossRef]
150. Okwako, O.E.; Lin, Z.-H.; Xin, M.; Premkumar, K.; Rodgers, A.J. Neural Network Controlled Solar PV Battery Powered Unified Power Quality Conditioner for Grid Connected Operation. *Energies* **2022**, *15*, 6825. [CrossRef]
151. Xu, B.; Heidari, A.A.; Zhang, S.; Chen, H.; Shao, Q. Extremal Nelder–Mead Colony Predation Algorithm for Parameter Estimation of Solar Photovoltaic Models. *Energy Sci. Eng.* **2022**, *10*, 4176–4219. [CrossRef]
152. Hafeez, M.A.; Naeem, A.; Akram, M.; Javed, M.Y.; Asghar, A.B.; Wang, Y. A Novel Hybrid MPPT Technique Based on Harris Hawk Optimization (HHO) and Perturb and Observer (P&O) under Partial and Complex Partial Shading Conditions. *Energies* **2022**, *15*, 5550. [CrossRef]
153. Yu, S.; Heidari, A.A.; He, C.; Cai, Z.; Althobaiti, M.M.; Mansour, R.F.; Liang, G.; Chen, H. Parameter Estimation of Static Solar Photovoltaic Models Using Laplacian Nelder-Mead Hunger Games Search. *Sol. Energy* **2022**, *242*, 79–104. [CrossRef]
154. López Santos, M.; García-Santiago, X.; Echevarría Camarero, F.; Blázquez Gil, G.; Carrasco Ortega, P. Application of Temporal Fusion Transformer for Day-Ahead PV Power Forecasting. *Energies* **2022**, *15*, 5232. [CrossRef]
155. Nassef, A.M.; Houssein, E.H.; Helmy, B.E.; Rezk, H. Modified Honey Badger Algorithm Based Global MPPT for Triple-Junction Solar Photovoltaic System under Partial Shading Condition and Global Optimization. *Energy* **2022**, *254*, 124363. [CrossRef]

156. Yu, Y.; Wang, K.; Zhang, T.; Wang, Y.; Peng, C.; Gao, S. A Population Diversity-Controlled Differential Evolution for Parameter Estimation of Solar Photovoltaic Models. *Sustain. Energy Technol. Assess.* **2022**, *51*, 101938. [CrossRef]
157. Houssein, E.H.; Nassef, A.M.; Fathy, A.; Mahdy, M.A.; Rezk, H. Modified Search and Rescue Optimization Algorithm for Identifying the Optimal Parameters of High Efficiency Triple-Junction Solar Cell/Module. *Int. J. Energy Res.* **2022**, *46*, 13961–13985. [CrossRef]
158. Xu, B.; Heidari, A.A.; Kuang, F.; Zhang, S.; Chen, H.; Cai, Z. Quantum Nelder-Mead Hunger Games Search for Optimizing Photovoltaic Solar Cells. *Int. J. Energy Res.* **2022**, *46*, 12417–12466. [CrossRef]
159. Al-Shammaa, A.A.; Abdurraqeeb, A.M.; Noman, A.M.; Alkuhayli, A.; Farh, H.M.H. Hardware-In-the-Loop Validation of Direct MPPT Based Cuckoo Search Optimization for Partially Shaded Photovoltaic System. *Electronics* **2022**, *11*, 1655. [CrossRef]
160. Premkumar, M.; Jangir, P.; Kumar, C.; Sundarsingh Jebaseelan, S.D.T.; Alhelou, H.H.; Madurai Elavarasan, R.; Chen, H. Constraint Estimation in Three-Diode Solar Photovoltaic Model Using Gaussian and Cauchy Mutation-Based Hunger Games Search Optimizer and Enhanced Newton–Raphson Method. *IET Renew. Power Gener.* **2022**, *16*, 1733–1772. [CrossRef]
161. Rangel-Heras, E.; Angeles-Camacho, C.; Cadenas-Calderón, E.; Campos-Amezcua, R. Short-Term Forecasting of Energy Production for a Photovoltaic System Using a NARX-CVM Hybrid Model. *Energies* **2022**, *15*, 2842. [CrossRef]
162. Shin, W.-G.; Shin, J.-Y.; Hwang, H.-M.; Park, C.-H.; Ko, S.-W. Power Generation Prediction of Building-Integrated Photovoltaic System with Colored Modules Using Machine Learning. *Energies* **2022**, *15*, 2589. [CrossRef]
163. Lin, H.; Ahmadianfar, I.; Amiri Golilarz, N.; Jamei, M.; Heidari, A.A.; Kuang, F.; Zhang, S.; Chen, H. Adaptive Slime Mould Algorithm for Optimal Design of Photovoltaic Models. *Energy Sci. Eng.* **2022**, *10*, 2035–2064. [CrossRef]
164. Hafdaoui, H.; Boudjelthia, E.A.K.; Chahtou, A.; Bouchakour, S.; Belhaouas, N. Analyzing the Performance of Photovoltaic Systems Using Support Vector Machine Classifier. *Sustain. Energy Grids Netw.* **2022**, *29*, 100592. [CrossRef]
165. Fentis, A.; Rafik, M.; Bahatti, L.; Bouattane, O.; Mestari, M. Data Driven Approach to Forecast the next Day Aggregate Production of Scattered Small Rooftop Solar Photovoltaic Systems without Meteorological Parameters. *Energy Rep.* **2022**, *8*, 3221–3233. [CrossRef]
166. Duman, S.; Kahraman, H.T.; Sonmez, Y.; Guvenc, U.; Kati, M.; Aras, S. A Powerful Meta-Heuristic Search Algorithm for Solving Global Optimization and Real-World Solar Photovoltaic Parameter Estimation Problems. *Eng. Appl. Artif. Intell.* **2022**, *111*, 104763. [CrossRef]
167. Rao Kasagani, D.V.S.K.; Manickam, P. Modeling of Solar Photovoltaic Power Using a Two-Stage Forecasting System with Operation and Weather Parameters. *Energy Sources Part A Recovery Util. Environ. Eff.* **2022**, 1–19. [CrossRef]
168. An, Q.; Tang, R.; Hu, Q. Large-Scale Photovoltaic System in Green Building: MPPT Control Based on Deep Neural Network and Dynamic Time-Window. *J. Intell. Fuzzy Syst.* **2022**, *43*, 4769–4787. [CrossRef]
169. Nejad, H.D.; Nazari, M.; Nazari, M.; Mardan, M.M.S.; Mohammadzadeh, A.; Vu, M.T.; Mosavi, A. Fuzzy State-Dependent Riccati Equation (FSDRE) Control of the Reverse Osmosis Desalination System with Photovoltaic Power Supply. *IEEE Access* **2022**, *10*, 95585–95603. [CrossRef]
170. An, Q.; Tang, R.; Su, H. Dynamic Model and Intelligent Maximum Power Point Tracking Approach for Large-Scale Building-Integrated Photovoltaic System. *Energy Technol.* **2022**, *10*, 2100493. [CrossRef]
171. Lodhi, E.; Wang, F.-Y.; Xiong, G.; Mallah, G.A.; Javed, M.Y.; Tamir, T.S.; Gao, D.W. A Dragonfly Optimization Algorithm for Extracting Maximum Power of Grid-Interfaced PV Systems. *Sustainability* **2021**, *13*, 10778. [CrossRef]
172. Zhang, W.; Liu, S.; Gandhi, O.; Rodríguez-Gallegos, C.D.; Quan, H.; Srinivasan, D. Deep-Learning-Based Probabilistic Estimation of Solar PV Soiling Loss. *IEEE Trans. Sustain. Energy* **2021**, *12*, 2436–2444. [CrossRef]
173. Fan, Y.; Wang, P.; Heidari, A.A.; Chen, H.; HamzaTurabieh; Mafarja, M. Random Reselection Particle Swarm Optimization for Optimal Design of Solar Photovoltaic Modules. *Energy* **2022**, *239*, 121865. [CrossRef]
174. Verma, A.; Upadhyay, K.G.; Tripathi, M.M. Development of Artificial Intelligence Techniques for Solar PV Power Forecasting for Dehradun Region of India. *J. Electr. Syst.* **2021**, *17*, 324–337.
175. David, T.W.; Soares, G.A.; Bristow, N.; Bagnis, D.; Kettle, J. Predicting Diurnal Outdoor Performance and Degradation of Organic Photovoltaics via Machine Learning; Relating Degradation to Outdoor Stress Conditions. *Prog. Photovolt. Res. Appl.* **2021**, *29*, 1274–1284. [CrossRef]
176. Wang, M.; Zhang, Q.; Chen, H.; Heidari, A.A.; Mafarja, M.; Turabieh, H. Evaluation of Constraint in Photovoltaic Cells Using Ensemble Multi-Strategy Shuffled Frog Leading Algorithms. *Energy Convers. Manag.* **2021**, *244*, 114484. [CrossRef]
177. Breesam, W.I. Real-Time Implementation of MPPT for Renewable Energy Systems Based on Artificial Intelligence. *Int. Trans. Electr. Energy Syst.* **2021**, *31*, e12864. [CrossRef]
178. Tavoosi, J.; Suratgar, A.A.; Menhaj, M.B.; Mosavi, A.; Mohammadzadeh, A.; Ranjbar, E. Modeling Renewable Energy Systems by a Self-Evolving Nonlinear Consequent Part Recurrent Type-2 Fuzzy System for Power Prediction. *Sustainability* **2021**, *13*, 3301. [CrossRef]

179. Wang, N.; Sun, Z.-L.; Zeng, Z.; Lam, K.-M. Effective Segmentation Approach for Solar Photovoltaic Panels in Uneven Illuminated Color Infrared Images. *IEEE J. Photovolt.* **2021**, *11*, 478–484. [CrossRef]
180. Aljanad, A.; Tan, N.M.L.; Agelidis, V.G.; Shareef, H. Neural Network Approach for Global Solar Irradiance Prediction at Extremely Short-Time-Intervals Using Particle Swarm Optimization Algorithm. *Energies* **2021**, *14*, 1213. [CrossRef]
181. Liu, Y.; Heidari, A.A.; Ye, X.; Chi, C.; Zhao, X.; Ma, C.; Turabieh, H.; Chen, H.; Le, R. Evolutionary Shuffled Frog Leaping with Memory Pool for Parameter Optimization. *Energy Rep.* **2021**, *7*, 584–606. [CrossRef]
182. Kumar, A.; Alaraj, M.; Rizwan, M.; Nangia, U. Novel AI Based Energy Management System for Smart Grid with RES Integration. *IEEE Access* **2021**, *9*, 162530–162542. [CrossRef]
183. Alden, R.E.; Gong, H.; Jones, E.S.; Ababei, C.; Ionel, D.M. Artificial Intelligence Method for the Forecast and Separation of Total and HVAC Loads with Application to Energy Management of Smart and NZE Homes. *IEEE Access* **2021**, *9*, 160497–160509. [CrossRef]
184. An, Q.; Tang, R.; Su, H.; Zhang, J.; Li, X. Robust Configuration and Intelligent MPPT Control for Building Integrated Photovoltaic System Based on Extreme Learning Machine. *J. Intell. Fuzzy Syst.* **2021**, *40*, 12283–12300. [CrossRef]
185. Liu, Y.; Chong, G.; Heidari, A.A.; Chen, H.; Liang, G.; Ye, X.; Cai, Z.; Wang, M. Horizontal and Vertical Crossover of Harris Hawk Optimizer with Nelder-Mead Simplex for Parameter Estimation of Photovoltaic Models. *Energy Convers. Manag.* **2020**, *223*, 113211. [CrossRef]
186. Abdel-Basset, M.; Mohamed, R.; Mirjalili, S.; Chakrabortty, R.K.; Ryan, M.J. Solar Photovoltaic Parameter Estimation Using an Improved Equilibrium Optimizer. *Sol. Energy* **2020**, *209*, 694–708. [CrossRef]
187. Jiao, S.; Chong, G.; Huang, C.; Hu, H.; Wang, M.; Heidari, A.A.; Chen, H.; Zhao, X. Orthogonally Adapted Harris Hawks Optimization for Parameter Estimation of Photovoltaic Models. *Energy* **2020**, *203*, 117804. [CrossRef]
188. Al-Majidi, S.D.; Abbod, M.F.; Al-Raweshidy, H.S. A Particle Swarm Optimisation-Trained Feedforward Neural Network for Predicting the Maximum Power Point of a Photovoltaic Array. *Eng. Appl. Artif. Intell.* **2020**, *92*, 103688. [CrossRef]
189. Ben Ammar, R.; Ben Ammar, M.; Oualha, A. Photovoltaic Power Forecast Using Empirical Models and Artificial Intelligence Approaches for Water Pumping Systems. *Renew. Energy* **2020**, *153*, 1016–1028. [CrossRef]
190. Ridha, H.M.; Gomes, C.; Hizam, H.; Mirjalili, S. Multiple Scenarios Multi-Objective Salp Swarm Optimization for Sizing of Standalone Photovoltaic System. *Renew. Energy* **2020**, *153*, 1330–1345. [CrossRef]
191. Zhang, H.; Heidari, A.A.; Wang, M.; Zhang, L.; Chen, H.; Li, C. Orthogonal Nelder-Mead Moth Flame Method for Parameters Identification of Photovoltaic Modules. *Energy Convers. Manag.* **2020**, *211*, 112764. [CrossRef]
192. Lee, H.; Ko, S.-K.; Lee, B.-T. Forecast Uncertainty-Based Performance Degradation Diagnosis of Solar PV Systems. *IET Renew. Power Gener.* **2020**, *14*, 968–975. [CrossRef]
193. Khatib, T.; Muhsen, D.H. Optimal Sizing of Standalone Photovoltaic System Using Improved Performance Model and Optimization Algorithm. *Sustainability* **2020**, *12*, 2233. [CrossRef]
194. Cortés, B.; Tapia Sánchez, R.; Flores, J.J. Characterization of a Polycrystalline Photovoltaic Cell Using Artificial Neural Networks. *Sol. Energy* **2020**, *196*, 157–167. [CrossRef]
195. Babu, P.C.; Prasanth, B.V.; Sujatha, P. Implementation of Anfis-Mptc for 20 Kwp Spv Power Generation and Comparison with Flmppt under Dissimilar Conditions. *Int. J. Ambient. Energy* **2022**, *43*, 1445–1455. [CrossRef]

processes

MDPI

Systematic Review

Emerging Advances in Sustainable Manufacturing

Samuel Polo, Eva María Rubio *, Jorge Ayllón and Beatriz de Agustina

Department of Manufacturing Engineering, Industrial Engineering School, Universidad Nacional de Educación a Distancia (National University of Distance Education), St/Juan del Rosal 12, E28040 Madrid, Spain; spolo64@alumno.uned.es (S.P.); jorge.ayllon@ind.uned.es (J.A.); bdeagustina@ind.uned.es (B.d.A.)
* Correspondence: erubio@ind.uned.es

Abstract

The present study undertakes a systematic review of the latest trends in sustainable manufacturing over the past five years, exploring future developments in technologies, methods, and strategies. Utilizing the PRISMA (Preferred Reporting Items for Systematic Reviews and Meta-Analyses) methodology, adapted to engineering, this research ensures transparency and reproducibility throughout the identification, screening, eligibility, and inclusion stages of the review process. A review of the extant literature identified several key terms that were frequently incorporated into research inquiries, including Industry 4.0, circular economy, additive manufacturing (AM), and the use of sustainable materials. The study's main findings highlight the transformative role of digital technologies in fostering closed-loop systems and resource efficiency. For industry professionals, this signifies concrete prospects for enhancing operational efficiency, mitigating waste, and shifting towards more robust production models. Nonetheless, this review discloses persistent challenges, particularly for small- and medium-sized enterprises (SMEs). These challenges encompass elevated implementation costs and the intricacy of advanced digital solutions. The study's findings indicate that while Industry 4.0 is a predominant catalyst for sustainable practices, its extensive adoption is impeded by substantial implementation costs and skill deficits. The study's concluding remarks outline strategic directions for future research and policy, advocating for scalable, inclusive solutions to facilitate a global transition toward sustainable industrial systems.

Keywords: sustainable manufacturing; literature review; Industry 4.0; circular economy

1. Introduction

The evolution of manufacturing has been marked by successive industrial revolutions, each bringing about profound transformations in production systems, economic structures, and societal dynamics. The advent of mechanized production, as exemplified by the First Industrial Revolution in the 18th century, signaled a transition from manual craftsmanship to large-scale manufacturing. While these advances catalyzed economic growth and urbanization, they also led to an increased demand for natural resources and significant environmental degradation. Concerns over resource depletion, pollution, and unsafe labor conditions gradually emerged, yet it was not until the late 20th century that sustainability became a central focus in industrial development [1,2].

A seminal moment in the evolution of this movement was the publication of the Brundtland Report in 1987. This report introduced the widely accepted definition of sustainable development as "meeting the needs of the present without compromising the ability of future generations to meet their own needs". This definition led to an increasing emphasis on environmental responsibility within manufacturing, initially focused on reducing emissions, waste, and pollution. As time progressed, however, a shift occurred, with sustainability in manufacturing transitioning from a reactive approach—centered on damage mitigation—to a proactive strategy that integrates sustainable practices into the fundamental operations of industry [2].

During the 2010s, the forces of globalization and mounting consumer pressure compelled numerous corporations to adopt more sustainable practices. This shift was accompanied by the introduction of mandatory regulations and certifications, which served to impose more stringent oversight of corporate activities. Consequently, sustainability has evolved from a mere regulatory compliance strategy to a critical business model for enhancing operational efficiency by reducing resource and energy consumption. This paradigm shift has led to the adoption of a holistic approach to sustainability, encompassing its three dimensions: economic, environmental, and social [3].

The evolution of sustainable manufacturing has been marked by a growing integration of technology and approach, driven by the paradigm of Industry 4.0. This digital transformation and business model innovation has enabled companies to enhance their sustainability. Through this holistic integration and the implementation of concepts such as the circular economy, which aims to close the life cycle of products, sustainable manufacturing seeks to adopt a restorative approach for the future. The overarching objective of sustainable manufacturing is to restore planetary boundaries to safe levels and ensure the long-term viability of life on Earth [4,5].

The integration of Industry 4.0 technologies and circular economy principles signifies a paradigm shift in sustainable manufacturing. The foundational element of Industry 4.0 is the digital infrastructure, which encompasses the Internet of Things (IoT), cyber–physical systems (CPS), and cloud computing. This infrastructure enables real-time data exchange, process automation, and dynamic optimization across manufacturing networks. These capabilities are critical for supporting circular economy strategies, which aim to regenerate natural systems by closing material loops through reuse, remanufacturing, and recycling [5]. The alignment between Industry 4.0 and the circular economy is particularly evident in the manner in which digital technologies facilitate actions such as product life extension, resource efficiency, and value recovery, which are considered to be the core pillars of circular innovation. By facilitating traceability, monitoring, and intelligent decision-making, Industry 4.0 technologies enable the implementation of closed-loop systems and reverse logistics networks that would otherwise be too complex or costly to manage [5]. Consequently, the integration of Industry 4.0 and the circular economy enables manufacturing systems to transition from linear to circular, resilient, and resource-efficient configurations.

Research Objective and Contribution

This document provides an overview that analyzes the latest research trends in sustainable manufacturing over the past five years. A systematic search of the literature on the subject was carried out using PRISMA (Preferred Reporting Items for Systematic Reviews and Meta-Analyses) to contextualize the research, and, considering global sustainability goals and the impact of industry on sustainable development, the study will focus on research, analyses, and advancements in sustainable manufacturing at an industrial level.

The primary objective of this review is to explore and synthesize the current advances and intersections of sustainable manufacturing strategies within the context of Industry 4.0, circular economy frameworks, and emerging biotechnologies. Specifically, the objective of this paper is to identify and critically examine how technological innovations—such as AM,

CPS, and bio-based materials—are converging to redefine sustainability across industrial systems. This integrative perspective is essential for guiding both academic discourse and practical implementation toward more resilient and resource-efficient production models.

This review makes a significant contribution to the extant literature by providing a comprehensive, interdisciplinary analysis that links technological enablers of Industry 4.0 with environmental and circular economic imperatives. Whereas extant studies address these themes in isolation, the present study proposes a unified framework that highlights operational, organizational, and ecological synergies. Furthermore, the study makes an empirical contribution through its mapping of emerging practices in industrial symbiosis and circular production loops. The integration of enabling technologies, such as the IoT, simulation tools, and reverse logistics, offers actionable insights for decision-makers seeking to optimize both environmental performance and economic viability. In summary, this paper provides a holistic and structured overview of sustainable manufacturing paradigms as they evolve under the influence of Industry 4.0. It offers a theoretical and empirical foundation for future research and practice, advocating for a transformation toward circular, bio-based, and intelligent production systems.

Notwithstanding the proliferation of studies on sustainable manufacturing, several crucial gaps persist, a subject this review aims to explicitly address. Firstly, the extant literature on this subject has a tendency to analyze Industry 4.0 technologies, circular economy models, and bio-based innovations separately, without exploring the complex synergies and trade-offs that arise at their intersection in depth. This review makes a significant contribution by addressing the issue of fragmentation through an integrative perspective that articulates technological, environmental, and organizational dimensions. Secondly, the extant literature has identified a paucity of attention towards the specific role of small- and medium-sized enterprises (SMEs), especially in terms of their capacity to adopt and implement these advanced technologies. The paucity of empirical studies exploring the particular barriers and opportunities faced by SMEs represents a significant gap that this work aims to address. Finally, although industrial symbiosis and circularity strategies have been extensively discussed at a conceptual level, there are still few studies that map their practical application and real impact in different industrial sectors and geographical contexts. Synthesizing scattered evidence and highlighting best practices, this review seeks to address these gaps, offering useful guidance for future research and supporting both professionals and policymakers in adopting more holistic and context-appropriate approaches to sustainable manufacturing.

2. Materials and Methods

As previously stated, the present document conducts a systematic review within the scope of the conceptual framework presented in the introduction, applying a PRISMA-based method of literature search that has been adjusted for the engineering field. This review follows the guidance provided by the PRISMA methodology in its latest 2020 revision. Moreover, the study by Blanco et al. in 2020 [6] is utilized as a point of reference, detailing the methodology to be employed to undertake systematic reviews with PRISMA. This includes an in-depth case study that focuses on the most recent trends in the utilization of magnesium, aluminum, and titanium in sectors such as the automotive and aeronautics sectors.

Although PRISMA is a strong basis for carrying out systematic reviews, it has certain inherent limitations. A salient drawback pertains to its reliance on the accessibility and caliber of the included papers, which can exhibit considerable variability based on factors such as database coverage or the precision of search terms. Furthermore, PRISMA does not assess the methodological soundness or relevance of the selected studies, leaving this crucial task to researchers. This delegation has the potential to introduce bias into the review process if not managed with rigor. Another notable limitation stems from the variability among the identified studies. These differences can arise in various aspects, including research methodologies, data collection approaches, and study types. This heterogeneity can impede the synthesis, analysis, and interpretation of results, thereby imposing an additional burden on researchers to ensure the coherence and reliability of their findings.

Consequently, the methodology employed in this study is built on a systematic approach that consists of several key steps. Initially, the research question was defined, with a focus on investigating the latest trends in sustainable manufacturing. Secondly, a reading quota of 15 articles was established, as it was expected that a substantial number of them would be found. The article seeks to identify the most relevant articles; therefore, the search could not be too specific. Following the PRISMA methodology, the articles with the most citations were selected as being the most relevant to the topic under investigation. Consequently, the inclusion and exclusion criteria were predetermined to screen the search results. These criteria were meticulously designed to encompass specific aspects, including publication mode and rigorous quality standards. Simultaneously, search criteria were established, consisting of carefully selected keywords and their synonyms, to ensure comprehensive coverage of the topic and to minimize bias in the search process. Once the initial set of articles was retrieved, a preliminary selection process of the 15 articles was conducted. This involved using database tools to apply filters based on the predefined criteria. The texts were then reviewed, and articles that did not align with the research scope or failed to meet the established quality standards were excluded. Consequently, in the event that 1 of the 15 articles identified was discarded, the subsequent article with the highest number of references was selected, and this process was repeated until the desired number of articles was obtained. The final selection of the literature was categorized and organized according to factors such as type, topics covered, and relevance, through detailed reading and statistical analysis supported by database tools. The process ensured a thorough analysis and facilitated the identification of valuable insights in the desired topic from the selected articles. Finally, the conclusions drawn from the analysis were synthesized in order to facilitate the development of the results and discussion sections of this article. The search and analysis strategies are shown in an outline in Figure 1.

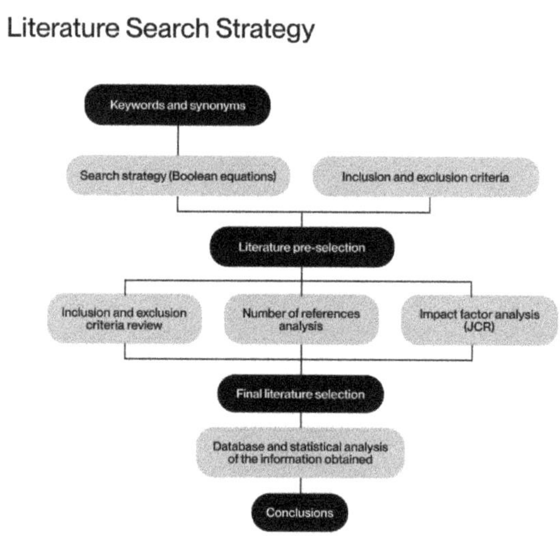

Figure 1. Literature search strategy based on Blanco et al. [6].

2.1. Inclusion and Exclusion Criteria Used in the Search

The preliminary definition of inclusion and exclusion criteria provides a solid foundation for the research and ensures that the results are consistent with the intended objectives. These criteria function not only to delineate the study's scope, but also to mitigate potential bias and ensure that the selected literature is both pertinent and representative for the subsequent analysis [6].

As stated earlier, this analysis focuses on studies published within the last five years to provide a thorough examination of recent trends in sustainable manufacturing techniques. The selection of studies was primarily focused on journal articles, as they are renowned for their meticulous methodologies and reliable findings. Additionally, review articles were included to offer a broad perspective on the current state of the field, highlighting key developments. Conference proceedings were also considered, given their significance in showcasing the latest advancements, innovations, and emerging trends. These sources are especially useful for identifying new approaches to sustainable manufacturing.

To refine the search within the defined topic, a set of carefully selected keywords and synonyms were structured into Boolean equations (see Figure 2). This methodological approach guaranteed extensive coverage, thereby minimizing the probability of overlooking pertinent studies.

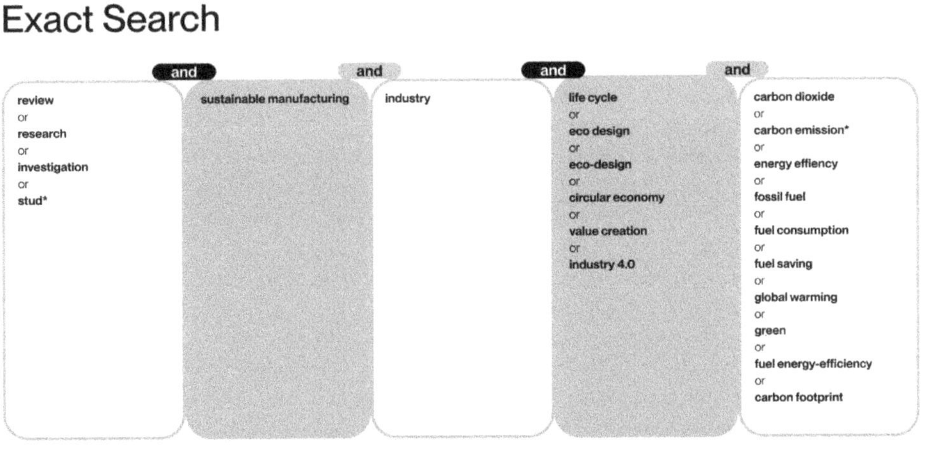

Figure 2. Terms and synonyms entered in this search. The asterisk (*) serves as an operator that captures any letters that follow it.

Web of Science (WoS) was selected as the primary bibliographic resource for this review because of its vast collection of high-quality, peer-reviewed publications and its sophisticated search and analysis tools. WoS provides access to reputable studies from top-tier journals, which adds credibility to the review. Its advanced filtering and Boolean search capabilities facilitate both comprehensive and precise literature searches, while its bibliometric tools enable trend analysis and citation tracking. The choice of WoS is in line with PRISMA's core principles of rigor and transparency, ensuring the validity of the review and the inclusion of influential studies on sustainable manufacturing.

The selection of publications was constrained to those of an open access nature, with the objective of ensuring global accessibility and thereby fostering collaboration in research. The application of strict quality criteria was a further essential element of the re-search process. The inclusion of articles was contingent upon their having undergone the process of peer review and having been published in journals that were ranked in the top quartiles (Q1 and

Q2) of the journal impact factor and journal citation reports. This ensured a high standard of reliability and rigor across the analyzed texts. These criteria are summarized in Table 1.

Table 1. Inclusion and exclusion criteria used in the search.

Inclusion/Exclusion Criteria	Feature
Publication period	From 1 January 2019 to 1 November 2024
Type of study	Articles: journals, reviews, and conference proceedings
Keywords and synonyms	See Figure 2
Information sources	Web of Science (WoS)
Databases in WoS	Option 'All databases'
Language of publication	English
Publication mode	Exclusively open access
Quality criteria required for each publication	Peer-reviewed articles included in WoS. Journal articles limited to Q1–Q2
Quality criteria reviewed for each publication	Reviewed by journal impact factor and journal citations reports

2.2. *Definition of Search Criteria*

The aim of this literature review is to explore the most significant studies on sustainable manufacturing, tracing its evolution from its inception to the present, while also considering potential future developments. Following the previously outlined search methodology, the study uses Boolean equations with keywords and their synonyms to effectively conduct the literature search.

The terms used in the search process are shown in Figure 2. The first column specifies the type of documents to be retrieved, including reviews, research papers, and studies, based on the predefined selection criteria. The second column refines the search to focus on sustainable manufacturing, ensuring that only relevant articles within this area are considered. The third column highlights the industrial applications of sustainable manufacturing, emphasizing its importance in advancing production and research in the manufacturing sector. The next column clusters key concepts related to sustainable manufacturing, such as product life cycle, eco-design, and circular economy, to increase the precision of the results and filter out unrelated studies. Finally, the last column includes terms related to sustainability and energy efficiency to maintain a strong focus on minimizing environmental impact. Given the broad nature of this topic and the likelihood of retrieving a large number of irrelevant articles, the advanced 'exact search' feature of WoS is used to ensure that each selected document contains at least one term from each column.

The Boolean equations used in the search are as follows: TS = (review OR research OR investigation OR stud*) AND TS = (sustainable manufacturing) AND TS = (industry) AND TS = (life cycle OR eco design OR eco-design OR circular economy OR value creation OR Industry 4.0) AND TS = (carbon dioxide OR carbon emission * OR energy efficient OR fossil fuel OR fuel consumption OR fuel saving OR global warming OR environment OR green OR carbon footprint)

With 'exact search' enabled in more options, the results of the WoS search, when entered with these Boolean equations, yielded a total of 1305 documents.

2.3. *Literature Selection*

This section describes the final selection process of the documents obtained in the previous search. First, the search was filtered according to the inclusion and exclusion criteria listed in Table 1. This filter was applied using the WoS search tools, including only open access and English language documents, which resulted in the exclusion of 697 (53.4%) documents from the search. The publication date filter was then applied, which excluded 171 (13.1%) documents from the search. With the exclusion of an additional 4 articles that were not in the English language, the result was a list of 522 articles to read.

In order to optimize document selection and ensure that the analysis was based on the most influential contributions in the field, it was decided to focus the study on the 15 most cited articles from the resulting set. This number was deemed sufficiently representative to encapsulate the most recent and pertinent trends in sustainable manufacturing, while maintaining an analysis on a manageable scale that would facilitate in-depth discussion. Moreover, the prioritization of the most cited works ensured that the selected articles have demonstrated substantial academic impact and recognition by the scientific community. This strategy enabled the study to concentrate on pivotal publications that influence the discourse and evolution of sustainable manufacturing paradigms, thereby ensuring that the results are grounded in well-established and widely recognized scientific contributions.

Therefore, the top 15 most cited articles underwent an initial review to apply the remaining inclusion and exclusion criteria. As a result, only one article (6.7%) was excluded because it did not meet the established quality standards. Specifically, the article was classified within the third quartile (Q3) of the journal citation reports, thereby falling short of the quality threshold defined by the review protocol, which required inclusion of only Q1 and Q2 publications. To maintain the selection criteria, the next most cited article was added to the final list. The article in question did satisfy all pre-established criteria for quality and credibility. The following flow chart, shown in Figure 3, summarizes the selection process for the literature found.

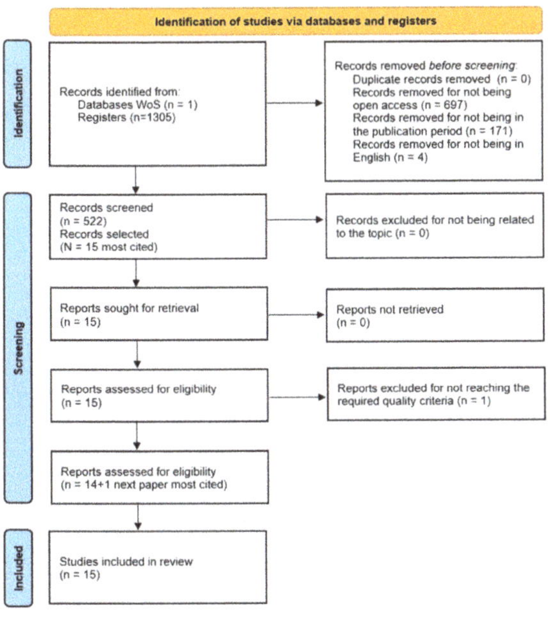

Figure 3. Flow chart of the process of selection of the documents found in the search.

2.4. Synthesis and Analysis of the Documents Included in the Literature Review

The 15 selected documents underwent an in-depth review, including a content analysis aimed at identifying methodologies, approaches, and findings. This process allowed for the identification of recurring patterns, key contributions to the field, and potential gaps in the existing research. To ensure a comprehensive understanding, the selected texts were thoroughly reviewed, resulting in the extraction of the data summarized in Table 2.

Table 2. Main analysis of data from chosen search documents. The symbol 'X' indicates that the documents considered refer to or contain content relevant to the specified category.

Documents	Number of References	Publication Type	State of the Art	Process Optimization	Experiments Design	Country	Publication Year
[1]	198	Q1		X		India	2020
[2]	198	Q1		X		India	2020
[3]	208	Q1		X		Austria	2020
[4]	198	Q1		X		India	2020
[5]	243	Q1		X	X	India	2020
[7]	157	Q2		X		France	2019
[8]	151	Q2			X	Italy	2019
[9]	169	Q2			X	Taiwan	2021
[10]	125	Q2		X		Portugal	2019
[11]	136	Q1		X		India	2020
[12]	178	Q1		X		Denmark	2020
[13]	224	Q1	X			Ireland	2020
[14]	304	Q2		X		Saudi Arabia	2020
[15]	135	Q1			X	United Kingdom	2020
[16]	162	Q1		X		Portugal	2020

As illustrated in Figure 4, a total of 14 documents (93.3%) were classified in business and engineering, which is a significant increase compared to other subject areas. In addition, 13 documents (86.7%) were associated with materials science, while 8 documents (53.3%) were associated with computer science and science technology. These results highlight the diversity and variation in the data collected, reinforcing their connection to the disciplines of engineering and economics.

Figure 4. Main research areas of the articles found in the search.

The years 2019 and 2020 represent the majority of the texts analyzed, as illustrated in Figure 5. The majority of citations are concentrated in these articles, attributed to the longevity of these publications, allowing for more citations to accumulate.

As mentioned above, the terms 'review', 'research', 'investigation', and 'study*' were included in the search criteria to identify articles on state-of-the-art advances, process optimization, and experimental design related to the topic. Of the analyzed texts, 1 (6.7%)

focused on state-of-the-art assessments, 11 (73.3%) explored process optimization, and 4 (26.7%) involved experiments aimed at improving or providing an alternative perspective on machining processes.

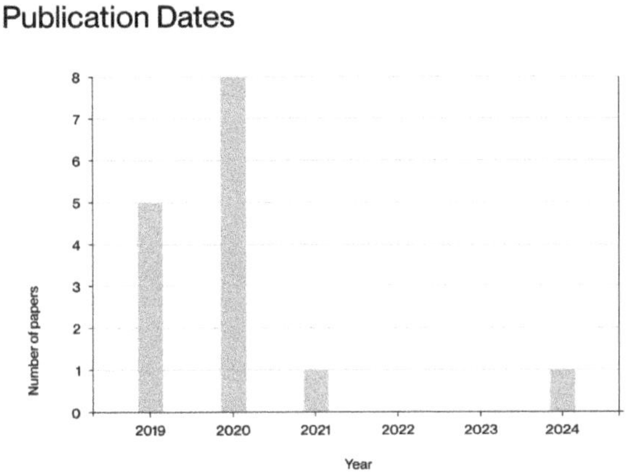

Figure 5. Date of publication of the documents selected from the search results.

In addition to the classification system outlined in Table 2, a further categorization of the selected articles was conducted based on their thematic content. This additional classification was performed with a particular emphasis on their contribution to the broader context of sustainable manufacturing and the emerging opportunities it offers for industrial, technological, and environmental progress. Accordingly, Table 3 below delineates the primary research themes that were identified across the articles that were analyzed. These themes represent the most prominent strategic frameworks that currently underpin the advancement of sustainable manufacturing.

Table 3. Analysis of key strategy themes from chosen search documents. The symbol 'X' indicates that the documents considered refer to or contain content relevant to the specified category.

Documents	Industry 4.0	Circular Economy	Environmental Impact Assessment	Additive Manufacturing (AM)	Sustainable Materials	Life Cycle Assessment (LCA)	Small and Medium-Sized Enterprises (SMEs) Applications
[1]	X	X	X	X	X		X
[2]	X	X		X		X	
[3]			X		X		
[4]	X						
[5]	X	X	X	X	X	X	
[7]	X						
[8]	X	X					X
[9]	X						
[10]	X	X	X				X
[11]	X		X		X		X
[12]	X	X	X	X	X	X	
[13]		X	X	X	X	X	
[14]				X	X	X	
[15]		X					X
[16]		X	X				X

A total of 10 articles (66.7%) concentrate on the implementation of Industry 4.0, with the objective of optimizing the adoption of sustainability measures in companies and organizations. This underscores the significance of Industry 4.0 in this domain, positioning it as a fundamental element for the effective development of the three dimensions of sustainability. The circular economy appears prominently in the reviewed literature, cited in 9 out of 15 articles (60%). It is presented as a key strategic framework that enables sustainable manufacturing by promoting material recirculation through reuse, remanufacturing, and recycling. Often integrated with Industry 4.0 technologies, the circular economy supports the transition from linear to closed-loop systems, enhancing both environmental performance and resource efficiency. Meanwhile, environmental impact assessment is addressed in 8 of the reviewed articles (53.3%), reflecting its importance as a methodological tool for assessing the environmental impact of manufacturing processes. Often linked to life cycle thinking, it provides a structured approach to quantifying emissions, resource use, and waste generation, thereby guiding more informed and sustainable decision-making in industrial contexts. Additive manufacturing (AM) is featured in 6 of the reviewed articles (40%) as a key enabler of sustainable production. Its integration with Industry 4.0 further enhances its potential within circular and low-impact manufacturing systems. A total of 7 of the 15 reviewed articles (46.7%) discuss sustainable materials, highlighting their role in reducing environmental impact through the use of bio-based, biodegradable, or recycled alternatives. These materials are frequently assessed via life cycle approaches and are regarded as pivotal components in promoting green manufacturing and attaining circularity in production systems. Therefore, Life Cycle Assessment (LCA) is employed in 5 out of the 15 examined (33.3%) articles as a core analytical instrument to quantify the environmental impacts of products and processes across their entire life cycle. The utilization of LCA fosters evidence-based decision-making and enhances the incorporation of sustainability criteria in material selection, process optimization, and product design. Finally, the reviewed articles address applications in SMEs in 6 articles (40%). These studies underscore the dualities of implementing sustainable manufacturing practices in settings characterized by resource constraints, accentuating the necessity for scalable technologies, financial support mechanisms, and targeted skill development to facilitate adoption.

3. Results and Discussion

As previously mentioned in the Introduction, the pursuit of sustainability has seen a marked increase in recent years. This endeavor, undertaken by companies, institutions, and organizations, has been propelled by the implementation of innovative technologies, various policies and regulations, and the adoption of sustainable practices and strategies, such as circular economy principles and the introduction of sustainable materials. A comprehensive analysis encompasses the three pillars of sustainability: economic, environmental, and social. The application of analytical methods and assessment approaches can serve to further strengthen these implementations.

The most frequently implemented approach in the analyzed articles was Industry 4.0, which appeared in 10 out of the 15 selected texts (66.67%). This prevalence is not surprising, as Industry 4.0 represents a significant transformation in companies and manufacturing operations, as well as in industrial production, through the integration of advanced technologies. This term was first introduced in Germany in 2011 to denote a contemporary era characterized by the convergence of digital, physical, and biological technologies. This paradigm encompasses a wide range of technological advances, including CPS, the IoT, big data analysis, artificial intelligence (AI), and cloud computing. This fluid communication between physical and digital components allows for the decentralization of decision-making processes and the development of highly efficient production environments. Consequently, Industry 4.0 not only transforms manufacturing operations, but also reshapes value creation processes and business strategies at a systemic level [5]. This paradigm shift establishes the foundation for novel production models and redefines industrial practices, underscoring the interconnection between technological innovation and sustainability.

The evolution of manufacturing systems within the Industry 4.0 paradigm necessitates a multidisciplinary theoretical framework to comprehend the interrelationships between technological innovation and sustainability. In this context, four key frameworks are of particular pertinence: environmental impact assessment, circular economy principles, AM, and sustainable materials science. Environmental impact assessment serves as the foundation for evaluating the sustainability of manufacturing processes and technologies. It encompasses the identification, quantification, and mitigation of environmental consequences related to resource consumption, energy use, emissions, and waste generation [5]. LCA, a widely utilized method for quantifying environmental impacts across the entire lifecycle of products and processes, further informs this evaluation [2]. Furthermore, the integration of real-time data analytics and embedded environmental indicators within Industry 4.0 systems provides novel capabilities to monitor and mitigate environmental impacts during production processes.

Concomitantly, the circular economy offers a systemic framework that redefines growth by decoupling economic activity from the consumption of finite resources. In contrast to the conventional "take–make–dispose" model, it advocates for the recirculation of materials through reuse, refurbishment, remanufacturing, and recycling strategies [1]. AM plays a critical role in this transition by enabling localized production, reducing energy use, minimizing material waste, and supporting mass customization. Furthermore, AM facilitates the use of recycled or renewable materials, aligning with circular economy and sustainable innovation principles [2]. Material sustainability has become pivotal in advancing green manufacturing. A mounting body of research underscores the imperative to substitute petroleum-based inputs with biodegradable, renewable, and low-impact alternatives. The selection of raw materials is guided by sustainable design principles, as addressed through LCA, circularity, and eco-innovation. These principles favor non-toxic constituents and encourage reuse and recycling strategies throughout the product's lifespan.

The relationship between sustainability and Industry 4.0 is of paramount importance in advancing sustainable practices within companies and organizations. The advent of Industry 4.0 technologies has engendered substantial opportunities to curtail environmental impacts, minimize material waste, and enhance energy efficiency. For instance, Kumar et al. [1] investigated the role of Industry 4.0 in enhancing the performance of SMEs in India through sustainable business models aligned with corporate social responsibility principles. Their findings underscore the potential of Industry 4.0 technologies to enhance the efficiency and energy savings of manufacturing processes. However, the study also notes that implementing these technologies remains particularly challenging for SMEs in developing countries due to resource limitations, identifying fifteen critical challenges that must be addressed to ensure successful integration. In a complementary line of research, Machado et al. [4] conducted a systematic review exploring how Industry 4.0 contributes to sustainable manufacturing. Utilizing bibliometric, social, and network analyses, they have provided quantitative insights into the development of research in this domain. While the correlation between Industry 4.0 and sustainability is recognized, the authors underscore that its integration remains inadequate and advocate for additional research grounded in practical case applications.

Additionally, Panetto et al. [7] analyzed the role of CPS in Industry 4.0 and highlighted key challenges in the transition toward the "industry of the future." The aforementioned framework encompasses the control of manufacturing plants via CPS, interoperability across systems of systems (SoS), interconnected networks, and data-driven decision-making processes. These systems contribute to sustainable manufacturing by integrating complementary technologies, such as collaborative robots (cobots) and augmented reality, thus improving production efficiency. Moreover, a number of studies have examined the practical applications of this phenomenon. For instance, Braccini et al. [8] examined the adoption of Industry 4.0 in a sanitary ceramics manufacturing company, noting the paucity of empirical research on the subject. The case study evaluated the implementation process by considering sustainability's three dimensions and employment variability. Contrary to concerns about job losses, the findings suggest that Industry 4.0 creates new roles while enhancing productivity, product quality, and workplace safety through continuous process improvement. Consequently, the study strongly recommends the adoption of these technologies to support sustainable manufacturing.

Despite the significant benefits that Industry 4.0 technologies can bring to sustainability and operational efficiency, SMEs face numerous and multifaceted barriers in their adoption. Key challenges include limited financial resources, a shortage of skilled personnel, underdeveloped digital infrastructure, and organizational inertia when it comes to embracing disruptive technologies [5]. Additionally, the substantial capital expenditure demanded by the integration of CPS, IoT platforms, and advanced data analytics tools poses a considerable obstacle for SMEs, particularly in developing economies [5]. These constraints are often compounded by difficulties in accessing external funding and the absence of tailored implementation roadmaps for smaller firms. To overcome these barriers, targeted policy measures and adaptive business strategies are essential. Governments and institutions can support SMEs through fiscal incentives, innovation grants, and low-interest financing mechanisms aimed at de-risking technological investments. Concurrently, capacity-building initiatives, including vocational training programs, digital literacy workshops, and industry–academia partnerships, can facilitate the acquisition of knowledge and skills. From a business perspective, the adoption of modular, interoperable, and scalable digital solutions enables SMEs to implement Industry 4.0 technologies incrementally, aligning with their operational and financial constraints. Furthermore, engagement in collaborative ecosystems, such as industrial clusters or digital innovation hubs, can facilitate access to shared infrastructure, technical expertise, and best practices, thereby reducing the barriers to successful digital transformation [5].

As previously mentioned, some of the reviewed articles explore additional concepts that reinforce the sustainable approach when implementing Industry 4.0. The studies by Tseng et al. [9] and Varela et al. [10] examine the integration of Industry 4.0 with lean manufacturing from a sustainability perspective. Lean manufacturing is a time-based philosophy that belongs to the set of "lean" techniques, focusing on creating value for the end consumer by eliminating activities that generate waste. In the initial study [9], bibliometric analyses were conducted to examine sustainable industry by linking lean manufacturing and Industry 4.0. The study concludes that further practical analyses of these cases are needed to strengthen the understanding of their combined impact. In the subsequent study [10], a structural equation model was proposed to quantitatively measure the effects of lean manufacturing and Industry 4.0, both with a focus on sustainability. To this end, a survey was administered in Spain and Portugal to assess the influence of these philosophies, and the resulting data were then analyzed both independently and in relation to one another. The resulting equation model was based on six hypotheses; however, it did not fully meet all the proposed criteria. The study ultimately suggests that the assumed variables differ from those commonly addressed in the analyzed literature, highlighting the need for further refinement and investigation in this area.

Another pivotal concept highlighted in several of the aforementioned articles is the circular economy. In the study published by Dev et al. [5], a pathway toward a sustainable supply chain is presented through the joint implementation of Industry 4.0 principles and the circular economy approach. Additionally, a lean approach and the implementation of reverse logistics are introduced as essential factors for improving sustainability. The study proposes a structured plan for this integrated implementation, outlining a comprehensive system that interconnects these elements. The study introduces optimized reuse and recycling processes facilitated by reverse logistics, designs for sustainable environments, and highlights key advantages such as cleaner and more efficient production.

AM, a process of paramount importance to the enhancement of sustainability in corporate and organizational entities, can also be regarded as a constituent element of Industry 4.0. In the study presented by Mehrpouya et al. [2], an analysis is conducted on the implementation of AM as a sustainable production process. The study underscores its substantial potential for sustainability improvements, as it facilitates the utilization of sustainable and recyclable materials, with some materials even being reusable. The study introduces key practices, including mass customization, a growing production model that prioritizes customer-specific considerations. It also introduces other essential concepts, such as eco-design and LCA. Eco-design ensures that every design process minimizes environmental impact, while LCA is a tool commonly used alongside eco-design to evaluate a product's environmental footprint throughout its entire life cycle, considering all stages of production, use, and disposal. The study concludes by emphasizing the potential of AM when combined with other Industry 4.0 technologies and its promising future in advancing sustainable manufacturing.

A critical complementary strategy to advance Industry 4.0 from a sustainability perspective involves the incorporation of sustainable materials to replace traditional, more polluting alternatives. This practice aligns closely with the principles of Industry 4.0, as it supports the integration of circular economy strategies aimed at reducing waste, extending product lifecycles, and enabling product designs that facilitate disassembly and recycling. The incorporation of bio-based, biodegradable, or recyclable materials within smart manufacturing environments signifies a systemic endeavor to mitigate environmental impact while preserving high efficiency and cultivating innovation. In this regard, Hynes et al. [11] examined the textile industry's use of dyes and chemicals, proposing sustainable alternatives and highlighting the role of Industry 4.0 technologies, including the IoT and smart sensors, to enhance management systems and improve sustainability performance. Fabris et al. [12] explored the industrialization and application of algae as a sustainable material in combination with Industry 4.0 technologies. The study's findings underscore the potential of algae-based solutions to provide cost-effective, technologically advanced alternatives that promote environmental responsibility in domains such as biomass generation and biotechnology. Notwithstanding the promising potential of algae, the authors note a paucity of extensive research concerning its industrial applications.

Concurrently, Jones et al. [3] investigated the potential of mycelium-based composites as substitutes for conventional thermal and acoustic insulation materials. These biofabricated composites exhibit reduced energy consumption during the manufacturing process and offer superior thermal, acoustic, and fire resistance properties. Despite the presence of certain limitations, the study emphasizes their substantial potential to propel the transition toward environmentally sustainable construction solutions. Additionally, RameshKumar et al. [13] have focused on the design of biodegradable polymers aimed at reducing plastic pollution by replacing conventional packaging materials. Their research addresses the balance between biodegradability and functionality, incorporating a LCA to evaluate the sustainable value chain of bioplastics. Finally, Shafey [14] examined biosynthesis processes for substituting metallic and metal oxide nanoparticles commonly used in cosmetics, pharmaceuticals, and personal care products. This study underscores the utilization of biodegradable polymers as a substitute for such nanoparticles, offering insights into fabrication techniques while maintaining a pronounced focus on sustainability and environmental responsibility.

It can be stated with complete certainty that the implementation of Industry 4.0 is a practice that enhances the sustainability of companies and organizations, providing improvements that extend beyond this domain. The integration of advanced technologies into business practices has been shown to enhance efficiency, reduce environmental impact, and develop products and services that are more sustainable. However, it is imperative to acknowledge the challenges that must be addressed, such as unemployment and data privacy, to ensure a just and advantageous transition across the three dimensions of sustainability: economic, environmental, and social.

The transition to a fully sustainable manufacturing paradigm is contingent upon the convergence of Industry 4.0 and the previously discussed pillars of sustainability. Of particular significance is the relationship between Industry 4.0 and the circular economy. The efficacy of circular economy strategies is significantly enhanced when supported by Industry 4.0 technologies, which facilitate the establishment of industrial ecosystems that enable the continuous repurposing of materials and components. This synergy not only mitigates environmental impact, but also enhances economic viability. AM further strengthens this connection by leveraging Industry 4.0 technologies to optimize design, production processes, and material efficiency. The application of AI-driven design has been demonstrated to play a complementary role in this regard, achieving a reduction in material usage while preserving structural integrity. This, in turn, has resulted in products that are lighter, stronger, and more sustainable. In this context, sustainable materials, such as biodegradable polymers, engineered mycelium composites, and recycled carbon fiber, are critical in bridging Industry 4.0-enabled manufacturing with circular economy objectives. These materials are instrumental in facilitating the development of resource-efficient, closed-loop production systems.

Smart factories, enabled by Industry 4.0 technologies, further reinforce this integration through the implementation of real-time monitoring systems that assess material properties and performance. This ensures that bio-based and recycled materials can be seamlessly incorporated into demanding applications without compromising quality. Concurrently, the implementation of predictive maintenance, driven by data analytics, has been shown to optimize AM processes, thereby extending the material's usability and reducing waste. The combination of these elements gives rise to a system in which the whole surpasses the sum of its parts. The advent of Industry 4.0 signifies more than merely a digital transformation instrument; it is the technological facilitator that enhances and integrates circular economy principles, additive manufacturing innovations, and sustainable material development. This integration is further advanced by real-time data.

On the other hand, despite the numerous advantages offered by the implementation of Industry 4.0, it is not a mandatory requirement for improving sustainable manufacturing in companies and organizations. In the study by Nana O. [15], the end-of-life applications of electric vehicle batteries are examined within a low-carbon circular economy. This research presents a sustainability approach aligned with the circular economy concept discussed earlier, but without considering Industry 4.0. The study, conducted in the United Kingdom, aims to reshape industrial systems. Initially, semi-structured surveys were carried out with various local companies, proposing closed-loop solutions, identifying challenges, and suggesting necessary actions. One key issue identified was the isolated implementation of closed-loop solutions, which should instead be applied collaboratively. The study also introduces the ReLiB Project, which aims to achieve nearly 100% battery mineral recycling. It is estimated that, through the implementation of regulatory measures such as these, a complete transition could be achieved by 2050 under a circular economy framework. Similarly, the study by Neves et al. [16] explores the implementation of industrial symbiosis, which enables different entities and companies—traditionally operating separately—to collaborate with one another. This article maps research trends in industrial symbiosis from a sustainability perspective, emphasizing how closing material cycles can significantly reduce previously generated waste. The study employs a case analysis method, gathering and qualitatively analyzing data to enhance understanding of existing cases. The findings highlight resource savings and economic benefits as the primary advantages of this approach.

A sustainability impact assessment was conducted to further elucidate the findings presented in the preceding sections. This assessment synthesizes the environmental, economic, and social contributions of the key technologies and strategies identified in the reviewed literature. The objective of this assessment is to provide a multifaceted perspective on the alignment of these innovations with the principles of sustainable manufacturing, which extends beyond their technical descriptions or conceptual functions. The resulting Table 4 classifies each element according to its systemic contributions and is based on the empirical evidence gathered from the selected studies included in this review.

Table 4. Sustainability impact assessment of the key technologies and strategies identified in the reviewed literature.

Technology or Strategy	Environmental Impact	Economic Impact	Social Impact
Industry 4.0	Enables real-time monitoring and energy optimization; reduces emissions and resource consumption.	Improves productivity and operational efficiency. Enables data-driven decision-making.	Transforms workforce dynamics, but requires upskilling and digital literacy training.
Circular economy	Promotes reuse, recycling, and reduced resource extraction; mitigates environmental degradation.	Reduces input costs through reuse and remanufacturing and enables circular business models.	Promotes responsible consumption and the creation of green jobs. Encourages stakeholder collaboration.
AM	Reduces waste and transportation emissions through localized production and enables efficient use of materials.	Reduces inventory and logistics costs; supports mass customization and on-demand production.	Creates local jobs and supports the development of digital manufacturing skills.
Sustainable materials	Replaces conventional materials with biodegradable or bio-based alternatives, reducing environmental footprint.	Potentially high cost, but leads to long-term savings from reduced resource consumption and environmental compliance.	Publicly favored. It promotes consumer awareness and responsible material use.
LCA	Supports holistic assessment of environmental impacts across product lifecycle stages.	Enables cost-effective environmental planning and resource efficiency strategies.	Raises awareness of sustainability in product design and policy; supports informed decision-making.
Industrial symbiosis	Minimizes waste by enabling resource sharing between industries. It also supports closed-loop systems.	Reduces raw material costs and creates synergies and new business opportunities.	Encourages inter-organizational collaboration; supports regional job growth.

As illustrated in Table 4, the reviewed literature elucidates the multifaceted contributions of each strategy and technology. The central enabling role of Industry 4.0 is evidenced by its integration of real-time data analytics, automation, and smart systems. Collectively, these technologies enhance energy efficiency, reduce waste, and support agile decision-making. Its economic relevance is particularly pronounced in the areas of operational optimization and cost reduction. On a social level, it demands a restructuring of workforce skills and digital literacy [4,7]. The circular economy has emerged as a system-level paradigm facilitating the continuous flow of materials through reuse, remanufacturing, and recycling. Its integration with digital tools enhances traceability and efficiency, while economically enabling new value creation models based on resource regeneration rather than extraction [1,5]. AM offers unique benefits by enabling localized, on-demand production with significantly reduced material waste. It facilitates customization, lightweight design, and reduced supply chain emissions. However, concerns regarding material recyclability and energy consumption during production underscore the need for future improvements [2]. The utilization of sustainable materials is imperative in the pursuit of decoupling manufacturing from its fossil-based inputs. These materials offer substantial environmental benefits but are still transitioning from niche applications to broader industrial scalability due to cost and durability constraints [3,13]. LCA serves as a methodological foundation for objectively evaluating sustainability outcomes across the entire lifespan of products [2]. Industrial symbiosis, on the other hand, is a strategic approach to enhancing resource efficiency at the system and product life cycle levels. It facilitates cross-sectoral material exchanges, thereby reducing collective waste [5,16]. The collective findings of the sustainability impact assessment underscore that technologies delivering the greatest sustainability gains are those that operate across multiple dimensions simultaneously—technological, environmental, economic, and social. This underscores the necessity of adopting holistic approaches when assessing the viability and strategic value of innovations within sustainable manufacturing systems.

Concurrently with technological and environmental innovations, global economic landscapes are undergoing profound transformations that are reshaping pathways toward sustainable manufacturing. In advanced economies, particularly in Western Europe and North America, there is a marked shift toward service-oriented and resource-regenerative business models. These regions are increasingly incorporating circular economy principles through the integration of advanced digital tools, Industry 4.0 technologies, and policy-driven incentives, including European directives and industrial symbiosis networks. Initiatives of this nature facilitate the establishment of closed-loop systems, real-time monitoring, digital twins, and collaborative platforms where firms exchange materials, energy, and information to optimize efficiency and minimize waste. This transition is facilitated by the presence of robust digital infrastructures and stringent environmental regulations, which have led to the development of sophisticated applications of sustainable manufacturing across various sectors [5,16].

Conversely, emerging and developing economies—particularly in Asia, Latin America, and select regions of Africa—are undergoing rapid industrialization, accompanied by increasing access to digital infrastructure and integration into global supply chains. In this context, the strategic implementation of Industry 4.0 technologies, reverse logistics, and AM is driven not solely by the pursuit of competitiveness, but also by the imperative to decouple economic growth from resource consumption. The aforementioned nations have advanced this agenda through the implementation of industrial symbiosis frameworks in eco-industrial parks. The promotion of shared use of materials, energy, and infrastructure is intended to achieve emissions reductions and cost savings. Moreover, additive manufacturing plays a pivotal role in the decentralization of production and the mitigation of transportation-related emissions, while facilitating the utilization of bio-based and biodegradable materials for mass customization [12,16].

However, considerable disparities persist in these regions due to varying levels of industrial maturity, regulatory support, and technological capabilities. The implementation of smart manufacturing remains uneven, particularly among SMEs, which face financial and technical challenges. Furthermore, while industrial symbiosis in advanced economies is often institutionalized and policy-driven, in developing contexts, it frequently emerges through informal or community-based initiatives influenced by resource scarcity and local socio-economic conditions [2,16].

When considered collectively, these region-specific trajectories indicate that sustainable manufacturing is not uniform. While global trends increasingly align around circularity, resilience, and digitalization, the operationalization of these concepts must be adapted to the unique economic, infrastructural, and regulatory realities of each region [2]. This underscores the necessity for context-sensitive strategies that ensure technological solutions are not only environmentally and socially beneficial, but also economically viable and locally relevant. Sustainable manufacturing, therefore, is best understood as a globally interconnected, yet regionally nuanced, transformation [13]. Nevertheless, significant disparities persist in these regions due to varying levels of industrial maturity, regulatory support, and technological capabilities. The implementation of smart manufacturing remains uneven, particularly among SMEs. These enterprises face financial and technical challenges. Furthermore, while industrial symbiosis in advanced economies is often institutionalized and policy-driven, in developing contexts, it frequently emerges through informal or community-based initiatives influenced by resource scarcity and local socio-economic conditions [1,2,16].

In light of these considerations, recent findings indicate that the future evolution of sustainable manufacturing will be influenced not only by regional adaptation and policy alignment, but also by the strategic incorporation of emerging technologies, bio-based innovations, and system-level circularity. A notable shift in manufacturing paradigms is evident, with a transition toward hybrid models that integrate digital intelligence with ecological functionality. For instance, the integration of Industry 4.0 tools—such as the IoT, real-time monitoring, digital twin simulations, and AI-driven decision-making—with bioengineered systems demonstrates a significant synergy between technological efficiency and biological regeneration [3,12]. These bio-digital systems possess the capacity to transform waste into value through decentralized, low-energy, and scalable processes that reduce environmental impact while enhancing material circularity. Concurrently with these innovations, the emergence of green nanotechnologies has opened new avenues for environmentally benign functional materials applicable in sectors ranging from electronics to water treatment [14]. Concurrently, systemic approaches, such as industrial symbiosis, are undergoing a transformation through the integration of advanced data infrastructures and supply chain analytics, thereby establishing interconnected networks for material and energy exchange. These digitally supported symbioses have the potential to enhance operational efficiency and to generate new business models grounded in eco-innovation and shared value creation [5,16]. These developments signify a fundamental shift in sustainable manufacturing, transitioning from isolated technological upgrades to comprehensive, interdisciplinary ecosystems capable of responding dynamically to local and global sustainability challenges. This paradigm shift underscores the imperative of integrating adaptability, resilience, and biological compatibility into the very fabric of industrial innovation strategies [12,14,16].

4. Conclusions and Future Research Directions

A thorough review of extant literature reveals that the integration of sustainable manufacturing processes within the industry engenders multiple advantages. These include the reduction in environmental impact and operational costs, while concomitantly enhancing production efficiency and overall machining performance. The following conclusions can be derived from this review:

- The advent of Industry 4.0 technologies has profoundly impacted the realm of sustainable manufacturing. These technologies serve as the foundational framework for real-time monitoring, predictive analytics, and resource optimization. By leveraging the capabilities of the IoT, AI, and blockchain, enterprises can enhance material traceability, reduce waste, and augment production efficiency. However, the widespread adoption of these technologies is impeded by several challenges. These include the high implementation costs, concerns regarding cybersecurity, and the necessity of technical expertise, particularly for SMEs.
- The circular economy is significantly enhanced through Industry 4.0 innovations, as digital systems facilitate closed-loop manufacturing, waste minimization, and improved material recovery processes. Nevertheless, the lack of standardized sustainability metrics and real-time LCA models limits the full exploitation of these benefits.
- AM offers significant potential for enhancing material efficiency and decentralized production. However, critical barriers persist, including high energy consumption, the recyclability of AM materials, and material degradation. To effectively integrate AM into circular systems, advancements in sustainable feedstocks, AI-driven design optimization, and automated end-of-life processing techniques are imperative.
- The utilization of sustainable materials, encompassing bio-based polymers, biodegradable composites, and engineered mycelium structures, is of paramount importance in the context of closing material loops. Concurrently, the industry continues to grapple with challenges pertaining to scalability, durability, and compatibility with prevailing manufacturing processes. The advent of innovations in nanotechnology-enhanced materials and AI-driven material selection holds immense potential in surmounting these impediments.

Despite the significant progress that has been made, future research must advance towards more rigorous and integrated methodologies to evaluate the real-world implementation of Industry 4.0 technologies in sustainable manufacturing environments. It is imperative to conduct longitudinal and mixed-method studies that can provide robust, context-sensitive insights into these transitions. A primary objective will be the formulation of comparative multi-case studies across diverse industrial sectors, with a particular emphasis on SMEs and high-impact value chains. These studies should integrate process tracing and realist evaluation approaches to elucidate the causal mechanisms underlying both successful and unsuccessful digital transformation pathways.

Furthermore, the integration of digital LCA models with real-time data streams sourced from IoT-enabled sensors, using platforms such as cloud-based Manufacturing Execution Systems (MES), will be vital. This dynamic coupling will enable near real-time monitoring of environmental performance indicators, thus enhancing precision and responsiveness in sustainability assessments.

Furthermore, it is recommended that subsequent research employ agent-based modeling techniques to simulate the adoption of circular economy practices fueled by Industry 4.0 technologies under diverse socioeconomic scenarios. These simulations should be enriched through participatory foresight workshops. Such workshops should include industry actors, policymakers, and supply chain partners. Together, they should co-create realistic and viable transition pathways. The integration of digital twin infrastructures with industrial symbiosis platforms, underpinned by semantic interoperability and ontology-based data integration frameworks, offers promising avenues for advanced coordination and decision-making across interconnected industrial ecosystems. The integration of digital LCA models with real-time data streams sourced from IoT-enabled sensors, using platforms such as cloud-based MES, will be vital. This dynamic coupling has the potential to facilitate the real-time monitoring of environmental performance indicators, thereby enhancing the precision and responsiveness of sustainability assessments.

Another critical area of inquiry pertains to the intersection of additive manufacturing, bio-based materials, and AI-powered decision support systems within circular economy loops. The employment of design science research methodologies will be instrumental in the iterative development and validation of decision-support tools that optimize material and resource flows, while effectively balancing trade-offs between economic, environmental, and social performance metrics.

The advancement of industrial symbiosis represents a significant opportunity to concurrently enhance sustainability and economic efficiency. The capacity of companies to exchange materials, energy, and by-products—repurposing waste streams from one sector as valuable inputs for another—has the potential to significantly reduce landfill waste, minimize resource extraction, and reduce production costs. Nevertheless, further research is necessary to effectively scale these models. Such research must address systemic enablers, including the development of robust regulatory frameworks, the standardization of waste classification, and the implementation of incentives for collaboration. Digital platforms and IoT-based monitoring architectures will also be essential to facilitate real-time tracking, ensure traceability, and enable seamless coordination across multiple industries.

Limitations of the Study

Notwithstanding the exhaustive nature of this systematic review and its contribution to the understanding of sustainable manufacturing trends, several limitations must be acknowledged. First and foremost, the study relies exclusively on previously published scientific monographs and peer-reviewed articles as primary data sources. While this approach ensures a certain degree of academic rigor, it may not encompass emerging, unpublished, or regionally specific practices that have yet to be documented in the literature. Consequently, the exclusive reliance on published monographs and articles constitutes an inadequate contribution to the formulation of a valuable article, as it curtails the diversity and immediacy of the knowledge base considered.

Furthermore, while the PRISMA methodology is widely accepted and systematic for literature reviews, it does possess inherent limitations. Specifically, PRISMA does not evaluate the methodological quality or practical relevance of the selected studies, placing the burden of assessment on the researchers themselves. This approach carries the potential for bias, particularly when the selected articles exhibit substantial heterogeneity in terms of methodology, data collection approaches, and industrial contexts.

Furthermore, the generalizability of the findings is constrained by the limited number of articles included in the final analysis (n = 15). While the selection of articles was based on factors such as citation count and relevance, this criterion may inadvertently favor older or more established research, potentially overlooking innovative but less-cited studies published more recently. Furthermore, the underrepresentation of certain technologies and sustainable manufacturing strategies can be attributed to their novelty and the limited number of comprehensive empirical validations available.

These limitations underscore the necessity for additional empirical research and the incorporation of a more extensive and diverse array of data sources in subsequent studies. The necessity of triangulating literature-based insights with field-level evidence is also underscored, as this approach is instrumental in developing more robust and context-sensitive conclusions.

Author Contributions: Conceptualization, S.P.; methodology, S.P.; validation, E.M.R., J.A. and B.d.A.; formal analysis, S.P.; investigation, S.P.; resources, S.P.; data curation, S.P. and E.M.R.; writing—original draft preparation, S.P.; writing—review and editing, S.P., E.M.R., J.A. and B.d.A.; visualization, S.P.; supervision, E.M.R.; project administration, E.M.R. All authors have read and agreed to the published version of the manuscript.

Funding: This research received no external funding.

Data Availability Statement: No new data were created or analyzed in this study.

Acknowledgments: The authors are grateful for the support of the Industrial Production and Manufacturing Engineering (IPME) Research Group, the Innovation and Teaching Group for Industrial Technologies in Productive Environments (TIA Plus UNED), and the Master of Manufacturing Advanced Engineering.

Conflicts of Interest: The authors declare no conflicts of interest.

References

1. Kumar, R.; Singh, R.K.; Dwivedi, Y.K. Application of industry 4.0 technologies in SMEs for ethical and sustainable operations: Analysis of challenges. *J. Clean. Prod.* **2020**, *275*, 124063–124076. [CrossRef] [PubMed]
2. Mehrpouya, M.; Dehghanghadikolaei, A.; Fotovvati, B.; Vosooghnia, A.; Emamian, S.S.; Gisario, A. The potential of additive manufacturing in the smart factory industrial 4.0: A review. *Appl. Sci.* **2019**, *9*, 3865. [CrossRef]
3. Jones, M.; Mautner, A.; Luenco, S.; Bismarck, A.; John, S. Engineered mycelium composite construction materials from fungal biorefineries: A critical review. *Mater. Des.* **2020**, *187*, 108397. [CrossRef]
4. Machado, C.G.; Winroth, M.P.; Ribeiro da Silva, E.H.D. Sustainable Manufacturing in Industry 4.0: An emerging Research agenda. *Int. J. Prod. Res.* **2019**, *58*, 1462–1484. [CrossRef]
5. Dev, N.K.; Shankar, R.; Qaiser, F.H. Industry 4.0 and circular economy: Operational excellence for sustainable reverse supply chain performance. *Resour. Conserv. Recycl.* **2020**, *153*, 104583. [CrossRef]
6. Blanco, D.; Rubio, E.M.; Marín, M.M.; De Agustina, B. Propuesta metodológica para revisión sistemática en el ámbito de la ingeniería basada en PRISMA. In Proceedings of the XXIII Congreso Nacional de Ingeniería Mecánica, Jaén, Spain, 21–23 October 2020.
7. Panetto, H.; Iung, B.; Ivanov, D.; Weichhart, G.; Wang, X.; Iung, B. Challenges for the cyber-physical manufacturing enterprises of the future. *Annu. Rev. Control* **2019**, *47*, 200–213. [CrossRef]
8. Braccini, A.M.; Margherita, E.G. Exploring organizational sustainability of Industry 4.0 under the triple bottom line: The case of a manufacturing company. *Sustainability* **2019**, *11*, 36–53. [CrossRef]
9. Tseng, M.L.; Tran, T.P.T.; Ha, H.M.; Bui, T.D.; Lim, M.K. Sustainable industrial and operation engineering trends and challenges Toward Industry 4.0: A data driven analysis. *J. Ind. Prod. Eng.* **2021**, *38*, 581–598. [CrossRef]
10. Varela, L.; Araújo, A.; Ávila, P.; Castro, H.; Putnik, G. Evaluation of the relation between lean manufacturing, industry 4.0, and sustainability. *Sustainability* **2019**, *11*, 1439. [CrossRef]
11. Hynes, N.R.J.; Kumar, J.S.; Kamyab, H.; Sujana, J.A.J.; Al-Khashman, O.A.; Kuslu, Y.; Ene, A.; Suresh Kumar, B. Modern enabling techniques and adsorbents based dye removal with sustainability concerns in textile industrial sector—A comprehensive review. *J. Clean. Prod.* **2020**, *272*, 122636. [CrossRef]
12. Fabris, M.; Abbriano, R.M.; Pernice, M.; Sutherland, D.L.; Commault, A.S.; Hall, C.C.; Labeeuw, L.; McCauley, J.I.; Kuzhiuparambil, U.; Ray, P.; et al. Emerging Technologies in Algal Biotechnology: Toward the Establishment of a Sustainable, Algae-Based Bioeconomy. *Front. Plant Sci.* **2020**, *11*, 279. [CrossRef] [PubMed]
13. RameshKumar, S.; Shaiju, P.; O'Connor, K.E.; Babu, R. Bio-based and biodegradable polymers-State-of-the-art, challenges and emerging trends. *Curr. Opin. Green Sustain. Chem.* **2020**, *21*, 75–81. [CrossRef]
14. El Shafey, A.M. Green synthesis of metal and metal oxide nanoparticles from plant leaf extracts and their applications: A review. *Green Process. Synth.* **2020**, *9*, 304–339. [CrossRef]

15. Bonsu, N.O. Towards a circular and low-carbon economy: Insights from the transitioning to electric vehicles and net zero economy. *J. Clean. Prod.* **2020**, *256*, 120659. [CrossRef]
16. Neves, A.; Godina, R.; Azevedo, S.G.; Matias, J.C.O. A comprehensive review of industrial symbiosis. *J. Clean. Prod.* **2020**, *247*, 119113. [CrossRef]

processes

MDPI

Article

Analyzing the Impact of Orifice Size and Retention Time in Private Tanks on Water Quality Indicators in Distribution Networks

Syed Rizvi * and Rabee Rustum *

Institute for Infrastructure and Environment, School of Energy, Geosciences, Infrastructure and Society, Heriot-Watt University, Dubai P.O. Box 38103, United Arab Emirates
* Correspondence: sar32@hw.ac.uk (S.R.); r.rustum@hw.ac.uk (R.R.)

Abstract

Chlorine decay in water distribution networks is significantly affected by the presence of private storage tanks, particularly due to the orifice size and retention time, which influence both hydraulic flow behavior and water residence time. This study introduces a novel simulation framework that integrates pressure-driven analysis with a first-order kinetic model for chlorine decay, implemented using the WQnetXL tool and validated through simulations in EPANET. Two schematic models, including a real-world case from Dubai, were analyzed under varying orifice sizes and retention times. Results show that larger orifices lead to higher initial chlorine concentrations during tank filling due to increased flow rates, but result in a rapid decline in chlorine levels once the tanks reach full capacity. In contrast, smaller orifices maintain more stable chlorine concentrations over time due to prolonged inflow durations. Extended retention times further delay tank filling and sustain higher chlorine levels until the system transitions to behavior typical of demand-driven analysis. A reliability assessment of the Dubai case study indicated that incorporating private tanks can result in deviations in chlorine concentration of up to 30 percent compared to conventional models. This approach addresses a key gap in conventional network modeling by quantifying the influence of decentralized storage on disinfection effectiveness and network reliability.

Keywords: orifice size; retention time; private tank; water supply; water quality

1. Introduction

Water quality analysis is essential for evaluating a network's ability to deliver clean water. This is done to avoid prolonged retention times within the network system, which can increase the water age and compromise quality. As such, chlorine is added to the effluent of the water treatment plant to ensure that no microorganisms are produced on the pipe's walls and that water remains drinkable for the end user [1]. However, chlorine concentration varies as it reacts with the pipe walls. Therefore, associated water quality models that perform first-order decay equations are used to observe the network's performance and whether it can maintain a set quantity of chlorine concentration prescribed by the Code of Practice [1–5].

Generally, to assess the performance of water distribution networks (WDNs), a certain value of known chlorine concentration is added to the reservoir of the network model, and based on the steady-state analysis, the chlorine concentration is calculated in the rest of the components using the Lagrangian Algorithm. This is based on fluid movement, which plays a vital role in the overall reaction of chlorine particles with the walls of the pipes. The concentration of chlorine changes over time based on the reaction of water parcels within the system using decay equations. Since integrating private tanks can affect the flow, it is important to analyze their effect on the chlorine concentration within the system and assess whether it remains under the prescribed threshold when associated with different orifice inlets and retention times.

Summary of the Literature

Based on the history of water quality modelling, many models have been developed that simulate the chlorine concentration in the nodes that decay over time as the water spreads into the network. The basis of modelling chlorine concentration comes from the algorithm that can solve the convenient transport problem in WDNs. Most researchers have worked on two techniques to evaluate concentration during water transport: steady-state analysis [6–13] and dynamic mathematical modelling [12,14,15]. Both of these techniques have shown different levels of success based on their capabilities, especially in the work of Liou and Kroon [15]. Their study introduced an algorithm known as the Lagrangian Time Driven Method (TDM).

In steady-state analysis, models use the law of conservation of mass to obtain the concentration distribution of the containment parcel that occurs once the WDN has approached the hydraulic equilibrium [12]. On the other hand, dynamic mathematical modelling is based on a system simulation approach to obtain the motion and progress of the containments under variable demand, continuous supply, and hydraulic conditions [12]. When comparing the two approaches, dynamic modelling tends to perform better because the system of WDN is based on the continuous process of water being supplied at every time interval. Therefore, dynamic modelling shows a more accurate and realistic representation of the time-varying interaction of hydraulic behavior and water quality.

Over the years, more research has been conducted to improve the limitations of dynamic modelling, including limitations such as long computational time, difficulty obtaining detailed network components, and model memory requirements. These problems were seen largely in networks with relatively long pipes with low velocities, which resulted in a loss of accuracy. As such, emphasis was placed on finding a more robust and efficient method.

To this end, the work of Boulos et al. [2] focused on employing a much more rigorous system simulation approach. The study derived the Lagrangian Event Driven Method (EDM) technique, which allows the network to simulate the containment–transport process based on the distribution system activities. The main merit of this method was that it allowed the dynamic modelling to become less sensitive to the structure of the network, along with decreased computational memory and system requirements. Later on, Boulos et al. [2] provided an extension for this method, which allowed EDM to handle time-varying hydraulic conditions by adopting a one-dimensional transport model that mixes the materials completely and instantaneously. Similarly, another study by Mau et al. [3] proposed a Linear Compartment Model (LCM) algorithm formulated analytically from the continuity and mass balance principles. This method also accounts for affective transport, the mixing process between the materials, and kinetic reactions by placing a decay coefficient.

The decay coefficient calculates the rate of pipe wall decay based on the difference in chlorine concentration between two points and estimates the water travel time between those two points. These decay rates are determined based on laboratory experiments that aim to reduce errors. First-order decay rate equations are affected by several factors, including pipe materials [16], initial chlorine concentration, pipe corrosion [17], and biofilm [18]. A review of these factors was investigated by Hallam et al. [4], who confirmed the influence of these factors in the actual water distribution networks.

As computational software improved, more researchers focused on water quality reliability. These indicators use a fraction of delivered quality to assess a network based on the simulation runs that kept the supplied concentration below the threshold [19]. Moreover, the same concept was applied to modelling the reliability of chlorine concentration, which is the ratio of days the residual fulfils the standards required for chlorine to the total number of days in the simulation [20]. Later, booster station operation was introduced to minimize the chlorine injection quantity [20–22]. Wang et al. [23] found the factors that affect the chlorine injection to WDNs by obtaining a chlorine decay coefficient in individual pipes in a network. This further validated that these coefficient values can be generalized to similar characteristics. Overall, modelling chlorine decay requires understanding the system and properly calibrating and validating tools for hydraulic modelling. This also includes accurate values for the kinetic constants required for the decay modelling. Therefore, such a technique was adopted in EPANET, which has a predefined set of water qualities. This allowed us to model chlorine decay based on a decay coefficient from the pipe wall and bulk composition.

In many types of research, the EPANET modelling tool was extended to water quality reliability for different real-time networks and extreme cases such as COVID-19. The methodology would often remain the same in these cases. However, the unique condition of each site with different decay constants, bulk decay rates, and calibration showed unique solutions for each case. For instance, research from Garcia-Avila et al. [1] has demonstrated that 45% of the nodal junctions in Azogues, Ecuador's water network failed to comply with the Ecuadorian standards during the pandemic. This was due to a rise in network demand, which led to changes in flow and velocities that eventually affected the network's bulk and wall decay coefficient. As such, the behavior of the network differentiated from the intended scenarios. This sudden change in demand also encouraged other tools for measuring water quality based on the risk index of disease occurrence [1]. Another instance took place in the city of Mohammedia, Morocco, where a study conducted by Belcaid et al. in 2023 [24] analyzed the city's WDN using chlorine decay modelling in EPANET to optimize the dosage of chlorine concentration by identifying the chemical components present in the network. Finally, the 2023 study by Giustolisi et al. [25] developed a solving tool known as "WQnetXL" that allows unlimited water parcels to be simulated in low-velocity conditions during water quality analysis using the Lagrangian approach. Their study allows the network to be simulated efficiently based on the memory of the computational system. The validity of these models was verified by field research studies such as the tracer-based measurement, although more emphasis on calibration was required to ensure higher accuracy is obtained from the water-based simulation tools [26,27].

Therefore, since many factors impact water quality modelling, any changes in the network's hydraulic behavior can affect chlorine's containment–transport process. Since it is already established that the integration of private tanks affects the flow and pressure of the network system, the present study observes the impact on chlorine concentration of the network for different orifice sizes and retention times when compared to the standard chlorine values proposed by the regulatory authority. As such, it would demonstrate the water quality reliability of the network.

Thus, this study addresses a significant gap in existing chlorine decay modelling by integrating private storage tanks and pressure-driven analysis (PDA) into a dynamic simulation framework using WQnetXL. While past studies have evaluated chlorine decay in standard looped networks, the unique influence of intermittent supply systems with decentralized storage has not been extensively explored.

By simulating water quality using Extended Period Simulation (EPS) and first-order kinetic equations in WQnetXL and validating outcomes against EPANET's chlorine decay model, this research offers a robust dual-model verification. The novel use of variable inflow curves derived from different orifice sizes and retention times within both platforms allows for a more realistic assessment of chlorine decay in networks with household-level tanks. This dual-layer modeling approach captures both hydraulic dynamics and water quality interactions more accurately, especially under non-ideal flow conditions.

2. Materials and Methods

2.1. Demand-Driven Analysis (DDA) vs. Pressure-Driven Analysis (PDA)

Hydraulic demand-driven analysis (DDA) assumes that the full consumer demand is always satisfied at each node in a water distribution network, regardless of the actual pressure available. It is a widely used and computationally efficient method, especially under normal operating conditions with adequate system pressure. DDA is commonly implemented in hydraulic simulation tools like EPANET (https://www.epa.gov/water-research/epanet, accessed on 21 May 2025), where it uses fixed demand values in the mass and energy balance equations to compute flows and pressures. However, this method cannot account for pressure-deficient scenarios or interruptions in supply, which limits its applicability in complex or non-ideal situations.

On the other hand, pressure-driven analysis (PDA) provides a more realistic approach by linking nodal outflows directly to available pressure at each junction. It allows demand to vary from zero up to the desired level depending on the system's hydraulic state, making it suitable for abnormal conditions such as pipe bursts, leakages, or low-pressure zones. Importantly, modelling private storage tanks in water distribution networks requires the use of PDA, as tank filling and emptying processes depend on pressure-dependent inflow and outflow rates. Since storage tanks interact dynamically with the network based on local pressure and volume, the fixed-demand assumptions used in DDA would not capture their behavior accurately.

2.2. Modelling with Integrated Private Tanks

To simulate a water distribution network (WDN) with local storage, the system's hydraulic behavior is modelled using momentum and continuity equations expressed in matrix form, as shown in Equation (1) [28].

$$\begin{aligned} \mathrm{A}_{pp}(t)\mathrm{Q}_p(t) + \mathrm{A}_{pn}\mathrm{H}_n(t) &= -\mathrm{A}_{p0}\mathrm{H}_0(t) \\ \mathrm{A}_{np}\mathrm{Q}_p(t) - \frac{\mathrm{V}_n(t,\mathrm{H}_n(t))}{\Delta T} &= 0_n \end{aligned} \tag{1}$$

In this formulation, $\mathrm{Q_p}(t)$ is the vector of unknown pipe flows, $\mathrm{H_n}(t)$ represents the unknown nodal heads, $\mathrm{H_0}(t)$ represents the known heads (e.g., reservoirs), and $\mathrm{V_n}(t,\mathrm{H_n}(t))$ is the vector of total outflows during time step ΔT. The matrices $\mathrm{A_{pn}}$, $\mathrm{A_{np}}$, and $\mathrm{A_{p0}}$ represent the network topology and relate links to nodes.

The second half of Equation (1) captures the mass balance at the nodes, where $\mathrm{V}_n = \mathrm{d}_n \cdot \Delta T$, and $\mathrm{d_n}$ represents pressure-dependent demands. These demands are broken down in Equation (2) into five components as follows:

$$\mathrm{V}_n(t,\mathrm{H}_n(t)) = \mathrm{V}_n^{cons}(t,\mathrm{H}_n(t)) + \mathrm{V}_n^{priv-tank}(t,\mathrm{H}_n(t)) + \mathrm{V}_n^{orif}(t,\mathrm{H}_n(t)) + \mathrm{V}_n^{tank}(t,\mathrm{H}_n(t)) + \mathrm{V}_n^{leak}(t,\mathrm{H}_n(t)) \tag{2}$$

where:

$V_n{}^{cons}$ $(t, H_n(t))$ denotes consumer demand, delivered fully if pressure meets minimum service levels; otherwise, calculated using Torricelli's law.
$V_n{}^{priv\text{-}tank}$ $(t, H_n(t))$ denotes the volume flowing into private tanks, depending on tank status and nodal pressure [29].
$V_n{}^{orif}$ $(t, H_n(t))$ denotes the flow rate through free orifices like hydrants or pipe bursts, either known or pressure-dependent.
$V_n{}^{tank}$ $(t, H_n(t))$ denotes the volume flowing into or out of urban tanks, linked to system-wide water balance.
$V_n{}^{leak}$ $(t, H_n(t))$ denotes leakage from the network, modelled as pressure-dependent outflows due to infrastructure flaws (e.g., cracks, joints).

2.3. The Setup of Private Tanks

Under the Dubai Electricity and Water Authority's (DEWA) regulations, every WDN connection must include a private storage tank sized to meet 24 h of consumption so that the supply remains uninterrupted during maintenance. In this study, private tanks were installed at each junction and sized according to DEWA's 24-h demand standards, with retention intervals of 6, 12, 18, and 24 h evaluated [30]. Although water is pumped from these underground tanks up to rooftop tanks in practice, only the underground tank outlet is modelled here. A locally derived water demand profile was applied to all residential nodes, and the dynamics of tank filling and emptying were incorporated using a pressure-driven approach as shown in Figure 1 [31].

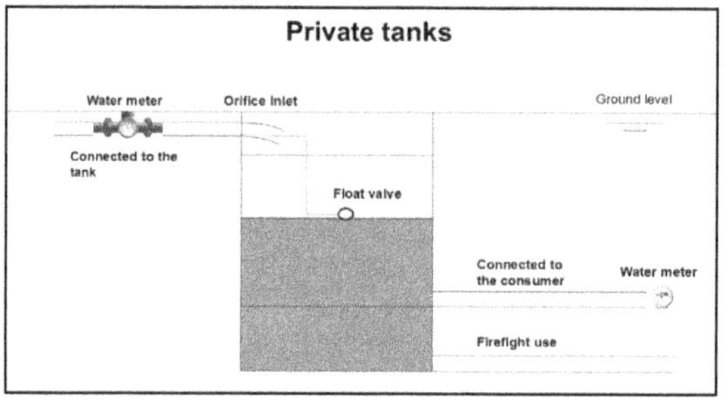

Figure 1. Schematic of a private tank at the junction of a UAE water distribution network.

Each tank receives flow through a small diameter orifice, which is a short pipe whose effective area varies with the stored volume and directly links to the network junction. Orifices are typically located at the tank's top, defining the junction's elevation head. Their diameters range from 2 cm to 10 cm depending on local pressure and demand, and float valves govern orifice opening and closing in response to the tank's water level based on the best practice from developers.

2.4. Modelling Flow to the Orifices of Private Tanks

The hydraulic model [29] used to simulate private storage tanks with a pressure-dependent demand function reflects tank filling and emptying based on orifice diameter, which changes depending on the tank's stored volume. When the orifice operates in a binary ON/OFF mode (either fully open or closed), the coefficient C(t) remains constant during the time step ΔT, provided the maximum tank volume $V_i{}^{max}$ is not exceeded. This allows the mass balance to be expressed as follows [29]:

$$V(t+\Delta T) = d^{fill}\Delta T - d^{act}\Delta T + V(t)\text{with } d^{fill}(t) = C^{max}\sqrt{P(t) - \Delta z^{orif}} \quad (3)$$

where d^{fill} (t), in cubic meter per second, is the average filling rate during ΔT; P(t), in meters, is the pressure head at time t; Δz^{orif}, in meters, is the elevation of the orifice inlet; and C^{max} is the maximum outflow coefficient of the orifice, which accounts for the orifice diameter via:

$$C^{max} = \frac{C_D \pi d^2 \sqrt{2g}}{4} \tag{4}$$

In Equation (4), C_D is the discharge coefficient, d is the orifice diameter, and g denotes gravitational acceleration.

The tank filling rate d_i^{fill} (t) and the actual customer demand $d_i^{act}(t)$ at node i are defined as follows [29]:

$$d_i^{fill}(P_i(t), V_i(t)) = \begin{cases} d_i^{act}(t) + \frac{V_i^{max} - V_i(t)}{\Delta T} & V_i(t+\Delta T) > V_i^{max} \\ C_i^{max}\sqrt{P(t)_i - \Delta z^{orif}} & V_i(t+\Delta T) \leq V_i^{max} \end{cases} \tag{5}$$

$$d_i^{act}(P_i(t), V_i(t)) = \begin{cases} d_i^{req,hum}(t) & V_i(t+\Delta T) > V_i^{max} \\ C_i^{max}\sqrt{P_i(t) - \Delta z^{orif}} + \frac{V_i(t)}{\Delta T} & V_i(t+\Delta T) \leq V_i^{max} \end{cases} \tag{6}$$

If the tank becomes empty within ΔT, then $V(t + \Delta T) = 0$, and:

$$d_i^{act}(t) = C_i^{max}\sqrt{P_i(t) - \Delta z^{orif}} + \frac{V_i(t)}{\Delta T} \tag{7}$$

This implies the tank completely empties, and although the orifice is fully open, the incoming flow is insufficient to meet consumer demand.

Conversely, when the tank becomes full during ΔT, i.e., $V_i(t + \Delta T) = V_i^{max}$, the inflow is assumed constant:

$$C_i^{max}\sqrt{P_i(t) - \Delta z^{orif}} \geq d_i^{act}(t) + \frac{V_i^{max} - V_i(t)}{\Delta T} \Rightarrow d_i^{fill}(t) = d_i^{act}(t) + \frac{V_i^{max} - V_i(t)}{\Delta T} \tag{8}$$

In this case, consumer demand is fully met, so $d_i^{act}(t) = d_i^{req,hum}(t)$, where $d_i^{req,hum}(t)$ is the expected demand from the demand profile at time t.

2.5. Water Quality Model

Most water quality models investigate numerous parameters such as disinfectant residual, water color, dissolved oxygen, smell, and turbidity. As such, chlorine residuals are the most commonly investigated parameter in these parameters.

The purpose of chlorine residuals is to ensure that any disinfectant formed is safely removed by reacting with chlorine particles. This ensures that the water quality remains high. The algorithms simulating chlorine decay within a WDN are classified as steady-state or dynamic approaches to mimic this process. Particularly, the dynamic approach is more accurate as it allows the spatial distribution of pollutants to change during the simulation steps under a time-dependent demand or water demand profile (WDP). These dynamic models can be classified into Eulerian and Lagrangian-based models [23]. The pipes are usually divided into equivalent segments in the former model and into variable segments in the latter.

Therefore, based on 1-D advection-diffusion, the chlorine residual models are quantified based on Equation (9) as follows:

$$\frac{dC_{k,t}}{dt} + v_k \frac{dC_{k,t}}{dx} + D_x \frac{d^2 C_{k,t}}{dx^2} + R(C_{k,t}) = 0 \tag{9}$$

where $C_{i,t}$ is the cross-sectional average chlorine concentration (mg/L) in pipe k as a function of both time t (s) and distance x (m); v_k represents the velocity (m/s) in pipe k; D_x is the diffusion coefficient (m^2/s) in the direction of the flow x; and $R(C_{k,t})$ is the reaction rate (s^{-1}) measured as a function of concentration.

The reaction rate also accounts for the total source-sink function and the bulk and wall reaction intensity between chlorine and other substances in the pipes. Therefore, the chlorine reaction rate can be further categorized using the first-order reaction in Equation (10) [23].

$$R(C_{k,t}) = -k_0 C_{k,t} \tag{10}$$

where k_0 is the first-order reaction rate (s^{-1}), the chlorine decay rate would be proportional to the chlorine concentration to the first power if k_0 is the first-order reaction rate (mg·s/L), represented in Equation (11) as follows:

$$k_0 = k_b + \frac{k_w k_f}{r_h\left(k_w + k_f\right)} \tag{11}$$

where k_b is the coefficient of bulk decay (s^{-1}), k_w is the coefficient of wall demand (m/t), r_h is the hydraulic radius (m), and k_f is the coefficient of mass transfer (m/s). Each of these factors needs to be taken by field calibration. However, they can be assumed based on the experience of an engineer [23,32]. Furthermore, there are other factors including pH level and temperature, that can impact the concentration of chlorine. However, they are assumed to remain consistent in the simulation to ensure a fair test when analyzing the different parameters of the private tank. Thus, the assumed pH value based on the characteristics of the network in Dubai was taken as 7.6, and the temperature of water was assumed to 22 °C [30]. Such models are incorporated into the simulation software to predict the chlorine residual in WDNs by considering their reaction with pipe walls and transportation.

Mathematical models incorporating chlorine kinetics, reaction rates, and transport phenomena are viable for simulating water chlorination. Water treatment systems can utilize models to forecast chlorine concentrations and disinfection outcomes. These models are established and authenticated through the application of actual data. The outcomes of the simulation aid in improving the chlorination process by enhancing the effectiveness of disinfection and reducing the production of disinfection by-products. The process of simulating water chlorination is commonly executed through the use of specialized software and tools. This enables water treatment experts to make informed decisions and ensure the provision of potable water that meets safety standards [25,32,33].

2.6. Sample Networks

The first sample WDN contains a single node connected directly to a reservoir by a single pump, as shown in Figure 2. The purpose of examining such a small network is to show the impact on different parameters of the network, such as pressure, flow, pump scheduling, aging, and chlorine concentration.

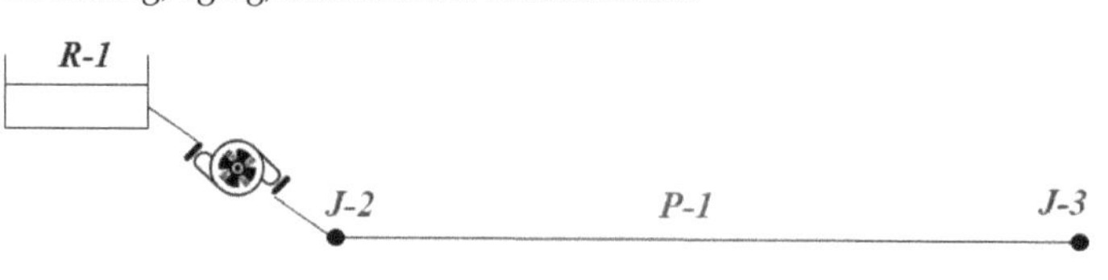

Figure 2. Schematics of Sample Network 1.

The second sample WDN contains a single source that supplies water using a single pump to a series of ten consecutive junctions, as shown in Figure 3. The purpose of examining this network configuration is to account for the head loss on the receiving parameters of the WDN. Furthermore, using this network allows us to see the downstream impacts due to the changes in the characteristics of the private tank's parameters at the upstream junction.

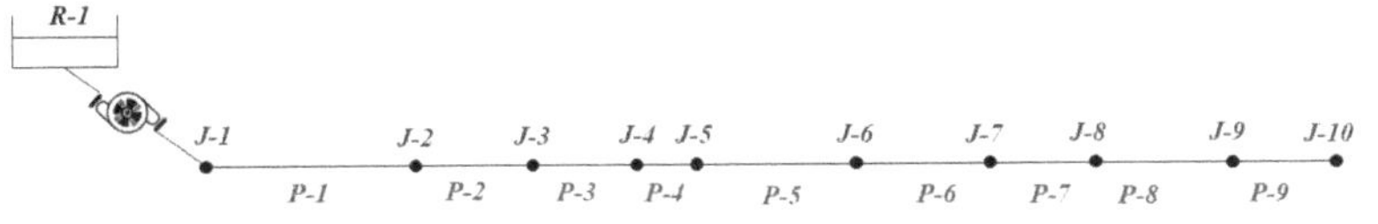

Figure 3. Schematics of Sample Network 2.

2.7. Real-World Network

The Dubai Silicon Oasis (DSO) network is typically a branched network comprising 2550 junctions and 2750 pipe links, see Figure 4 [31]. Water is sourced from a singular reservoir and is distributed through three parallel pumps. The range of pipe diameters falls within the range of 150 mm to 900 mm, whereas the daily per capita demand is approximately 300 liters per day [30]. The projected user count is approximately 165,000 individuals. The region's topography exhibits altitude fluctuations, with limited regions surpassing the 30-m elevation of the reservoir's apex. The network's nodal characteristics exhibit variability due to the diverse structures present within the area, including residential buildings, villas, commercial buildings, educational institutions, governmental offices, and medical facilities. As such, the water quality analysis would use this network to test the reliability when integrated with private tanks.

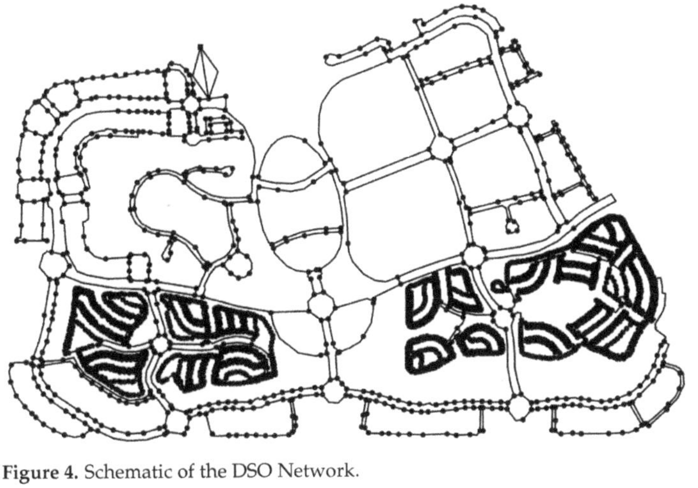

Figure 4. Schematic of the DSO Network.

2.8. Calibration Process

The calibration process in this study was tailored to its specific purpose: hydraulic and water quality modelling, since calibrated models are not universally applicable. Calibration involved reconstructing network geometry and estimating key parameters using often incomplete or erroneous data, necessitating field verification, which is limited by DEWA regulations. Integrating Geographic Information System (GIS) data helped identify missing information by comparing pipe layouts with road networks. While most physical parameters like pipe diameters and pump characteristics were stable, pipe roughness required calibration through flow rate comparisons in the literature due to its variability with age.

Water demand data, sourced from well-maintained DEWA meters, did not require calibration, but tank sizes were verified against developer-provided data. For the chlorine concentration models, wall decay coefficients were calibrated using the literature and lab data, although the high-quality desalinated water in Dubai reduced the need for changes. The calibration process also included aligning model pipes with GIS topography and verifying tank sizes at 12 locations per network. Variations were minor (under 4 percent), largely due to differences in building types and layouts. An error analysis was conducted to quantify uncertainties in tank sizes, pipe roughness, and chlorine concentrations using statistical measures like standard deviation. A summary of the calibration process is provided in Figure 5.

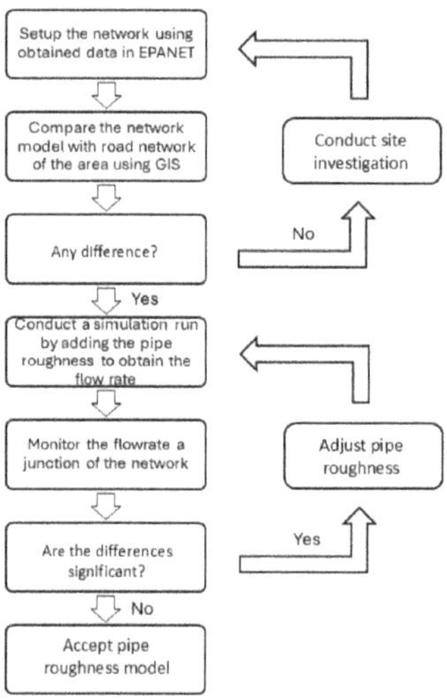

Figure 5. Calibration process flowchart.

2.9. Procedure of the Application

The approach for adopting this technique is based on the EPS of the first-order kinetic equation and chlorine decay for a network with integrated private tanks using PDA in WQnetXL [25]. Furthermore, to validate the findings from WQnetXL, EPANET analysis was used to determine the chlorine concentration at the junctions. Like water age analysis, the inflow curve obtained from different orifice sizes and retention times was plugged into the EPANET model for water quality analysis. The process adopted in both approaches is based on the Lagrangian Transport Algorithm (LTA), which utilizes Equations (9) and (11) to simulate the chlorine concentration at each part of the junction.

The same methodology has been adopted to facilitate the integration of private tanks. As for water quality models, the principle of PDA with LTA utilizes the sizing of a segment carrying discrete water units. As the simulation continues, the size of upstream nodes can increase with more water entering the pipe, while having an equal loss in the size of the downstream segment when water leaves the pipe. However, the overall volume of every pipe segment would not change; therefore, the segments' size of the leading and trailing segments would also remain unchanged. For the LTA model to perform quality analysis, the following actions were performed at each time step:

1. The value for chlorine concentration was updated to reflect any reaction that has occurred over the time interval.
2. For each upstream junction, the water from the leading segments of the pipe with the flow was mixed to calculate the new value of chlorine concentration. The volume contributed from each segment would equal the product of the pipe flow and the time interval. With the upcoming flow, if the volume exceeded that of the segment, then the segment was removed, and the next one behind it started contributing to the volume.
3. The new quality was calculated for every junction based on the total inflow mass divided by the total inflow volume.
4. The concentration at the junction was changed based on the contribution of external water quality sources, such as networks with two reservoirs.
5. Finally, a new segment was built in the pipes, and the flow came out of the node. The volume of this new segment was again calculated using the product of the flow rate and time interval. Its quality was equated to the new quality found for the junction.

In this algorithm, the number of segments was limited for quicker analysis; new segments were only produced if the new junction's quality differed more than a user-specified tolerance from the quality of the previous segment. This process was repeated for each time step.

2.10. Simulation

For this simulation, the first step was to compare the performance of DDA models with that of integrated private tanks. The next step was to increase the orifice diameter and retention time to observe their impact on the concentration of chlorine. As such, any external factor, including pipe wall reaction and bulk decay rate, was not considered, since the main emphasis was on observing private tank integration's impact on simple networks. The initial quality of the reservoir was kept at 3 mg/L for each network, because this comes under the range of chlorine concentration specified by the DEWA. The simulation ran based on the following steps for 72 h during weekdays:

1. Networks 1, 2, and DSO were run using DDA in EPANET, followed by the integration of private tanks using different orifice sizes and retention times under PDA.
2. The water quality parameters (wall coefficient, bulk coefficient, limiting concentration and wall correlation) were selected based on this region.
3. The selected range of orifice size varied from 2 cm to 5 cm; the retention time varied from 6 h to 24 h.
4. Each variation of orifice size was tested using LTA while keeping the retention time constant.
5. Then, the same water quality simulation was performed by changing the retention time of the private tanks whilst keeping the orifice size the same.
6. A comparison of chlorine concentrations over time was graphed for the different scenarios.

The next part of the simulation considers the water quality reliability indicators that assess the junction's chlorine concentration. These indicators calculate the network's performance by using a ratio of junctions above the chlorine concentration level to the network's overall junctions. The chlorine concentration level indicated by DEWA suggests that the chlorine concentration should always be greater than 0.3 mg/L at each junction. Therefore, a reliability assessment was carried out by changing the tank parameters of the network and comparing them with the design set up by engineers (DDA models). This reliability assessment followed the same steps mentioned previously to obtain the chlorine concentration. Afterward, the reliability indicator used Equation (12) to determine whether the junctions have passed or failed the assessment. For each retention time and orifice size, the reliability indicators in Equation (13) were used to calculate the percentage of junctions that have passed.

$$f_{t,Q} = \begin{cases} 1, & if\ C_{i,t} > 0.3\ \text{mg/L} \\ 0, & if\ C_{i,t} \leq 0.3\ \text{mg/L} \end{cases} \quad (12)$$

$$R_Q = \left(\frac{\sum_{t=1}^{N} f_{t,Q}}{N} \right) \% \quad (13)$$

where $f_{t,Q}$ denotes the failure event/condition at each time step t for the quality indicator (chlorine concentration) and R_Q represents the percentage of passing and other variables.

3. Results and Discussion

For the first part of the following analysis, the comparison was graphed to establish the variation of chlorine level in the junction, with different orifice sizes and retention times.

3.1. Impact of Orifice Size on Chlorine Concentration

As a result of the simulation of the DDA model of Network 1, Figure 6 shows the variation of the chlorine level through time. The demand profile is depicted in Figure 7.

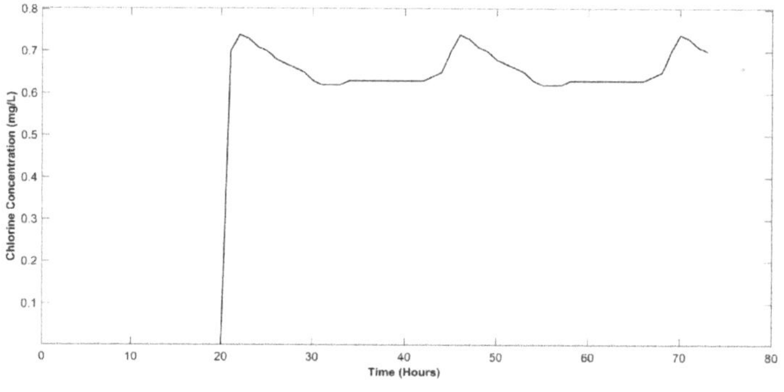

Figure 6. Chlorine concentration over time at the single junction in Network 1.

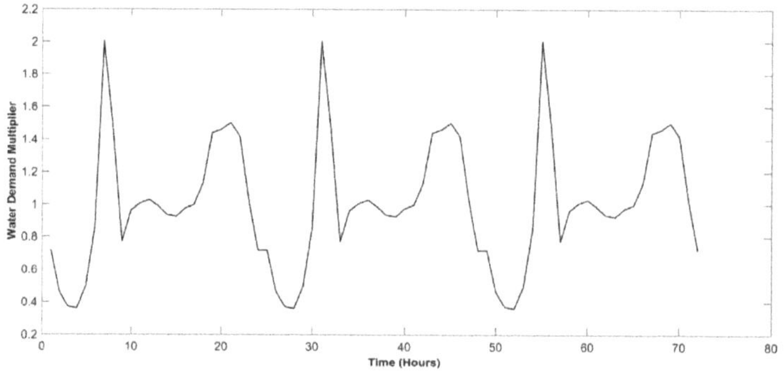

Figure 7. WDP in terms of multiplier for the 3-day simulation in Network 1.

It can be observed in Figure 6 that the chlorine level fluctuates continuously and periodically from 22 h to 72 h. The periods where the chlorine level drops are due to high demand in the network, which increases the flow rate, as it coincides with the high demand at 32 h in Figure 7. As a result, the chlorine level drops during the peak demand period. This impact was also validated by [34], who stated that an increased flow rate would reduce the chlorine concentration.

By comparing the DDA model to the model of integrated private tanks with different orifice sizes, Figure 8 shows the fluctuation level of chlorine at the junction for the network over the span of 48 h (a time span of 72 h was ignored because the behavior of chlorine concentration repeated). Furthermore, Figure 9 depicts how the inflow at the orifice inlet coincides with the quality level and the flow rate.

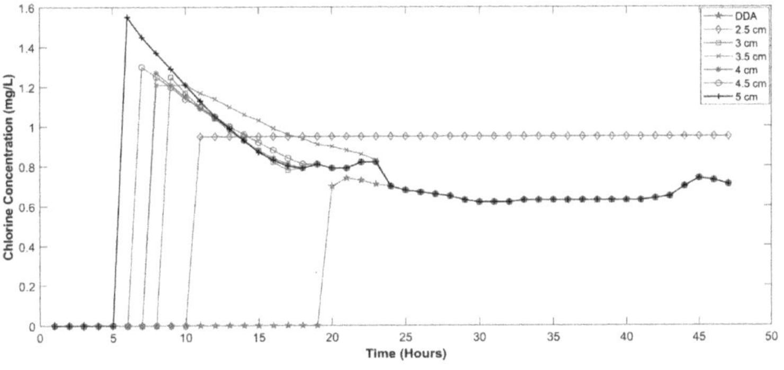

Figure 8. Chlorine concentration over time for different orifice sizes, compared to the DDA model for the junction in Network 1.

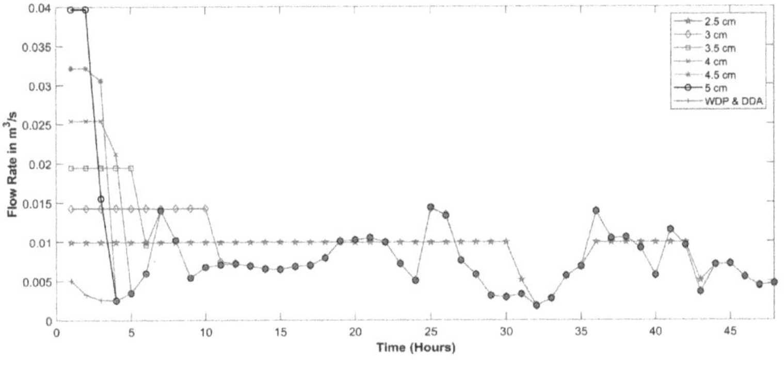

Figure 9. Flow rate variation over time for the junction in Network 1 for different orifice sizes, compared to the DDA model.

Based on the results shown in Figure 8, it can be observed that the chlorine concentration reaches a maximum level for the highest orifice inlet since it has the highest flow. As such, the chlorine concentration reaches the maximum value of 1.6 mg/L at a time of 6 h, compared to other inlet sizes that take another hour or more to reach the maximum value. Secondly, as soon as the tank is filled, the flow behavior is similar to DDA in each case of orifice size from 3 cm to 5 cm. Afterward, the chlorine concentration decreases based on the flow rate. Since this happens very quickly, the chlorine concentration drops immediately because the flow rate becomes very similar to that of DDA. However, based on observation after 12 h, it can be noted that the orifice size of 2.5 cm was not completely filled throughout the simulation; as a result, the chlorine concentration remained constant. Furthermore, the chlorine concentration of the orifice inlet of 2.5 cm remained higher than DDA and other cases after the time step of 15 h, as the flow rate for the orifice inlet of

2.5 cm remained higher than any of the other cases (because the volume of the private tank did not reach maximum capacity). This indicates that larger orifice sizes can increase the chlorine concentration because they are able to reach the junction quickly without reacting to the boundary wall. As a result, increasing the orifice size results in an increase in the chlorine levels at the junction.

Similar results were observed for Network 2, which can be seen in Figure 10, which compares different orifice inlets with the DDA model for junction 8.

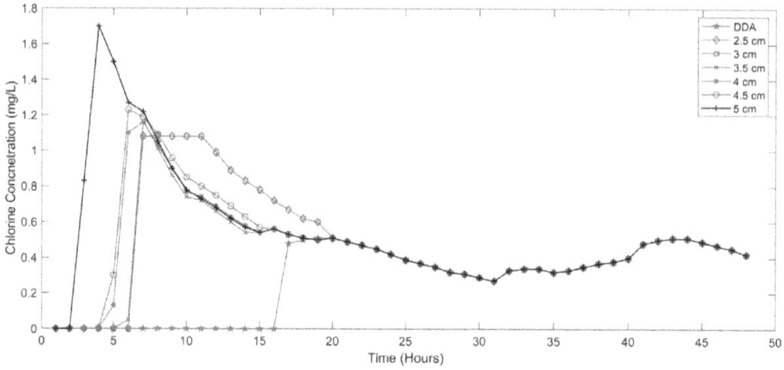

Figure 10. Chlorine concentration over time at junction 8 for different orifice sizes, compared to the DDA model in Network 2.

This network also validates that increasing the orifice size would increase the chlorine level due to the high flow rate until the water level in the tank does not reach its full capacity. However, higher orifice sizes fill up the tank quickly, decreasing the system's flow rate and ultimately decreasing the concentration. For instance, in Figure 11, the chlorine level has dropped from 1.7 mg/L to 0.6 mg/L in a time span of 9 h for the orifice inlet of 5 cm. This decrease in chlorine level is the highest when compared to another orifice inlet, since the flow decreases rapidly when the orifice inlet closes. This is evident from Figure 11, which showcases the drastic change in flow when the orifice size is decreased from 2 cm to 5 cm. The higher corresponding flow for orifice sizes of 5 cm resulted in higher chlorine concentration in the pipe in Figure 10. As a result, this can impact the chlorine concentration level in the network when the network has a much lower demand. Hence, lower orifice sizes are again preferred since they retain higher chlorine concentration for a long period of time, since the tank does not reach its maximum capacity.

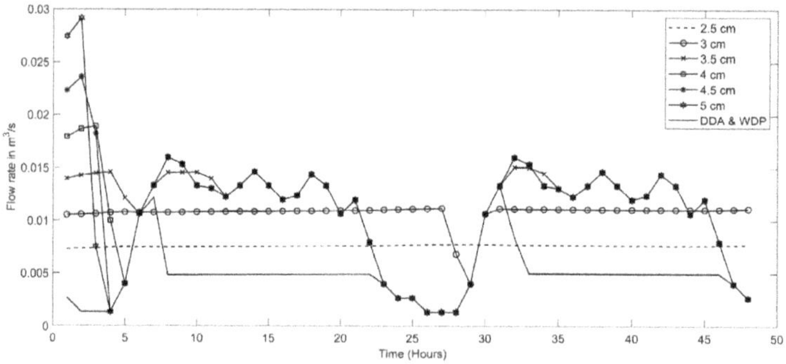

Figure 11. Flow rate over time at junction 8 in Network 2 for different orifice sizes, compared to the DDA model.

3.2. Impact of Retention Time on Chlorine Concentration

The impact of retention time relatively affects flow based on the water level in the tank, which can ultimately open or close the orifice. Under optimal orifice settings, the behavior of filling and emptying is regularly encouraged, which remains the same regardless of the tank's retention time change. As a result, the only notable impact of changing retention time is observed when the orifice is sufficient to satisfy demand and discourage any drop in the water level. Such a scenario is observed when a float valve is placed in tanks that control the opening and closing of the orifice based on the water level. As such, the orifice sizes are kept larger to encourage quick opening and closing when water drops below a certain set level.

Therefore, for Network 1, the orifice level was kept at a constant value of 2.5 cm, while the retention time was changed to observe the impact on chlorine concentration. As a result, Figure 12 shows the chlorine level for the junction over time for the DDA and integrated tank models with different retention times.

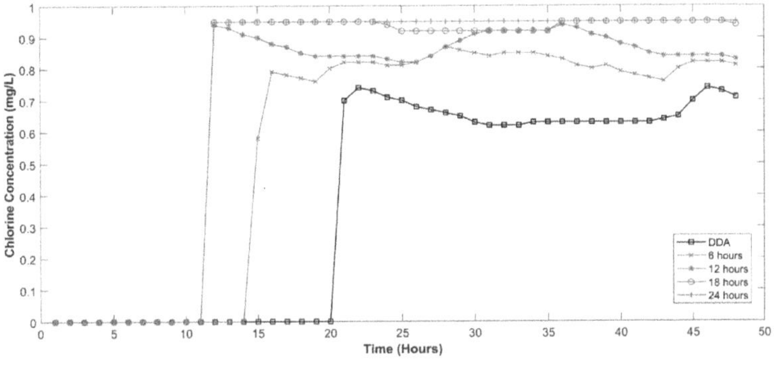

Figure 12. Chlorine concentration over time at the junction for different retention times, compared to the DDA model in Network 1.

From Figure 12, it can be observed that increasing the retention times of the private tank causes the chlorine concentration in junctions to increase. This is because the time the orifice inlet remains open increases, which causes more flow to enter the system. As expected from Figure 12, the private tank retention time of 6 h tends to fill first, followed by 12 h, then 18 h and 24-h retention time, which causes the chlorine concentration to increase over time. As soon as the private tank is full, the graph of chlorine concentration starts to become similar to that of the DDA model. As a result, the lower retention time of 6 h has behavior similar to that of the DDA model. It is also important to note that the maximum difference between the peak chlorine level of the DDA model and the model with a retention size of 24 h is 0.3 mg/L. This difference validates that inserting a certain storage volume tends to impact concentration because more water is added to the system.

Another observation from Figure 12 is that the variation in chlorine concentration remains similar for all the retention sizes. The changes in the chlorine level variation only happen when the tank's water level reaches the maximum capacity. For instance, when observing the chlorine level from Figure 12 for the retention times of 18 h and 24 h, it can be observed that the variation of chlorine concentration remained similar since the tanks for both retention times did not reach the maximum capacity. There is a slight variation at 24 h, where the tank size with a retention time of 18 h reaches maximum capacity, which shows a small drop in chlorine concentration of 0.02 mg/L. Before this time, both curves were similar due to the longer opening of the orifice.

For Network 2, the same simulation was carried out to produce the chlorine concentration level for each junction. As such, the downstream junction (J-8) results are presented in Figure 13, which shows the chlorine level pattern for the DDA model and integrated private tanks with different retention times.

The results from Figure 13 validate the findings observed in the Network 1 simulation, as the chlorine levels tend to increase with increasing retention times. The model shows a relatively large difference in concentration levels because more junctions are present in the network, allowing more flow to be added. The maximum difference between the smallest retention time of 6 h and the largest 24 h retention time was close to 0.4 mg/L. Another observation based on Figure 13 shows that the chlorine concentration takes more time to increase for higher retention times. This is due to more flow being utilized by upstream nodes.

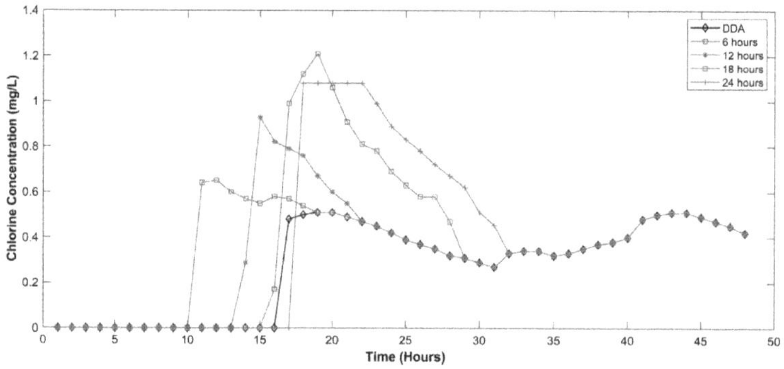

Figure 13. Chlorine concentration over time at junction 8 for different retention times, compared to the DDA model in Network 2.

3.3. *Reliability Analysis for the Real-Time Network*

As observed for the cases of Network 1 and Network 2, the impact of both orifice sizes and retention time of the private tank tends to impact the chlorine concentration level due to the changes in flow. Therefore, a reliability analysis of the DSO network was conducted to account for whether these changes in chlorine concentration result in any failure when the network is set up without integrating private tanks (as done in the design stage). The analysis results presented in Table 1 show the network's performance with different orifice inlet sizes and retention times.

Table 1. Percentage of junctions that passed the assessment based on different orifice sizes and retention times for the DSO network.

Retention Time	Orifice Sizes					
	2.5 cm	3 cm	3.5 cm	4 cm	4.5 cm	5 cm
6 h	91%	92%	92%	94%	94%	95%
12 h	94%	95%	95%	96%	96%	97%
18 h	97%	98%	98%	98%	98%	98%
24 h	98%	99%	99%	100%	100%	100%

It can be observed from Table 1 that increasing the orifice inlet increases the network's reliability because the system's flow rate increases. Even when keeping an optimal orifice size of 2.5, the percentage of the junction with chlorine level higher than the standard level by DEWA is close to 98% for a retention time of 24 h. For an orifice size that goes up to 5 cm (retention time of 24 h), the optimal percentage of junction that had passed in the entire simulation was 100%, because larger orifice sizes cause a higher flow rate, resulting in increased chlorine concentration. Failure in the system largely occurs due to smaller retention times, which indicates a change in the initial concentration of the chlorine level, since the network would not be able to satisfy the demand quality for the consumer. As such, the initial chlorine dosage had to be increased by 50% for the smallest orifice size and retention time to achieve full network reliability without any failure.

In Table 1, it can also be observed that increasing the retention time increases the network's reliability more effectively than changing orifice sizes. The percentage of passing increases from 94% to 100% when changing the retention time from 6 h to 24 h for an orifice size of 4 cm. As validated before from the simple networks, this is due to the system's increased flow, which increases the chlorine concentration level. Moreover, most of these failures only occur in the simulation's beginning time step, after which they become more similar. Furthermore, it implies again that the initial chlorine dosage in the reservoir has to increase to satisfy the water quality parameters.

Finally, the results of chlorine concentrations for different orifice sizes in the DSO network are showcased using flow path C shown in Figure 14. The pressure distribution of the network is depicted in Figure 15.

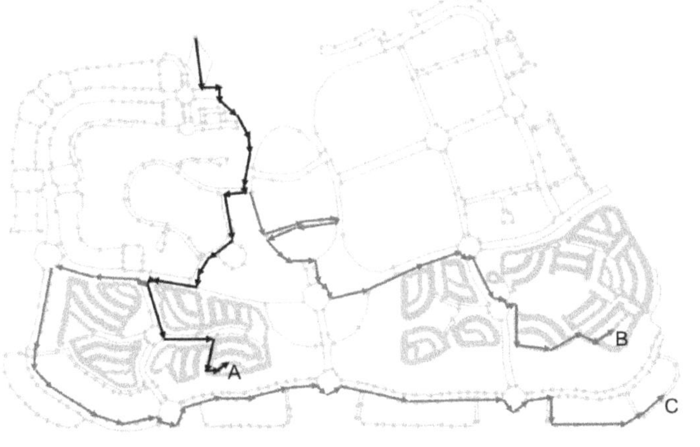

Figure 14. Flow path for 3 stages, out of which flow path C was used.

Figure 15. Pressure distribution across the DSO network.

Based on the selected flow path C, the chlorine concentration obtained along the junctions from the reservoir to Junction C for different orifice sizes is presented in Figure 16.

From Figure 16, it can be noted that increasing the orifice size results in an increase in the chlorine concentration due to high flow. As such, the observed chlorine concentration for the last junction is closer to 0.97 mg/L for the orifice size of 5 cm. In comparison to the smallest orifice size, the chlorine concentration has decreased by 30% during the filling period of private tanks. However, this impact is only observed during the filling time of the tank; thereafter, once the tank has reached its full capacity, the observed concentration gradually decreases. Moreover, as the water flows through the pipe, the chlorine concentration decreases in the junction due to lower pressure, as seen in Figure 15, which decreases the overall flow of the pipe.

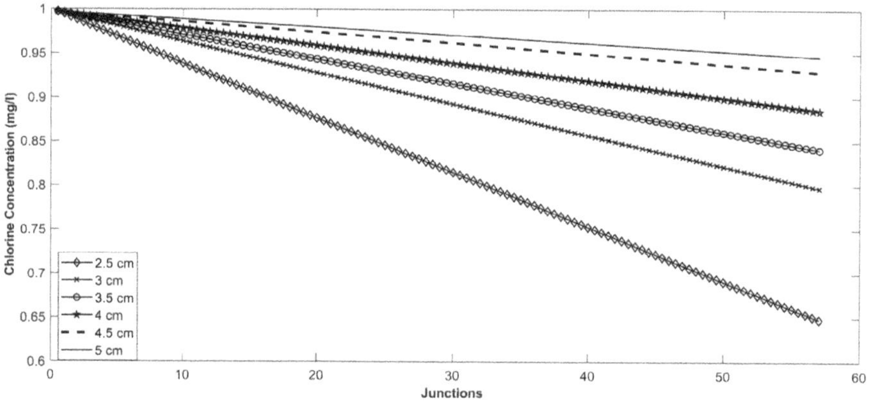

Figure 16. Chlorine concentration for different junctions in flow path C at a time step of 4 h.

At the end of the analysis, it is also worth pointing out the limitations of the network; the uncertainty of WDP remains a factor due to the integration of private tanks. Secondly, the analysis assumes that the flow rate at each time step remains constant. While this assumption is quite helpful for quick analysis, it dampens some of the peaks that can be observed during a relative time step. As such, it increases the uncertainty of the reliability indicators as it does not account for any changes in demand during the time step. Finally, due to the restriction in accessing real-time networks, the coefficients were based on the literature rather than actual field tests for the DSO model. Data from a field test could have improved the accuracy of the result; however, since each condition had an equal fair test, it would not have impacted the findings.

From the results, it was noticed that larger orifice sizes were initially able to provide high chlorine concentration throughout the network, but were not able to maintain it since the tank became full and the flow in the system dropped. Ideally, the best case observed in this study suggested controlling the orifice area based on the pressure-demand relation [29]. This would allow continuous flow within the system without the water parcel being halted, thereby improving the water quality in the system.

Similarly, when the tank is being filled, the chlorine concentration would remain quite high for larger tank sizes compared to smaller tank sizes. In fact, the concentration of chlorine was 12% higher for large tanks (24 h retention time) when compared with the smallest tank size with a retention time of 6 h. The presence of more chlorine allows the water quality to be maintained as it reacts with any contaminants to achieve the standard required of drinking water. Similarly, the reliability with tank parameters was quite high, as large orifice sizes and retention times were able to achieve the highest reliability. In terms of scale, a study by Garcia-Avila et al. [26] mentions that 45% of the junctions were below index if they had unaccounted design constraints. The failure percentage for this analysis was significantly lower than the addition of private tanks on the system, and higher pipe diameters are advised in the Code of Practice [30].

3.4. Error Analysis

As observed during the investigation of water age, the impact of WDP is limited. Thus, when extended towards water quality, similar observations were made. However, in the chlorine concentration parameter, another key parameter is the wall reaction coefficient factor, which directly impacts the chlorine concentration. Since there was a limitation during the calibration of the wall decay coefficient parameters, due to restrictions on accessing the network, the range of values that was considered can be found in the Code of Practice [30]. The impact of different decay values (0.5 to 1) was implemented in the model to see the changes in the chlorine concentration. Table 2 shows the differences in the value of chlorine concentration when using different values of the wall decay coefficient.

Table 2. Change in chlorine concentration of the networks at peak hour for the DDA model due to different values of decay coefficient.

			Sample Network	**Real-World Network**
		1	2	DSO
Chlorine Concentration (mg/L)	Maximum Change	0.12	0.24	0.54
	Average Change	0.12	0.15	0.29
	SD	0	0.45	0.87

Based on the Code of Practice, the range of wall coefficient lies between 0.5 to 1, which shows the percentage uncertainty of chlorine concentration to range from 7% to 11% from the result in Table 2 throughout the difference junction for each network. These changes go higher in areas with lower flow rates, where maximum change in concentration was observed at 0.54 mg/L. While these differences are acceptable, such differences in values can increase or decrease the actual concentration of chlorine when applied in a real-life network. However, the maximum absolute uncertainty of chlorine concentration of 0.54 mg/L was concerning. However, in real life, the network can be regulated by performing calibration by doing laboratory testing for chlorine. Therefore, these errors can be accepted and are consistent for different orifice sizes and retention times, which does not impact the overall investigation of the result.

4. Conclusions

This study demonstrates the critical role of private storage tank parameters, namely, orifice size and retention time, in influencing chlorine concentration and network reliability in water distribution systems. Simulation results revealed that larger orifice sizes, such as 5 cm, led to higher initial chlorine concentrations, reaching peak values of up to 1.6 mg/L within 6 h. However, once the tank reached capacity, the flow rate dropped sharply, causing the chlorine concentration to decrease to as low as 0.6 mg/L over the next 9 h. In contrast, smaller orifice sizes, particularly 2.5 cm, maintained more stable concentrations throughout the simulation, as the tank did not fill completely and thus sustained a higher flow rate over time.

Retention time also played a significant role in water quality performance. Increasing retention time from 6 to 24 h improved chlorine concentration, with differences as high as

0.3 to 0.4 mg/L between the shortest and longest durations. Moreover, network reliability improved with both larger orifice sizes and longer retention times. For instance, the percentage of junctions meeting quality standards rose from 91 percent to 100 percent when the orifice size increased from 2.5 cm to 5 cm and retention time increased from 6 to 24 h.

These findings emphasize the need to account for tank-specific parameters in network design and quality analysis. Future work should consider integrating real-time smart meter data to improve the accuracy and efficiency of demand profiling. Additionally, collecting data on pipe wall decay from actual networks would improve the precision of chlorine decay modeling.

Author Contributions: R.R. initiated the conceptual design, after which the analysis and write-up were conducted by S.R. The entire paper was supervised by R.R. and necessary improvements were made again at each section by R.R. All authors have read and agreed to the published version of the manuscript.

Funding: This research did not receive any external funding.

Data Availability Statement: The data for this research can be made available upon request as it is unavailable to the public.

Acknowledgments: The WDNetXL system used to run the simulation of all WDN models was provided by IDEA-RT s.r.l. (www.idea-rt.com).

Conflicts of Interest: The authors declare no conflicts of interest.

Abbreviations

The following abbreviations are used in this manuscript:

WDN	Water Distribution Network
DSO	Dubai Silicon Oasis
LTA	Lagrangian Time-Driven Method
PDA	Pressure-Driven Analysis
DDA	Demand-Driven Analysis
EDM	Lagrangian Event Driven Method
WDP	Water Demand Profile
DEWA	Dubai Electricity and Water Authority
GIS	Geographic Information System
LCM	Linear Compartment Model
EPS	Extended Period Simulation

References

1. García-Ávila, F.; Avilés-Añazco, A.; Ordoñez-Jara, J.; Guanuchi-Quezada, C.; Flores del Pino, L.; Ramos-Fernández, L. Modeling of residual chlorine in a drinking water network in times of pandemic of the SARS-CoV-2 (COVID-19). *Sustain. Environ. Res.* **2021**, *31*, 12. [CrossRef]
2. Boulos, P.F.; Altman, T.; Jarrige, P.A.; Collevati, F. Discrete simulation approach for network-water-quality models. *J. Water Resour. Plan. Manag.* **1995**, *121*, 49–60. [CrossRef]
3. Mau, R.E.; Boulos, P.F.; Bowcock, R.W. Modelling distribution storage water quality: An analytical approach. *Appl. Math. Model.* **1996**, *20*, 329–338. [CrossRef]
4. Hallam, N.B.; West, J.R.; Forster, C.F.; Powell, J.C.; Spencer, I. The decay of chlorine associated with the pipe wall in water distribution systems. *Water Res.* **2002**, *36*, 3479–3488. [CrossRef]
5. Chhipi-Shrestha, G.; Mohammadiun, S.; Ishaq, S.; Hu, G.; Mian, H.; Pokhrel, S.; Hewage, K.; Sadiq, R. Modeling hydraulics and water quality in distribution networks: A review of existing mathematical techniques and software. In *Water Engineering Modeling and Mathematic Tools*; Elsevier: Amsterdam, The Netherlands, 2021; pp. 187–212.

6. Chun, D.G.; Selznick, H.L. Computer modeling of distribution system water quality. In *Computer Applications in Water Resources*; ASCE: Reston, VA, USA, 1985; pp. 448–456.
7. Males, R.M.; Grayman, W.M.; Clark, R.M. Modeling water quality in distribution systems. *J. Water Resour. Plan. Manag.* **1988**, *114*, 197–209. [CrossRef]
8. Murphy, S.B. Modeling Chlorine Concentrations in Municipal Water Systems. Doctoral Dissertation, College of Engineering, Montana State University-Bozeman, Bozeman, MT, USA, 1985.
9. Shah, M.; Sinai, G. Steady state model for dilution in water networks. *J. Hydraul. Eng.* **1988**, *114*, 192–206. [CrossRef]
10. Wood, D.J.; Ormsbee, L.E. Supply identification for water distribution systems. *J.-Am. Water Work. Assoc.* **1989**, *81*, 74–80. [CrossRef]
11. Fowler, A.G.; Jones, P. Simulation of water quality in water distribution systems. In Proceedings of the Water Quality Modeling in Distribution Systems 1991, Cincinnati, OH, USA, 4–5 February 1991; pp. 4–5.
12. Rossman, L.; Boulos, P.; Altman, T. Discrete Volume-Element Method for Network Water-Quality Models. *J. Water Resour. Plan. Manag.* **1993**, *119*, 505–517. [CrossRef]
13. Clark, R.M.; Grayman, W.M.; Males, R.M.; Hess, A.F. Modeling contaminant propagation in drinking-water distribution systems. *J. Environ. Eng.* **1993**, *119*, 349–364. [CrossRef]
14. Liou, C.P.; Kroon, J.R. Modeling the propagation of waterborne substances in distribution networks. *J.-Am. Water Work. Assoc.* **1987**, *79*, 54–58. [CrossRef]
15. Grayman, W.M.; Clark, R.M.; Males, R.M. Modeling distribution-system water quality; dynamic approach. *J. Water Resour. Plan. Manag.* **1988**, *114*, 295–312. [CrossRef]
16. Tansley, N.S.; Brammer, L.F. Chlorine residual modelling in distribution: The improvement of taste and the maintenance of effective disinfection. In *Integrated Computer Applications in Water Supply (Vol. 2) Applications and Implementations for Systems Operation and Management*; John Wiley and Sons Ltd.: West Sussex, UK, 1994; pp. 111–126.
17. Kiéné, L.; Lu, W.; Lévi, Y. Relative importance of the phenomena responsible for chlorine decay in drinking water distribution systems. *Water Sci. Technol.* **1998**, *38*, 219–227. [CrossRef]
18. Lu, W.; Kiéné, L.; Lévi, Y. Chlorine demand of biofilms in water distribution systems. *Water Res.* **1999**, *33*, 827–835. [CrossRef]
19. Ostfeld, A.; Kogan, D.; Shamir, U. Reliability simulation of water distribution systems–single and multiquality. *Urban Water* **2002**, *4*, 53–61. [CrossRef]
20. Gupta, R.; Dhapade, S.; Ganguly, S.; Bhave, P.R. Water quality based reliability analysis for water distribution networks. *ISH J. Hydraul. Eng.* **2012**, *18*, 80–89. [CrossRef]
21. Tryby, M.E.; Boccelli, D.L.; Uber, J.G.; Rossman, L.A. Facility location model for booster disinfection of water supply networks. *J. Water Resour. Plan. Manag.* **2002**, *128*, 322–333. [CrossRef]
22. Kang, D.; Lansey, K. Real-time optimal valve operation and booster disinfection for water quality in water distribution systems. *J. Water Resour. Plan. Manag.* **2010**, *136*, 463–473. [CrossRef]
23. Wang, Y.; Zhu, G.; Yang, Z. Analysis of water quality characteristic for water distribution systems. *J. Water Reuse Desalin.* **2019**, *9*, 152–162. [CrossRef]
24. Belcaid, A.E.; Taghlabi, F.; Tounsi, A. Chlorine Decay Modeling in a Water Distribution Network in Mohammedia City, Morocco. *Ecol. Eng. Environ. Technol. (EEET)* **2023**, *24*, 269–278. [CrossRef]
25. Giustolisi, O.; Ciliberti, F.G.; Berardi, L.; Laucelli, D.B. Leakage Management Influence on Water Age of Water Distribution Networks. *Water Resour. Res.* **2023**, *59*, e2021WR031919. [CrossRef]
26. García-Avila, F.; Asitimbay-Barbecho, G.; Espinoza-Bustamante, M.; Valdiviezo-Gonzales, L.; Sánchez-Cordero, E.; Cabello-Torres, R.; Gutiérrez-Ortega, H. Water age in drinking water distribution systems: A case study comparing tracers and EPANET. *Case Stud. Chem. Environ. Eng.* **2024**, *10*, 100817. [CrossRef]
27. Li, Z.; Liu, H.; Zhang, C.; Fu, G. Real-time water quality prediction in water distribution networks using graph neural networks with sparse monitoring data. *Water Res.* **2024**, *250*, 121018. [CrossRef] [PubMed]
28. Giustolisi, O.; Berardi, L.; Laucelli, D. Generalizing WDN simulation models to variable tank levels. *J. Hydroinform.* **2012**, *14*, 562–573. [CrossRef]
29. Giustolisi, O.; Berardi, L.; Laucelli, D. Modeling Local Water Storages Delivering Customer Demands in WDN Models. *J. Hydraul. Eng.* **2014**, *140*, 89–104. [CrossRef]
30. DEWA. *Guide to Water Supply Regulations-Issue 5*; Dubai Electricity and Water Authority: Dubai, United Arab Emirates, 2020.
31. Rizvi, S.; Rustum, R.; Giustolisi, O.; Wright, G.; Arthur, S.; Berardi, L. Effects of Orifice Diameter and Retention Time of Local Tanks on the Reliability and Carbon Footprint of Water Distribution Networks. *J. Water Resour. Plan. Manag.* **2021**, *147*, 05021023. [CrossRef]

32. Slavik, I.; Oliveira, K.R.; Cheung, P.B.; Uhl, W. Water quality aspects related to domestic drinking water storage tanks and consideration in current standards and guidelines throughout the world–a review. *J. Water Health* **2020**, *18*, 439–463. [CrossRef]
33. Cvelihárová, D.; Pauliková, A.; Kopilčáková, L.; Eštoková, A.; Stefanova, M.G.; Dománková, M.; Šutiaková, I.; Kusý, M.; Moravčíková, J.; Hazlinger, M. Optimization of the Interaction Transport System—Transported Medium to Ensure the Required Water Quality. *Water* **2023**, *15*, 2573. [CrossRef]
34. Liu, B.; Reckhow, D.A.; Li, Y. A two-site chlorine decay model for the combined effects of pH, water distribution temperature and in-home heating profiles using differential evolution. *Water Res.* **2014**, *53*, 47–57. [CrossRef]

processes

MDPI

Article

Multi-Objective Large-Scale ALB Considering Position and Equipment Conflicts Using an Improved NSGA-II

Haiwei Li [1,2], Yanghua Cao [3,*], Fansen Kong [1], Xi Zhang [2] and Guoqiu Song [2]

[1] School of Mechanical and Aerospace Engineering, Jilin University, Jilin 132022, China
[2] Shenyang Aircraft Corporation, Shenyang 110850, China
[3] School of Economics and Management, Shenyang Aerospace University, Shenyang 110136, China
* Correspondence: caoyanghua@sau.edu.cn

Abstract

On large-scale product assembly lines, such as those used in aircraft manufacturing, multiple assembly positions and devices often coexist within a single workstation, leading to complex task interactions. As a result, the problem of parallel task execution within workstations must be effectively addressed. This study focuses on positional and equipment conflicts within workstations. To manage positional and equipment conflicts, a multi-objective optimization model is developed that integrates assembly sequence planning with the first type of assembly line balancing problem. This model aims to minimize the number of workstations, balance task loads, and reduce equipment procurement costs. An improved NSGA-II algorithm is proposed by incorporating artificial immune algorithm concepts and neighborhood search. A selection strategy based on dominance rate and concentration is introduced, and crossover and mutation operators are refined to enhance search efficiency under restrictive parallel constraints. Case studies reveal that a chromosome concentration weight of about 0.6 yields superior search performance. Compared with the traditional NSGA-II algorithm, the improved version achieves the same optimal number of workstations but provides a 5% better workload balance, 2% lower cost, a 76% larger hyper-volume, and a 133% increase in Pareto front solutions. The results demonstrate that the proposed algorithm effectively handles assembly line balancing with complex parallel constraints, improving Pareto front quality and maintaining diversity. It offers an efficient, practical optimization strategy for scheduling and resource allocation in large-scale assembly systems.

Keywords: assembly line balancing; improved NSGA-II algorithm; multi-objective optimization; restrictive parallel constraints

1. Introduction

As one of the most representative modes of production organization in modern manufacturing, assembly lines have been extensively applied in industries such as automotive, electronics, and aerospace [1]. Owing to the high degree of specialization in labor division and collaboration, assembly lines can substantially enhance production efficiency while reducing manufacturing costs. A typical assembly line comprises a series of workstations, each responsible for designated assembly tasks [2]. These tasks are typically executed by one or more workers operating the relevant equipment in accordance with a predetermined process sequence. Upon completion, the product is transferred to the next workstation for subsequent assembly, continuing in this manner until the entire process is completed. In this context, the scientific and rational allocation of tasks to workstations directly influences assembly line efficiency, production cycle time, and both construction and operational costs [3].

The Assembly Line Balancing (ALB) problem aims to allocate assembly tasks in such a way that objectives such as minimizing the number of workstations, optimizing cycle time, or maximizing production efficiency can be achieved [4,5]. Traditional studies on ALB are often based on the following assumptions: each workstation is operated by a single worker, tasks must strictly follow the predefined precedence constraints, and no parallel tasks are allowed. Under these assumptions, research mainly focuses on optimizing task allocation under precedence constraints [6,7]. However, real-world assembly processes are usually more complex. In particular, in two-sided or multi-sided assembly line settings, although parallel tasks have been partially considered, most studies still focus primarily on precedence relationships, while constraints arising from workers, equipment, and spatial positions have not been sufficiently addressed [8].

Large-scale, complex product assembly lines, particularly those in the aerospace industry, constitute a typical ALB problem characterized by restrictive parallel constraints [9]. In such contexts, numerous tasks must be executed in parallel. A single workstation often encompasses multiple workers, specialized equipment, and several assembly positions [10]. However, tasks within the same workstation may contend for identical resources; for instance, multiple tasks may simultaneously require the same worker, the same equipment, or the same spatial location. Since a worker, a piece of equipment, or a specific position can engage in only one task at a given time, such resource conflicts substantially limit task parallelism. Consequently, the feasibility of parallel task execution is determined not only by precedence constraints but also by conflicts involving personnel, equipment, and workstation positions. If such parallel conflicts are not appropriately resolved, they may result in decreased assembly efficiency, resource wastage, and production delays [11].

Simultaneously, the construction and operation of large-scale assembly lines generally necessitate substantial investment, particularly for high-precision assembly lines that heavily depend on expensive tooling and automated or semi-automated equipment [12]. Against this backdrop, minimizing the number of workstations and reducing equipment investment costs are commonly considered primary objectives by enterprises. Furthermore, as each workstation is typically operated by a team, task allocation must also account for workload balance among teams to ensure fairness and enhance overall assembly performance while satisfying production capacity requirements [13,14].

Motivated by these considerations, this paper investigates the ALB problem for large-scale product assembly lines characterized by restrictive parallel constraints. Unlike traditional approaches that solely consider precedence relations, this study explicitly integrates conflicts associated with positions, equipment, and workers into the optimization model. The objective is to achieve cost reduction and efficiency enhancement via multi-objective optimization. To this end, an improved Non-dominated Sorting Genetic Algorithm II (NSGA-II) is proposed to optimize task allocation and assembly sequencing, thereby effectively mitigating parallel conflicts and enhancing the overall performance of the assembly line.

The remainder of this paper is organized as follows. Section 2 reviews the related literature on assembly line balancing and assembly sequence planning. Section 3 provides a detailed description of the problem. Section 4 presents the proposed improved NSGA-II algorithm. Section 5 illustrates the solution process and results through a case study. Finally, Section 6 concludes the paper and outlines future research directions.

2. Literature Review

The ALB problem has been extensively investigated, with primary objectives typically encompassing the minimization of the number of workstations under a given cycle time [15–17] or the minimization of the cycle time under a fixed number of workstations [18–20]. In addition, some studies have focused on maximizing production efficiency while minimizing costs [21–23].

From the perspective of constraints, ALB models are generally formulated based on factors such as cycle time, task precedence relations, workstation capacity, task attributes, and worker skills. For instance, Mukund N. J. et al. investigated the robotic ALB problem by incorporating constraints such as robot assignment, the number of tasks per robot, cycle time, and task precedence, thereby addressing the impact of task execution sequences on processing time [24]. Junior P. C. M. et al. proposed an exact constraint programming (CP) method capable of handling a wide range of real-world constraints and formulated a general and unified model [25]. Abdous A. M. et al. introduced a scenario-based optimization model that integrates task time variability and its associated probabilities across different scenarios to mitigate worker fatigue [26]. Huang Y. Z. formulated a mixed-integer programming model to address ALB problems with multiple constraints, including setup time, positional restrictions, zoning limitations, and synchronization requirements [27]. Amir N. et al. proposed a constraint programming model to address practical issues such as varying skill levels of workers and robots, energy requirements, tool changes, and spatial zoning [28].

Classical ALB research predominantly assumes that tasks within a workstation are executed sequentially, whereas the introduction of parallel tasks substantially increases problem complexity. In recent years, researchers have increasingly focused on parallel assembly line balancing (Parallel ALB), emphasizing task allocation and worker collaboration to enhance workstation utilization [29–33]. However, existing studies primarily address scenarios with unrestricted parallelism, whereas research on parallel tasks under constraints such as limited space and equipment remains relatively scarce. As the number of parallel tasks increases, the complexity of resource allocation within workstations intensifies, and more stringent requirements are imposed on assembly sequence planning.

Assembly sequence planning (ASP) seeks to improve the efficiency and stability of the assembly process by systematically determining the order in which components are assembled. With respect to constraints, ASP models typically consider both physical limitations—such as geometric feasibility and tool availability—and logical constraints, including precedence relations and process requirements [34–37]. From an algorithmic perspective, intelligent optimization techniques such as genetic algorithms, particle swarm optimization, and ant colony optimization have been widely employed to address ASP problems.

For instance, Zhao R. H. et al. [38] proposed a synergistic optimizer, termed the "Self-Learning Reinforced Multi-guided Genetic and Differential Evolution Operator (SL-MGDO)," to address multi-objective and multi-constraint human–robot collaborative assembly tasks, thereby effectively handling the complexity of task planning. Hwai-E T. et al. [39] applied genetic algorithms to ASP optimization, elaborating on the design of genetic operators and the formulation of fitness functions. Wan X. et al. [40] enhanced the traditional genetic algorithm to mitigate its tendency to fall into local optima, thereby enabling global search under machining constraints and obtaining the optimal assembly sequence. Sheng Z. Y. et al. [41] proposed a symbiotic organisms search (SOS) algorithm that integrates robotic kinematics analysis with diversification strategies, thereby satisfying the requirements of efficiency and stability in space missions. Keijiro W. et al. [42] employed an improved Q-learning search algorithm that explores assembly sequences through reinforcement learning and trial-and-error, thus eliminating the need for detailed pre-investigation of product structures required in earlier studies.

In recent years, several studies have sought to integrate ALB with ASP for joint optimization. For example, Rashid A. F. F. M. et al. formulated the objectives of minimizing tool orientation changes and tool substitutions in ASP, as well as reducing production cycle time, the number of workstations, and workload variation in ALB [43]. Under the constraints of maximum workstation cycle time and operation precedence, they employed an improved artificial bee colony algorithm to achieve joint optimization. Subsequently, Rashid A. F. F. M. et al. incorporated task similarity and operation precedence as constraints, and devised a multi-objective discrete particle swarm optimization approach with minor solution adjustments to attain optimal or near-optimal performance [44]. Abdolmajid Y. et al. proposed a hybrid genetic algorithm to jointly address ALB and ASP, explicitly considering the impact of task execution order on processing time [45]. Their approach reduced assembly line idle time and enhanced production efficiency through rational task allocation and sequencing. Gunji B. M. et al. reformulated ASP as a discrete optimization problem and applied an improved cuckoo search algorithm to shorten planning time, alleviate bottleneck operations, and enhance ALB performance [46]. S. Ehsan Hashemi P. et al. adopted an "adversarial approach" to assembly line design, employing local search to generate product sequences and achieve dynamic task allocation, thereby overcoming the scalability limitations of mixed-integer linear programming models [47]. Nevertheless, most existing studies have primarily concentrated on integrating the objectives and conventional constraints of ALB and ASP, whereas limited attention has been devoted to more complex scenarios involving restrictive parallel constraints, assembly costs, and positional conflicts. Ramezanian R. et al. introduced the Imperialist Competitive Algorithm and Genetic Algorithm into ALB to enhance the effectiveness of multi-product supply chain network design and assembly line balancing [48]. Nowicki E. et al. applied the Tabu Search Algorithm to minimize the makespan in job shop scheduling [49].

The main contributions of this study are summarized as follows:

1. Modeling restrictive parallel constraints: This study explicitly incorporates restrictive parallel constraints, including conflicts associated with workers, equipment, and assembly positions, thereby providing a more realistic representation of large-scale product assembly lines.
2. Multi-objective optimization framework: A multi-objective model is formulated with the aims of minimizing the number of workstations, reducing equipment investment costs, and balancing workloads among teams, thereby aligning with practical industrial needs.
3. Improved NSGA-II algorithm: An enhanced version of the NSGA-II is developed to solve the proposed model, enabling efficient handling of complex constraints and improvement of solution quality.
4. Case study validation: The effectiveness and superiority of the proposed approach are validated through a real-world-inspired case study, demonstrating its potential to reduce costs and enhance production efficiency in large-scale product assembly.

In summary, existing research on the integration of ALB and ASP remains limited, particularly in the context of flow-line design for large-scale products subject to restrictive constraints. In studies addressing multi-operator workstations, practical issues such as positional conflicts and equipment interference have likewise received limited attention. Therefore, this study addresses the challenge of parallel constraints in real-world production by explicitly considering factors such as positional and equipment conflicts. By systematically integrating ALB and ASP, this study investigates assembly line balancing under restrictive parallel constraints, aiming to enhance the design efficiency of assembly lines for large-scale products.

3. Problem Description

3.1. Restrictive Parallel Constraints and Balance Objectives

For the ALB problem of large-scale products with restrictive parallel constraints, this study considers not only conventional constraints, such as production takt time and task precedence, but also emphasizes two types of restrictive parallel constraints: position conflicts and equipment conflicts. Incorporating these constraints enables the assembly line balancing scheme to more accurately reflect real production conditions, preventing interruptions or efficiency losses caused by position or equipment conflicts, thereby enhancing the feasibility, stability, and overall production efficiency of the assembly line.

(1) Equipment Conflicts

In parallel assembly tasks of large-scale products, equipment conflicts are both complex and critical. Such assembly operations often necessitate dedicated tooling or specialized equipment. Due to the presence of multiple assembly positions within a workstation and considerable distances between them, sharing equipment across positions is often infeasible. Consequently, in most cases, each assembly position must be provided with the necessary tools and equipment to accomplish its tasks. For large-scale products with high precision or complex assembly processes (e.g., aircraft assembly), certain equipment may entail substantial single-unit procurement costs. As a result, typically only a single unit of the relevant equipment is installed in a workstation, with devices such as rails employed to perform tasks across multiple positions. Consequently, even when task precedence permits parallel execution, tasks requiring the same limited equipment must still be executed sequentially, as shown in Figure 1a.

(2) Position Conflicts

Although large-scale product assembly can accommodate multiple workers performing parallel tasks simultaneously, spatial limitations constrain the number of workers able to operate concurrently. This implies that even if tasks can theoretically be executed in parallel according to task precedence, they cannot run concurrently when requiring the same position, resulting in position conflicts. The aforementioned restrictive parallel constraints are prevalent in large-scale product assembly, particularly within the aerospace industry. Other constraints, including task posture conflicts and personnel skill conflicts, may also be addressed using analogous methods. This study primarily focuses on equipment and position conflicts in assembly line balancing, making it particularly suitable for the design of new production lines. The approach begins with production capacity requirements, derives the corresponding takt time, and conducts assembly line balancing under the specified takt, as shown in Figure 1b.

Based on the characteristics of large-scale product assembly lines, this study aims to optimize three primary objectives:

(1) Minimizing the Number of Workstations

This represents a conventional objective in assembly line balancing and is particularly critical for large-scale products. Workstations for large products occupy substantial space; for instance, assembling a small fixed-wing UAV may necessitate approximately 15 m $\times$ 10 m. More complex products, such as fighter jets or transport aircraft, demand even larger areas. Excessive workstations reduce factory space utilization, increase logistics costs, and complicate management; hence, minimizing the number of workstations is essential.

(2) Balancing Workload Among Workstations

For assembly lines with numerous parallel tasks, each workstation is typically staffed by a team of workers. Production output is initially allocated among workstations and subsequently distributed within the team according to individual task assignments. Therefore, balancing workloads among workstations is critical for efficient line operation and effective performance management.

(3) Minimizing Equipment Procurement Costs

Large-scale product assembly, particularly in aerospace, necessitates high-precision equipment, which is frequently costly. For example, an automated riveting and drilling machine may cost up to 70 million RMB. If equipment cannot be relocated or shared, multiple units may be required for different positions within the same workstation; if the equipment is movable, only a single unit is needed per workstation. Since equipment is generally non-shareable across workstations, assembly line balancing significantly impacts equipment procurement costs.

These three objectives collectively contribute to optimizing the cost and efficiency of large-scale product assembly lines. Other objectives, such as maximizing personnel utilization, may also be considered; however, they fall beyond the scope of this study, as they involve personnel assignment.

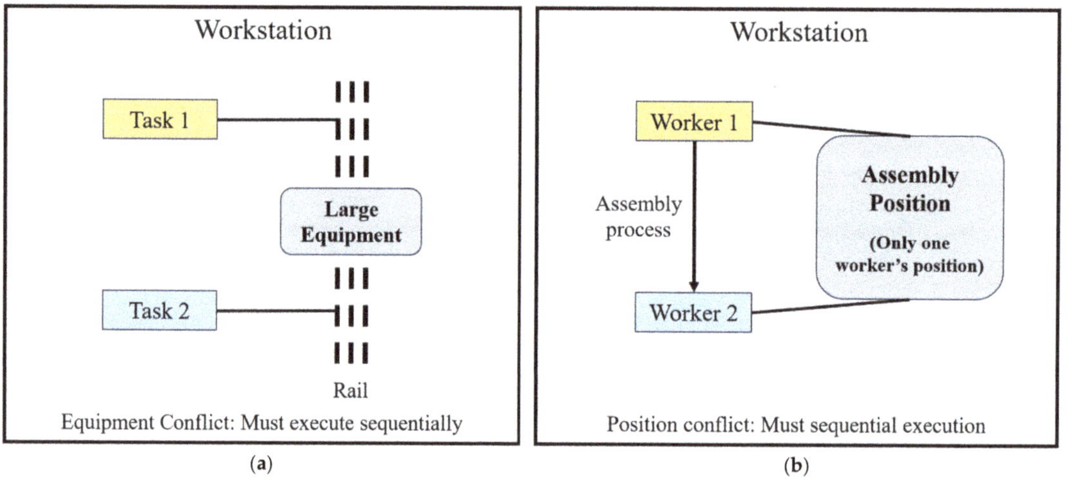

Figure 1. Conflict illustrations. (**a**) Equipment Conflicts; (**b**) Position Conflicts.

3.2. Assumptions

The following assumptions are made concerning ALB with restrictive parallel constraints:

1. The assembly line produces only a single type of product.
2. The assembly line has a known and predetermined takt time requirement.
3. The precedence relationships among tasks are known.
4. The workstation positions and the equipment required for each assembly task are known.
5. The processing time for each assembly task is predetermined and independent of the assigned workstation.
6. Each workstation position or piece of equipment can perform only one task at a time.
7. Workstations are arranged in series, and no parallel workstations are considered.

8. Parallel tasks within the same workstation are scheduled as early as possible to maximize the number of tasks that can be executed within the workstation.

3.3. Notation and Variables Used in the Proposed Model

The notations and variables commonly used in the proposed model are summarized in Table 1.

Table 1. Notation used in the proposed model.

Symbol	Description
V	Set of all tasks, $V = \{1, 2, \ldots, n\}$.
CT	Takt time.
t_i	Duration-time for task i.
K	Set of all workstations.
N	Total number of workstations, $N = \|K\|$.
C_j	Purchase price of one device j.
N_s	Number of devices s required for the entire assembly line.
P	Set of all assembly positions.
S_F	Set of fixed device types used in the entire assembly line.
S_M	Set of movable device types used in the entire assembly line.
S	Set of all device types used in the entire assembly line, $S = S_F \cup S_M$.
$Pred_i$	Set of immediate predecessor tasks for task i.
T_k	Cycle time of the k-th workstation
TT_k	Total duration time of all tasks at the k-th workstation
R_k	Set of tasks in the k-th workstation, $R_k = \{i \mid x_{ik} = 1, i = 1, 2, \ldots, n\}$
A_k	0 or 1 parameter. 1 indicates that at least one task has been assigned to the k-th workstation, otherwise it is 0
S_{jk}	0 or 1 parameter. 1 means that the k-th workstation uses the mobile device j, otherwise it is 0.
VM_j	Set of tasks using mobile device j.
VF_s	Set of tasks using fixed device s
pos_{ip}	0 or 1 parameter. 1 indicates that task i uses position p, otherwise it is 0.
eq_{is}	0 or 1 parameter. 1 indicates that task i uses device s, otherwise it is 0.
E_s	Number of tasks using device s.
H_i	Set of tasks that are in the same workstation as task i and have conflicts with task i in terms of position, $H_i = \left\{ j \middle\| pos_{jp} = pos_{ip} \ and \ x_{jk} = x_{ik}, j \in V, p \in P, k \in K \right\}$.
D_i	Set of tasks that are in the same workstation as task i and have conflicts with task i in terms of devices, $D_i = \left\{ j \middle\| eq_{js} = eq_{is} \ and \ x_{jk} = x_{ik}, j \in V, s \in S_M, k \in K \right\}$.
M	A sufficiently large constant.
pos_{kps}	0 or 1 parameter. If there is a task using fixed device s at position p in the k-th workstation, it is 1, otherwise it is 0.
N_{sk}	Number of fixed devices s required for the k-th workstation.
V_{kp}	Set of tasks assigned to position p of the k-th workstation.
W_l	Task with the minimum end-time.
W_i	Task corresponding to the i-th gene locus in the chromosome.
Q_k	Set of assigned tasks for workstation k.
T_{W_i}	Set of tasks that have parallel conflicts with W_i in current workstation.

Table 2 presents the decision variables employed in the proposed model.

Table 2. Variables used in the proposed model.

Symbol	Description
st_i	Start-time for task i.
et_i	End-time for task i.
x_{ik}	0 or 1 variable. 1 indicates that task i is assigned to the k-th workstation, otherwise it is 0.
y_{ij}	0 or 1 variable. 1 indicates that task i is executed before task j, otherwise it is 0.

3.4. Mathematical Model

Based on the constraints and objectives outlined in this study, the mathematical model for ALB is formulated as follows:

$$Minimize\ obj1 = \sum_{k \in K} A_k, \quad (1)$$

$$Minimize\ obj2 = 1 - \frac{\sum_{i=1}^{n} t_i}{N * \underset{k \in K}{Max}\{TT_k\}}, \quad (2)$$

$$Minimize\ obj3 = \sum_{j \in S} C_j N_j, \quad (3)$$

$$\sum_{k \in K} x_{ik} = 1, \forall k \in K, \quad (4)$$

$$\sum_{i=1}^{n} x_{ik} \leq nA_k, \forall k \in K, \quad (5)$$

$$\sum_{k \in K} (kx_{ik} - kx_{jk}) \geq 0, (\forall i \in V, j \in Pred_i), \quad (6)$$

$$st_i \leq st_j + M(1 - y_{ij}) - t_i, \forall i \in V, j \in (H_i \cup D_i), \quad (7)$$

$$st_i \geq st_j - M \cdot y_{ij} + t_j, \forall\ i \in V, j \in (H_i \cup D_i), \quad (8)$$

$$et_i = st_i + t_i, \forall i \in V, \quad (9)$$

$$T_k = \max\{et_i | i \in R_k\}, \forall k \in K, \quad (10)$$

$$T_k \leq CT, \forall k \in K, \quad (11)$$

$$TT_k = \sum_{i \in R_k} t_i, \forall k \in K, \quad (12)$$

$$E_s = \sum_{i=1}^{n} eq_{is}, \forall s \in S, \quad (13)$$

$$\sum_{i \in VM_j} x_{ik} \leq E_j S_{jk}, \forall j \in S_M, k \in K, \quad (14)$$

$$N_j = \sum_{k \in K} S_{jk}, \forall j \in S_M, \quad (15)$$

$$\sum_{i \in VF_s} x_{ik} \cdot pos_{ip} \leq E_s \cdot pos_{kps}, \forall k \in K, p \in P, s \in S_F, \quad (16)$$

$$N_{sk} = \sum_{p \in P} pos_{kps}, \forall s \in S_F, k \in K, \quad (17)$$

$$N_j = \sum_{k \in K} N_{jk}, \forall j \in S_F. \quad (18)$$

Equations (1)–(3) define the objective functions of the model, which correspond to minimizing the number of workstations, achieving the most balanced workload distribution across workstations, and minimizing the total cost of equipment procurement, respectively. Equations (4)–(18) specify the constraints of the model. Equation (4) ensures that each task is assigned to exactly one workstation. Equation (5) requires that each workstation be assigned at least one task. Equation (6) enforces that task assignments adhere to precedence constraints. Equations (7) and (8) state that if two tasks within the same workstation conflict in terms of position or equipment usage, they cannot be executed in parallel. Equation (9) specifies the end-time constraint for each task. Equation (10) imposes the cycle time constraint for the k-th workstation. Equation (11) imposes the takt time constraint, requiring that the cycle time of any workstation must not exceed the takt time. Equation (12) defines the workload of a workstation as the sum of the processing times of all tasks assigned to it. Equation (13) specifies the total number of tasks that utilize a particular device across the assembly line. Equation (14) ensures that if the k-th workstation includes a task requiring a movable device j, then one unit of device j must be allocated to that workstation. Equation (15) calculates the total number of movable devices j across the assembly line as the sum of devices allocated to each workstation. Equation (16) requires that if a task at a specific position in a workstation utilizes a fixed device s, then that position must be equipped with device s. Equations (17) and (18) determine the total number of fixed devices s required for each workstation and for the entire assembly line, respectively.

3.5. ASP Methods Within the Workstation

Due to the presence of numerous parallel tasks as well as position and device conflicts within the workstation, ASP must be considered in the ALB model proposed herein. It is necessary to determine which tasks can be parallelized and the appropriate initiation time for parallelization. This has a significant impact on attaining line balance, as shown in Figure 2.

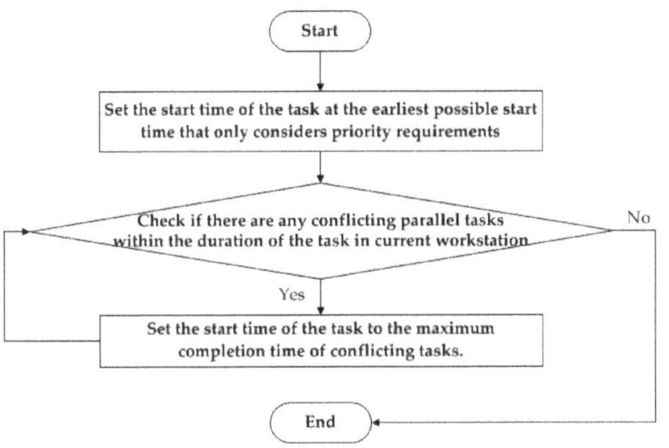

Figure 2. Workflow diagram of the ASP method.

ASP itself has numerous optimization objectives and constitutes an NP-hard problem. Based on the balance objective proposed in Section 3.1 and to simplify the calculation, this study proposes the parallel task processing principle of "parallelize as soon as possible if feasible" at the workstation level. Although this principle may not be optimal for ASP, it is advantageous for attaining ALB objectives, and there remains scope for further ASP optimization at the workstation level after assembly line balancing:

(1) The parallel task processing principle of "parallelize as soon as possible if feasible" ensures that the initiation time of parallel tasks at the workstation is advanced to the earliest feasible point. This approach partially reduces the workstation cycle time and ensures that the workstation can accommodate as many tasks as possible under the given task assignment sequence, thereby minimizing the number of workstations and device procurement costs.

(2) After the assembly line is balanced, the cycle time of the bottleneck workstation serves as the workstation cycle time constraint, while other workstations retain some idle capacity. When considering the goals of balancing workload within and between workstations, there remains potential for further optimization in ASP, which can also be addressed through worker reallocation across workstations. These optimizations do not affect the attainment of the balance objective in this study. Therefore, this study focuses on ALB, and it is appropriate to adhere to the parallel task processing principle of "parallelize as early as feasible" in ASP.

4. Improved NSGA-II Algorithm

4.1. NSGA-II Algorithm

The NSGA-II algorithm is a widely adopted meta-heuristic method for solving multi-objective optimization problems [50,51]. Due to the complex assembly processes and numerous tasks of large-scale products, the resulting chromosome representation tends to be lengthy. To enhance algorithmic efficiency, drawing upon the principles of the artificial immune algorithm [52] and the neighborhood search algorithm [53], this study improves the traditional NSGA-II algorithm from two perspectives: initial population generation and genetic operator design. By introducing affinity-based selection and neighborhood search mechanisms, the local and global refinement capabilities of the genetic algorithm are enhanced. This effectively alleviates the tendency of the NSGA-II algorithm to become trapped in local optima when dealing with lengthy chromosomes.

4.2. Chromosome Coding

This study adopts an integer encoding method, wherein integers serve as the gene values of chromosomes, and each gene value uniquely corresponds to a task on the assembly line. The ordering of tasks within a chromosome represents the task assignment sequence. For example, the chromosome "1 2 3 4 5 6 7 8 9" represents a sequence of integers ranging from 1 to 9, where each integer uniquely corresponds to a task and occurs only once within the chromosome, thereby ensuring that each task is assigned exactly once. The ordering of this chromosome directly reflects the task assignment sequence, namely, task 1 is assigned first, followed sequentially by tasks 2 through 9.

4.3. Chromosome Decoding

(1) Feasibility Testing and Repair of Chromosomes

Owing to precedence constraints in assembly, infeasible task assignment sequences may arise during chromosome generation and genetic operator operations. This study employs the following procedures to conduct feasibility testing and repair of infeasible chromosomes:

Step 1: Starting from the first gene, check whether the corresponding task is assignable (a task is assignable when all its immediate predecessors have been assigned). If it is, mark the task as assigned and proceed to Step 2; otherwise, skip the gene and examine the next one until a feasible task is identified, then mark it as assigned and proceed to Step 2.
Step 2: Insert the task corresponding to the gene into the last position of the new chromosome. If there are tasks preceding it, shift them forward by one position and proceed to Step 3.
Step 3: Delete the gene from the original chromosome and shift subsequent genes forward by one position. Repeat Steps 1–3 until all genes in the original chromosome have been deleted, then proceed to Step 4.

Step 4: Replace the original chromosome with the newly formed one to complete feasibility testing and repair.

(2) Decoding of Chromosomes

The decoding process entails sequentially assigning tasks according to the gene sequence of the chromosome. Depending on whether positional or device conflicts exist between the task being assigned and those already allocated to the current workstation, and whether the production takt time constraint is violated, it is determined whether the current task should be assigned to the current workstation or to a newly initiated workstation. Simultaneously, the initiation time of the task within the current workstation is established. By continuously repeating these task assignment steps until all tasks are allocated, the full task assignment corresponding to the chromosome is ultimately derived.

For assembly lines with numerous restrictive parallel tasks, the decoding process must determine the feasible parallelization of tasks. In accordance with the objective outlined in this study, the initiation time of a parallel task is determined using the method presented in Section 3.5. The decoding procedure proceeds as follows, where W_i denotes the task associated with the i-th gene position:

Step 1: Initialization. Let $i = 0$, $Q_k = \varphi$, $k = 1$; initialize the first workstation of the assembly line, and proceed to Step 2 (Q_k is the set of tasks assigned to the k-th workstation,φ is an empty set).
Step 2: Initialize $i = i + 1$, indicating preparation to handle the corresponding task in the first gene sequence, and proceed to Step 3.
Step 3: Determine whether is $Q_k \cap Pred_{W_i} = \varphi$ true. If it is, proceed to Step 4; otherwise, proceed to Step 11.
Step 4: Because $Q_k \cap Pred_{W_i} = \varphi$, it indicates the absence of an immediate predecessor task of i in the currently assigned tasks within the k-th workstation. Initialize $st_{W_i}\prime = 0$, $T_{W_i} = (H_{W_i} \cup D_{W_i}) \cap Q_k$, turn to Step 5.
Step 5: Determine whether $st_{W_i}\prime + t_{W_i} \leq CT$ is true. If it is, proceed to Step 6; otherwise, proceed to Step 10.
Step 6: Determine whether T_{W_i} is empty. If it is, proceed to Step 8; otherwise, proceed to Step 7.
Step 7: Determine whether $st_{W_i}\prime + t_{W_i} \leq min(st_{W_l})$ is true, where $W_l \in T_{W_i}$. If it is, proceed to Step 8; otherwise, proceed to Step 9.
Step 8: Initialize $st_{W_i} = st_{W_i}\prime$, $Q_k = Q_k \cup W_i$, turn to Step 17.
Step 9: Find the task W_l with the minimum end-time and its corresponding end-time in T_{W_i}, denote them as W_l and , respectively, let $st_{W_i}\prime = et_{W_l}$, and remove W_l from T_{W_i}. Then turn to Step 5.
Step 10: Initialize $k = k + 1$; initialize the subsequent workstation. Proceed to Step 3.
Step 11: Initialize $st_{W_i}\prime = max(et_{W_j})$, $W_j \in (Q_k \cap Pred_{W_i})$, turn to Step 12.
Step 12: Initialize $T_{W_i} = (H_{W_i} \cup D_{W_i}) \cap Q_k$, turn to Step 13.
Step 13: Determine whether $st_{W_i}\prime + t_{W_i} \leq CT$ is true. If it is, proceed to Step 14; otherwise, proceed to Step 10.
Step 14: Remove tasks W_l with end-time less than $st_{W_i}\prime$ in T_{W_i}, then proceed to Step 15.
Step 15: Determine whether T_{W_i} is empty. If it is, proceed to Step 8; otherwise, proceed to Step 16.
Step 16: Determine whether $st_{W_i}\prime + t_{W_i} \leq min(st_{W_l})$ is true, where $W_l \in T_{W_i}$. If it is, proceed to Step 8; otherwise, proceed to Step 17.
Step 17: Find the task W_l with the minimum end-time in T_{W_i}, let $st_{W_i}\prime = et_{W_l}$, and remove W_l from T_{W_i}. Then turn to Step 13.

Step 18: Verify whether all tasks have been assigned. If so, proceed to Step 19; otherwise, return to Step 2.
Step 19: Output the final task assignment results.

The detailed flowchart illustrating the decoding steps is presented in Figure 3.

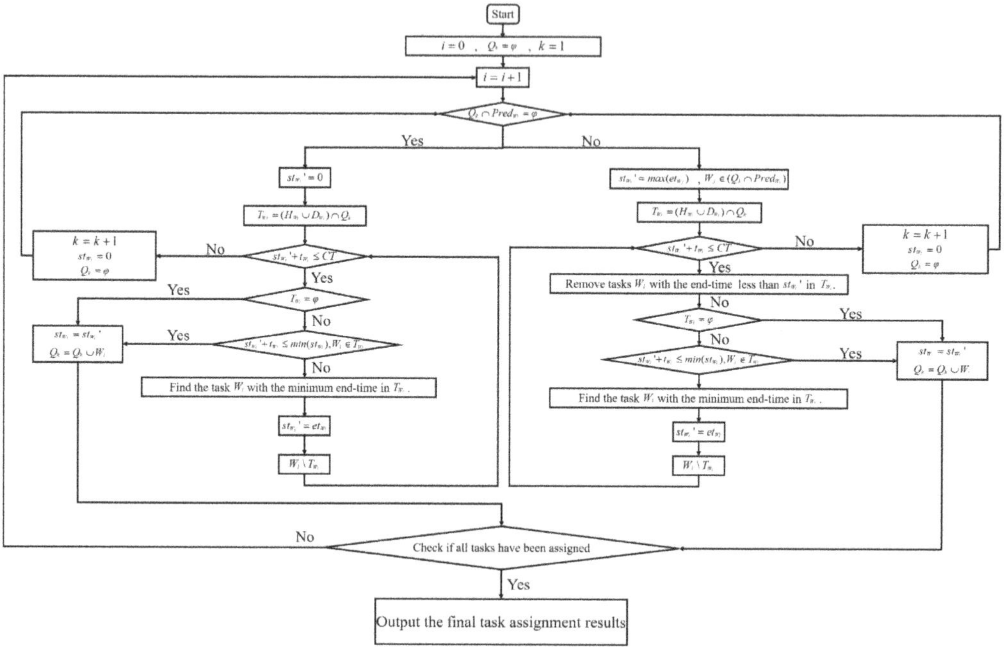

Figure 3. Decoding flowchart.

4.4. Population Initialization

In the NSGA-II algorithm, chromosomes within the initial population are typically generated randomly. However, the task priority sequence imposes constraints on their positions within each chromosome, so randomly generated chromosomes frequently produce infeasible solutions. Therefore, to ensure the feasibility of the initial chromosomes, minimize the chromosome repair workload, and preserve population diversity, this study proposes the following procedure for generating the initial population:

Step 1: Construct the task precedence relationship matrix. Based on the precedence relationships among tasks, define the relationship matrix PR such that if task i is the immediate predecessor of task j, then $PR_{ij} = 1$, otherwise, $PR_{ij} = 0$.
Step 2: Initialize $i = 1$.
Step 3: Randomly select a task from the set of tasks whose corresponding column sum in the matrix PR equals 0, and assign it to the i-th gene position of the chromosome. Remove the corresponding row and column of this task from matrix PR, update PR accordingly, and increment i by 1.
Step 4: Repeat Step 3 until matrix PR becomes empty, thereby forming a feasible chromosome.
Step 5: Repeat Steps 2–4 until the entire initial population is generated.

The flowchart of population initialization is shown in Figure 4.

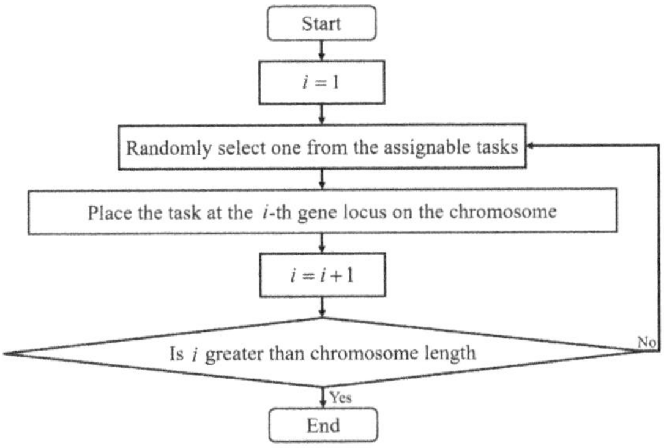

Figure 4. Workflow diagram of the population initialization.

4.5. Genetic Operators

4.5.1. Selection Operator

To ensure algorithmic convergence, this study constructs a separate elite pool and selects chromosomes according to an elite retention strategy. To balance global and local search and enhance search efficiency, this study does not employ the hierarchical selection strategy of the NSGA-II algorithm, which relies on crowding distance and hierarchical ranking. Instead, this strategy applies crowding-based selection only at the final layer that satisfies the population size requirement. For long chromosomes and large-scale populations, this strategy may fail to maintain population diversity during early iterations and is prone to convergence to local optima. Therefore, this study proposes an affinity-based selection strategy, which preserves population diversity during early iterations while promoting population convergence in later iterations.

(1) Elite Pool Update: Identify chromosomes on the Pareto front in the current population, merge them with the existing elite pool, and select the Pareto-optimal chromosomes from the merged set. Add these selected chromosomes to the elite pool and remove others to complete the update.
(2) Population Update: Inspired by the artificial immune algorithm [54], population updates are performed based on affinity. The affinity of a chromosome x is defined as a function of its dominance rate and concentration:

$$aff(x) = \lambda \frac{1}{1.0 + c(x)} + (1 - \lambda)\rho(x), \tag{19}$$

where $c(x)$ is the dominance rate of the chromosome x, $\rho(x)$ is the concentration of the chromosome x, and λ is a variable weight coefficient that ensures that the concentration effect gradually decreases as the iteration progresses, its maximum and minimum values are limited to λ_{max} and λ_{min}, respectively. The above variables are calculated by the following formulas:

$$c(x) = \frac{|\{y|y < x, \forall y \in X\}|}{|X|}, \tag{20}$$

$$\rho(x) = \frac{\underset{\forall y \in X}{min} d(x,y)}{\underset{\forall y,z \in X}{max} d(y,z)}, \tag{21}$$

$$\lambda = \lambda_{max} - (\lambda_{max} - \lambda_{min})/n, \tag{22}$$

where $y \prec x$ indicates individual y dominating individual x, $|X|$ indicates the number of individuals in the population, $d(x,y)$ indicates the Euclidean distance between the objective functions of chromosome x and y, and n is the current iteration algebra. To eliminate the influence of dimensionality, $d(x,y)$ is calculated by the following formula:

$$d(x,y) = \sqrt{\sum_{i=1}^{m} d_i(x,y)}, \tag{23}$$

$$d_i(x,y) = \begin{cases} \frac{(f_{ix}-f_{iy})^2}{\max\limits_{\forall x_1,x_2 \in X}(f_{ix_1}-f_{ix_2})^2}, \max\limits_{\forall x_1,x_2 \in X}(f_{ix_1}-f_{ix_2})^2 > 0 \\ 0, \max\limits_{\forall x_1,x_2 \in X}(f_{ix_1}-f_{ix_2})^2 = 0 \end{cases}, i = 1,2,\ldots,m, \tag{24}$$

where m denotes the number of objectives, f_{ix} represents the objective function value of chromosome x on the i-th objective, and X denotes the chromosome population.

The selection process adopts the binary tournament method: two individuals are randomly chosen from the population, and their affinity values are compared. The individual with the higher affinity value is retained for subsequent operations, while both individuals are returned to the original population. If their affinity values are equal, one is randomly selected. This selection and comparison process is repeated until a sufficient number of individuals are obtained to construct the next-generation population.

4.5.2. Crossover Operator

To enhance search efficiency, this study designs two crossover operators, and the optimal crossover result is determined by comparing their outcomes.

(1) Crossover Based on Sequence Features. This operator inherits the sequence features of the crossover fragments from the parent generation while ensuring the feasibility of the task assignment sequence after crossover. Consequently, no chromosome repair is required, which substantially reduces the algorithm's computational workload. The specific crossover process is illustrated in Figure 5.

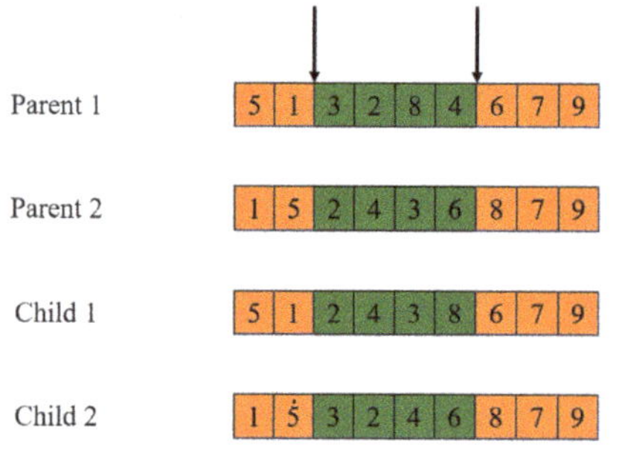

Figure 5. Crossover based on sequence features.

Two identical crossover positions are randomly selected in Parent 1 and Parent 2, dividing each parent into three segments: head, middle, and tail. To construct offspring Child 1, the head and tail of Parent 1 are directly inherited while preserving the original task sequence. The middle segment is reassigned internally according to the task assignment sequence of Parent 2. Offspring Child 2 is obtained in a similar manner. This crossover method retains the sequence features of the parental crossover segment while ensuring the feasibility of the chromosome task assignment sequence. However, when a large number of parallel tasks are present, the solution quality after crossover often remains unchanged, thereby diminishing the significance of crossover.

(2) Crossover Based on Full Preservation of Crossover Segments. In this method, the offspring fully retain the parental crossover segment, although the positions of the segment and other genes may vary. This operator is capable of inheriting validated, effective local structures from the parent individual. The specific crossover process is illustrated in Figure 6.

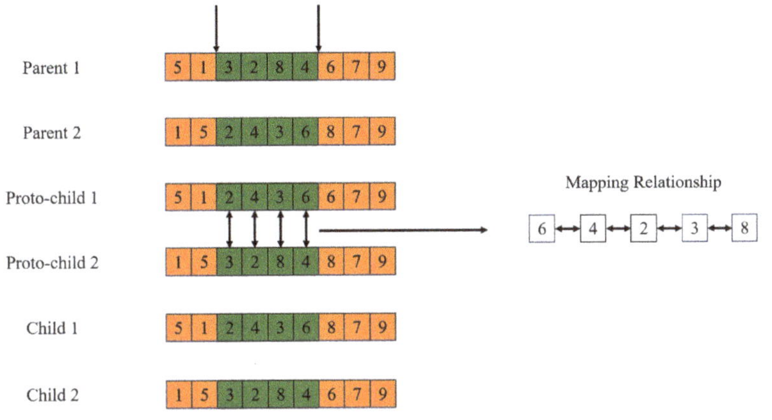

Figure 6. Crossover based on full preservation of crossover segments.

After selecting the crossover point, the corresponding middle segments of the parent chromosomes are exchanged. If repetitive tasks occur between the head or tail segments and the exchanged middle segment, these are replaced according to the mapping relationship between the tasks in the middle segments of the two parent chromosomes. If the chromosome resulting from crossover is infeasible, the crossover segment (middle segment) is treated as an indivisible unit for repair. The specific repair process is described in Section 4.3. This crossover pattern effectively preserves gene adjacency relationships, which is critical for maintaining high-quality patterns.

The two crossover methods employ different mechanisms for preserving crossover segments, and the resulting solutions may also differ. To improve iteration efficiency, the offspring generated by the two crossover methods are compared based on their objective function values. The non-dominated solution is selected as the effective offspring; if no dominant relationship exists between the offspring, the final effective offspring is determined through random selection.

4.5.3. Mutation Operator

To ensure mutation quality, this study employs a neighborhood search-based mutation strategy.

(1) Feasible mutation space of task i

To maintain chromosome feasibility, the feasible mutation space of task i is defined by the maximum position of all immediate predecessor tasks and the minimum position of all immediate successor tasks. An illustrative example of a feasible mutation space is provided in Figure 7.

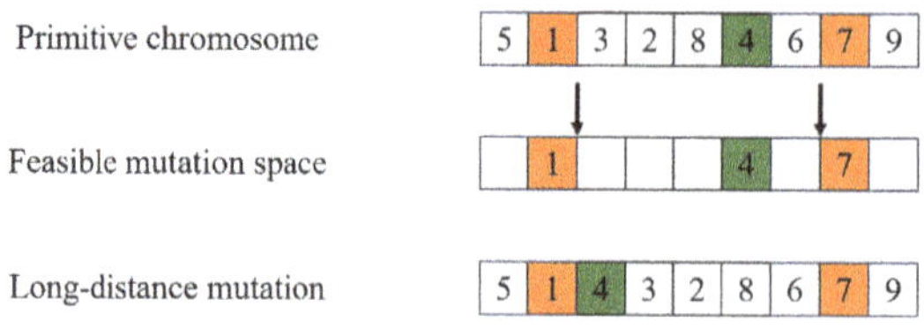

Figure 7. Long-distance mutation method.

Assume that task 4 is selected for mutation. Tasks 5 and 1 are its immediate predecessors, while tasks 7 and 9 are its immediate successors. The interval between the positions of tasks 1 and 7 defines the feasible mutation space of task 4.

(2) Mutation method

Given the presence of numerous parallel tasks, and to maximize changes in the objective values after decoding the mutated chromosome, this study adopts the long-distance mutation method. Specifically, the point within the feasible mutation space farthest from the original position of the task is selected, and insertion-based mutation is performed. For example, in Figure 7, task 4 is inserted between task 1 and task 3.

(3) Mutation neighborhood search

To further enhance mutation quality, this study applies neighborhood search. The mutation neighborhood of task i is defined as the space beyond the farthest point within the feasible mutation space. Considering computational cost, the maximum number of neighborhood searches is limited to five. The method for selecting the optimal chromosome after mutation is consistent with that used for crossover.

4.6. Iteration Termination Conditions

The algorithm terminates when the maximum number of iterations is reached. Additionally, if more than 90% of individuals in the elite pool recur over 10 consecutive generations, the algorithm is also terminated.

5. Case Study

5.1. Case Introduction

To evaluate the effectiveness of the proposed model and algorithm, this study compares the improved NSGA-II with the traditional NSGA-II using a practical case. The case concerns the assembly line of an aviation manufacturing enterprise, with a designed takt time of 20 h and a total of 50 tasks. The task priority sequence is illustrated in Figure 8.

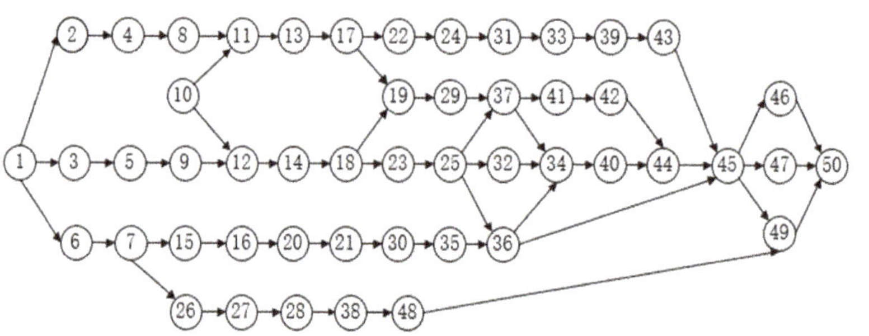

Figure 8. Priority sequence diagram.

The product requires five assembly positions, which do not interfere with each other. A total of eight types of devices are required: seven are fixed within the assembly positions and cannot be moved, while the remaining one is movable and can be used across multiple assembly positions. The task details are provided in Table 3.

Table 3. Basic task information.

Task	Task Position	Device Required	Device Property	Device Cost (Yuan)	Duration Time (Hours)	Task	Task Position	Device Required	Device Property	Device Cost (Yuan)	Duration Time (Hours)
1	p1	A	fixed	5000	3	26	p1	A	fixed	5000	2
2	p2	B	fixed	3000	2	27	p1	B	fixed	3000	3
3	p3	B	fixed	3000	2	28	p1	E	fixed	2000	1
4	p2	C	fixed	1500	1	29	p4	H	fixed	1500	5
5	p3	C	fixed	1500	1	30	p5	D	fixed	4000	3
6	p1	D	fixed	4000	4	31	p2	H	fixed	1500	2
7	p1	E	fixed	2000	2	32	p3	H	fixed	1500	2
8	p2	F	fixed	1000	3	33	p2	None	-	0	3
9	p3	F	fixed	1000	3	34	p3	None	-	0	3
10	p4	G	movable	60,000	2	35	p5	F	fixed	1000	3
11	p2	G	movable	60,000	3	36	p5	C	fixed	1500	1
12	p3	G	movable	60,000	3	37	p4	G	movable	60,000	4
13	p2	H	fixed	1500	1	38	p1	E	fixed	2000	2
14	p3	H	fixed	1500	1	39	p2	None	-	0	1
15	p5	A	fixed	5000	5	40	p3	None	-	0	1
16	p5	D	fixed	4000	4	41	p4	H	fixed	1500	5
17	p2	B	fixed	3000	2	42	p4	D	fixed	4000	2
18	p3	B	fixed	3000	2	43	p2	B	fixed	3000	2
19	p4	C	fixed	1500	3	44	p3	B	fixed	3000	2
20	p5	E	fixed	2000	2	45	p5	None	-	0	3
21	p5	G	movable	60,000	3	46	p5	None	-	0	1
22	p2	F	fixed	1000	4	47	p5	None	-	0	1
23	p3	F	fixed	1000	4	48	p1	A	fixed	5000	2
24	p2	C	fixed	1500	2	49	p1	F	fixed	1000	3
25	p3	C	fixed	1500	2	50	p1	E	fixed	2000	2

5.2. Calculation Results

The calculations were performed on a computer equipped with an Intel i7-1365HX processor and 24 GB of RAM. The minimum value of parameter λ was set to 0.6 and the maximum to 1. The crossover probability was set to 0.9, the mutation probability to 0.3, the population size to 150, and the maximum number of iterations to 100. The parameters of the NSGA-II algorithm were selected based on previous literature and further refined through empirical tuning and preliminary optimization experiments to ensure robust convergence and solution diversity.

Figure 9 presents the Pareto front solutions obtained by running both the improved NSGA-II algorithm and the traditional NSGA-II algorithm in MATLAB R2021b (as shown in the Appendix A). Since the number of workstations corresponding to the Pareto front solutions obtained by both algorithms is identical (three stations), for clarity, only a comparison of solutions on the two-dimensional Pareto front formed by the remaining two objectives is presented. The figure clearly shows that the improved NSGA-II algorithm demonstrates superior overall performance. Specifically, the improved NSGA-II algorithm yields a greater number of frontier solutions, a more uniform distribution, and higher-quality outcomes.

From the perspective of workload balance rate, the quality of the single-objective optimal solution on the Pareto front obtained by the improved NSGA-II algorithm is superior to that of the traditional NSGA-II algorithm. Figure 10 presents the task allocation schemes corresponding to the single-objective optimal solutions on each Pareto front for the two algorithms, while Table 4 reports the task load statistics of the corresponding solutions. These results indicate that the improved algorithm offers significant advantages.

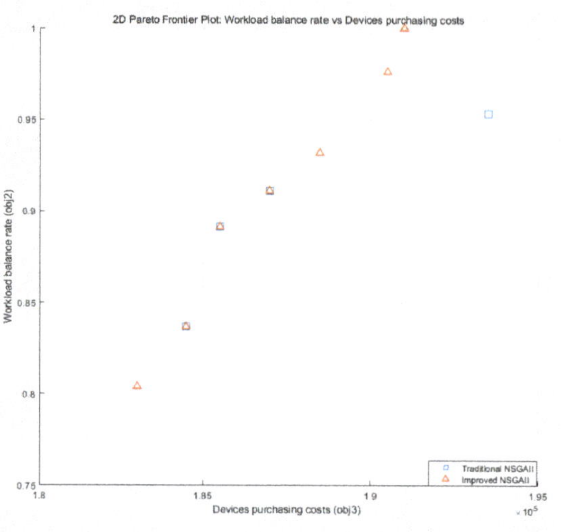

Figure 9. Comparison of Pareto frontiers between traditional NSGA-II and improved NSGA-II algorithm.

Table 4. Comparison of Workload between Traditional NSGAII and Improved NSGAII Algorithm.

Workstation	Traditional NSGA-II Algorithm Workload (Hours)	Improved NSGA-II Algorithm Workload (Hours)
Workstation 1	43	41
Workstation 2	43	41
Workstation 3	37	41

From the perspective of equipment procurement costs, the quality of the single-objective optimal solution on the Pareto front obtained by the traditional NSGA-II algorithm is also superior to that obtained by the improved NSGA-II algorithm. Figure 11 presents the task plans corresponding to the single-objective optimal solutions on each Pareto front of the two algorithms, while Table 5 reports the equipment procurement plans and cost statistics for the corresponding solutions. These findings further demonstrate that the improved algorithm provides clear advantages.

Table 5. Comparison of Cost between Traditional NSGAII and Improved NSGAII Algorithm.

Algorithm	Workstation	Number of Devices								Cost (Yuan)
		A	B	C	D	E	F	G	H	
Traditional NSGA-II algorithm	Workstation 1	1	3	2	1	1	2	1	2	184,500
	Workstation 2	1	0	3	2	1	1	1	3	
	Workstation 3	0	2	1	0	1	2	0	0	
Improved NSGA-II algorithm	Workstation 1	2	3	2	1	1	2	1	1	183,000
	Workstation 2	0	1	3	1	1	1	1	3	
	Workstation 3	0	1	1	1	1	2	0	0	

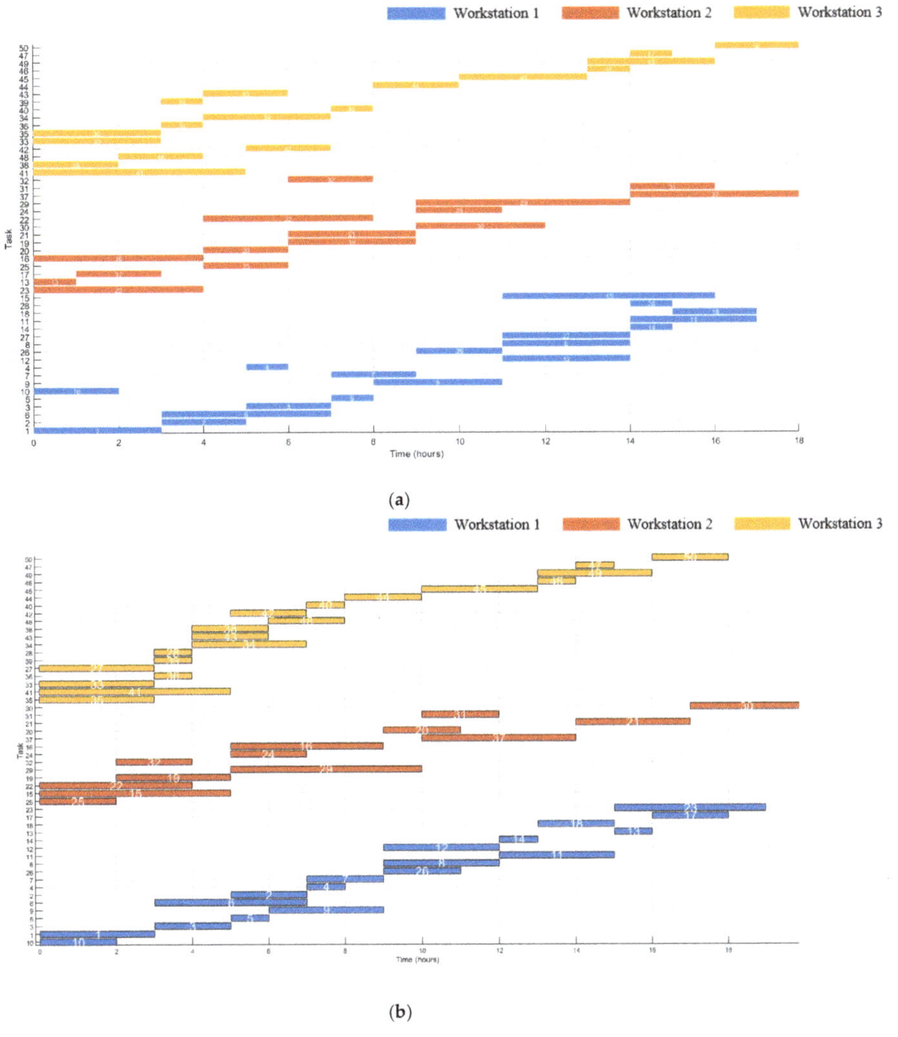

Figure 10. Comparison of Pareto front optimal solutions of two algorithms on workload balance rate. (**a**) Traditional NSGA-II algorithm; (**b**) Improved NSGA-II algorithm.

Hyper-volume is a key metric for evaluating the performance of multi-objective optimization algorithms. A larger hyper-volume indicates a wider coverage of the target space by the Pareto-optimal solution set and a higher solution quality. Based on the calculation method proposed by Wang et al. [55], Table 6 indicates that the improved NSGA-II algorithm achieves a significantly larger hyper-volume than the traditional NSGA-II algorithm, further confirming the superiority of the improved algorithm in terms of solution coverage and quality from a quantitative perspective.

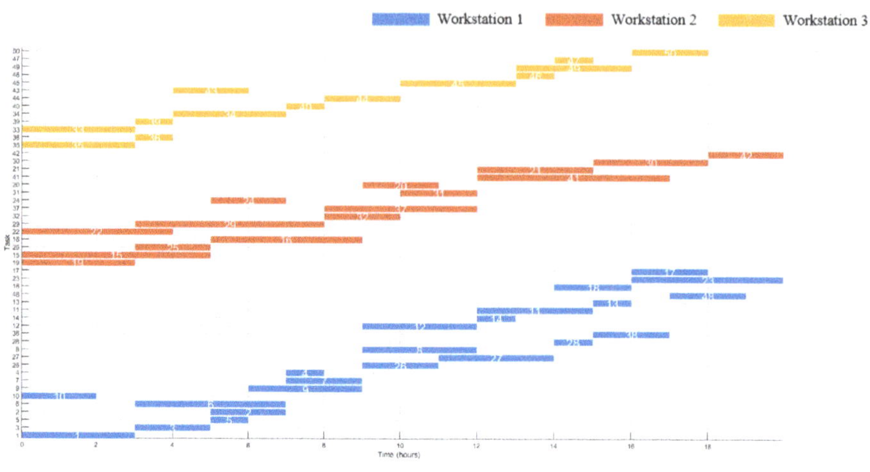

(**a**)

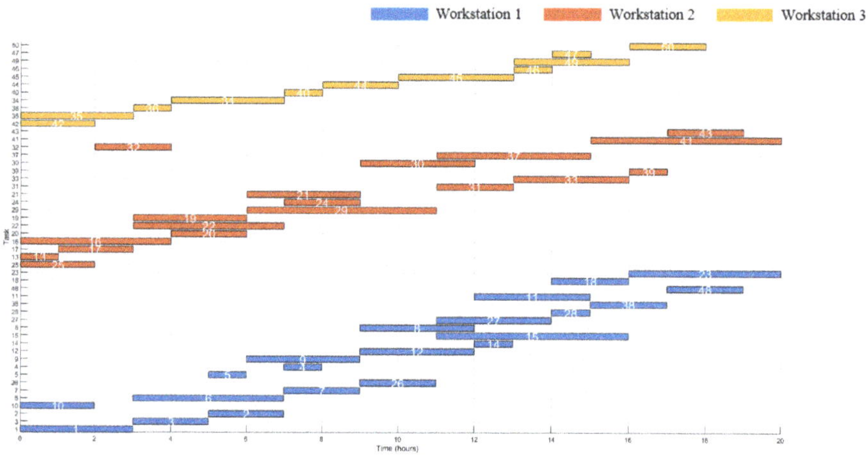

(**b**)

Figure 11. Comparison of Pareto front optimal solutions of two algorithms on equipment procurement costs. (**a**) Traditional NSGA-II algorithm; (**b**) improved NSGA-II algorithm.

Table 6. Comparison of Running Time and Hyper-volume between Traditional NSGAII and Improved NSGAII Algorithm.

	Run Time	Normalized Hyper-Volume (Mean)
Traditional NSGA-II	3:11:02	2.7599
Improved NSGA-II	4:45:47	4.8474

Furthermore, the improved algorithm produces a more balanced performance of solutions on the Pareto front across multiple objectives. The Pareto front obtained by the traditional NSGA-II algorithm may include solutions that perform well on one objective but poorly on others; the improved algorithm better balances the relationships among objectives, yielding solutions with higher overall performance.

Regarding running time, the traditional NSGA-II algorithm exhibits faster performance (Table 6). This is because the improved NSGA-II algorithm incorporates additional operational steps and more complex computational logic, such as refined selection strategies and more sophisticated mutation methods, which increase the computational load and result in longer running times. However, given today's advanced computing capabilities, from an application perspective, the running time for a single instance is no longer a major limitation for ALB.

Additionally, to assess the stability of the proposed model, ten independent experimental runs were conducted, and the standard deviation (SD) of the obtained results was calculated. The results demonstrated that the SD of the improved NSGA-II algorithm was 0.151, compared with 0.686 for the traditional NSGA-II algorithm. Furthermore, a two-sample t-test was performed to determine whether the difference in SDs between the two algorithms was statistically significant. The test results indicated a statistically significant difference ($p < 0.01$), confirming that the improved algorithm achieved more stable performance. Moreover, the 95% confidence interval (CI) for the mean difference was [1.54, 2.63], further supporting that the proposed improvements effectively maintained consistent optimization performance across different execution conditions.

To investigate the influence of the initial (minimum) value of parameter λ in the affinity calculation on the performance of the improved NSGA-II algorithm, this study conducted a further analysis based on the case study, with the results presented in Figure 12.

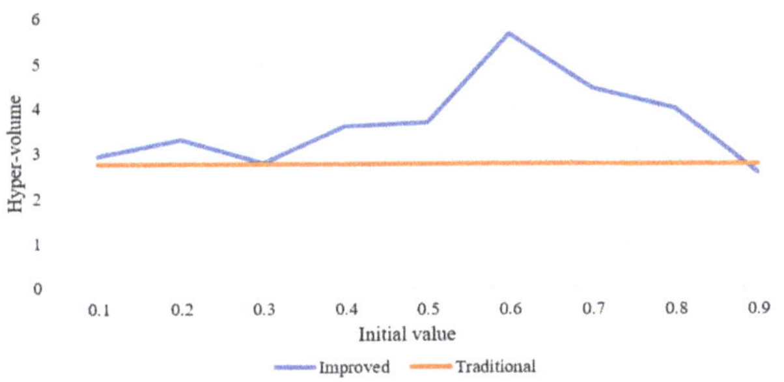

Figure 12. The influence of the initial value of parameter λ on the quality of the improved NSGA-II algorithm.

It can be observed that, in terms of hyper-volume, the quality of the final solution varies with the initial value of parameter λ. In most cases, the improved NSGA-II algorithm outperforms the traditional NSGA-II algorithm, particularly when the initial value is approximately 0.6. The hyper-volume of the Pareto front solutions obtained by the improved NSGA-II algorithm is significantly larger than that of the traditional NSGA-II algorithm. This trend is especially evident in assembly line balancing problems with a large number of restrictive parallel constraints. By optimizing selection and mutation strategies, the improved NSGA-II algorithm is able to generate a more balanced workload distribution among workstations, achieve a more rational allocation of workstation numbers, and reduce equipment procurement costs, all while satisfying the design takt time requirements. Consequently, the algorithm enhances both the coverage of the Pareto front and the hypervolume indicator.

6. Conclusions

This study investigates assembly line balancing (ALB) problems with a large number of restrictive parallel constraints. Tasks are assigned to appropriate positions within suitable workstations. The assignment process considers both positional conflicts and device usage conflicts. Task start times are calculated according to the principle of parallelism, aiming to begin as early as possible. While satisfying the design takt time requirements, the approach aims to achieve three objectives: minimizing the number of workstations, balancing the workload among workstations, and minimizing device acquisition costs.

The ALB problem with restrictive parallel constraints is classified as NP-hard. The NSGA-II algorithm is a widely used method for solving NP-hard multi-objective optimization problems, and its effectiveness has been well established. However, due to the characteristics of the model proposed in this study, the traditional NSGA-II algorithm experienced premature convergence. Drawing on concepts from artificial immune algorithms and neighborhood search algorithms, this study improves the traditional NSGA-II algorithm from two perspectives: initial population generation and genetic operator design.

Case studies indicate that (1) the improved NSGA-II algorithm outperforms the traditional NSGA-II algorithm in overall solution quality; however, the redesign of genetic operators increases computational complexity, resulting in longer computation times compared to the traditional algorithm. (2) The initial value of parameter λ in the affinity calculation significantly affects performance, with solution quality being higher when the initial value is approximately 0.6.

The proposed model effectively addresses the balancing problem of large-scale product assembly lines with numerous restrictive parallel constraints. Given the limited literature on the type of assembly line balancing problem addressed in this study and the scarcity of corresponding case sets, further comparative studies of algorithms are needed.

This study addresses the line balancing problem in large-scale product assembly lines. For the first time, practical constraints—including intra-station position conflicts and equipment conflicts—are comprehensively incorporated into the traditional assembly line balancing model. A multi-objective optimization model is developed with the objectives of minimizing the number of workstations, achieving a balanced task load across stations, and reducing equipment procurement costs, thereby simultaneously addressing production efficiency, cost control, and team management considerations. Regarding algorithm design, for large-scale parallel-task assembly line balancing problems, the NSGA-II algorithm is enhanced by integrating affinity and concentration selection strategies inspired by artificial immune algorithms, while neighborhood search is applied to optimize crossover and mutation operators, improving both solution efficiency and convergence performance. This study not only enhances the practical applicability of the model but also provides theoretical and methodological guidance for the optimized design of high-precision, large-scale product assembly lines, demonstrating substantial practical significance and innovation.

This study has certain limitations. Due to the limited availability of public benchmark cases for assembly line balancing problems with numerous restrictive parallel constraints, the number of case studies used for algorithm validation is relatively small, and therefore, the generalizability and practical applicability of the proposed method require further validation. Moreover, the algorithm's performance is highly sensitive to the initial parameter settings, and it currently lacks a systematic self-adaptive adjustment mechanism. Additionally, several practical production factors—such as variations in worker proficiency, equipment maintenance schedules, and logistics constraints—were simplified during model formulation, potentially leading to deviations from real-world production scenarios. Future research should focus on expanding the benchmark case set, optimizing parameter configurations, reducing algorithmic complexity, and integrating more realistic production constraints to further enhance the robustness and applicability of the proposed model and algorithm.

Author Contributions: Conceptualization, H.L. and Y.C.; methodology, H.L.; software, H.L.; validation, Y.C., X.Z. and F.K.; formal analysis, G.S.; investigation, X.Z.; resources, F.K.; data curation, G.S.; writing—original draft preparation, H.L.; writing—review and editing, Y.C.; visualization, Y.C.; supervision, Y.C.; project administration, Y.C.; funding acquisition, Y.C. All authors have read and agreed to the published version of the manuscript.

Funding: This research was funded by the Aeronautical Science Foundation, grant number 20230009054001.

Data Availability Statement: Given the ongoing research value of the data, scholars may reach out to the author to obtain access to relevant data.

Conflicts of Interest: Authors Haiwei Li, Xi Zhang and Guoqiu Song were employed by the company Shenyang Aircraft Corporation. The remaining authors declare that the research was conducted in the absence of any commercial or financial relationships that could be construed as a potential conflict of interest.

Appendix A

Sample Calculation

```
for gen = 1:max_gen
    offspring = generate_offspring(pop_size, population, crossover_rate, mutation_rate, predecessors, ...
                                   durations, m, conflict_jobs, required_equipment, ...
                                   equipment_types, equipment_cost_per_type);
    combined_population = [population; offspring];
    for i = 1:size(combined_population, 1)
        combined_population(i, :) = repair_chromosome(combined_population(i, :), predecessors);
    end
    combined_objectives = calculate_objectives(combined_population, predecessors, durations, m, conflict_jobs, required_equipment, equipment_types, equipment_cost_per_type);
        [fronts, affinities] = non_dominated_sort_with_affinity(combined_objectives, gen);
      population = tournament_selection(combined_population, affinities, pop_size);
    current_combined_pareto = fronts{1};
    current_combined_elite = combined_population(current_combined_pareto, :);
    current_combined_obj = combined_objectives(current_combined_pareto, :);
    if isempty(current_elite_set)
            merged_elite = current_combined_elite;
        merged_obj = current_combined_obj;
    else
            merged_elite = [current_elite_set; current_combined_elite];
        merged_obj = [current_elite_objectives; current_combined_obj];
                [~, unique_indices] = unique(merged_elite, 'rows');
        merged_elite = merged_elite(unique_indices, :);
        merged_obj = merged_obj(unique_indices, :);
    end
```

```
    [merged_fronts, ~] = non_dominated_sort_with_affinity(merged_obj, 1);
  new_pareto_indices = merged_fronts{1};
  current_elite_set = merged_elite(new_pareto_indices, :);
  current_elite_objectives = merged_obj(new_pareto_indices, :);
  elite_count = size(current_elite_set, 1);
  if ~isempty(current_elite_objectives)
      min_stations = min(current_elite_objectives(:,1));
      min_cost = min(current_elite_objectives(current_elite_objectives(:,1) ==
min_stations, 2));
      max_load_balance = max(current_elite_objectives(current_elite_objectives(:,1)
== min_stations & current_elite_objectives(:,2) == min_cost, 3));
  else
              current_obj = calculate_objectives(population, predecessors,
durations, m, conflict_jobs, required_equipment, equipment_types,
equipment_cost_per_type);
      min_stations = min(current_obj(:,1));
      min_cost = min(current_obj(current_obj(:,1) == min_stations, 2));
      max_load_balance = max(current_obj(current_obj(:,1) == min_stations &
current_obj(:,2) == min_cost, 3));
  end
    similarity_ratio = 0;
  if gen > 1 && ~isempty(previous_elite_set) && elite_count > 0
      similarity_ratio = calculate_elite_similarity(current_elite_set,
previous_elite_set);
  end
          consecutive_similar_count = 0;
  end
  previous_elite_set = current_elite_set;
end
```

References

1. Aguilar, H.; García-Villoria, A.; Pastor, R. A survey of the parallel assembly lines balancing problem. *Comput. Oper. Res.* **2020**, *124*, 105061. [CrossRef]
2. Schmid, N.A.; Montreuil, B.; Limère, V. Modeling and solving integrated assembly line balancing, assembly line feeding, and facility sizing problems. *Int. J. Prod. Econ.* **2024**, *277*, 109354. [CrossRef]
3. Gianassi, M.; Leoni, L.; Fantozzi, I.C.; De Carlo, F.; Tucci, M. Mixed-model and multi-model assembly lines: A systematic literature review on resource management. *J. Manuf. Syst.* **2025**, *82*, 632–657. [CrossRef]
4. Şahin, M.C.; Tural, M. Robotic stochastic assembly line balancing. *Flex. Serv. Manuf. J.* **2023**, *35*, 1076–1115. [CrossRef]
5. Tiacci, L. Simultaneous balancing and buffer allocation decisions for the design of mixed-model assembly lines with parallel workstations and stochastic task times. *Int. J. Prod. Econ.* **2015**, *162*, 201–215. [CrossRef]
6. Hou, W.; Zhang, S. Assembly line balancing and optimal scheduling for flexible manufacturing workshop. *J. Mech. Sci. Technol.* **2024**, *38*, 2757–2772. [CrossRef]
7. Kheirabadi, M.; Keivanpour, S.; Chinniah, Y.A.; Frayret, J. Human-robot collaboration in assembly line balancing problems: Review and research gaps. *Comput. Ind. Eng.* **2023**, *186*, 109737. [CrossRef]
8. Huang, D.; Mao, Z.; Fang, K.; Fu, E.; Pinedo, M.L. An Improved Combinatorial Benders Decomposition Algorithm for the Human-Robot Collaborative Assembly Line Balancing Problem. *INF. J. Comput.* **2024**, *37*, 1267–1283. [CrossRef]
9. Wang, Q.; Wang, X.; Luo, H.; Xiong, J. An Improved Multi-Objective Evolutionary Approach for Aerospace Shell Production Scheduling Problem. *Symmetry* **2020**, *12*, 509. [CrossRef]
10. Zhang, Y.; Gu, C.; Wang, J.; Zhang, L.; Hu, X.; Ma, X. Dynamic Modeling Method of Mission Reliability of a Multi-State Manufacturing System with Multiple Production Lines. *IEEE Access* **2020**, *8*, 57012–57023. [CrossRef]

11. Telles, E.S.; Lacerda, D.P.; Morandi, M.I.; Piran, F.A. Drum-buffer-rope in an engineering-to-order system: An analysis of an aerospace manufacturer using data envelopment analysis (DEA). *Int. J. Prod. Econ.* **2020**, *222*, 107500. [CrossRef]
12. Zhang, H.; Feng, L.; Wang, J.; Zhu, T.; Jin, L. Investigating product innovation pathway from a modular standpoint: A case study of large aircraft assembly line. *Heliyon* **2023**, *10*, e23356. [CrossRef] [PubMed]
13. Johnson, R.V. Optimally balancing large assembly lines with 'FABLE'. *Manag. Sci.* **1988**, *34*, 240–253. [CrossRef]
14. Martignago, M.; Battaïa, O.; Battini, D. Workforce management in manual assembly lines of large products: A case study. *IFAC-Pap.* **2017**, *50*, 6906–6911. [CrossRef]
15. Liao, S.; Sang, C.; Liu, A.; Liu, H. Solving Type-I unpaced synchronous mixed-model two-sided assembly line balancing problem using a genetic algorithm. *Comput. Oper. Res.* **2025**, *184*, 107257. [CrossRef]
16. Yin, X.; Yang, Y.; Li, X.; Wu, R.; Guo, A.; Zhao, Q. Research on the balancing problem of human-robot collaborative assembly line in SMEs considering ergonomic risk and cost. *Comput. Ind. Eng.* **2025**, *204*, 111091. [CrossRef]
17. Jiao, Y.; Wang, Y.; Su, X.; Wang, F. An ant colony hybrid simulated annealing algorithm for collaborative optimization of robotic mixed-model parallel two-sided assembly lines balancing. *Comput. Oper. Res.* **2025**, *182*, 107113. [CrossRef]
18. El Machouti, S.; Hlyal, M.; Babay, A.; Alami, J.E. Optimisation of Simple Assembly Line Balancing Problem Type E: A Systematic Literature Review. *Manag. Syst. Prod. Eng.* **2024**, *32*, 162–173. [CrossRef]
19. Aguilar, H.; García-Villoria, A.; Pastor, R. Heuristic and metaheuristic procedures for the buffer sizing problem in parallel assembly lines balancing problem with multi-line workstations and different cycle times. *Comput. Oper. Res.* **2023**, *157*, 106285. [CrossRef]
20. Lai, T.; Sotskov, Y.N.; Dolgui, A.; Zatsiupa, A. Stability radii of optimal assembly line balances with a fixed workstation set. *Int. J. Prod. Econ.* **2016**, *182*, 356–371. [CrossRef]
21. Chao, Y.; Chen, X.; Chen, S.; Yuan, Y. An improved multi-objective antlion optimization algorithm for assembly line balancing problem considering learning cost and workstation area. *Int. J. Interact. Des. Manuf. (IJIDeM)* **2025**, *19*, 6691–6705. [CrossRef]
22. Tiacci, L. Combining balancing, sequencing and buffer allocation decisions to improve the efficiency of mixed-model asynchronous assembly lines. *Comput. Ind. Eng.* **2024**, *194*, 110357. [CrossRef]
23. Demiralay, Y.D.; Kara, Y. A Cost-Effective Balancing Model for Human-Robot Collaborative Assembly Lines. *IEEE Int. Symp. Multimed.* **2024**, *253*, 2766–2775. [CrossRef]
24. Janardhanan, M.N.; Li, Z.; Bocewicz, G.; Banaszak, Z.A.; Nielsen, P. Metaheuristic algorithms for balancing robotic assembly lines with sequence-dependent robot setup times. *Appl. Math. Model.* **2019**, *65*, 256–270. [CrossRef]
25. Junior, M.C.; Michels, A.S.; Magatão, L. An exact constraint programming method for the multi-manned assembly line balancing problem with assignment restrictions. *Expert Syst. Appl.* **2025**, *259*, 125294. [CrossRef]
26. Abdous, M.; Delorme, X.; Battini, D.; Sgarbossa, F. Scenario-based optimization and simulation framework for human-centered Assembly Line Balancing. *Int. J. Prod. Econ.* **2025**, *282*, 109513. [CrossRef]
27. Yang, H. Balance of mixed flow assembly line based on industrial engineering mathematics and simulated annealing improved algorithm. *Results Eng.* **2024**, *22*, 102071. [CrossRef]
28. Nourmohammadi, A.; Arbaoui, T.; Fathi, M.; Dolgui, A. Balancing human-robot collaborative assembly lines: A constraint programming approach. *Comput. Ind. Eng.* **2025**, *205*, 111154. [CrossRef]
29. Zamzam, N.; El-Kharbotly, A. Balancing two-sided multi-manned assembly line under time and space constraint. *Ain Shams Eng. J.* **2024**, *15*, 102464. [CrossRef]
30. Nourmohammadi, A.; Fathi, M.; Ng, A.H. Balancing and scheduling human-robot collaborated assembly lines with layout and objective consideration. *Comput. Ind. Eng.* **2023**, *187*, 109775. [CrossRef]
31. Alhomaidi, E.; Askin, R.G. Parallel Assembly Line Balancing Model with Tooling Consideration and Demand Fulfilment (ALBPTD). *IFAC-Pap.* **2022**, *55*, 103–108. [CrossRef]
32. Alhomaidi, E.; Askin, R.G. Exact and approximation heuristic of mixed model assembly line balancing with parallel lines and task-dependent tooling consideration. *Comput. Ind. Eng.* **2024**, *193*, 110265. [CrossRef]
33. Aguilar, H.; García-Villoria, A.; Pastor, R. Heuristic and metaheuristic procedures for the Parallel Assembly Lines Balancing Problem with multi-line workstations and buffer sizing. *Comput. Oper. Res.* **2024**, *166*, 106596. [CrossRef]
34. Zhang, H. Assembly sequence planning: A Review. *Recent Adv. Comput. Sci. Commun.* **2022**, *15*, 531–539. [CrossRef]
35. Kang, M.; Seo, J.; Chung, H. Ship block assembly sequence planning considering productivity and welding deformation. *Int. J. Nav. Archit. Ocean Eng.* **2018**, *10*, 450–457. [CrossRef]
36. Chen, Z.; Adel, A. Advancing robotic assembly in construction: Innovations, challenges, and opportunities. *Autom. Constr.* **2025**, *178*, 106370. [CrossRef]

37. Parsi, S.S.; Whittaker, A.S.; Sivaselvan, M.V.; Velez-Lopez, E.; Stewart, W.R.; Shirvan, K. Experimental investigations into the rocking response of graphite-block assemblies in a HTGR core under horizontal earthquake shaking. *Nucl. Eng. Des.* **2025**, *443*, 114300. [CrossRef]
38. Zhao, R.; Tao, S.; Li, P. Human-centric assembly planning framework for human-robot collaborative systems with efficient reinforcement self-learning multi-objective evolutionary optimizer. *Adv. Eng. Inform.* **2025**, *67*, 103508. [CrossRef]
39. Tseng, H.; Wang, W.; Shih, H. Using memetic algorithms with guided local search to solve assembly sequence planning. *Expert Syst. Appl.* **2007**, *33*, 451–467. [CrossRef]
40. Wan, X.; Liu, K.; Qiu, W.; Kang, Z. An Assembly Sequence Planning Method Based on Multiple Optimal Solutions Genetic Algorithm. *Mathematics* **2024**, *12*, 574. [CrossRef]
41. Sheng, Z.; Chen, W.; Chen, Z.; Wen, H. Sequence planning for on-orbit robotic assembly based on symbiotic organisms search with diversification strategy. *Acta Astronaut.* **2024**, *219*, 941–951. [CrossRef]
42. Watanabe, K.; Inada, S. Search algorithm of the assembly sequence of products by using past learning results. *Int. J. Prod. Econ.* **2020**, *226*, 107615. [CrossRef]
43. Ab. Rashid, M.F.; Nik Mohamed, N.M.; Mohd Rose, A.N. A modified artificial bee colony algorithm to optimise integrated assembly sequence planning and assembly line balancing. *J. Mech. Eng. Sci.* **2019**, *13*, 5905–5921. [CrossRef]
44. Rashid, M.F.; Tiwari, A.; Hutabarat, W. Integrated optimization of mixed-model assembly sequence planning and line balancing using Multi-objective Discrete Particle Swarm Optimization. *Artif. Intell. Eng. Des. Anal. Manuf.* **2019**, *33*, 332–345. [CrossRef]
45. Yolmeh, A.; Kianfar, F. An efficient hybrid genetic algorithm to solve assembly line balancing problem with sequence-dependent setup times. *Comput. Ind. Eng.* **2012**, *62*, 936–945. [CrossRef]
46. Murali, G.B.; Deepak, B.B.; Biswal, B.B.; Mohanta, G.B.; Rout, A. Robotic Optimal Assembly Sequence Using Improved Cuckoo Search Algorithm. *Procedia Comput. Sci.* **2018**, *133*, 323–330. [CrossRef]
47. Petroodi, S.E.H.; Thevenin, S.; Kovalev, S.; Kovalev, S.; Dolgui, A. Robust multi-manned mixed-model assembly line balancing with dynamic task assignment considering product mix uncertainty. *Int. J. Prod. Econ.* **2025**, 109756. [CrossRef]
48. Ramezanian, R.; Khalesi, S. Integration of multi-product supply chain network design and assembly line balancing. *Oper. Res. Int. J.* **2021**, *21*, 453–483. [CrossRef]
49. Nowicki, E.; Smutnicki, C. A Fast Taboo Search Algorithm for the Job Shop Problem. *Manag. Sci.* **1996**, *42*, 797–813. [CrossRef]
50. Yijie, S.; Shen, G. Improved NSGA-II Multi-objective Genetic Algorithm Based on Hybridization-encouraged Mechanism. *Chin. J. Aeronaut.* **2008**, *21*, 540–549. [CrossRef]
51. Hao, J.; Yang, X.; Wang, C.; Tu, R.; Zhang, T. An Improved NSGA-II Algorithm Based on Adaptive Weighting and Searching Strategy. *Appl. Sci.* **2022**, *12*, 11573. [CrossRef]
52. Li, L.; Lin, Q.; Ming, Z. A survey of artificial immune algorithms for multi-objective optimization. *Neurocomputing* **2022**, *489*, 211–229. [CrossRef]
53. Sener, A. Hybrid large neighbourhood search algorithm for capacitated vehicle routing problem. *Expert Syst. Appl.* **2016**, *61*, 28–38. [CrossRef]
54. Hao, Y.; Zhao, C.; Zhang, Y.; Cao, Y.; Li, Z. Constrained multi-objective optimization problems: Methodologies, algorithms and applications. *Knowl.-Based Syst.* **2024**, *299*, 111998.
55. Wang, L.P.; Ren, Y.; Qiu, Q.C.; Qiu, F.Y. Survey on performance indicators for multi-objective evolutionary algorithms. *Chin. J. Comput.* **2021**, *44*, 1590–1619.

MDPI AG
Grosspeteranlage 5
4052 Basel
Switzerland
Tel.: +41 61 683 77 34

Processes Editorial Office
E-mail: processes@mdpi.com
www.mdpi.com/journal/processes